Synthesis of Cannabinoid Ligands
– Novel Compound Classes, Routes and Perspectives

Zur Erlangung des akademischen Grades eines

DOKTORS DER NATURWISSENSCHAFTEN

(Dr. rer. nat.)

von der KIT-Fakultät für Chemie und Biowissenschaften

des Karlsruher Instituts für Technologie (KIT)

genehmigte

DISSERTATION

von

M. Sc. Thomas Hurrle
aus Gernsbach

KIT-Dekan: Prof. Dr. Reinhard Fischer

1. Referent: Prof. Dr. Stefan Bräse

2. Referent: Prof. Dr. Michael A. R. Meier

Tag der mündlichen Prüfung: 14.12.2017

Band 73
Beiträge zur organischen Synthese
Hrsg.: Stefan Bräse

Prof. Dr. Stefan Bräse
Institut für Organische Chemie
Karlsruher Institut für Technologie (KIT)
Fritz-Haber-Weg 6
D-76131 Karlsruhe

Bibliographic information published by the Deutsche Nationalbibliothek

The Deutsche Nationalbibliothek lists this publication in the Deutsche Nationalbibliografie; detailed bibliographic data are available in the Internet at http://dnb.d-nb.de .

ISBN 978-3-8325-4777-6
ISSN 1862-5681

Logos Verlag Berlin GmbH
Comeniushof, Gubener Str. 47,
10243 Berlin
Tel.: +49 030 42 85 10 90
Fax: +49 030 42 85 10 92
INTERNET: http://www.logos-verlag.de

Die vorliegende Arbeit wurde im Zeitraum vom 15. November 2014 bis 6. November 2017 am Institut für Organische Chemie und dem Institut für Toxikologie und Genetik, der Fakultät für Chemie und Biowissenschaften am Karlsruher Institut für Technologie (KIT) unter der Leitung von Prof. Dr. Stefan Bräse angefertigt. Ein Teil der praktischen Arbeit wurde im Rahmen des „BioInterfaces in Technology and Medicine" Programms (BIFTM) als JOINT Projekt mit dem Titel "Coumarin derivatives and microparticulate carriers for immunomodulation" zwischen dem 1. Februar 2016 und dem 31. März 2016 am Helmholtzzentrum Geesthacht, Standort Teltow unter der Leitung von Dr. Christian Wischke durchgeführt.

The present thesis was realized at the Karlsruhe Institute of Technology, Faculty of Chemistry and Biosciences, Institute of Organic chemistry and Institute of Toxicology and Genetics, between the November 15th 2014 and the November 6th 2017 under the supervision of Prof. Dr. Stefan Bräse. A part of the practical part was carried out as a JOINT-project "Coumarin derivatives and microparticulate carriers for immunomodulation" in the context of the BIFTM program. This part was realized between the February 1st 2016 and the March 31st 2016 at the Helmholz center Geesthacht, Campus Teltow and was supervised by Dr. Christian Wischke.

Hiermit versichere ich, die vorliegende Arbeit selbstständig verfasst und keine anderen als die angegebenen Quellen und Hilfsmittel verwendet sowie Zitate kenntlich gemacht zu haben. Die Dissertation wurde bisher an keiner anderen Hochschule oder Universität eingereicht.

Table of contents

Abstract

Since the discovery of the endocannabinoid system, cannabis and cannabinoids have received a lot of attention. The elucidation of the biochemical machinery responsible for cannabinoid effects has led to the identification of several targets that can be addressed by cannabinoid analogs. The most prominent targets are the cannabinoid receptors CB_1 and CB_2. The endocannabinoid system acts as a versatile regulating system found in mammals that is involved in a broad range of physiological and neuronal processes. Accordingly, cannabinoid ligands have been proposed for the treatment of a large number of diseases and symptoms, including cancer, multiple sclerosis and AIDS. Selective ligands, that can address specific targets in the endocannabinoid system, are of high interest and new compounds are regularly reported.

In the present thesis, the synthesis and development of several novel potentially active cannabinoid compound classes are described. The goal is to identify selective ligands and implement close modifications based on structure-activity relationships. Based on the lead structure of the potent cannabinoid analogs, 3-benzylcoumarins, several novel compound classes have been designed and more than one hundred new compounds have been synthesized.

Furthermore, the formal total synthesis of tetrahydrocannabinol (THC) that has been developed in our group was completed using a novel and modular synthesis strategy. The key step involves the DIELS-ALDER-reaction of a functionalized diene that can be synthesized in a six-step sequence. The DIELS-ALDER reaction was implemented to synthesize six exemplary compounds to emphasize the synthetic opportunities. Proceeding from the maleic acid anhydride adduct, a six-step protocol generated a nabilone derivative that concluded the formal synthesis of THC.

Additionally, a project dedicated to the investigation of the potential of drug delivery and controlled release systems in the context of cannabinoids was pursued. Thereby, several micro- and nanoparticulate carrier systems have been prepared and some of the most potent compounds from preceding work have been encapsulated. The achieved particles were capable to release their cannabinoid payloads over hours, days, weeks and even months. Traditional application methods of cannabinoids in general lead to systemic distribution of the active ingredients. The herein described microparticulate carrier systems are able to achieve site specifity and constitute the opportunity to generate stable long-term drug levels.

Kurzzusammenfassung/ Abstract in German

Cannabis und Cannabinoid Analoga haben seit der Entdeckung des Endocannabinoid-Systems eine Menge Aufmerksamkeit erfahren. Durch die Aufklärung des biochemischen Wirkmechanismus konnten zahlreiche Targets identifiziert werden, zu denen auch die cannabinoiden Rezeptoren CB_1 und CB_2 gehören, welche durch Cannabinoid-Liganden adressiert werden. Das Endocannabinoid-System kommt in allen Säugetieren vor und ist an einer Vielzahl physiologischer und neuronaler Prozesse beteiligt. Entsprechend vielfältig sind auch die potentiellen Anwendungsgebiete für Cannabinoid-Liganden. Die Behandlung verschiedenster Krankheitsbilder und Symptome, z. B. Krebs, multiple Sklerose und AIDS, mit Cannabinoiden wird angestrebt. Dabei sind selektive Liganden, die spezifische Targets innerhalb des Endocannabinoid-Systems beeinflussen können, weiterhin gefragt.

In der vorliegenden Arbeit ist die Synthese und Entwicklung neuartiger Substanzklassen beschrieben, welche potentielle Cannabinoid-analoge Eigenschaften zeigen sollen. Das Ziel ist die Identifikation selektiver Liganden, welche im Anschluss basierend auf Struktur-Aktivitäts-Beziehungen modifiziert werden. Ausgehend von der Leitstruktur der Cannabinoid-aktiven 3-Benzylcumarine wurden verschiedene neue Substanzklassen entworfen und mehr als hundert neue Verbindungen synthetisiert. Des Weiteren konnte eine Strategie für die formale Totalsynthese von THC umgesetzt werden. Diese modulare Synthesestrategie wurde in unserer Gruppe seit längerem verfolgt und involviert als Schlüsselschritt eine DIELS-ALDER-Reaktion. Das entsprechende Dien konnte in einer sechsstufigen Synthesesequenz erhalten und in beispielhaften Reaktionen umgesetzt werden. Die sechs erhaltenen Produkte unterstreichen die synthetische Relevanz der Strategie. Das Produkt aus dem Dien mit Maleinsäureanhydrid konnte in der Folge in sechs weiteren Schritten zu einem Nabilon-Derivat umgesetzt werden. Diese Verbindung stellt die Vollendung der formalen Totalsynthese dar, da alle weiteren notwendigen synthetischen Stufen zum THC bereits publiziert sind.

Außerdem wurden verschiedene Micro- und Nanopartikelsysteme untersucht, welche Wirkstofftransport und -freisetzung in einer kontrollierten Art und Weiße ermöglichen sollten. Dazu wurden einige der vielversprechendsten Cannabinoid-Liganden erfolgreich in verschiedenen Polymerpartikeln verkapselt und anschließend das Freisetzungsverhalten untersucht. Es konnte dabei gezeigt werden, dass die hergestellten Partikeln in der Lage sind ihre Ladung kontrolliert über Stunden, Tage, Wochen oder gar Monate freizusetzen. Die beschriebenen Systeme könnten zudem für ortsaufgelöste Wirkstoffapplikationen verwendet werden und um konstante Wirkstoffkonzentrationen zu erhalten.

1. Introduction

1.1 Cannabis: The Dualism of Illicit Drug and Remedy[i]

The intoxicating effects of the cannabis plant, e.g. *Cannabis sativa* and *Cannabis indica*, have been known by mankind for several thousands of years.[2] First reports of the psychotropic effects can be found in the oldest medicinal transcript, the medicinal book of PEN TS'AO CHING, based on lore from the time of emperor SHEN-NUNG around 2000 BC. Transcripts dating back to ancient Egypt and China around 1700 BC report the medical applications of cannabis for treatment of glaucoma and infectious diseases.[3] Throughout human development, reports of cannabis can be found. Applications are described e.g. in ayurvedic medicine (300-350 BC) or the use for the treatment of asthma in Europe during the modern era.[4-5] The first chemical efforts on cannabis began in the 1840s. In the late 19th-century the crystal structure of cannabinol was elucidated and the first clinical research was undertaken.[6] At the beginning of the 20th-century, cannabis based drugs were widely prescribed for the treatment of numerous afflictions and conditions.[7] However, the establishment of modern medicine heralded the demise of cannabis treatment. The lack of standards and quality controls often failed to allow a precise dosage. Furthermore, public opinion was tainted by the high abuse-, addiction- and risk-potential of medicine like cocaine and morphine, which was also linked to cannabis. The medical descent reached its apex when cannabis use was made illegal in the US in 1937[7-8] followed by many countries throughout the world.[4]

The prohibition of cannabis did little to diminish the recreational use of the plant. Next to tobacco and alcohol, two socially approved and legal intoxicants, the use of cannabis is widespread in most countries of the world.[9-10] Since 1997, the United Nations Office on Drugs and Crime (UNODC) monitors the development of cannabis cultivation, trafficking and use alongside opioids and amphetamines.[11] In their reports the global number of cannabis users in 2015 was given as 183 million which equals around 3.8% of the world population and has stagnated since 1998.[12] This makes cannabis the most common illicit drug and as up to date the cultivation and consummation of cannabis remains illegal in large parts of the world.

Initiated by the breakthrough discovery of the cannabinoid receptors CB_1 and CB_2,[13] and the associated endocannabinoid system,[14-15] interest in cannabinoids for medical application has found new vigor. The renewed popularity of medical applications in modern medicine has led

[i] This chapter is based on sections from the own master thesis.[1]

to a change in public perception and more and more exception rules for the medical use of either plant preparations, extracts, or synthetic cannabinoid analogs were created.[11-12,16-17] The general trend leans towards decriminalization and often prosecution is only partly or not at all enforced. Uruguay, Bangladesh, North Korea, the Netherlands and several US-states have established regulations to allow acquisition and consume for recreational use with certain restrictions to amount and quality.[17] In countries that stand on the brink of legalization, e.g. Canada, the debate continues. It is often complicated to make educated guesses about the impact of legalization on public health, due to several factors ranging from incomplete scientific evidence to missing experience with large scale legalization.[18] In 2016 the peer reviewed open access journal 'Cannabis and Cannabinoid Research' was dedicated to give a platform to discuss all aspects of the cannabinoids.[19]

1.2 Chemical Aspects and Terminology[ii]

The structure of (–)-Δ^9-tetrahydrocannabinol (THC, **1**, Figure 1) was first elucidated by GANOI and MECHOULAM in 1964.[20] THC (**1**) is responsible for the infamous psychoactive effects of *C. sativa*. However, THC (**1**) is only one of around 60 secondary metabolites found in the plant that are structurally related. To mention only a few, cannabidiol (CBD, **2**),[21] cannabigerol (CBG, **3**)[22] and cannabichromene (CBC, **4**)[23] were among the first compounds identified in the extracts of *C. sativa*.

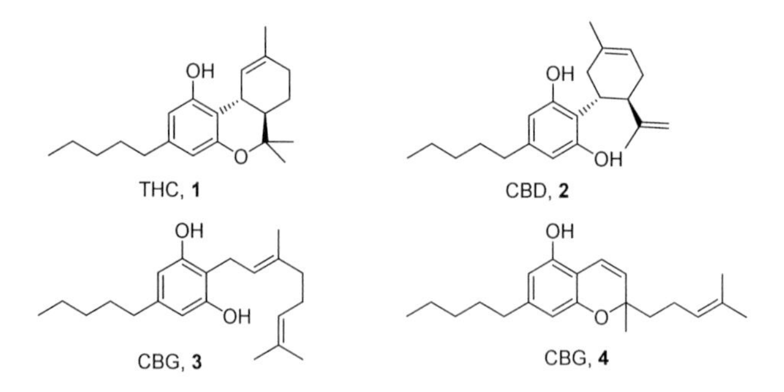

Figure 1: Examples for secondary metabolites from *C. sativa*.

[ii] This chapter is based on sections from the own master thesis.[1]

The term "cannabinoid" was shaped to refer to these secondary metabolites as they are almost unique to the various strains of *C. sativa*.[6,24] The terminology however is often used in a broader sense with no clear definition, including the later found endocannabinoids as well as synthetic cannabinoids, and does not imply a biological activity. It is therefore vital to use a clearly defined terminology, as has been advocated by DI MARZO *et al.*[24] In the following, definitions of the more important terms are given. The endocannabinoid system is defined as a signaling system that consists of the '*cannabinoid CB1 and CB2 receptors, of endogenous compounds known as endocannabinoids that can target these receptors, of enzymes that catalyze endocannabinoid biosynthesis and metabolism, and of processes responsible for the cellular uptake of some endocannabinoids*'.[25] Phytocannabinoids refer to the '*plant-derived secondary metabolites capable of either directly activating* CB_1 *and* CB_2 *receptors or sharing chemical similarity with cannabinoids, or both*'.[26] The term 'synthetic cannabinoids' describes compounds both naturally and non-naturally occurring with cannabinoid-like structures. This however, excluded the large variety of synthetic compounds without any structural relationship to cannabinoids, which cause THC-like (or inverse) effects. The overarching term 'synthocannabinoids' has been proposed[24] and is synonymously used with the term 'cannabinoid analogs'.

In Figure 2, several examples of the above defined compound classes are given. The first discovered and best studied endocannabinoids are anandamide (**5**) and 2-arachidonylglycerol (2-AG, **6**).[27-28] Both are highly lipophilic structures that are derived from arachidonic acid. The compounds HU-210 (**7**) and its reduced form HU-211 (**8**), which is still presumed to be the most potent cannabinoid known, are examples for synthetic cannabinoids.[29] Another compound derived from the cannabinoid structure is nabilone (**9**)[30] which was developed by ELI LILLY & CO. and is marketed under the name Cesamet®. Examples of synthocannabinoids (or cannabinoid analogs) that are structurally not related, and therefore non-classical cannabinoids, are rimonabant (**10**), WIN 55,212-2 (**11**) and AM-678 (**12**).[31] Examples of phytocannabinoids are given above in Figure 1.

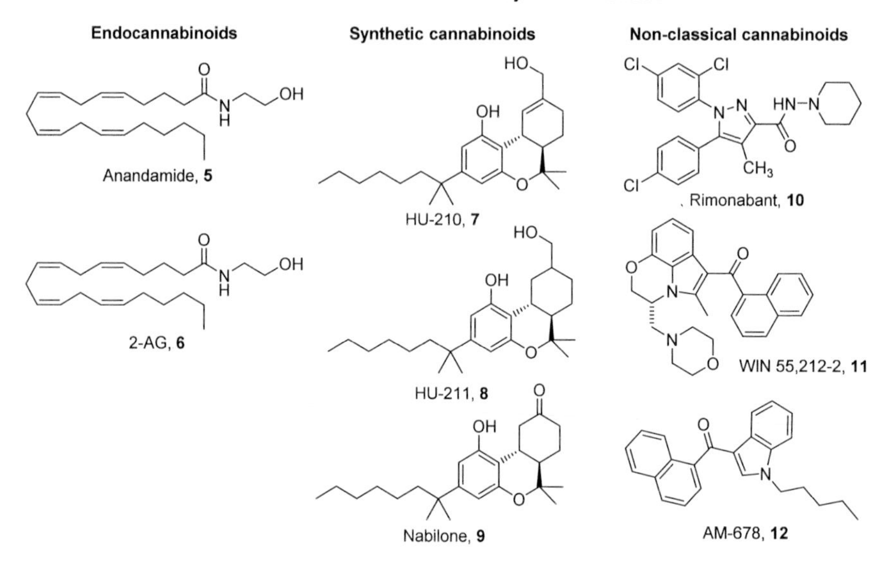

Figure 2: Examples for endocannabinoids and synthocannabinoids.

1.2.1 THC Synthesis

On industrial scale, THC (**1**) is acquired by extraction of *C. sativa*. As regulations (e.g. in Germany) prohibit the cultivation of cannabis strains with a high THC-content (around 15–20%), semi-synthetic pathways have been developed. Strategies of pharmaceutical companies such as BIORNICA ETHICS and THC PHARM are based on the extraction of CBD (**2**) from hemp strains[32] with a THC (**1**) content of less than 0.2%. CBD (**2**) is transformed to THC (**1**) in an acid-catalyzed cyclization reaction.[33-35] This semisynthetic THC (**1**) is sold under the name Dronabinol.

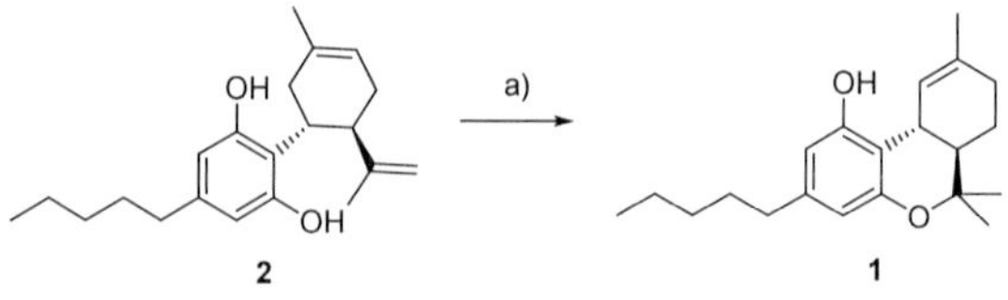

Scheme 1: Acid-catalyzed cyclization reaction of CBD (**2**) to THC (**1**): *a) BF_3 or Ethanol/HCl.*

The biosynthesis pathway of THC (**1**) is shown in Scheme 2. The metabolic precursor olivetolic acid (**13**) is formed via the polyketide pathway[36] and prenylated with geranyldiphosphate under enzymatic catalysis through a prenyltransferase to yield in cannabigerolic acid (**14**, CBG-acid).[37] A second enzyme, THCA-synthase, initiates a cyclization to tetrahydrocannabinolic acid (THCA, **15**) that is subsequently decarboxylated to the secondary metabolite THC (**1**).

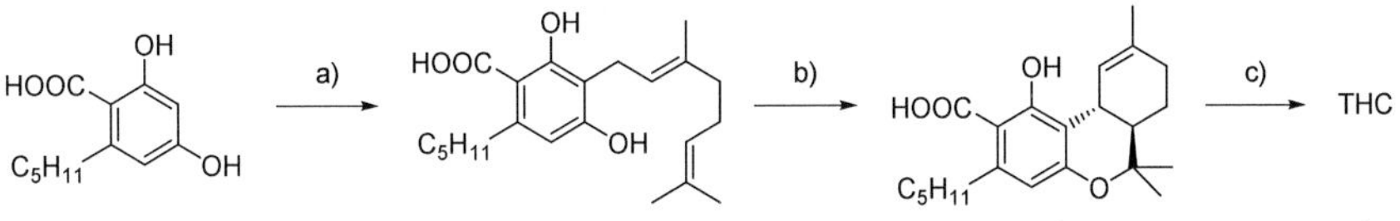

Scheme 2: Biosynthesis of cannabinoids from olivetolic acid **13**: *a) Geranyldiphosphat (GPP), prenyltransferase; b) THCA synthase; c) Decarboxylation.*[37]

Despite this effective access to THC (**1**), more than 20 total syntheses have been published.[38-57] MECHOULAM *et al.* published the first racemic total synthesis as early as 1965.[38] Two years later, they also established an enantioselective route to obtain (–)-Δ^9-*trans*-Tetrahydrocannabinol (**1**), starting from (–)-α-pinene and olivetol.[41,58]

In the following, the racemic total synthesis of *dl*-THC (*dl*-**1**), published 1966 by FAHRENHOLTZ *et al.*,[39] is described in detail (Scheme 3). The total synthesis starts with a VON PECHMANN condensation[59] on olivetol (**16**) by which coumarin **18** is formed. An intramolecular cyclization results in the tricycle **19**. The resulting keto function is protected as an acetal **20** by heating of ketone **19** in ethylene glycol in the presence of an acid. Then the methyl groups are introduced with methyl magnesium iodide and an acidic workup results in the deprotection of the acetal with a simultaneous isomerization of the double bond to the α,β-unsaturated ketone **21**. The ketone is reduced with lithium in liquid ammonia to ketone **22**. Acetylation and subsequent GRIGNARD-reaction with an acidic workup resulted in tertiary alcohol **24**. The alcohol is then transformed via LUCAS-reagent ($ZnCl_2$ in HCl) to chloride **25** and then treated with sodium hydride resulting in a dehydrohalogenation to the racemic *dl*-THC (*dl*-**1**).

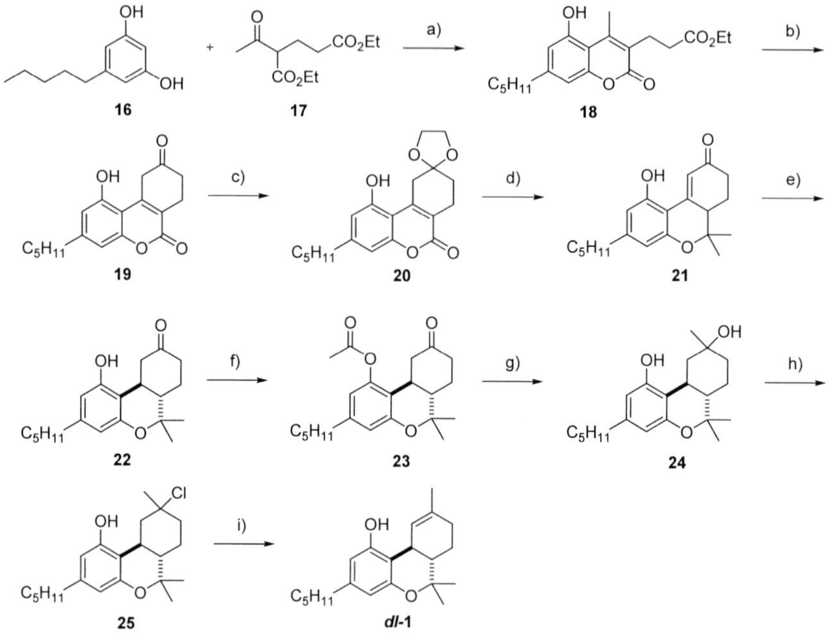

Scheme 3: Total synthesis of *dl*-THC (*dl*-**1**) by FAHRENHOLTZ *et al.*:[39] *a) $POCl_3$, r.t., 10 d, 84%; b) NaH, DMSO, –5 °C, 16 h, 78%; c) ethylene glycole, p-TsOH, benzene, reflux, 16 h, 94%; d) MeMgI, diethyl ether, reflux, 2 d, $HCl_{(aq.)}$, 70%; e) $Li/NH_{3(aq.)}$, –70 °C, 5 min, 53%; f) pyridine, acetic anhydride, 100 °C, 15 min; g) MeMgI, diethyl ether, reflux, 1 h, 88%; h) $ZnCl_2$, HCl, AcOH, r.t., 3 h, 74%; i) NaH, THF, reflux, 16 h, 73%.*

1.3 Principles of Cannabinoid Pharmacology[iii]

For a long time it was assumed that the pharmacological attributes of the cannabinoids originated in their interaction with the phospholipid membrane structure.[60] This assumption was based on the lipophilic properties of the compound class. First, the discovery of the G-Protein coupled receptors (GPCR) and the receptor concept in the seventies[61-63] initiated the search for cannabinoid receptors. The breakthrough discovery of the cannabinoid receptors CB_1 and CB_2 in 1988[13] led to the elucidation of the cannabinoid signal system. The cannabinoid receptors belong to the biggest protein superfamily, the GPCRs.[64]

[iii] This chapter is based on sections from the own master thesis.[1]

The layout of all GPCRs is similar. In Figure 3 (A) the crystal structure of the β-adrenergic receptor (β-AR) is depicted in its active state. The 2012 Nobel prize for chemistry was awarded to LEFTKOWITZ and KOBILKA for the identification and reconstruction of the gene sequence of the receptor.[65-67] The crystal structure of β-AR[68-71] shows the receptor (blue) with its seven transmembrane helices embedded in the cell membrane (grey). Adrenaline (yellow) can bind to the extracellular binding site. On the intracellular side, a heterotrimeric G-Protein (red and orange) with its three α-, β- and γ- subunits is bound. In the last year, crystal structures of the CB_1-receptor have been published.[72-73] The crystal structure shown in Figure 3 (B), shows CB_1 in its inactivated form with a tightly bound ligand. The seven transmembrane matrices are drawn in orange, the intracellular G-protein binding site in blue, and the ligand is depicted in green.

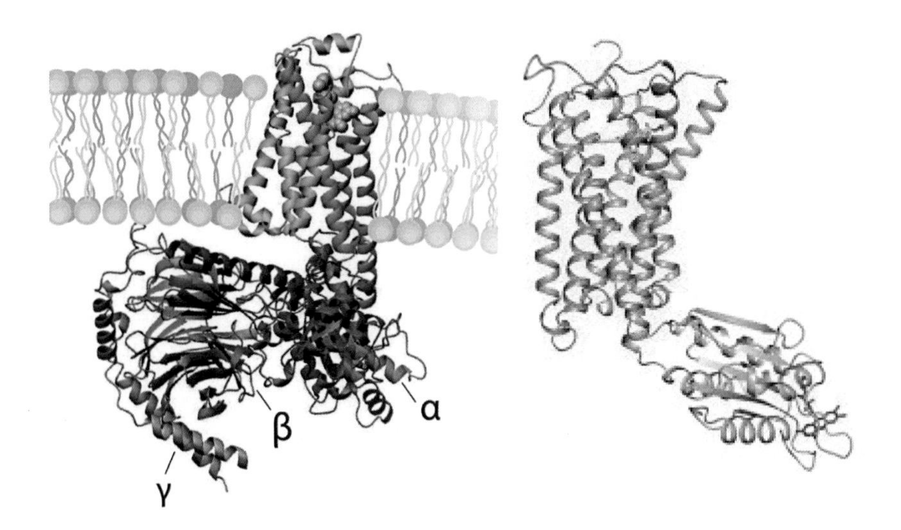

Figure 3: A: Crystal structure of the β-adrenergic receptor (blue) in its active state. The activation is a consequence of a signal molecule (i.e. a hormone, yellow) binding to the receptor on the extracellular site. The trimeric G-protein (red) with its α-,β- *and* γ- subunits can then be activated by the receptor. Adapted from The Nobel Prize in Chemistry 2012 - *Popular Information* © **(2014)** Johan Jarnestad/The Royal Swedish Academy of Sciences.[74-75] **B**: Crystal structure of the cannabinoid receptor CB_1. Adapted from *Crystal Structure of the Human Cannabinoid Receptor CB_1*, Hua *et al.* Cell 167 (3) **2016** 750-762. © Elsevier[72]

Receptors act by transducing an external stimulus into the cell. The schematic mode of action is depicted in Figure 4. When a ligand, e.g. a hormone, binds to the extracellular binding site of the receptor (A), a change in conformation is triggered. In this active conformation (B) the intracellular part of the receptor can bind and activate a G-protein. The receptor acts as guanine nucleotide exchange factor, exchanging guanosine diphosphate (GDP) to guanosine triphosphate (GTP). The thereby activated G-protein breaks into three subunits (C). The α-, β- and γ- subunits can then trigger intracellular signaling pathways (C). The GPCR stays in its activated form as long as the ligand is bound and can activate hundreds of G-proteins (D).[76]

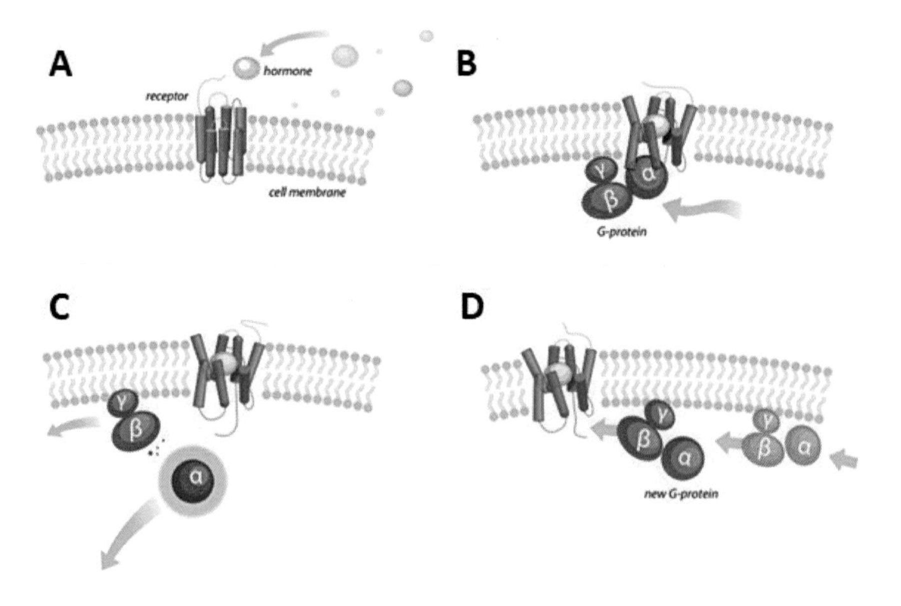

Figure 4: Mechanism of GPCR mediated signal transduction: **A**: A ligand such as a hormone binds to the receptor; **B**: As an effect, the receptor changes its shape into an activated state and G-protein is bound and activated; **C**: The activated trimeric G-protein breaks down into subunits, the α-subunit triggers a signaling cascade in the cell; **D**: The activation of G-Proteins continues until the receptor returns to its inactive state. Adapted from The Nobel Prize in Chemistry 2012 - *Popular Information* © (**2014**) Johan Jarnestad/The Royal Swedish Academy of Sciences.[75]

In general, three types of ligands are distinguished according to their effects on the receptors. The above described cycle is only activated by agonistic ligands, which cause a positive conformational change of the GPCR and thereby initiate the signal transduction. The opposite effect is caused by inverse agonists. They lead to the invert pharmacological effects when binding to the receptor. The last class are the antagonistic ligands (or competitive inhibitors), that dock onto the binding site but do not initiate a change into an active state. They block the signal transduction by denying the agonistic ligands. Their pharmacological effect is caused by inhibition of endogenous ligands.[77-78]

1.3.1 The Endocannabinoid System

The concept of the endocannabinoid signaling system was developed after the CB_1 and CB_2 receptors had been found.[13] The system is found in mammals and acts as a versatile regulator in both neuronal and non-neuronal cells and organs. The endocannabinoid system is comprised of the cannabinoid receptors, endogenous ligands and the metabolic machinery for endocannabinoid synthesis and degradation.[79-80]

In humans, CB_1-receptors are found in the brain, predominantly in the hippocampus, cerebral cortex, cerebellum and basal ganglia.[14,81-83] These areas are instrumentally involved in perception, short-term memory, motoric processes and pain reception. Most of the characteristic effects of cannabinoids (i.e. THC), such as modulation of pain perception, hallucinations, motoric restrictions and forgetfulness, can be attributed to influence on the CB_1-receptor. Besides the central nervous system (CNS), CB_1 protein expression was observed in peripheral organs and tissues, leucocytes, muscles spleen, heart, bone marrow and parts of the reproduction-, urinary-, cardiovascular- and gastrointestinal tracts. Their existence in small numbers at the peripheral nerve endings led to the assumption that they are involved in regulation of neuro transmitter release.[14,80,84]

The protein sequences of the CB_2-receptor coincide to only 48% with CB_1. In contrast to CB_1, no expression of CB_2 is reported in the CNS and is predominantly expressed in immune system associated tissue. Expression was observed in leucocytes, spleen, pancreas, thymus gland, bone marrow and tonsils.[64] It is presumed that they are involved in immune regulation, however the scientific groundwork is less solid when compared to CB_1.

More recent studies suggest that more receptors such as GPR55 and GPR18 are related to the endocannabinoid system.[80,85-87] The attribution is based on the interaction of cannabinoid-agonists such as THC (**1**) with the receptors. Research on GPR18 is still limited with more data for GPR55 being available. GPR55 has been found in human heart,[88] brain and cancer cells[89] and it is assumed that the receptor is involved inflammatory events, neural pain and angiogenesis (blood vessel formation).[90] The relationship between the cannabinoid receptors and GPR55 has been established through the activation of GPR55 by several cannabinoid ligands. Yet GPR55 has a protein sequence identity with the cannabinoid receptors of only 13–14%. Therefore, the binding pocket shows significant differences and particular CB_1/CB_2 agonists can have an antagonistic effect on GPR55.[89-90] For example in pain regulation, CB_1

and CB_2 agonists suppress the pain perception while GPR55 antagonists have the same effect and show a potential target for the therapy of neuropathic pain and inflammation.[91-93]

Thus, the endocannabinoid system constitutes a target for the treatment of numerous diseases such as diabetes, Parkinson, chronic pain or cancer[94-95] and numerous applications are already employed in modern therapy. Selective ligands, that can discriminate between the receptors have been in the focus of scientific efforts in the last decades.[96]

1.3.2 Medical Applications

A variety of applications for cannabinoids has been established. THC (**1**) is marketed under the name dronabinol and is a major constituent in several formulations. UNIMED PHARMACEUTICALS sells Marinol®, a formulation of dronabinol in sesame oil in retard capsules. In Canada and the USA it is approved for the treatment of anorexia in AIDS-patients,[97] as well as the nausea as a side effect of chemotherapy.[98] Additionally, all cannabinoid agonists exhibit analgesic properties and thereby contribute to the comfort of patients.[81-82]

ELI LILLY & CO marketed the synthetic cannabinoid nabilone (**9**) under the name Cesamet®.[99] It is also used for the treatment of nausea as a side effect of chemotherapy and is currently in clinical trials as an alternative to opioid based analgesics in patient with fibromyalgia, a chronic pain disorder.[100]

A combination of THC (**1**) and CBD (**2**) from *C. sativa* extracts is sold by BAYER HEALTH CARE under the trade name Sativex®. It is applied as a spray via trans mucosal resorption and constitutes a treatment option for spastics in multiple sclerosis.[100-101]

Rimonabant (**10**) has been developed for the treatment of chronic overweight. Rimonabant shows an antagonistic effect on the CB-receptors. While CB-agonists upregulate appetite, as it is the case for the previously described applications, CB-antagonists have an opposing effect and can consequently help in cases of extreme obesity.[102] However, the trials on this endocannabinoid receptor antagonists have been suspended due to severe psychiatric side effects.[103-104]

1.4 Cannabinoid Analogs based on the Coumarin Motif[iv]

BRÄSE and MÜLLER *et al.* were able to show in 2009 that 3-benzyl substituted coumarins **26**, can act as cannabinoid analogs.[105-106] The structural resemblance becomes apparent when the compounds are examined next to each other (Figure 5). It was also assumed that smaller substitutions in 3-position such as shown in 3-methylcoumarin **27** could lead to an antagonistic effect and therefore make them a pharmacologically interesting lead structure.

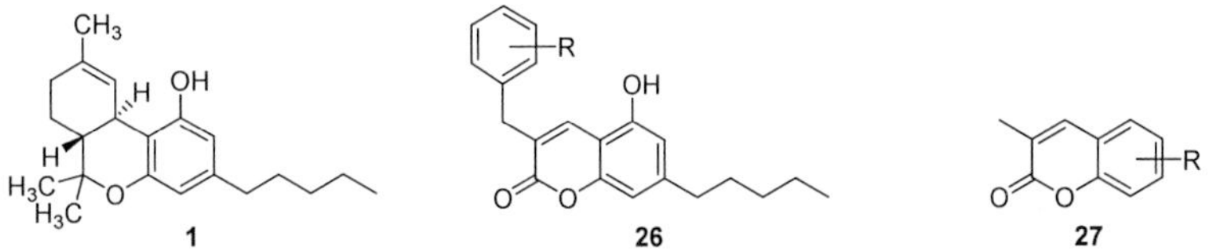

Figure 5: Structural comparison between THC (**1**), 3-benzylcoumarins **26** and a 3-methylcoumarin **27**.

Structure-activity relationship (SAR) studies on cannabinoids show that receptor binding correlates to *in vivo* activity.[107] Extensive SAR studies on classical and non-classical cannabinoids have been carried out and led to the identification of several pharmacophoric sites which are shown on CP-55.940 (**28**, Figure 6): (i) northern aliphatic hydroxy (NAH); (ii) phenolic hydroxyl (PH); (iii) lipophilic alkyl side chain (SC) and (iv) southern aliphatic hydroxyl (SAH).[96,108-109]

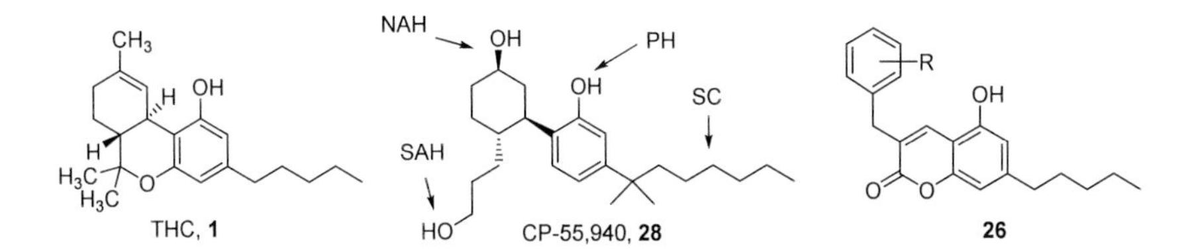

Figure 6: Pharmacophoric sites in classical and non-classical cannabinoids[96] compared to 3-benzylcoumarins. Pharmacophores are shown on CP-55.940: *northern aliphatic hydroxy (NAH); phenolic hydroxyl (PH); lipophilic alkyl side chain (SC) and southern aliphatic hydroxyl (SAH).*

The NAH in THC (**1**) is generated by metabolic processes, i.e. oxidation of the methyl group to 11-hydroxy-THC.[110] Other polar groups in this position have a similar effect, e.g. the ketone group of nabilone. In the 3-benzylcoumarin system, polar groups can be incorporated by the

[iv] This chapter is based on sections from the own master thesis.[1]

choice of the cinnamaldehyde. The PH was long thought to be essential for cannabinoid activity, as the absence of the PH leads to an extreme decrease of CB_1 activity. After the discovery of the CB_2-receptor it was found that a compound retains its CB_2 activity when a methyl ether is introduced in this position, but CB_1 activity decreases sharply. Compounds with a methyl ether at the PH site are therefore the first examples for CB_2 selective ligands.[96] SAR studies on the 3-benzylcoumarins led to a similar conclusion for this compound class.[111] The SC is essential for pharmacophoric activity of cannabinoid. It has been shown that short side chains (3 or less C-atoms) show a drastically reduced receptor affinity and 1,1'-dimethylheptyl side chains as shown in CP-55,940 **28** lead to an optimized affinity.[96] Furthermore, the exchange of the 1,1'-dimethyl part of the side chain with cyclic alkyl chains led to an even more increased selectivity.[112-113] This structural feature was already implemented into the 3-benzylcoumarins **26** by the synthesis of corresponding salicylic aldehydes to increase the affinities.[111,114] SHARMA *et al.* also showed that polar groups in the SC are tolerated and can even lead to increased receptor affinity. In their system, they incorporated ester functions into the SC.[115] The SAH is the only pharmacophore that is not found in classical cannabinoids such as THC (**1**).[96] In the 3-benzylcoumarins, the polar carbonyl function of the coumarin could have a similar binding effect.

1.4.1 Coumarin Synthesis

The synthesis of 3-benzylcoumarins **26** is performed via a method that was developed in our group.[105] The coumarin framework is fabricated by the reaction of salicylic aldehydes **29** with cinnamic aldehydes **30** and is facilitated through the presence of an ionic liquid (IL, **31**), that acts as a *N*-heterocyclic carbene (NHC). Functional groups are introduced through the choice of the starting materials and a broad scope of substrates is accepted.

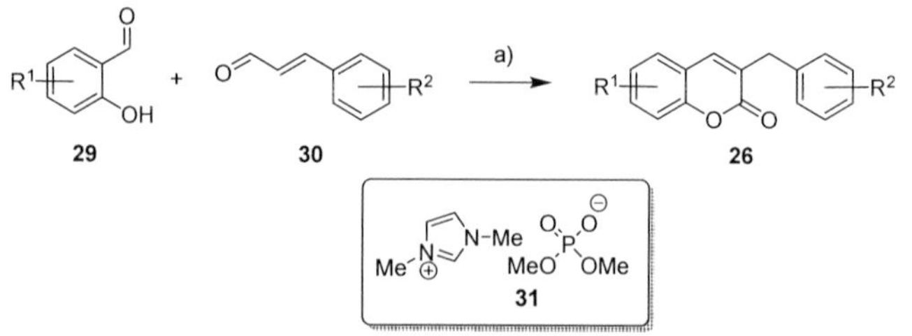

Scheme 4: NHC mediated coumarin synthesis developed by BRÄSE *et al.*[105]: *a)IL **31**, K_2CO_3, toluene, 110 °C, MWI (max. 200 W, 7 bar), 50 min.*

The proposed mechanism for the coumarin synthesis is shown in Scheme 5. The IL **31** is deprotonated to an NHC **32**. The NHC then reacts with the α,β- unsaturated aldehyde **30** and forms an adduct **33**. A proton shift generates enamine **34a** which is in a tautomeric equilibrium with the enol form **34b**. Protonation in β-position leads to the tautomeric forms **35a** and **35b**. Salicylic aldehyde **29** can now react with **35b** to form the intermediate **37** under reformation of the NHC mediator **32** or is quenched by water to the side product **36**. The intermediate **37** then undergoes an intramolecular condensation reaction to yield coumarin product **26**.

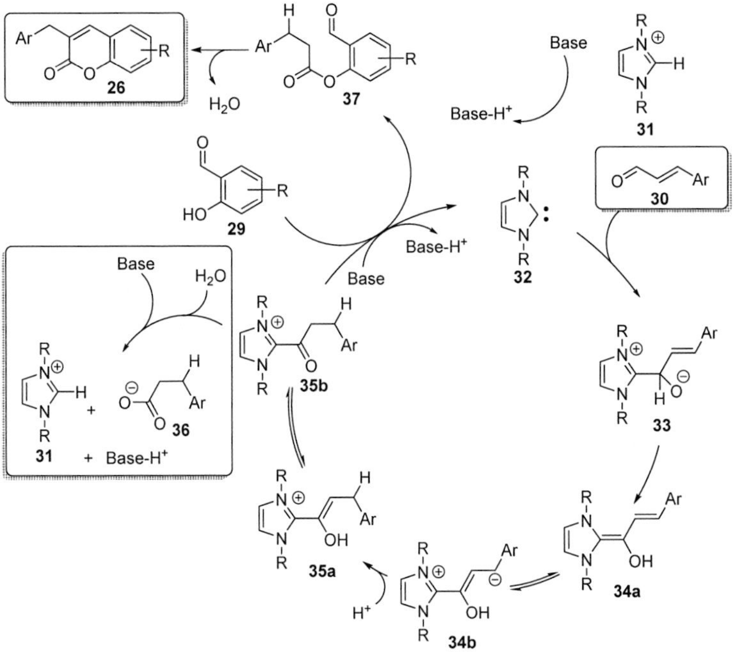

Scheme 5: Proposed mechanism for the NHC mediated 3-benzylcoumarin synthesis according to BRÄSE *et al.*[105,116]

1.5 Drug Development

The process of drug development is costly and can take more than two decades from initial discovery to market launch of a product[117] and estimated costs range between roughly \$ 100 million to \$ 1.5 billion for a single drug.[118] Drug development is a huge interdisciplinary field that employs a terminology that is attempted to be clarified in the following explanations.

At the beginning stands the discovery of a hit, i.e. that a compound is capable to manipulate a biological target. Targets are biomolecules that are involved in metabolic processes. The manipulation of a target can consequently trigger a change in the biological processes. Targets include enzymes, nucleic acids, ion channels and GPCRs.[119-120] A comprehensive study from 1996 estimates the total number of targets addressed by all drugs marketed at the present time to around 500, with 45% of the target structures being GPCRs. Often one target can be addressed by multiple drugs, e.g. the glucocorticoid receptor has 61 approved drugs. Newer analysis estimate the number of protein targets alone at around 700 and significant growth is expected as the knowledge of biological processes develops.[121]

New hits are often generated by high throughput screens of chemical compound libraries and initiate the process of Hit-to-Lead development (Figure 4).[120] When a hit is observed the structure is examined closely and a lead structure is developed.[119] Lead structures are generated by comprehensive assessment of the initial hit compound(s). SAR, chemical integrity and synthetic accessibility are crucial for promising lead structures. A rigorous attention to bio-physical and chemical requirements is necessary in early development to prevent costly late stage attrition. Awareness of drug likeness and pharmacokinetic parameters should be considered early on.

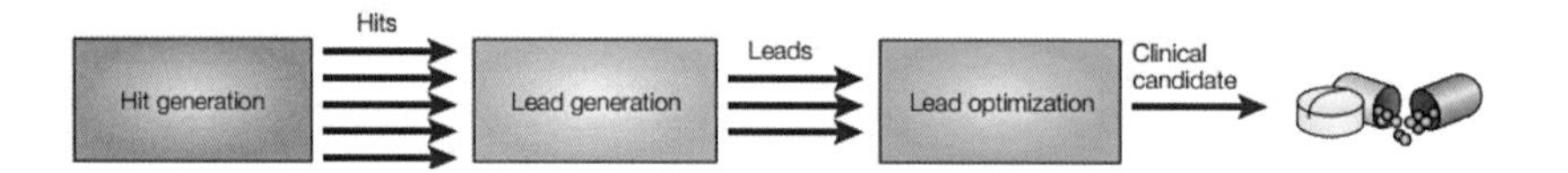

Figure 7: Stages from hit to lead. **Red**: A hit is often generated by high throughput screening of compound libraries. **Yellow**: The hit structure is evaluated, lead structures generated. **Green**: Lead structures are optimized and in the best case a drug candidate for clinical trials is found. Adapted with permission from *Hit and Lead Generation: Beyond High-Throughput Screening* Bleicher *et al.*, *Nat. Rev. Drug Discovery* **2003** (2) 369-378 © Nature Publishing Group.[119]

To assess the drug likeness of a compound, several parameters can be considered. One of the most famous rule-of-thumb approaches is LIPINSKYs rule-of-five (RO5).[122-123] It was one of the first attempts to break down the likeliness of a compound to become a drug. The principal statement is that a drug candidate should not violate more than one of the following rules: (i) the number of hydrogen bond donors is below 5, (ii) the number of hydrogen bond acceptors does not exceed 10, (iii) molar mass should be below 500 daltons, (iv) the octanol-water partition coefficient Log P does not exceed 5. The RO5 was later refined by GOHSE *et al.* through comprehensive assessment of successful drugs.[124] The modified rules are the following: (i) the calculated log P (octanol-water partition coefficient) should be in the range of -0.4 and 5.6, (ii) the molecular weight should be between 160 and 480 g/mol, (iii) the molar refractivity (total polarizability) should lie between 40 and 130, (iv) the total number of atoms should be in the range of 20–70. Both rules help to predict a preferential pharmacokinetic behavior for orally administrable drugs, i.e. absorption, distribution, metabolism and excretion (ADME). However, they do not help to predict if a compound is biologically active and often drugs are developed that are not orally administrable.

1.6 Drug Delivery and Controlled Release Systems

Drug delivery is a fast growing and productive research area with a high paper output.[125] When a drug is developed, a suitable way of application must be found. The preferred noninvasive techniques such as oral, transdermal, transmucosal and inhalation are not always feasible because of resorption or stability problems. Drugs that cannot be administered via these routes are often administered by injection or lead to the failure of a potential drug. It is often necessary to enhance the systemic distribution, manage the temporal progression of drug levels and find suitable administration techniques. Therefore, formulations and techniques have been developed to allow temporal and distributional control[126] and to assist pharmacokinetics, bioavailability, and minimize adverse effects.[125] Historically, the development of controlled drug delivery systems dates back to the 1960s where a sealed silicon tube containing a drug was first proposed as implantable drug delivery system.[127-128] In the 1970s and 1980s, macroscopic devices and implants had been developed. Delivery was achieved by mucosal inserts (e.g. in the eye), implants, ingestible formulations and topical patches. Over time, the systems decreased in size and the first microscopic and nanoscopic polymer systems were introduced.[129]

In the following, the two major principles of drug delivery are outlined.

1.6.1 Temporal Control of Drug Levels

The control of drug levels can help to improve efficacy, toxicity, patient compliance and convenience.[130] To retain constant drug levels in the therapeutic window, i.e. concentration at which only beneficial but no harmful effects occur, a temporal management of drug resorption or administration is necessary. This is especially important for drugs with a low stability in a biological system. After administration, e.g. oral or by injection in the blood stream, drug levels reach a maximum shortly after administration and decrease depending on the metabolic properties.[129]

In Figure 8, an example for control of drug levels is given. The aim is to retain a constant drug level in the therapeutic window (shaded in grey). For a standard therapy, by injecting the drug every six hours, the drug level rises after every injection (dashed line) and a rapid decrease is observed (thin line). Therefore, periodically new injections become necessary and sub therapeutic drug levels are observed for several periods. A controlled release system can achieve constant drug levels over longer periods of time (bold line). After an initial lag phase where the drug accumulates, an equilibrium state between depletion and release is reached.[130]

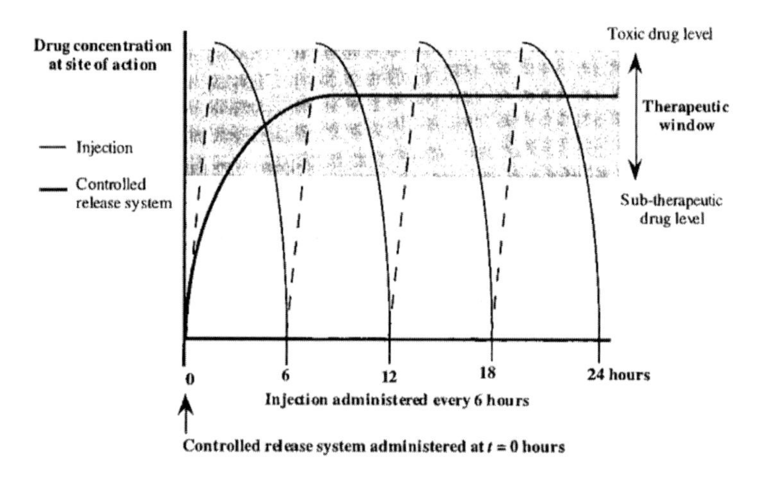

Figure 8: Control of drug concentrations at therapeutic site after delivery as conventional injection (thin line) and when administered through temporal controlled release system (bold line). Reprinted with permission from (*Polymeric Systems for Controlled Drug Release* Kathryn E. Uhrich, Scott M. Cannizzaro and, Robert S. Langer*, and Kevin M. Shakesheff *Chem. Rev.* **1999** *99* (11), 3181-3198 © (1999) American Chemical Society.

For orally administrable drugs, one solution to retain more constant drug levels are depot formulation solutions. The active substance is here covered with functional coatings that can release their payload after predefined time periods controlled by e.g. thickness of the dissolving coating or type of the coating. One popularly known example would be depot capsules for long term supply of vitamin C and zinc.[131]

For drugs that have to be injected into the bloodstream because of poor oral availability, a continued administration via technical devices is possible. One example where constant levels of a drug is achieved, is a morphine pump.[132] Either external or implanted, this pump gives a continuous flow of drug into the bloodstream in a rate that counters the metabolic decrease of the drug level. The greatest advantage of such systems is that the drug level can be adjusted manually if a medication in intervals is needed, e.g. insulin, or the drug level can be adjusted according to the need.[133-134] However, the technical requirements are high and often involve a stationary observation or surgical intervention. Therefore, novel techniques for controlled release have been developed.

Implants that can slowly release drugs have already been proposed in the 1960s. In the beginning, especially macro implants were developed. Famous examples are contraceptive implants such as Norplant® or Vaginal Ring®. The disadvantage of such macroscopic systems is that they often involved surgical intervention both on implantation and extraction.[129]

1.6.2 Distributional Control of Drug Levels

Controlled release devices can also be employed when a distributional control of a drug is preferable or high systemic drug levels would have adverse effects. In Figure 9, the effects of distributional control are outlined. In this example, a controlled release system is employed to allow a high concentration at the desired site of action (bold line) while the systemic drug levels (thin line) are kept low. Thereby, the systemic drug concentration can be kept below a certain threshold at which side effects would occur (dashed line).[130] As implied, a perfect site specific drug release is hard to achieve if not impossible as diffusion occurs and active mechanisms can lead to systemic distributions.

One example for an application where distributional control is necessary is the treatment of cancers with chemotherapy. The employed drugs often show severe systemic side effects and localized application helps to prevent or reduce them.[130]

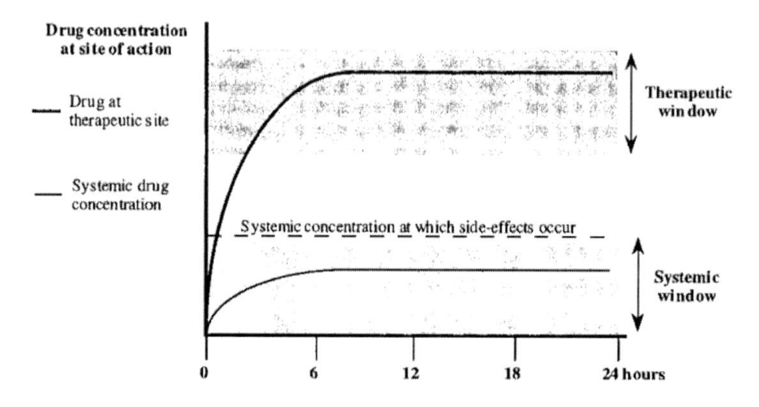

Figure 9: Distribution control of a drug level for an ideal controlled release system. **Bold line**: Drug concentration at site of therapeutic action. **Thin line**: Systemic drug levels. Reprinted with permission from (*Polymeric Systems for Controlled Drug Release* Kathryn E. Uhrich, Scott M. Cannizzaro and, Robert S. Langer*, and Kevin M. Shakesheff *Chem. Rev.* 1999 *99* (11), 3181-3198 © (**1999**) American Chemical Society.

1.6.3 Micro and Nanoparticulate Carrier Systems

Controlled delivery systems have been miniaturized ever since their first introduction.[128-129] In present days, the nano- and micro scale biodegradable microparticulate carriers are of prime interest. A drug is encapsulated in a matrix polymer that can be injected into tissue where it releases the compound, degrades and is ultimately resorbed by the organism. The release is controlled by diffusion of the drug through the matrix and is assisted by swelling and degradation effects that both accelerate the release of the payload compound.[135]

The broad scope and need for drug delivery applications has led to many inventions in the field.[136-144] Ranging from simple setups where biodistribution is controlled by different surface properties of particles[145] to magnetic drug delivery systems that can be directed by external magnetic fields.[138] Nanocarriers that are stimuli responsive have been developed and can target their release according to external factors, i.e. higher pH-levels in tumor tissue or higher temperatures in inflamed sites.[136]

Setups of micro- and nanoparticulate polymer carrier materials impress through simple preparation methods[146] and a broad scope of applications.[135] Size, surface to volume ratio, surface structure and release rate can be adjusted according to the planned application.[146-147] In contrast to macroscopic implants, micro- and nanoparticles are suitable for a syringe based injection without the need for surgical intervention.[148] Often employed are therefore biodegradable and biocompatible polymers such as polyesters, i.e. polylactic acid (PLA) and poly(lactic-co-glycolic acid) (PLGA).[135,149-153] Especially PLGA is often used as it shows a fast and traceless hydrolytic degradation over the course of several weeks.

Another advantage of PLGA/PLA micro- and nanospheres is the good distribution in tissue and good biocompatibility. After injection several factors can be observed that contribute to the tissue response. Wound healing, inflammation and foreign body responses are triggered, that are tissue-, organ- and species-dependent.[149]

Highly interesting is also the passive targeting function of competent cells that are able to internalize the particles via phagocytosis/endocytosis, eventually leading to intracellular drug release.[154-155] This behavior was previously shown for microparticles in the range of 5–10 μm.[135,156]

The expression of cannabinoid receptors in immune cells makes them interesting targets to either trigger or suppress immune responses on a cellular level. A study published in 2013 found that human peripheral blood leukocytes (PBL) express CB_2 receptor protein, but could only find it on the extracellular membrane of human B-cells, whereas the distribution in T-cells and monocytes is predominantly intracellular.[157] However, the role of the intracellular receptor protein remains unclear. The combination of microparticulate carriers in a size range that allows phagocytosis and a payload of CB-modulating substances might help to get insight in this function.

2. Aim of the Project

Cannabinoid research has received extensive attention over the last decades.[6,158] Cannabinoids and cannabinoid analogs are marketed as drugs for a broad range of diseases. However, demand for novel compounds remains high. New developments and insights in the field are still unfolding, such as the recent discovery of novel receptors (GPR55/GPR18).[80,85-87] Especially compounds that effect only a selective target within the mechanism of action of the cannabinoids are of primary interest.

The here presented work covers several approaches to achieve selectivity and generate novel analogs. One objective is the formal total synthesis of THC (**1**). Though there are more than 20 total syntheses routes already available,[38-57] a number of interesting intermediates can be accomplished by our route. The most challenging part is the inclusion of a DIELS-ALDER reaction in the synthesis protocol (Scheme 6). The modular set up allows the generation of multiple target molecules and the introduction of synthetic working handles to further diversify the structures.

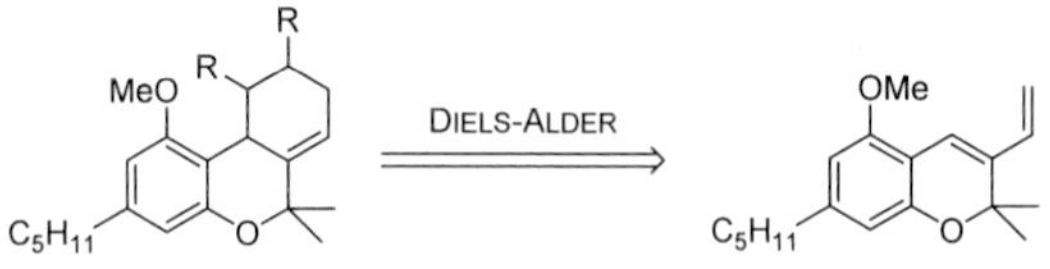

Scheme 6: DIELS-ALDER access to THC-like structures.

In addition to the work on the THC-framework, synthetic cannabinoids based on the 3-benzylcoumarin **26** motif are targeted (Figure 10). Previous work in the field identified 3-benzylcoumarins with various substitution patterns as cannabinoid ligands.[105-106] Commitment to the project includes the obligation to resynthesize compounds when requested by biological partners.

Furthermore, the synthesis of 3-arylcoumarin **40**, 3-styrylcoumarin **41** and 3-alkylcoumarin **27** libraries was a designated goal of the thesis. Aryl and styryl substitution on the coumarin core could result in compounds that act as cannabinoid analogs in respect to the CB-receptors and have an increased fluorescence when compared to the benzyl substituted compounds. Alkyl substitution on the other hand promised to deliver compounds that act CBD (**2**) analog.

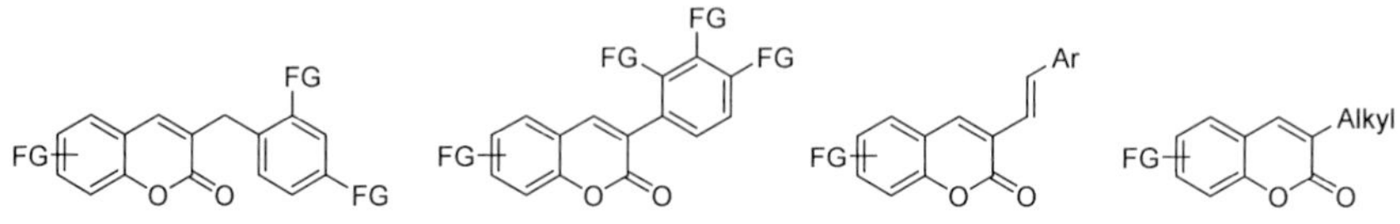

Figure 10: Objective molecule classes with potential cannabinoid activity. *FG = Functional Group.*

Scientific effort towards the synthesis and development of novel potential cannabinoid structures was not restricted to the abovementioned compound classes. Ideas to broaden the substrate scope or the implementation of novel motives were considered to be a part of the work.

The third part of the thesis was planned as a research stay at the Helmholtz-Zentrum Geesthacht (HZG), campus Teltow, at the WISCHKE group as part of the JOINT-project „Coumarin derivatives and microparticulate carriers for immunomodulation". Some of the most promising compounds from preceding work **T1-T3** were selected to be encapsulated in polymer carriers. These carrier systems have the capacity to entrap and protect bioactive molecules and can deliver their payload over a prolonged period. In addition, a site specificity can be achieved by injection of the carriers and a passive targeting function is expected as the polymer particles can enter competent cells via phagocytosis.[135,154-156]

The challenge addressed in this work was development of methods that are suitable to encapsulate the selected compounds. Encapsulation of the compounds in PLGA-based biodegradable microparticles by emulsion-based methods allows to prepare particles in different sizes including the phagocytizable range.[135,149-153] The work package included the characterization of thermal and morphological properties of the particles and the development of analytical methods for quantification of the compounds. With the completed methods in hand, the release behavior from the particles can then be studied for different particle sizes and polymer properties.

Ultimately the preparation of adequate amounts of compounds and compound loaded particles for biological testing was envisaged. The *in vitro* behavior of cells when exposed to extracellular compound solution, constant compound levels from controlled release and intracellular impact after a potential intracellular release from phagocytized particles should further be examined by biological partners.

3. Results and Discussion

The overarching topic of this work are cannabinoids and the focus lies on their synthesis. The present thesis is split into three major parts. First the work on the formal total synthesis of THC is discussed. The second part discusses the efforts to synthesize and develop novel coumarin based cannabinoids. The last part covers an excursion where previously evaluated cannabinoids were encapsulated in polymer carrier systems.

3.1 Work on the THC-Framework[v]

3.1.1 Retrosynthetic Aspects

One of the main goals of this work was the completion of a formal total synthesis of THC that has been developed in our group. More than 20 total synthesis approaches towards the THC-framework have been published before.[38-57] In Figure 11 a) the three-ring THC motif is shown. The established synthetic approaches are rather divers, but one common motif is the formation of ring B in either one or two steps from building blocks where ring A and C are already present (Figure 11, b). In contrast our strategy consists of a stepwise approach where ring B and C are synthesized consecutively (Figure 11, c).

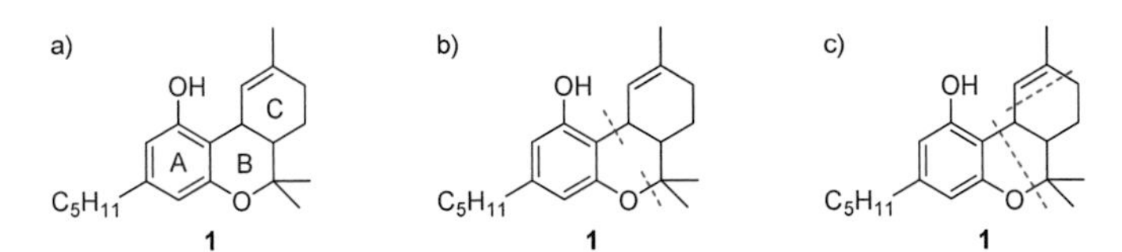

Figure 11: Conceptual approach towards THC.

Our strategy is centered on the synthetic transformation of diene **39** in a DIELS-ALDER-(DA)-reaction (Scheme 7). The third ring (Figure 11 a) ring-C) of THC-like structures can thereby be built up in a modular fashion. In theory, any electron deficient dienophile should be suitable for the reaction.

---

[v] This chapter is based on preliminary work performed by BERNHARD LESCH,[159-160] MANUEL C. BRÖHMER[161-162] and FRANZISKA GLÄSER.[114,162] and the own master thesis.[1]

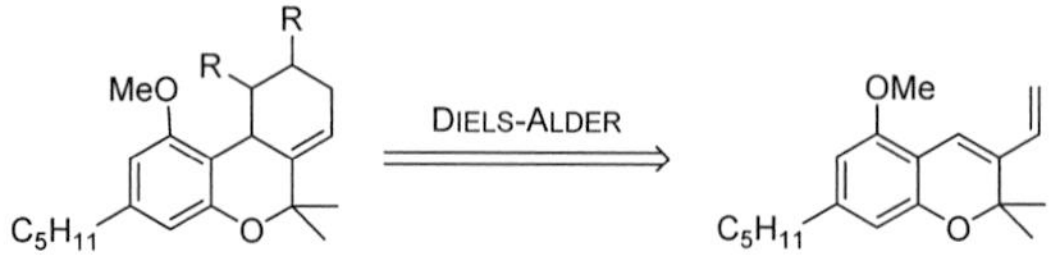

Scheme 7: Key step for the modular construction of the third ring of the THC-motif.

Proof of the concept was provided by the group of MIANMI *et al.*, who showed that it is possible to use the DIELS-ALDER reaction to access cannabinoid structures (Scheme 8).[163] They used the vinyl chromene **42** that retains a free hydroxy group. The substrate had to be prepared *in situ* and was considered labile. The high temperatures necessary for the reaction led to isomerization of the double bond and the overall yield was not satisfactory (36%).

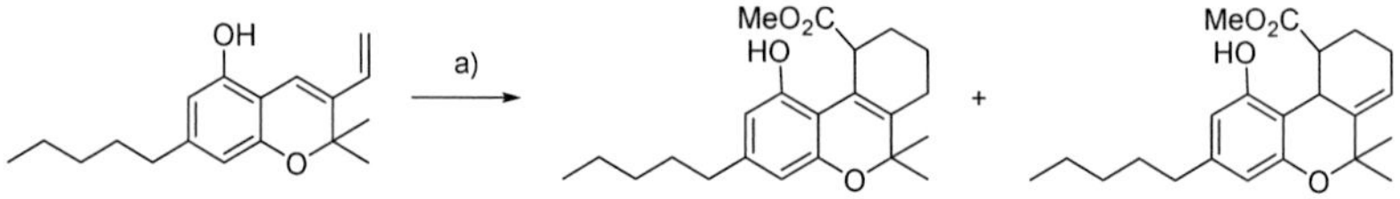

Scheme 8: Diels-Alder-concept by MIANMI *et al.*: *a) 10 equiv. acrylic acid ethyl ester, 1 equiv. DBU, DMF, 100 °C.*[163]

Regardless, including this crucial step in the total synthesis appeals since it would allow a modular synthesis and constitutes an extensive synthetic opportunity. The application of a wide range of dienophiles should give access to new potentially cannabinoid active compounds. Also, the different possible substitution patterns represent synthetic leverage for further diversifications, which has been shown on a model system (Scheme 9).[162] I.e. DA products that have two different substituents are interesting, as they present individual working points for synthetic transformations. With the model system it was shown, that aldehyde and ester function could be independently processed in further synthetic steps.[162]

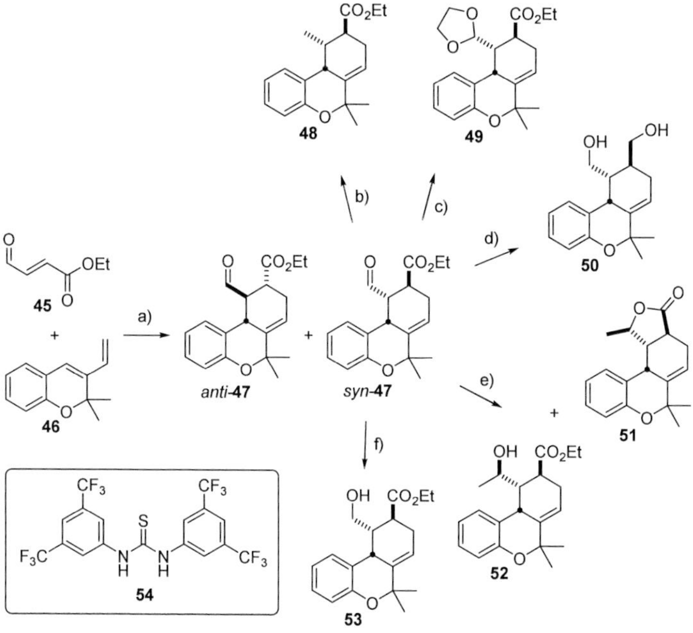

Scheme 9: Post-DIELS ALDER modification potential as shown on a model system:[162] *a)* ***54*** *(20 mol%), DCM, r.t., 12 h, 72% (11:1, syn:anti); b)* H_2, PtO_2, *conc. acetic acid, 32%, c) ethylene glycol, p-TsOH (1 mol%), toluene, water separator, reflux, 48 h, 49%; d)* $LiAlH_4$, *diethyl ether, r.t., 2 h, reflux, 4 h, 30%; e) MeMgBr, diethyl ether, –5 °C, 2.5 h, 49%* ***51****, 27%* ***52****; f)* $NaBH_4$, *methanol, r.t., 15 min, >98%.*

Scheme 10 shows the retrosynthetic approach towards THC developed in our group. The concept starts from *α,β*-unsaturated ketone **55**, where the additional steps for the transformation to THC have already been developed by HUFFMAN[164-165] (deprotection) and FAHRENHOLZ.[40] Ketone **55** should be available by an epoxide opening of **56** followed by an oxidation. The epoxide can in turn be generated from alkene **57** which is accessible by oxidative bisdecarboxylation of the hydrolyzed form of anhydride **58**. The anhydride could be generated by above mentioned DIELS-ALDER reaction of diene **39** with maleic anhydride and constitutes the key-step around which the described pathway is centered. Diene **39** should be accessible via WITTIG reaction from *α,β*-unsaturated aldehyde **59** that could be achieved via SWERN-oxidation on the hydroxy methyl side chain of diol **60** involving the elimination of water. Diol **61** should in turn be accessible by reduction of ketone **62** which can be generated through hydroxy methylation and prior protection from chromene **63**. The synthesis of chromene **63** can be achieved by literature procedures from the commercially available olivetol.[166]

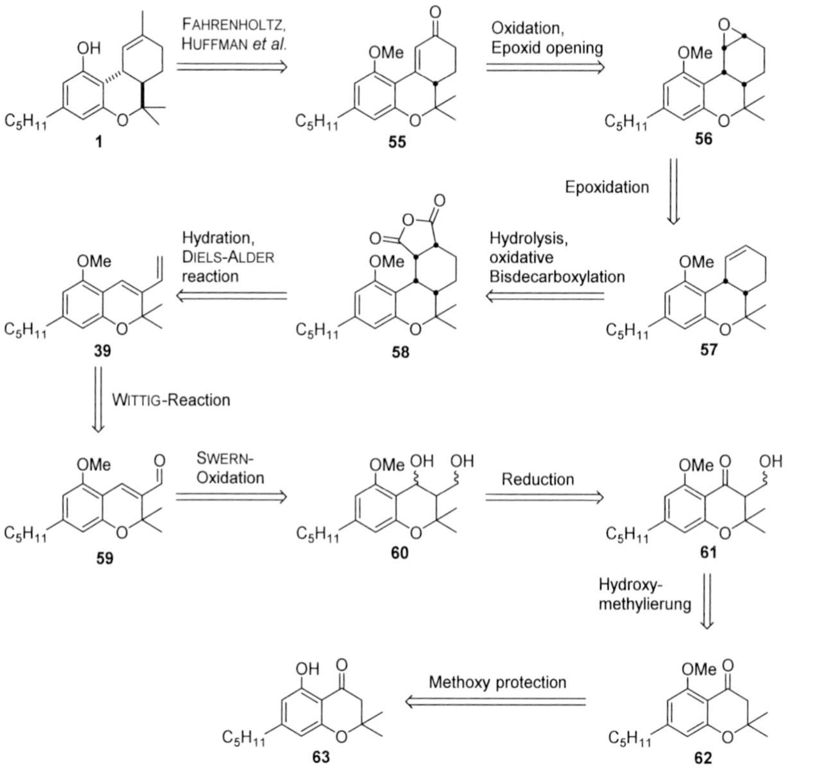

Scheme 10: Retrosynthetic approach centered around the DIELS-ALDER reaction.[1,114,161]

3.1.2 Synthetic Work on the THC Framework

Following the retrosynthetic strategy shown in Scheme 10 the synthesis follows the preliminary work of MANUEL C. BRÖHMER[161] and FRANZISKA GLÄSER,[114] and work of my master thesis.[1]

Starting from chromanone **63** that is synthesized via established routes by BIRNBERG *et al.*[166] the first synthetic step is a methylation (Scheme 11). This methylation to methoxy ether **62** is necessary for subsequent transformations, as a number of reactions would be influenced by the protic character of the hydroxyl group. Starting from chromene **62**, a lithiation and subsequent quenching with *N*-hydroxy methylene phthalimide, generates hydroxy methyl chromanone **61**, which is then reduced to racemic diol **60**. Subsequently, diol **60** is used without purification to synthesize *α,β*-unsaturated aldehyde **59** under SWERN conditions, a mild oxidation method. A WITTIG reaction leads then to the key compound, diene **39**. The diene **39** is stable enough for analysis and can be stored for several days under argon atmosphere in a freezer but a conversion to more stable intermediates is recommended as dimerization or polymerization of the compound is possible. It has been observed that a prolonged idle time between the synthesis and purification leads to a severe loss of the product yield as degradation of the product in the crude mixture occurs. Also, the removal of volatiles at room temperature is preferable for this compound and the column chromatography had to be performed as quickly as possible. If handled correctly between 89% and up to quantitative yield of **39** can be achieved.

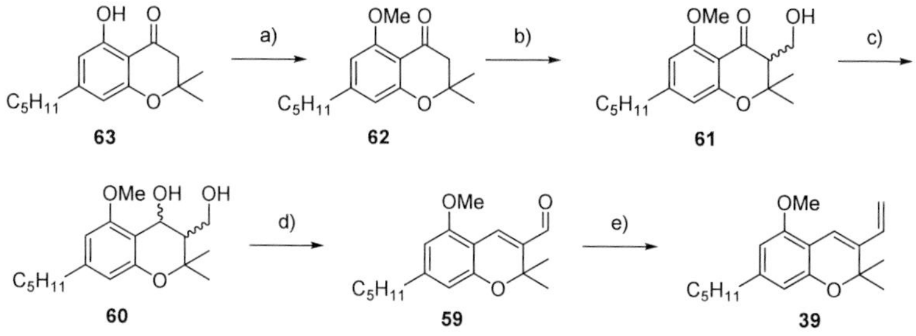

Scheme 11: Synthetic steps towards diene **39**:[114,161] *a) Me_2SO_4, acetone, K_2CO_3, reflux, 4 h, 97%; b) TMP, n-BuLi, N-hydroxy methylene phthalimide, THF, –78 °C, 6 h, 82%; c) $LiAlH_4$, THF, 0 °C, 30 min, r.t. 2 h, quant.; d) SO_2Cl_2, DMSO, NEt_3, DCM, –55 °C to r.t., 3 h, quant.; e) Ph_3PCH_3Br, tBuOK, THF, –78 °C to r.t., 2 h, 89%.*

To show the feasibility of creating novel compounds using the DIELS-ALDER reaction, a range of dienophiles were applied to the diene **39**, under conditions that were similar to the optimized conditions for maleic anhydride (Scheme 12, a)). Scheme 12 shows the results of unoptimized reactions. Compared to the reaction with maleic anhydride (top row), that proceeds with quantitative yield after 16 h at room temperature, the other dienophiles show slower conversions. Monitoring of the reactions via TLC showed that after 16 h starting material was left in the reaction mixtures and more time was allowed for the reactions to proceed. A main difference in the reaction is also that for the reaction of maleic anhydride no further purification is needed since there is only a single product **38a** formed. In reactions b) to d) regioisomers can form and have been isolated for b) (**38b** and **38c**) and d) (**38f** and **38g**). The reaction mixtures from reaction b) to d) were consequently purified via column chromatography and the polar character of the anhydrides and the possibility of hydrolytic degradation on the stationary phase might explain the significantly lower yields when compared to a). This assumption is supported by the difference in mass between the crude mixture and the mass of the combined fractions found after column chromatography as well as the observation of a baseline spot on TLC and column that could not be moved even with polar solvents. Secondly, as discussed before, diene **39** showed to be a rather delicate compound and long storage times had to be prevented if possible. Samples that had been kept for prolonged periods of time showed signs of degradation. Therefore, it is possible that the longer reaction times necessary for reactions b) to d) (Scheme 12) allowed parts of starting material to degrade and consequently reduce the yield of the reaction. The observed formation of unidentifiable side products, that could be dimerized or polymerized starting material, as well as recovered dienophile starting material support this hypothesis.

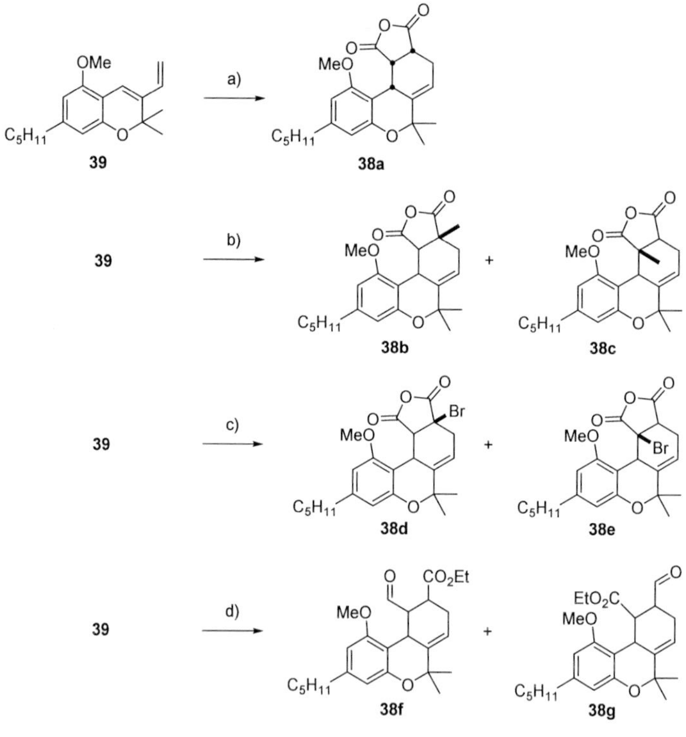

Scheme 12: DIELS-ALDER reactions with diene **39**: *a) maleic anhydride, ACN, 0 °C to r.t., 16 h, **38a**, quant.; b) methyl maleic anhydride, ACN, 0 °C to r.t., 60 h, **38b** 33%, **38c** 10%; c) bromo maleic anhydride, ACN, 0 °C to r.t., 36 h, **38d** 17%, **38e** not isolated; d) ethyl (E)-4-oxobut-2-enoate, DCM, 0 °C to r.t., 36 h, **38f** 14%, **38g** 7%.*

To increase the yields a reaction optimization for every dienophile might be necessary. The conversion could also be accelerated by the addition of a catalyst or an increase in reaction temperature. However, as was shown by MIANMI *et al.* especially higher reaction temperatures might allow even more side reactions such as isomerization of the resulting double bond.[163]

The synthetic transformation of anhydride **38a**, to alkene **57** followed an established optimized route by F. GLÄSER (Scheme 13). In her preliminary work she was able to determine the absolute configuration of anhydride **58** via X-ray crystallography.[114] First the double bond is hydrated with palladium on activated charcoal in a pressure device until quantitative conversion is observed. The anhydride is then exposed to hydrolyzing conditions, resulting in dicarboxylic acid **64**. Followed by an oxidative decarboxylation in the presence of lead acetate. Alkene **57** can be isolated in moderate yields of up to 43%. Again, the absolute configuration of **57** has been determined via X-ray crystallography.[114] It was found that the alkene is relatively stable at –20 °C and under argon atmosphere but shows signs of degradation at room temperature either dry or when dissolved in various solvents. The most likely explanation for this observation would be an oxidation to the diene and ultimately to an aromatic ring, but this theory could not be proven via analytical experiments since the isolation of single compounds after degradation was unsuccessful so far.

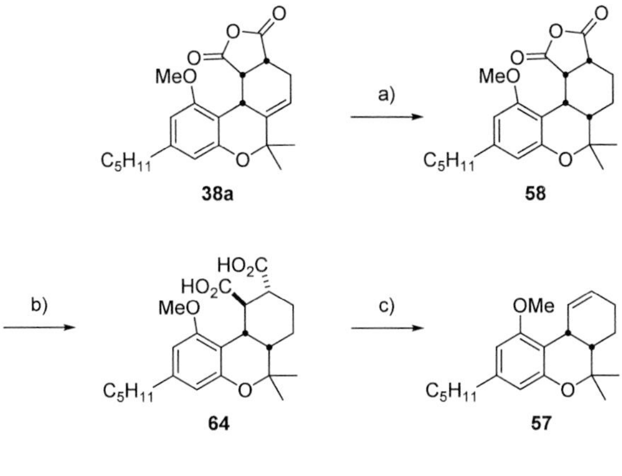

Scheme 13: Synthetic steps towards alkene **57**:[114] *a) 25 mol% Pd/C, 20 bar H_2, EtOAc, 16 h, 99%; b) $NaHCO_3$-solution, 80 °C, 3 h, 0 °C, HCl, 98%; c) $Pb(OAc)_4$, pyridine, 55 °C, 2 h, 43%.*[114]

To follow the retrosynthetic approach (Scheme 10) the next synthetic transformation is the epoxidation of alkene **57**. Several epoxidation methods were tested and are summarized in Table 1. Using standard epoxidation techniques with *m*-CPBA the product was obtained in low yield (2%, Entry 1). When *m*-CPBA was purified prior to use,[167] the yield could be increased to 9% (Entry 2). However, TLC and GCMS monitoring of the reaction showed the formation of several side products whose structure could not be elucidated. A system of trifluoroacetic anhydride (TFAA) and urea-hydrogenperoxide led to the loss of the starting material without the identification of the desired product. Hydrogenperoxide in *iso*-propanol in various concentrations showed no turnover and the starting material was recovered (Entry 5). Finally, the employment of dimethyl dioxirane (DMDO, **65**), a potent but mild epoxidation agent, led to the product in acceptable yield of 23% (Entry 5). The yield was optimized to 28% using 1.50 equiv. of DMDO (**65**) and by removal of the volatiles at 0 °C by condensation into a liquid nitrogen cooling trap under high vacuum (Entry 6).

Table 1: Summary of tested epoxidation reactions with alkene **57**: *A: m-CPBA (70%), DCM, 0 °C to r.t., 12 h; B: m-CPBA (99%), DCM, 0 °C to r.t., 12 h; C: TFAA, $CO(NH_2)_2$•H_2O_2, DCM, 0 °C, 2 h; D: 2.00 equiv. H_2O_2, NaOH, iPrOH, 0 °C, 2 h, r.t., 12 h; E: 1.00 equiv. DMDO (**65**), acetone, 0 °C, 2 h, r.t., 12 h. F: 1.50 equiv. DMDO (**65**), acetone, 0 °C, 2 h, volatiles removed in high vacuum with cooling trap.*

Entry	Conditions	Oxidant	Yield 56 [%]
1	A	*m*-CPBA 70%	2
2	B	*m*-CPBA 99%	9
3	C	$CO(NH_2)_2$•H_2O_2	–[a]
4	D	H_2O_2	–[b]
5	E	DMDO (**65**)	23
6	F	DMDO (**65**)	28

[a] no product, starting material was lost, [b] no turnover, partial recovery of starting material.

Since there is no commercial source for DMDO (**65**) it had to be freshly prepared according to a simplified method provided by TABER *et al.* (Scheme 14).[168] Small amounts of the reagent are generated by suspending potassium peroxymonosulfate (tradename Oxone®) with potassium carbonate in a water/acetone mixture and slow distillation into a bump trap. The concentration is determined prior to use by oxidation of a measured amount of phenyl methyl sulfide to phenyl methyl sulfoxide in deuterated acetone and comparison of the ^{1}H NMR spectra. The concentrations achieved varied from 40-80 mM DMDO in acetone.

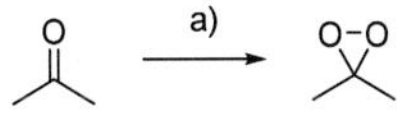

Scheme 14: Simplified DMDO (**65**) synthesis according to TABER *et al.*:[168] *a)* H_2O/*acetone,* K_2CO_3, $KHSO_5$ *0 °C to r.t., 1 h.*

Absolute *cis*-configuration of the epoxide was already determined in the master thesis[1] via NOESY(*nuclear overhauser enhancement and exchange spectroscopy*) NMR spectroscopy (Figure 12). NOESY spectroscopy allows to determine coupling between protons that are spatially close in a molecule. The intensity of the coupling decreases with increasing distance (proportionally to $I \sim \frac{1}{r^6}$). The observed coupling between the protons at 9b and 9c can only occur for the *cis*-, but not the *trans*- configuration.

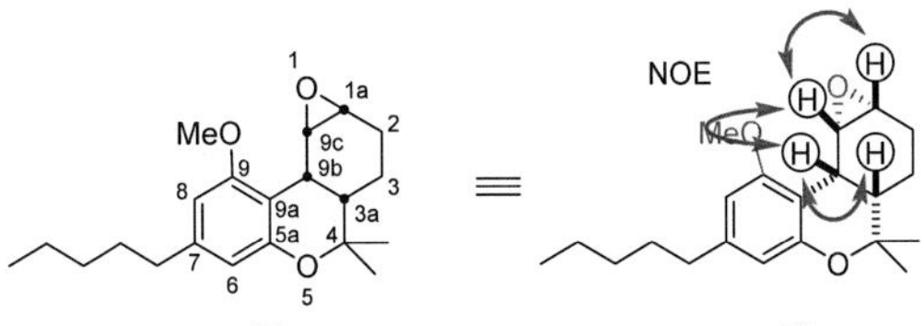

Figure 12: NOESY-spectroscopy was used to determine the absolute configuration of epoxide **56**, the red arrows on the right visualize the important NOE couplings that were observed.[1]

In addition to the epoxidation several alternative pathways to modify alkene **57** were investigated. Several ideas were tested and are shown in Scheme 15. The first idea was to brominate the double bond with molecular bromine to result in either dibrominated **67** or after dehydrohalogenation in monobromide **68**. The reaction mixture was analyzed by TLC and GCMS and revealed that a series of side reactions happened. The mixture could not be separated by chromatographic purification and the reaction was discarded. The second idea, the lithiation of alkene **57** and subsequent quenching with a methyl electrophile (methyl iodide) showed no conversion at all. The initial objective to introduce the methyl group and target THC (**1**) in one step was a speculative option from the start. There is no guaranty that the lithiation takes place in the right position as the estimated pKs values for other protons in the molecule are lower. Initially the idea was inspired by the work of NIKAS *et al.* who reported the *in situ* formation of a similar lithiated species.[53] In their system, they used a hydrazine precursor which afforded the organo lithium species upon treatment with *n*-butyl lithium and thus directed the position of the lithiation.

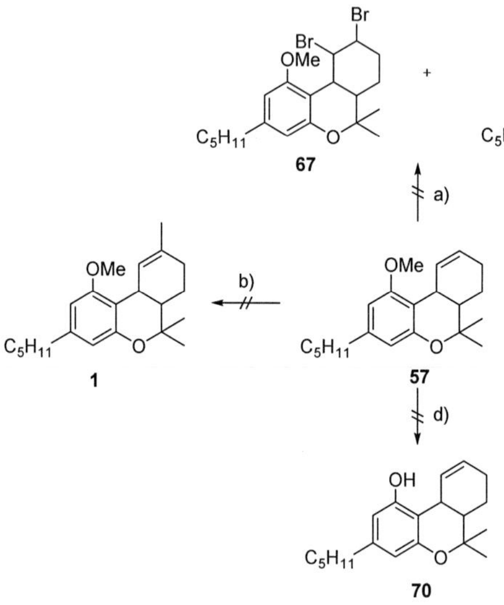

Scheme 15: Reactions performed on alkene **57**: *a) 1.00 equiv. Br_2, $CHCl_3$, 2 h r.t.; b) 2.30 equiv. n-BuLi, TMEDA, hexane, –78 °C, 15 min, 0 °C, 20 min, MeI, 10 min; c) 2.10 eq, Cu/Zn, 4.10 equiv. CH_2I_2, diethyl ether, reflux, 15 h.; d) 10 equiv. sodium thiomethoxide, DMF, 140 °C, 3 h.*

A third option for the modification of alkene **57** was explored with the attempted cyclopropanation. Even though GCMS analysis hints that traces of the product had been formed, no product was isolated after several chromatographic attempts. The effort was dropped as it became apparent that extensive optimization would be necessary and not enough alkene was available for this purpose. However, cyclopropane **69** embodies a very interesting compound that should have analogue properties to THC and constitutes a promising target for further research.

As empirical data for synthetic cannabinoids shows, the free hydroxy group found in THC (**1**) has severe influence on the affinity towards the cannabinoid receptors (compare Chapter 1.4).[96,108-109] Therefore, a deprotection of the methoxy ether was attempted. The standard method to deprotect a phenol with boron tribromide leads as expected to side reactions and ultimately destroys the compound. A milder procedure that is reported for the deprotection of THC[56] was tested but initial results were negative and further investigations have not been performed.

The synthetic transformations of alkene **57** are limited as relatively harsh conditions must be applied to process the double bond. Furthermore, analysis is complicated by the possible stereo configurations in the product and the overall lability of the starting material. However, the successful epoxidation shows persistence might allow further transformations. The synthetic effort in this work was focused towards the formal total synthesis progressing from epoxide **56**.

The epoxide **56** was treated with methyl lithium to achieve a selective ring opening to alcohol **71** which was isolated in 64% yield. Methyl lithium abstracts the 9b proton and the formation of the double bond should proceed while retaining the stereochemistry. Therefore, the *R*-configuration of the 1a-postion in the epoxide is preserved for the hydroxy group in the product. Not observed is an nucleophilic attack of a methyl anion which would have introduced a methyl group in 1a or 9c position of the epoxide **56**.

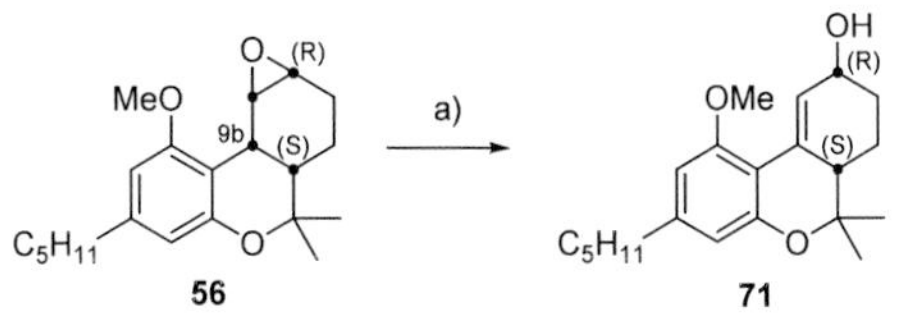

Scheme 16: Epoxide opening to *α,β*-unsaturated alcohol **71**: *a) THF, MeLi, –70 °C, 2 h, 64%.*

The oxidation of *α,β*-unsaturated alcohol **71** with Dess-Martin periodinane at room temperature proceeds in mediocre yield (Scheme 17). The formation of *α,β*-unsaturated ketone **55** symbolizes the formal total synthesis of THC (**1**). The necessary steps are first a deprotection of the methoxy functionality according to HUFFMAN *et al.*[164-165] followed by several steps according to the total synthesis by FAHRENHOLZ *et al.*[40] (compare Chapter 1.2.1).

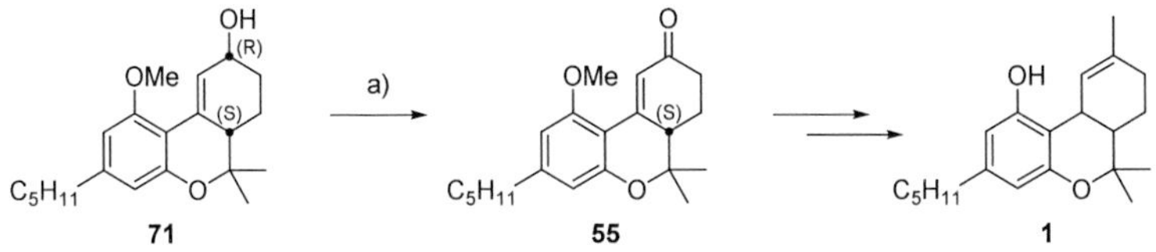

Scheme 17: Oxidation to *α,β*-unsaturated ketone **56** and formal total synthesis of THC: *a) DCM, DESS-MARTIN-Periodinane, r.t., 4 h, 49%.*

3.2 Synthetic Cannabinoid Analogs based on the Coumarin Motif

Interest of our group in cannabinoid analogs began with the first positive results for 3-benzylcoumarins **26** regarding their affinity towards the cannabis receptors. Coumarin **26a** (Figure 13) is one of the first compounds, synthesized by Dr. JAKOB TORÄNG, with high affinity and K_i-values in the high nanomolar range.[169] At the onset of the project, Dr. NICOLE VOLZ began to design novel compounds with the focus on enhanced affinity and selectivity between the two main receptors CB_1 and CB_2.[111] An example of such an selective ligand is **26b**, which shows a 100fold higher affinity towards CB_2 when compared to CB_1. Since the discovery of a novel cannabinoid receptor GPR55,[85] the coumarins have been tested for their affinity towards this receptor. The pharmacophoric requirements of GPR55 differ slightly from CB_1 and CB_2 and led to the inclusion of smaller substituents.[170] An example for such a GPR55-active coumarin is given with 5-isopropyl-8-methyl benzylcoumarin **26c**.

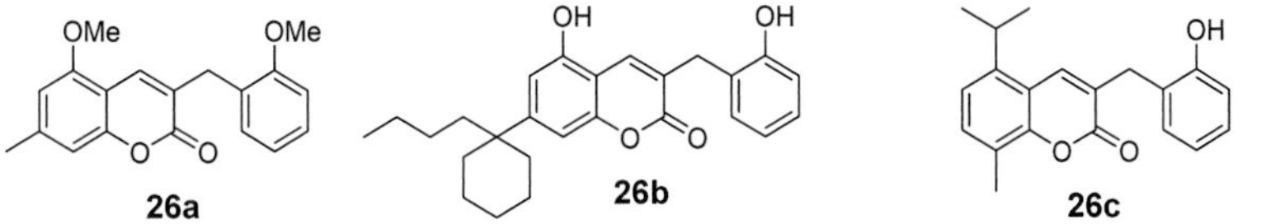

OMe OMe O O
26a
K_i (hCB_1) = 738 nM
K_i (hCB_2) = 944 nM

OH OH O O
26b
K_i (hCB_1) = 4890 nM
K_i (hCB_2) = 49 nM

OH O O
26c
IC_{50} (hGPR55) = 1.77 μm
K_i (hCB_1) = 738 nM
K_i (hCB_2) = 944 nM

Figure 13: Cannabinoid lead structures **26a** - **26c**.[111,169]

SAR studies on an increasing number of coumarins allowed the assessment of pharmacophoric sites that comply mostly with those on the THC scaffold (Chapter 1.4).[111,114,169] The most important pharmacophores are the aliphatic substitution in the 7-position, phenolic hydroxy in the 5-position and the substitution of the 3-benyzl moiety. On the other hand it was found that, contrary to initial assumptions, small substituents (e.g. methyl, isopropyl, methoxy) lead to increased GPR55-interaction.[106] This led to the inclusion of smaller substituents in positions 5-8 to the lead structures. In addition, it was found that some compounds act as CBD (**2**) analogues in non-CB receptor related studies. CBD analogues present a non-psychoactive category of compounds that is highly interesting for therapeutic approaches to a broad range of diseases.[171] In summary, this led to new target structures that seek broader diversification (Figure 14). Especially the potential of 3-alkyl, 3-styryl and 3-arylcoumarins was of interest and synthesis methods for these compounds were devised.

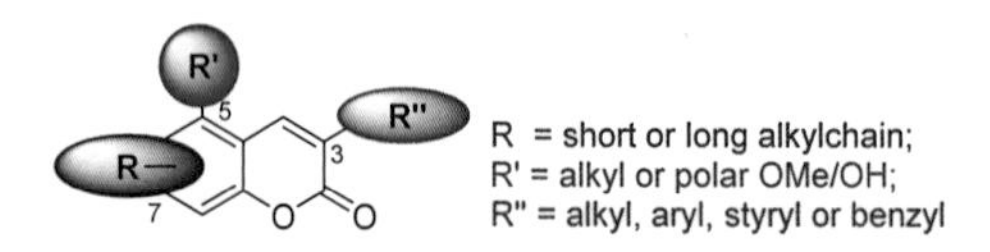

Figure 14: Proposed variations of substitution to generate cannabinoid active compounds from coumarin.[114]

In this work, several of the more potent compounds from previous screens where resynthesized, when demand by our biological partners arose, i.e. when initial findings required larger amounts for verification or advanced screenings were proposed. Close modifications of the required compounds were implemented to gather more information on SAR. The focus, however, was shifted towards the expansion of the existing libraries by pushing the established methods to the limits, as well as the development of new potential analogues by rational design. For several substance classes new synthetic strategies were developed. As the common motif for all novel compounds the coumarin core structure was retained, which allows for robust compounds and modular syntheses.

3.2.1 Synthesis of Salicylic Aldehydes

The preparation of coumarins in a modular fashion begins with the synthesis of the building blocks. The synthetic strategies have been developed, so that a mutual origin, i.e. salicylic aldehydes **29**, built the foundation of the strategies.

This subchapter summarizes the synthesized salicylic aldehydes **29**, that were later used for the generation of coumarin libraries. Several salicylic aldehyde syntheses were reproduced according to previous work and have been supplemented by the synthesis of novel compounds.

Pentyl substituted salicylic aldehyde **29a** (Scheme 18) was synthesized according to an established protocol.[114] Starting from olivetol **72**, a methylation reaction with dimethyl sulfate achieves dimethoxy ether **73** in good yield. Lithiation and quenching with dimethylformamide (DMF) results in aldehyde **74** in excellent yield, which is then selectively mono-deprotected to salicylic aldehyde **29a**.

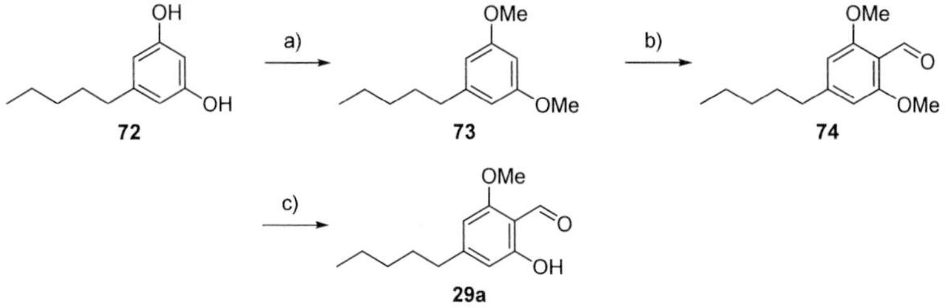

Scheme 18: Synthesis of pentyl-substituted salicylic aldehyde **29a**: *a)* Me_2SO_4, K_2CO_3, *acetone, reflux, 4 h, 82%; b) 1) N,N,N',N'-Tetramethylethane-1,2-diamine (TMEDA), diethyl ether, 0 °C, n-BuLi, 2 h, r.t., 2) 0 °C, DMF, 4 h, r.t., 95%; c)* $AlCl_3$, *NaI, ACN, DCM, 1.5 h, r.t., 80%.*

The synthesis sequence for salicylic aldehydes bearing a 1-butylcyclopentyl **29b** or 1-butylcyclohexyl side chain **29c** is likewise following an established protocol (Scheme 19).[111] Proceeding from nitrile **75**, different cyclic alkyl chains can be introduced under strong basic conditions. In the described reaction sequence, dibromobutane and dibromopentane were employed (Scheme 19, reaction a). However, other cyclic chains can be installed in a similar manner.[112] In the next step (b), the nitriles are reduced to the aldehydes **77** which can then undergo WITTIG-reactions (c) with *n*-propyl triphenylphosphonium bromide. The resulting alkenes **78** are reduced with hydrogen in the presence of palladium (d) and an aldehyde is introduced via lithiation and quenching with DMF (e). Selective mono-deprotection (f) results in the desired salicylic aldehydes **29b** and **29c**.

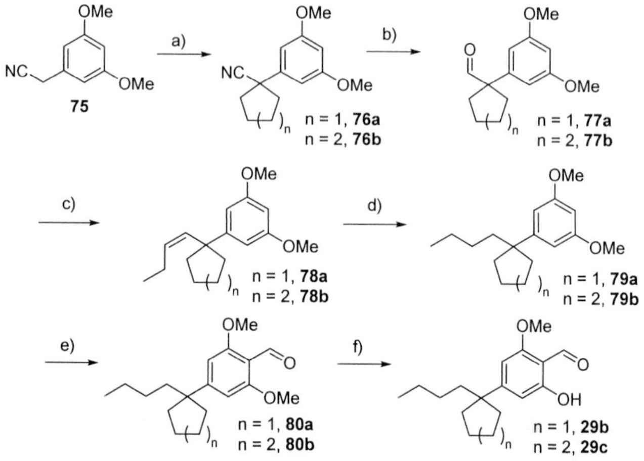

Scheme 19. Synthetic route towards salicylic aldehydes **29b** and **29c**: *a) KHMDS, THF, dibromo alkane, –16 °C to r.t., 16 h,* ***76a****, 88%,* ***76b****, 78%; b) DIBAL-H, DCM, –78 °C, 1 h,* ***77a*** *86%,* ***77b*** *92%; c) 1) n-propyl triphenyl phosphonium bromide, KHMDS, THF, 0 °C to 10 °C, 30 min, 2)* ***77a*** *or* ***77b****, 10 °C, 1 h,* ***78a*** *99%,* ***78b*** *99%; d) Pd/C, H_2-atmosphere, ethyl acetate, 24 h,* ***79a*** *99%,* ***79b*** *97%; e) 1.) TMEDA, diethyl ether, 0 °C, n-BuLi, 2 h, r.t., 2.) 0 °C, DMF, 4 h, r.t.,* ***80a*** *n.d.,* ***80b*** *91%; g) $AlCl_3$, NaI, ACN, DCM, 1.5 h, r.t.,* ***29a*** *69%,* ***29b*** *80%.*

In addition to the established salicylic aldehydes, derivatives with an ether function in 7-position were proposed (**29d-f**, Scheme 20). The idea behind these structures were to shorten the sequences, i.e. for cyclopentyl substituted salicylic aldehydes, as well as a modular access to the side chains. Besides, as described in Chapter 1.4, the side chains in 7-position that contain polar moieties should still be pharmacologically active.[115] The length of the aliphatic chain of the ether was chosen to result in the same chain length as the aliphatic salicylic aldehydes, in order to give comparable results in biological studies.

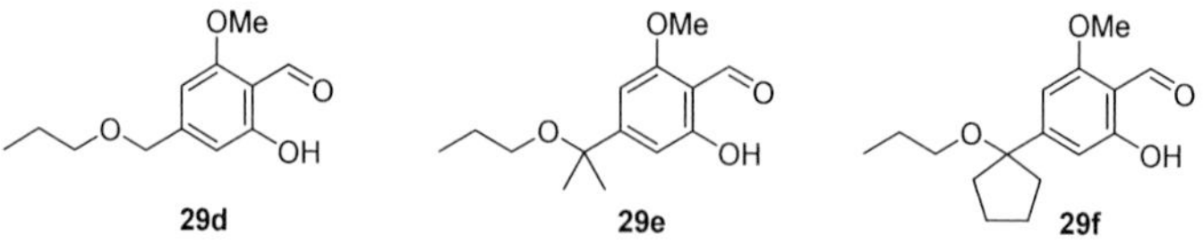

Scheme 20. Proposed structures for salicylic aldehydes **29e-f** with an ether function.

The strategy for the synthesis for the salicylic aldehydes is based on benzyl alcohols **81a-c** (Scheme 21). Starting from these benzyl alcohols **81a-c**, a WILLIAMSON ether synthesis with bromopropane should result in benzyl ethers **82a-c**. Subsequently, a formylation reaction ought to allow access to aldehydes **93a-c**. Followed by a selective mono-deprotection, the desired salicylic aldehydes **29d-f** could then be obtained.

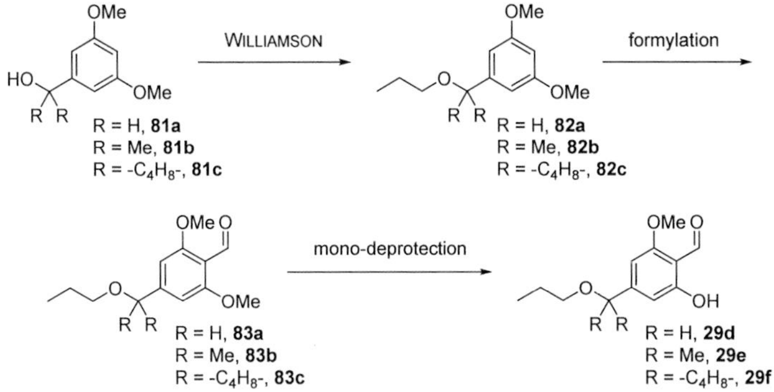

Scheme 21. Synthesis strategy for benzyl ether salicylic aldehydes **29d-f**.

The developed strategy was executed in a bachelor project by SARAH AL MUTHAFER.[172] While benzyl alcohols **81a** and **81b** can be purchased, the cyclopentylbenzyl alcohol **81c** had to be synthesized (Scheme 22). This was achieved starting from 1-bromo-3,5-dimethoxybenzene (**84**). First, the compound was *in situ* transferred with magnesium to the GRIGNARD reagent **85**, which is then quenched with cyclopentanone to give **81c** in 76% yield.

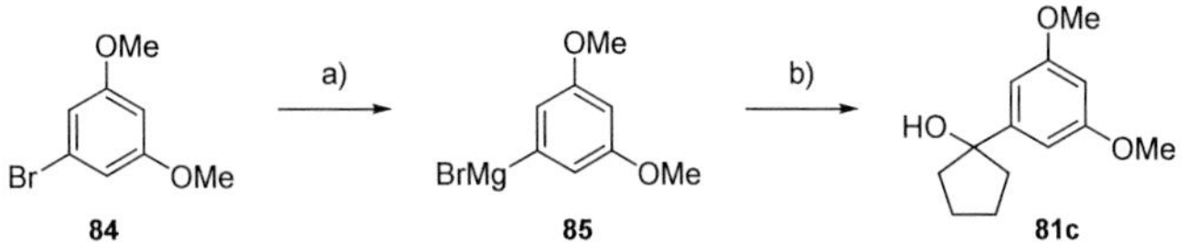

Scheme 22. Synthesis of cyclopentylbenzyl alcohol **81c**: *a) Mg, THF, r.t., 1 h; b) cyclopentanone, r.t., 1 h, 76%.*

With all three benzyl alcohols **81a-c** in hands, the syntheses of the salicylic aldehydes were carried out. The results of the individual synthesis steps are summarized in Table 2. The yields for the WILLIAMSON ether synthesis decrease with increasing steric demand of the benzyl alcohols. While the unsubstituted alcohol **81a** gives an excellent yield of 99%, the dimethyl-substituted alcohol **81b** allows only a moderate yield of 54%. The cyclopentyl-substituted alcohol **81c** results in good of 85% yield which matches the expectations since the steric demand of the cyclopentane ring is lower than that of the two methyl groups. The yields for the second reaction step range between 43 and 86%. The lower yield for unsubstituted ether **81a** can be explained through possible side reactions due to the strong base, *n*-butyl lithium employed in the reaction. Benzyl ethers have been known to be labile in the presence of organo lithium-species, or could undergo a WITTIG rearrangement due to the abstraction of an α-proton.[173-176] In the final reaction step, the selective mono-deprotection, salicylic aldehydes **29d** and **29e** were obtained in good yields. The last salicylic aldehyde **29f** was only found in traces and an elimination product was isolated as a main compound.

Table 2: Summary of benzyl ether salicylic aldehydes **29d-f**: *a) n-bromopropane, NaH, THF, reflux, o.n.; b) 1) TMEDA, diethyl ether, 0 °C, n-BuLi, 2 h, r.t., 2) 0 °C, DMF, 4 h, r.t.; c) AlCl₃, NaI, ACN, DCM, 1.5 h, r.t.*.[172]

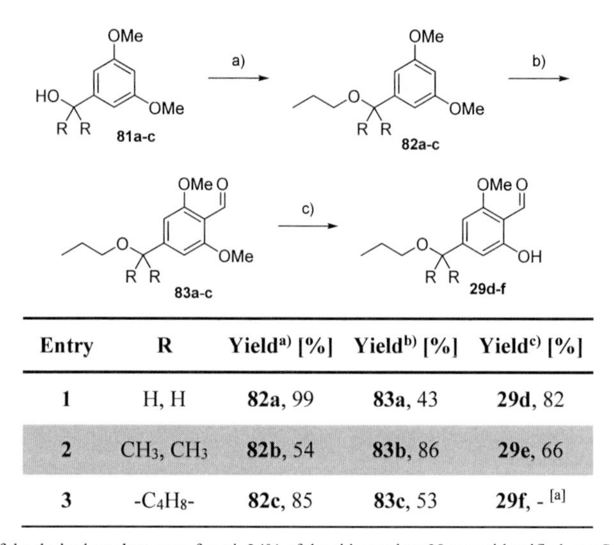

Entry	R	Yield$^{a)}$ [%]	Yield$^{b)}$ [%]	Yield$^{c)}$ [%]
1	H, H	**82a**, 99	**83a**, 43	**29d**, 82
2	CH_3, CH_3	**82b**, 54	**83b**, 86	**29e**, 66
3	-C_4H_8-	**82c**, 85	**83c**, 53	**29f**, - [a]

[a]only traces of the desired product were found, 24% of the side product **29g** was identified, see Scheme 23.

The side product for the mono-deprotection of **83c** was identified as elimination product **29g** and has been isolated in 24% yield (Scheme 23). There are several conceivable mechanistic pathways that could lead to the elimination. The most plausible mechanism seems to be a LEWIS acid-catalyzed E_1-elimination. The intermediary tertiary carbocation would be stabilized through the high substitution, benzylic position and no base is present in the reaction mixture.

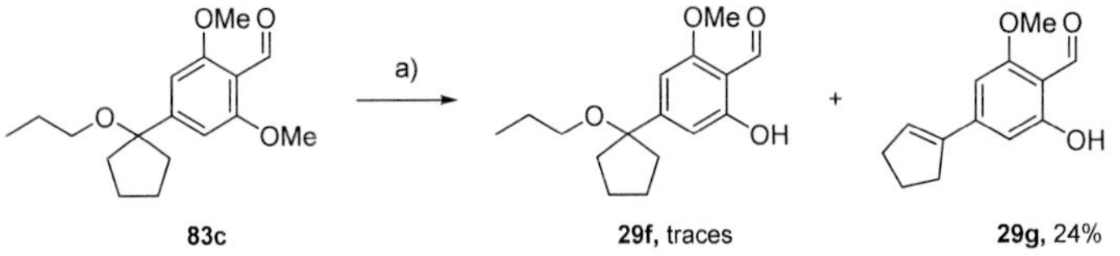

Scheme 23. Identified side product for mono deprotection on **83c**: *a) $AlCl_3$, NaI, ACN, DCM, 1.5 h, r.t., **29f** traces, **29g** 24%.*

In the next sequence, salicylic aldehydes with small substituents were the center of synthetic ambitions. As was shown for several examples in preliminary work, coumarins with small aliphatic substituents on the benzene moiety of the coumarin core exhibit antagonistic effects on GPR55.[170]

Therefore, trimethyl-substituted salicylic aldehyde **29h** was prepared. In the original protocol, the methylation is performed as the first step, followed by formylation. By adjusting the pathway and utilization of an alternative formylation method the overall yield was improved from 23%[114] to 28%. Additionally, the idea occurred, that the resulting salicylic aldehyde **87** from the first reaction step can be used in the coumarin syntheses to provide a free hydroxy group in one step. This proved to be impractical as the free hydroxy group interferes with the reaction conditions of the condensation reactions or leads to unwanted by-products.

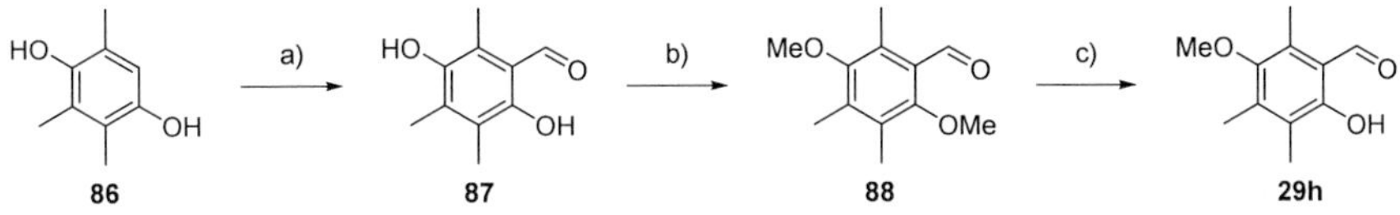

Scheme 24. Synthesis of salicylic aldehyde **29h**. *a) DCM, $TiCl_4$, 1,1-dichlorodimethylether, 0 °C to r.t. 15 h, 42%; b) K_2CO_3, dimethyl sulfate, acetone, 12 h, reflux, 75%; c) ACN, DCM, $AlCl_3$, NaI, 1 h, r.t., 89%.*

In previous work, a salicylic aldehyde based on the natural product carvacrol **89a** has been employed for the synthesis of biological active 3-benzylcoumarins.[111] This led to the idea to employ thymol **89d**, the regio isomer of carvacrol and also a natural product, to synthesize a salicylic aldehyde. JAN-SIMON FRIEDRICHS synthesized several derivatives (Scheme 25) and generated a small library of salicylic aldehydes with them during his bachelor work.[177] Starting from carvacrol **89a** and thymol **89d**, a bromination reaction led to the derivatives **89b** and **89e**. An additional transformation via an ULLMANN-like reaction on **89b**, with sodium methanolate and copper iodide as a catalyst, gave the third derivative **89c** according to a procedure of FIELDS *et al.*[22]

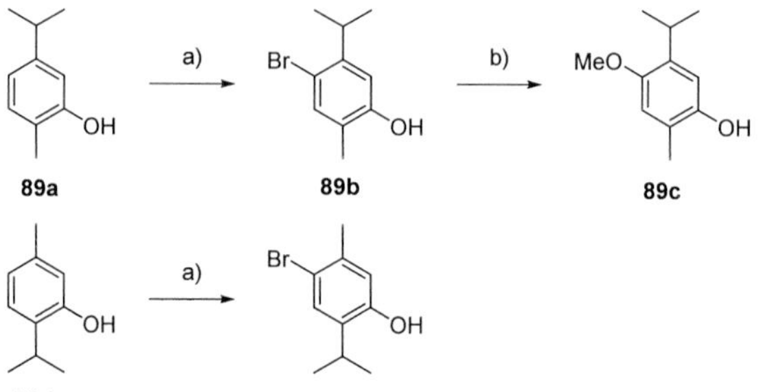

Scheme 25. Synthesis of salicylic aldehyde precursors **89c, 89e** and **89b**: *a) Br_2, AcOH, 0 °C to r.t., 3 h, **89c**, 51%, **89e**, 28%; b) CuI, NaOMe, DMF, 110 °C, 48 h, 70%.*[177]

The phenol derivatives **89a-e** were then formylated according to an established protocol with paraformaldehyde, magnesium chloride and triethylamine (Table 3).[178] The synthesis of the bromo substituted salicylic aldehydes **29k** and **29m** showed low yields and the respective phenols **89b** an **89e** might not be suitable for these reaction conditions. In the product fraction of **29j**, 6% of a side product was identified as methoxymethyl ether, also known as MOM protecting group. The formation of this by-product has been reported for other derivatives,[178] and should not interfere in subsequent reactions. Therefore, **29j** was used without further purification. Similarly, the product fraction of **29m** contained 12% of dibromo thymol that was already present as impurity in the starting material, where it was not detected by NMR-spectroscopy but could be identified later via GCMS. The higher concentration in the product fraction of the next step is explained by the low yield and parallel elution of the impurity in column chromatography. However, the impurities should not affect subsequent reactions and the salicylic aldehydes were used without further purification steps. In summary, all five phenol derivatives have been successfully synthesized in low to moderate yield (6–55%).

Table 3: Formylation reactions on carvacrol derivatives **29i-m**: *a) $(CH_2O)_n$, $MgCl_2$, NEt_3, ACN, 4-16 h, reflux.*

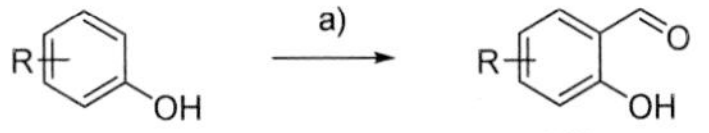

a)

89a-e → **29i-m**

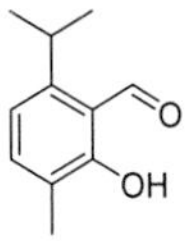

29i, 55%	**29j**, 24%[a]	**29k**, 10%	**29l**, 19%	**29m**, 6%[b]

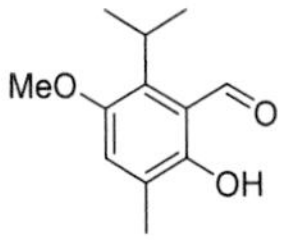

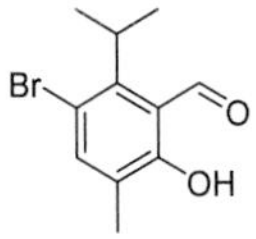

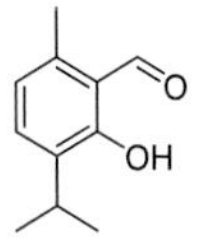

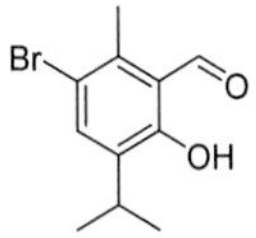

[a] Yield calculated by ^{1}H NMR, contains 6% 1-isopropyl-2-methoxy-5-(methoxymethoxy)-4-methylbenzene; [b] yield calculated by ^{1}H NMR, contains 12% of dibromo phenol by-product

The prepared salicylic aldehydes were used in the syntheses of the coumarins that will be discussed in the following sections.

3.2.2 3-Benzylcoumarins

The synthesis of cannabinoid analogs in our group began, when 3-benzylcoumarins were identified to have cannabinoid active properties.[111] A number of biological results are already available for this compound class and deductions could be made according to this data. The compounds mentioned in this chapter are in part resynthesized compounds, that had been specifically requested in bigger quantities by our cooperation partners, or variations of compounds that have been the most promising compounds in biological assays. The synthesis followed the method of Bräse *et al.*[105] described in detail in Chapter 1.4.1.

Table 4: Summary of the library synthesis of 3-benzylcoumarins **26k-m**: *a) IL **31**, K_2CO_3, toluene, 110 °C, MWI (max. 200 W, 7 bar), 50 min; b) BBr_3, DCM, 30 min, –78 °C, overnight, r.t.*

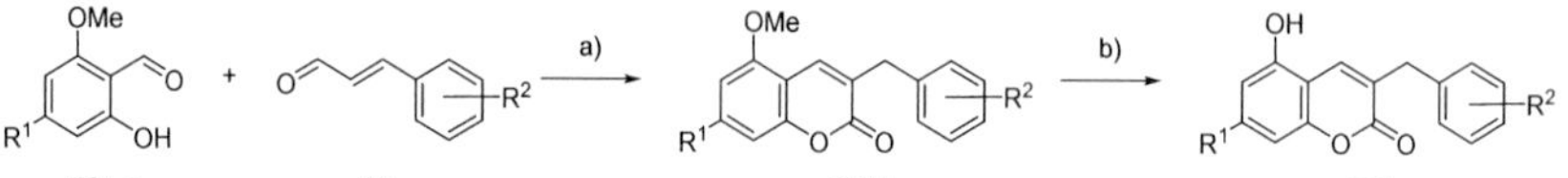

Entry	R^1	R^2	Yield[a)] [%]	Yield[b)] [%]
1	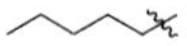	H	23, **26d**	–
2	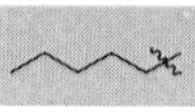	*o*-OMe	57, **26e**	–
3	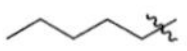	*o*-Me	67, **26f**	–
4	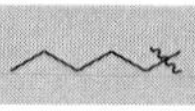	*p*-NMe_2	46, **26g**	63, **26k**
5	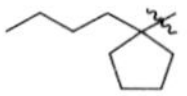	*o*-OMe	51, **26h**	79, **26l**
6	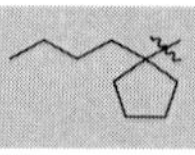	*p*-NMe_2	43, **26i**	49, **26m**
7	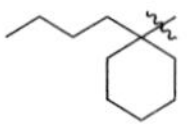	*p*-F	25, **26j**	–

Some of the cannabinoid compounds showed promising results as potential inhibitors of melanoma cell metastasis and tumor initiation comparable to CBD (**2**) (Cooperation with the SLEEMAN Group, Heidelberg University).[179] A few selected coumarins were requested in big quantities for *in vivo* studies (10 g/55 mmol scale). The necessary amounts that were previously required have never topped a quantity that had not been practicable to synthesize via scale-out methods. Therefore, the coumarin synthesis, i.e. the reaction shown in Scheme 26, has been performed in 10 g scale to investigate if the upscaling is feasible. Salicylic aldehyde **29i** was selected since it is accessible in a one-step synthesis and data for the formation of the coumarin is available for the default scaled reaction which is ~ 0.5 - 1.0 mmol. For the reaction a 30-fold upscaling was necessary to yield 10 g of the product **26c**. Although, the established protocol works well with microwave irradiation, it has been originally developed with standard heating conditions. Since our microwave does not meet the required dimensions, we decided to return to a classical set up and use a flask with a reflux condenser heated by an oil bath, which required a prolonged reaction time of 24 h.

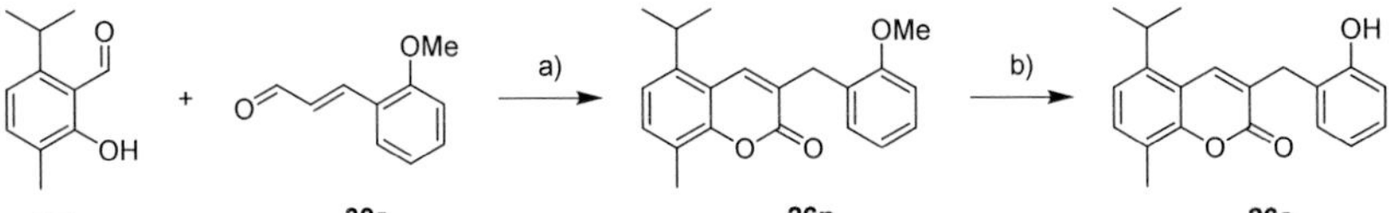

Scheme 26: Upscaling of the coumarin synthesis: *a) IL 31, K_2CO_3, toluene, reflux,24 h, 62%; b) BBr_3, DCM, 30 min, –78 °C, 16 h, r.t., 96%.*

The structure of **26c** was confirmed *via* X-ray analysis (Figure 15).

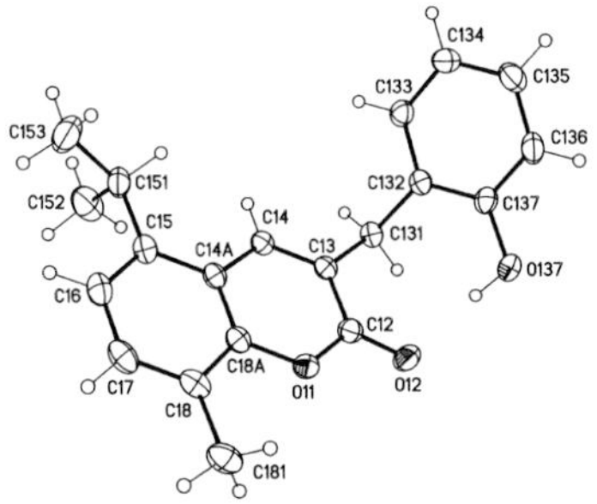

Figure 15: Molecular structure of **26c**. (X-ray image is depicted with displacement paraeters drawn at 50% probability levels).

The reaction proceeded equivalent to the small-scale microwave conditions. Unpolar and poorly soluble by-products that are also observed in the small scale ensured, a challenging purification solely due to the amount and a short filter-column was pre- run to the proper chromatography. Ultimately, the pure product was obtained in 62% yield after washing several times with 1 M sodium hydroxide-solution. The yield of the upscaled process is thereby even superior to the microwave reaction, where only 55% were obtained for the given salicylic aldehyde and cinnamic aldehyde combination. In conclusion, this result shows, that our method is indeed capable of producing high quantities, of 3-benzylcoumarins.

Furthermore, a small library of coumarins **26o-r** has been synthesized, where the isopropyl group and the methyl group on the coumarin moiety are inverted (Table 5). The structures show similarity to the 3-benzylcoumarin **26c** (Scheme 26) and were specifically designed to clarify structure activities when compared to the established compound.

Table 5: Summary of the library synthesis of 3-benzylcoumarins **26o-r** from salicylic aldehyde **29l**: *a) IL **31**, K_2CO_3, toluene, 110 °C, MWI (max. 200 W, 7 bar), 50 min.*

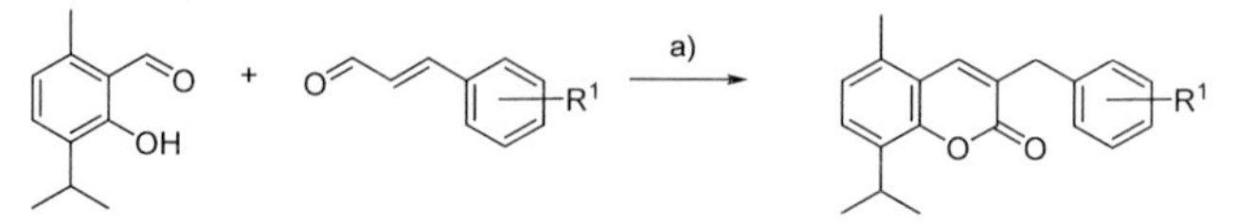

Entry	R^1	Yield [%]
1	H	41, **26o**
2	*o*-OMe	28, **26p**
3	*p*-Me	40, **26q**
4	*p*-Cl	24, **26r**

As mentioned before, the novel salicylic aldehydes that bear a benzyl ether moiety have been designed for the synthesis of coumarins. The available aldehydes were combined with several

Table 6: Summary of the library synthesis of 3-benzylcoumarins **26s-z**: *a) IL **31**, K_2CO_3, toluene, 110 °C, MWI (max. 200 W, 7 bar), 50 min.*

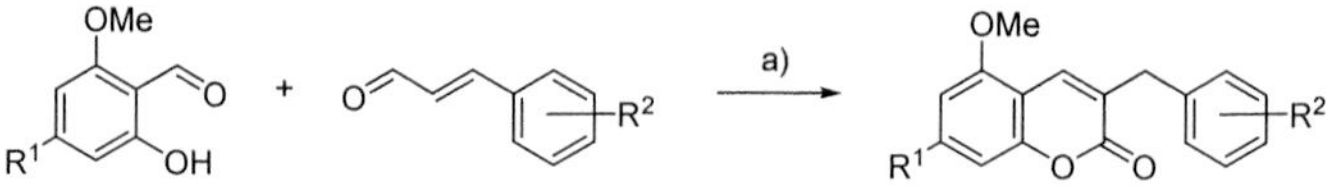

Entry	R^1	R^2	Yield [%]
1	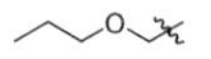	H	40, **26s**
2	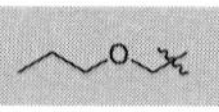	*o*-OMe	37, **26t**
3	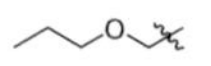	*p*-OMe	41, **26u**
4	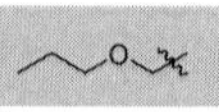	*p*-Cl	34, **26v**
5	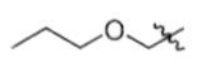	*o*-Me	81, **26w**
6	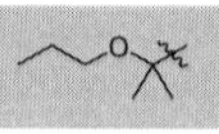	*o*-OMe	60, **26x**
7	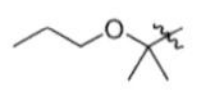	*o*-Me	46, **26y**
8	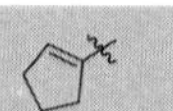	*o*-Me	31, **26z**

cinnamic aldehydes. The resulting coumarins are summarized in Table 6. The yields for the reactions are reasonable when compared to those regularly achieved in the reaction setup.

Pushing the substrate spectrum of the established 3-benzylcoumarin syntheses, heteroaromatic moieties should be introduced via the NHC mediated method. Therefore, the applicable cinnamic aldehyde derivatives had to be synthesized.

3.2.3 3-Pyridylmethylcoumarins

The first attempt for the synthesis of heteroaryl substituted cinnamic aldehyde derivatives, namely pyridyl acrylaldehydes **92**, led to the desired molecules but failed to achieve the compounds as pure products. Fortunately, the impurity was identified to be triphenylphosphine oxide **93**. While the compounds were obtained in the necessary quantities, several purification steps failed to yield the pure compounds for *m*-pyridyl acrylaldehyde **92a** and *p*-pyridyl acrylaldehyde **92c**. Therefore, an alternative pathway was investigated.

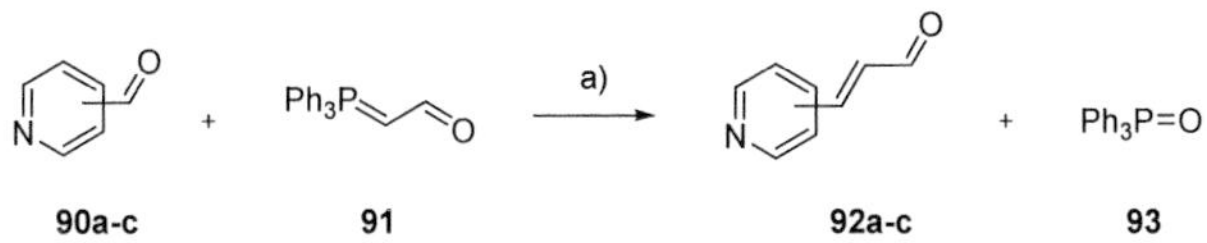

90a-c + Ph_3P=CH–CHO **91** —a)→ **92a-c** + Ph_3P=O **93**

Scheme 27: Synthesis of pyridyl acrylaldehyde **92a-c** *via* WITTIG-reaction: *a) nitrobenzene, **91**, 24 h, r.t.; **92a** 46%, **92b** 43% (60:40 mixture with Ph_3PO), **92c** 43% (64:36 mixture with Ph_3PO).*

In the alternative pathway (Scheme 28), a HECK reaction was employed. Pyridyl acrylaldehyde **92b** was obtained as pure compound in 71% yield, while the other two regioisomers **92a** and **92c** could not be isolated.

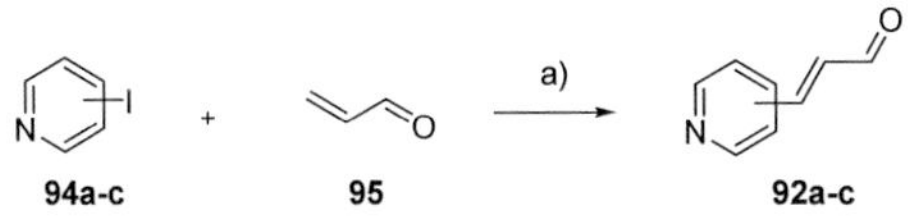

94a-c + **95** —a)→ **92a-c**

Scheme 28: Synthesis of pyridyl acrylaldehyde **92a-c** *via* HECK-reaction: *a) $Pd(OAc)_2$, n-Bu_4NI, $NaHCO_3$, DMF, 60 °C, 24 h. **92a** n.d., **92b** 70%, **92c** n.d..*

To test the influence of triphenylphosphine oxide, the two acrylaldehyde fractions from both synthetic pathways for *m*-pyridyl acrylaldehyde **92b**, were converted in a test reaction (Scheme

29). One batch of **92b** contained triphenylphosphine oxide (**93**) while the other one was pure. The isolated yield for both reactions was almost identical, which showed that the coumarin synthesis is not significantly impaired by the presence of triphenylphosphine oxide (**93**). Since the reaction appeared unaffected, the acrylaldehyde fractions still containing triphenylphosphine oxide (**93**) were employed for further transformations.

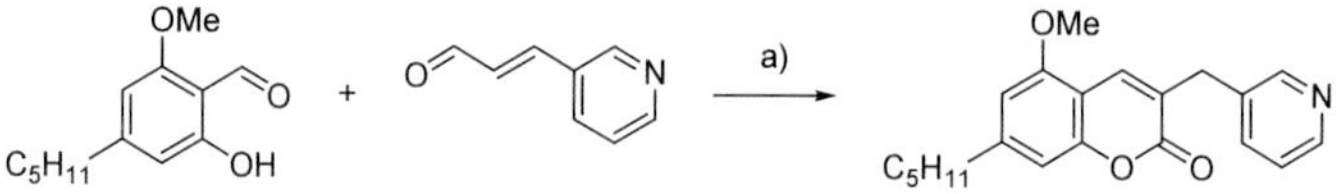

Scheme 29: Test reaction, performed to see the influence of triphenylphosphine oxide (**93**) in the reaction system: a) *IL 31, K_2CO_3, toluene, 110 °C, MWI (max. 200 W, 7 bar), 50 min; with $P(Ph_3)O$, 17%, without $P(Ph_3)O$, 17%.*

In the next step, a library of 3-pyridylmethylcoumarins **96** was synthesized (Table 7). The three salicylic aldehydes were chosen because comparable data with 3-benzyl analogues was already available. The yields for the synthesized 3-pyridylmethylcoumarins **96-Me** were low ~ 15–26%. The collected empirical data for the applied coumarin syntheses indicates that electron-deficient cinnamic aldehydes result in lower yields than electron-rich cinnamic aldehydes.[111,114] The rather small electronic pull caused by the nitrogen atom explains that behavior. Additionally, the pyridine moiety leads to a more difficult purification of the compounds, as the product fractions are distributed in bigger eluent volumes compared to other benzyl substituted compounds. The successful synthesized 3-pyridylmethyl compounds **96-Me** were then deprotected to further expand the compound library.

Table 7: Summary of the library synthesis of 3-pyridylmethylcoumarins **96-Me** and their deprotected derivativs **96-H**: *a) IL 31, K_2CO_3, toluene, 110 °C, MWI (max. 200 W, 7 bar), 50 min; b) BBr_3, DCM, 30 min, –78 °C, overnight, r.t.*

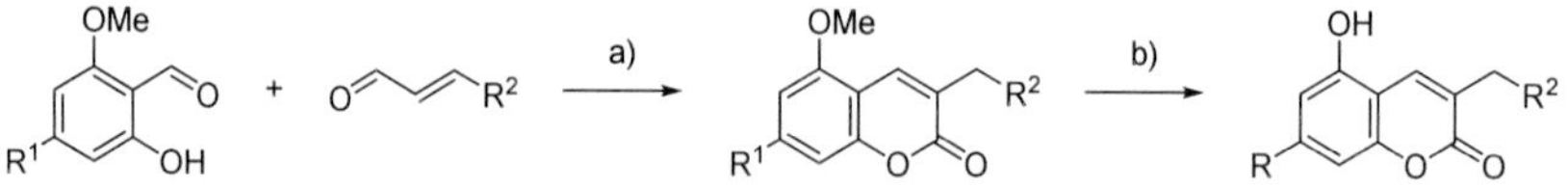

Entry	R¹	R²	Yield[a)] [%]	Yield[b)] [%]
1	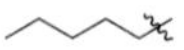	*o*-pyridyl	18, **96a-Me**	21, **96a-H**
2	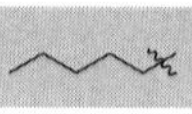	*m*-pyridyl	18, **96b-Me**	52, **96b-H**
3	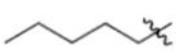	*p*-pyridyl	23, **96c-Me**	45, **96c-H**

Entry	R¹	R²	Yield[a)] [%]	Yield[b)] [%]
4	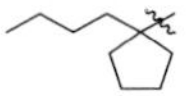	*o*-pyridyl	25, **96d-Me**	60, **96d-H**
5	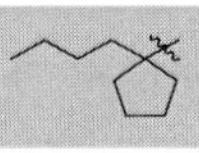	*m*-pyridyl	18, **96e-Me**	52, **96e-H**
6	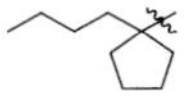	*p*-pyridyl	24, **96f-Me**	42, **96f-H**
7	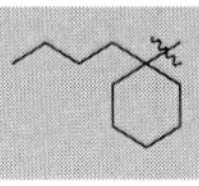	*o*-pyridyl	19, **96g-Me**	49, **96g-H**
8	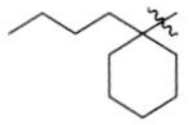	*m*-pyridyl	15, **96h-Me**	88, **96h-H**
9	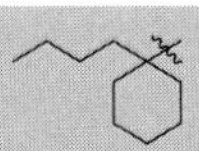	*p*-pyridyl	26, **96i-Me**	39, **96i-H**

In addition to 3-pyridylmethyl substitution, a reaction with the commercially available 2-furyl acrylaldehyde (**97**) was performed (Scheme 30). The 3-furylmethylcoumarin **98** was obtained in 30% yield, proving the transferability of the concept towards other heteroaryl substituted cinnamic aldehyde analogs. Depending on biological results for these compounds the enlargement of these libraries should be feasible. In contrast, the deprotection of coumarin **98** was not possible via the standard protocol involving BBr_3 (Scheme 30). A milder method for the deprotection is already in development in succeeding work.

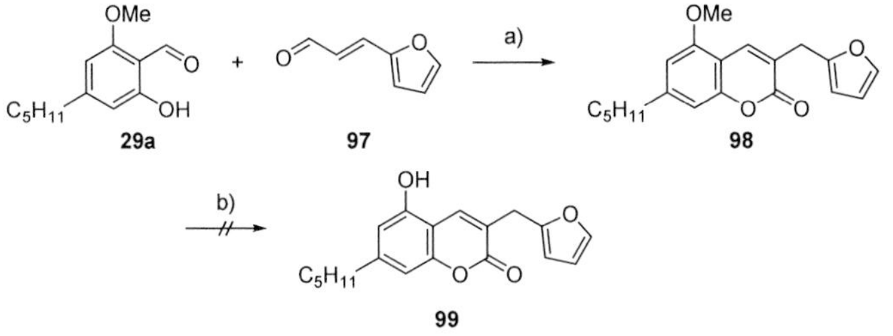

Scheme 30: Synthesis of furylmethyl coumarin **98**: *a) IL **31**, K_2CO_3, toluene, 110 °C, MWI (max. 200 W, 7 bar), 50 min, 30%; b) BBr_3, DCM, 30 min, –78 °C, overnight, r.t., no product obtained.*

3.2.4 3-Alkylcoumarins

Another compound class that was synthesized are 3-alkylcoumarins **27**. The substitution with small side chains promises to yield antagonistic ligands for the cannabinoid receptors while longer side chains might approximate in their behavior with endocannabinoids like anandamide. The idea behind this is that the long flexible side chains should be able to blend in the binding pockets. Aside from that they have been identified as lead structures for biological assays in which they act as CBD (**2**) analogs. They showed promising results in preliminary studies by our collaboration partners in the SLEEMAN group acting as CBD (**2**) analogs[179] and have potential for GPR55 activity.

The synthesis of 3-alkylcoumarins **27** is achieved via PERKIN reaction (Scheme 26). W. H. PERKIN first synthesized coumarin in 1868 by heating the sodium salt of salicylic aldehyde with acetic anhydride. He also synthesized derivatives by deploying carboxylic acids with longer side chains, which resulted in 3-alkyl substitution.[180] The mechanism of the PERKIN-reaction starts with a base-catalyzed acylation of salicylic aldehyde **29** with acid anhydride **100**. The resulting acyl species **101** is in equilibrium with the enolate form **102**, which can perform an intramolecular aldol addition to the intermediate **103**, followed by the elimination of water to result in the desired coumarin **27**.

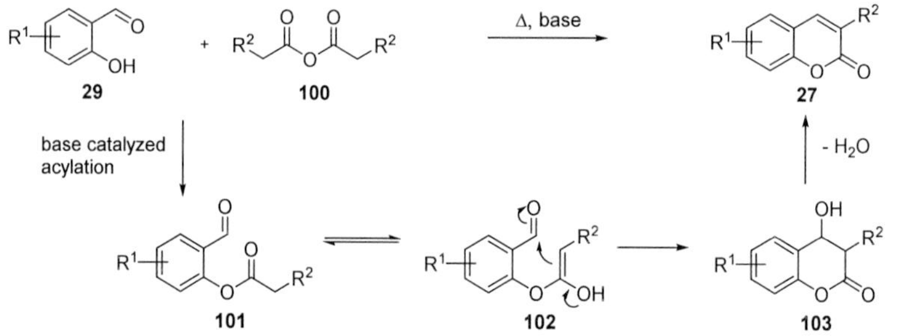

Scheme 31: PERKIN synthesis[180] and its mechanism.[181]

In practice, a reaction protocol by GÁPLOVSKÝ *et al.* was adapted. The protocol has been optimized for PERKIN reactions under microwave irradiation and allows to shorten reaction times while simultaneously optimizing reaction yields under solvent free conditions. The conducted experiments have been summarized in Table 8 and have been performed in part (entries 3 and 4, Table 8) by JAN-SIMON FRIEDRICHS during his bachelor project.[177]

Table 8: Summary of the isolated yields [%] for the library synthesis of 3-alkylcoumarins **27**: *a) K_2CO_3, 180 °C, MWI (max. 300 W, 15 bar), 1 h.*

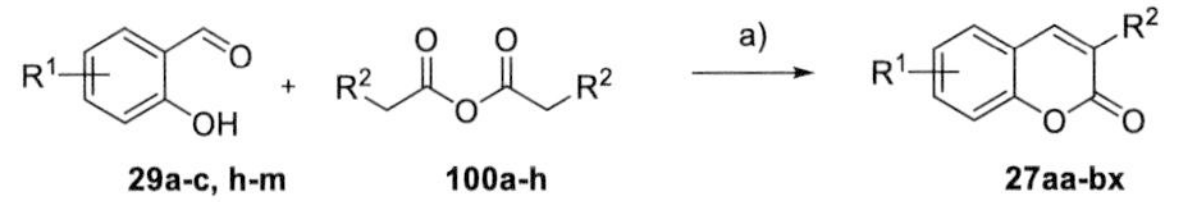

29a-c, h-m + **100a-h** —a)→ **27aa-bx**

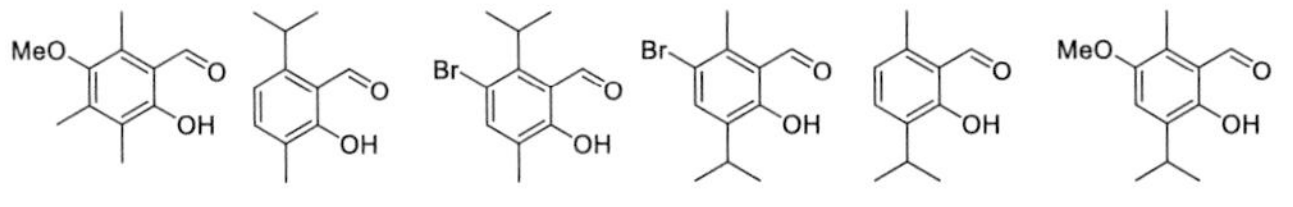

29h **29i** **29k** **29m** **29l** **29j**

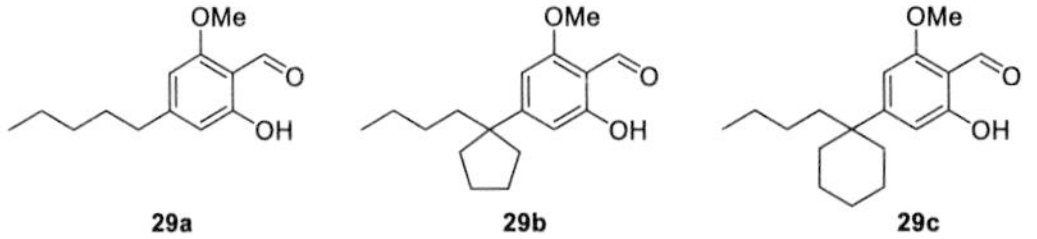

29a **29b** **29c**

Entry	*Salicylic aldehyde*	*Acetic anhydride* **100a**	*Propionic anhydride* **100b**	*Butyric anhydride* **100c**	*Valeric anhydride* **100d**	*Hexanoic anhydride* **100e**	*Heptanoic anhydride* **100f**	*Octanoic anhydride* **100g**	*Isovaleric anhydride* **100h**
		R^2=H	CH_3	C_2H_5	C_3H_7	C_4H_9	C_5H_{11}	C_6H_{13}	*i*Pr[b]
1	**29h**	82	89	–	–	–	–	–	–
2	**29i**	quant.	82	quant.	84	72	42	57	78
3[177]	**29k**	quant.	quant.	quant.	95%	n. i.[c]	n. i.[c]	86	71
4[177]	**29m**	85	80	73	13	n. i.[c]	n. i.[c]	n. i.[c]	n. i.[c]
5	**29l**	–	quant.	–	–	–	–	–	–
6	**29j**	95	74	83	55	57	82	61	quant.
7	**29a**	67[a]	80	63	81	68	–	–	–
8	**29b**	93	86	87	92	31	–	–	–
9	**29c**	56	77	93	68	82	–	–	–

[a]15% of OMe → OAc side product obtained, [b]reaction time 3 h; the reaction didn't proceed to completion after one hour and an intermediate ester was found, [c]due to purification issues these compounds have not been isolated (n. i.), a repetition of the experiments was not conducted so far.

The missing entries (entries 1 and 5) show experiments that have not been carried out so far. The two salicylic aldehydes had not been synthesized in adequate quantities, but the established protocols should allow to fill out the blanks efficiently and will be performed in succeeding works. For entries in 7 to 9 the permutation with carbonic acid anhydrides with more than six carbon atoms was not performed as the calculated log *P* and Clog *P* values make the compounds rather uninteresting. This is a result of the already implemented hydrophobic side chains in the 7-position.

In total 44 new 3-alkylcoumarins **27aa-27bx** were successfully synthesized. The entries are numbered according to their appearance in Table 8. The yields for the synthesis of 3-alkylcoumarins via PERKIN reaction were overall good. Where the yields are modest, issues with the purification were encountered. Entries 3 and 6 show several positions where the yield could not be determined, because no satisfactory purity was reached after repeated (2-3) chromatographical steps. There are two probable factors playing a role in this. Firstly, the presence of bromide in the salicylic aldehyde seems to facilitate the formation of side products that could not be identified at this stage. Secondly, we assume that the residual carbonic acids that inevitably form during the reaction could not be separated via column chromatography when their elution behavior is too similar to the desired compounds. The latter issue was solved later for entries 2 and 6 (Table 8) by including extensive washing steps in the purification set up, i.e. washing the crude mixtures with sizeable amounts (500 – 1000 mL) of sodium carbonate and sodium hydroxide solutions prior to chromatography.

Where applicable (Table 8, Entries 6-9) deprotections were carried out to further extend the library and alter the physicochemical properties. The deprotection was performed analog to the conditions described in previous chapters. The resulting compounds are summarized in Table 9. In total 22 deprotected 3-alkylcoumarins **27bb-H – 27bx-H** were isolated.

Table 9: Summary of the yields [%] for methoxy deprotections of 3-alkylcoumarins **27**: *a) BBr_3, DCM, 30 min, –78 °C, overnight, r.t.*

a)

27bb-bx → **27bb-H - 27bx-H**

27A **27B** **27C** **27D**

Entry	Core structure	Substituent R						
		H	CH_3	C_2H_5	C_3H_7	C_4H_9	C_5H_{11}	C_6H_{13}
1	**27A**	89	98	90	70	quant.	88	75
2	**27B**	69	86	89	89	78	–	–
3	**27C**	90	87	77	97	85	–	–
4	**27D**	87	92	91	90	92	–	–

3.2.5 3-Arylcoumarins

The first attempts for the synthesis of 3-arylcoumarins **40** were based on the idea to start from salicylic aldehydes **29**. That way, the modular character of the library synthesis could be retained and a vast number of compounds would be accessible, starting from already synthesized salicylic aldehydes **29**. Therefore, two different synthetic strategies involving phenylacetic acids **104** were tested on a substituted salicylic aldehyde **29a**. The tested reaction conditions are summarized in Table 10.

Table 10: Summary of attempted 3-arylcoumarin synthesis reactions from salicylic aldehyde **29a** and phenylacetic acid **104**: *A: 2 equiv. p-methoxy-phenylacetic acid (**104**), 6 equiv. DABCO, 90 min, 180 °C MWI (max. 200 W, 15 bar); B: 2 equiv. p-methoxy-phenylacetic acid (**104**), 6 equiv. DABCO, 15 min, 180 °C MWI (max. 200 W, 15 bar), C: 2 equiv. p-methoxy-phenylacetic acid (**104**), 6 equiv. DABCO, anisole, 16 h, reflux. D: 2 equiv. p-methoxy-phenylacetic acid (**104**), 26.5 equiv. acetic anhydride, 4.4 equiv. NEt_3, 120 °C, 8 h.*

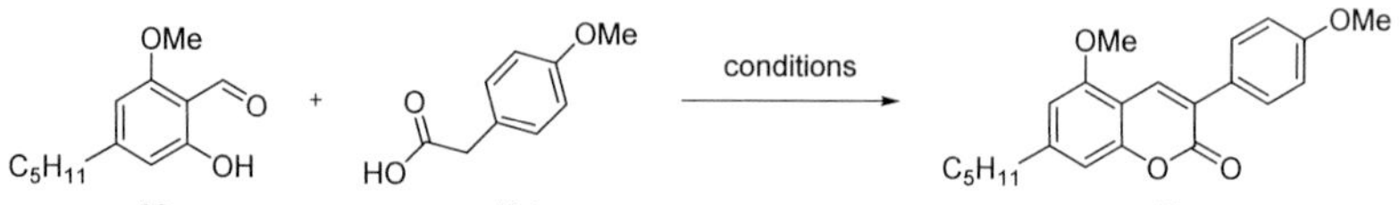

Entry	Conditions	Yield [%]
1	A	traces
2	B	traces
3	C	traces
4	D	60[a]

[a]difficult purification

The first attempts (Table 10, Entries 1-3) were founded on a 3-arylcoumarin synthesis reported by RAHMANI-NEZHAD *et al.*[182] In their synthesis, 1,4-diazabicyclo[2.2.2]octane (DABCO) catalyzed the reaction at 180 °C under solvent free conditions. However, for our substitution pattern, after 90 min microwave irradiation (Entry 1), the formation of many side products was observed. The reduction of the reaction time down to 15 min (Entry 2) showed similarly no improvement, as well as the implementation of anisole as a solvent and 16 h of heating at reflux in an oil bath. Only the employment of an alternative synthesis strategy by PU *et al.*,[183] that employs acetic acid anhydride and triethylamine gave the product in 60% yield. However, the purification involved repetition of several purification steps and another test reaction with a different salicylic aldehyde **29b** showed similar issues (Scheme 32).

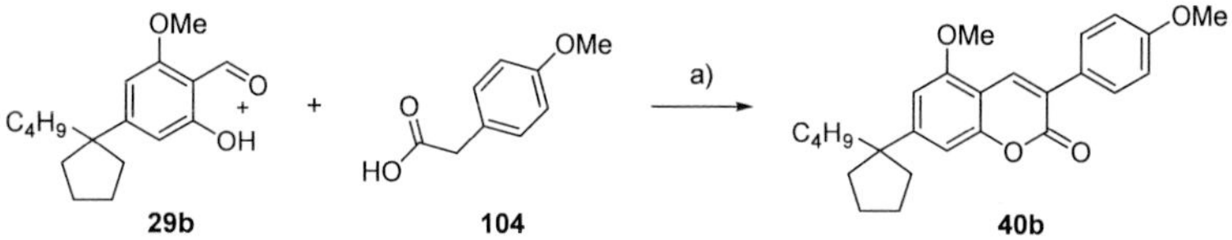

OMe C4H9 O + OH 29b + O HO OMe 104 a) C4H9 OMe O O OMe 40b

Scheme 32: Synthesis of 3-arylcoumarin **40b**: *2 equiv. p-methoxy-phenylacetic acid (**104**), 26.5 equiv. acetic anhydride, 4.4 equiv. NEt$_3$, 120 °C, 8 h, 62%.*

At this point, the synthesis strategy from salicylic aldehydes **29** was discontinued and we began to search for alternative pathways. The implementation of a SUZUKI-cross coupling, inspired by the work of BÄUERLE *et al.*, was taken into consideration. BÄUERLE *et al.* created a 3-arylcoumarin library containing 151 compounds via post condensation modification.[184] For the realization of this strategy, a 3-bromo substituted coumarin **105** was necessary. Therefore, 3-unsubstituted coumarin **27bo** was synthesized via PERKIN-reaction, as described in Chapter 3.2.4. Due to limitations in the experimental set up, the reaction is not readily upscalable, however, a scale-out-approach by repeating the same reaction several times resulted in adequate amounts for further investigations. Coumarin **27bo** was then brominated (Scheme 33) in a good yield of up to 83% according to a procedure developed by KIM *et al.*[185] The transformation proceeds via an intermediate dibromo compound, that eliminates under basic conditions to result in 3-bromocoumarin **105**.

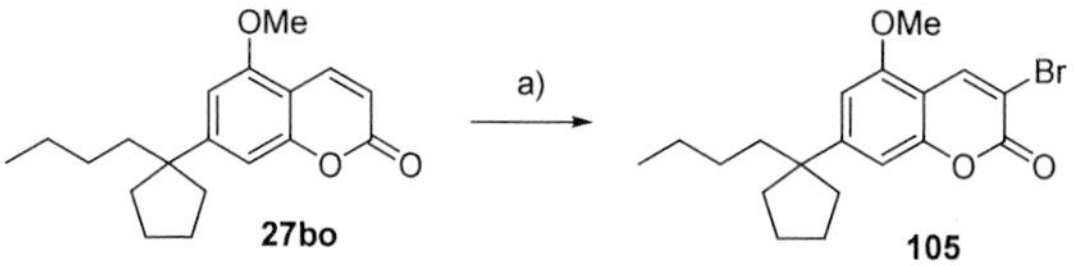

OMe O O 27bo a) OMe Br O O 105

Scheme 33: Synthesis of 3-bromocoumarin **105**: *a) 1.) Oxone®, HBr, DCM, 2 h, r.t., 2.) NEt$_3$, 15 min, r.t., 83%.*

The SUZUKI-reaction conditions had been optimized by BÄUERLE *et al.* for the use of boronic esters. The conditions were not optimal for the use of boronic acids **106a**, but a change of the base from cesium fluoride towards cesium carbonate resulted in an 15% increase of the yield in our reaction setting (Scheme 34) and allowed the isolation of **40c** in 82%.

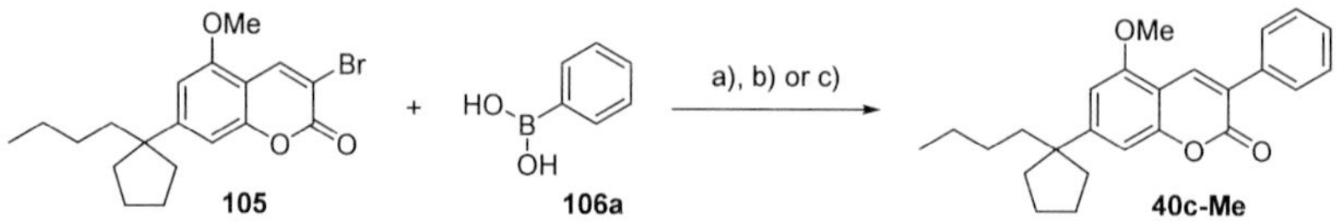

Scheme 34: Test reactions for SUZUKI-cross coupling: *a) 2 equiv. boronic acid* ***106a****, 2 equiv. CsF, 0.05 equiv.* $Pd(PPh_3)_4$*, degassed dioxane, 90 °C, 16 h, 65%; b) 2 equiv. arylboronic acid* ***106a****, 2 equiv. CsF, 0.05 equiv.* $Pd(PPh_3)_4$*, dioxane, 90 °C, 16 h, 67%, side products formed; c) 2 equiv. boronic acid, 2 equiv.* Cs_2CO_3*, 0.05 equiv.* $Pd(PPh_3)_4$*, degassed dioxane, 90 °C, 16 h, 82%.*

With the optimized reaction conditions, a library of 13 3-arylcoumarins **40-Me** was synthesized (Table 11). The selection of the boronic acids **106** was done mainly due to the comparability of the resulting structural features with 3-benzylcoumarins previously synthesized in our group.

For the reaction of *o*-fluorine boronic acid, no product was found; the reaction has not been repeated yet and the explanation for the failed reaction might also be a decomposed starting material. Apart from that, except for *p*-pyridyl boronic acid (Entry 10), all yields were fair or good (47-90%). The poor yield of 23% for the 3-pyridyl coumarin is at least partly caused by the purification via chromatography which is distinctly more challenging due to the high polarity.

However, further expansion of the library should be feasible if desired and is only limited by the number of boronic acids that are commercially available. In this, the system bears another advantage when compared to the initially developed methods. Boronic acids are, due to their widespread use in synthetic chemistry, obtainable with a wide variety. On the contrary, only a small number of phenylacetic acids, necessary for the alternative synthesis, would be available.

Table 11: Overview of synthesized 3-arylcoumarins **40-Me**: *a) 2 equiv. arylboronic acid* ***106****, 2 equiv.* Cs_2CO_3*, 0.05 equiv.* $Pd(PPh_3)_4$*, degassed dioxane, 90 h, 16 h.*

Entry	R	Yield [%]	Product
1	H	82	**40c-Me**
2	*o*-Me	90	**40d-Me**
3	*m*-Me	67	**40e-Me**
4	*p*-Me	73	**40f-Me**
5	*o*-OMe	63	**40g-Me**
6	*m*-OMe	83	**40h-Me**
7	*p*-OMe	79	**40b-Me**
8	*m*-NMe_2	63	**40i-Me**
9	*p*-NMe_2	90	**40j-Me**
10	*p*-pyridyl	23	**40k-Me**
11	*o*-F	-[a]	**40l-Me**
12	*m*-F	77	**40m-Me**
13	*p*-F	47	**40n-Me**
14	*p*-CF_3	56	**40o-Me**

[a] no product found.

The resulting 3-arylcoumarin library was subsequently treated with boron tribromide to generate free hydroxy groups from the methoxy ethers (Table 12). The two compounds *p*-pyridyl **40k-Me** and *p*-fluorine **40m-Me** were not deprotected, because not enough material was obtained in the SUZUKI-coupling.

Table 12: Overview of deprotection reactions on 3-arylcoumarins **40**: *a) 5 equiv. BBr_3, DCM, 30 min 0 °C, 16 h, r.t.*.

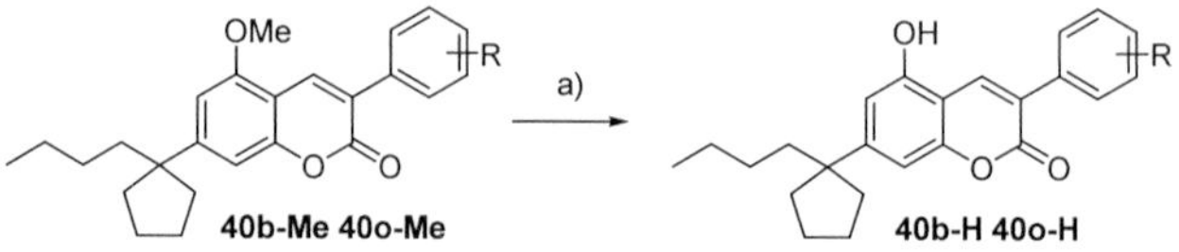

Entry	Starting Material	R	Yield [%]	Product
1	**40c-Me**	H	85	**40c-H**
2	**40d-Me**	*o*-Me	73	**40d-H**
3	**40e-Me**	*m*-Me	24	**40e-H**
4	**40f-Me**	*p*-Me	93	**40f-H**
5	**40g-Me**	*o*-OH	97	**40g-H**
6	**40h-Me**	*m*-OH	61	**40h-H**
7	**40b-Me**	*p*-OH	90	**40b-H**
8	**40i-Me**	*m*-NMe_2	29	**40i-H**
9	**40j-Me**	*p*-NMe_2	84	**40j-H**
11	**40m-Me**	*m*-F	64	**40m-H**
13	**40o-Me**	*p*-CF_3	79	**40o-H**

Interestingly, the yields of the deprotection for *m*-substituted 3-arylcoumarins is consistently lower than that for the respective *o*- and *p*-substituted derivatives. However, the amount of data is not sufficient to make an assumption about the cause.

The successfully synthesized compounds were handed over for biological screening. Depending on the results, SARs can be deduced and a directed synthesis towards relevant compounds can be carried out. Notably is the structural comparison between the 3-benzyl and 3-aryl compounds. The benzyl CH_2-group leads to a more flexible structure when compared to the rigid structure of the 3-arylcoumarins and can have significant influence on the receptor affinity. Additionally, the aromatic ring is expected to be out of plane in relation to the coumarin structure (Scheme 35).

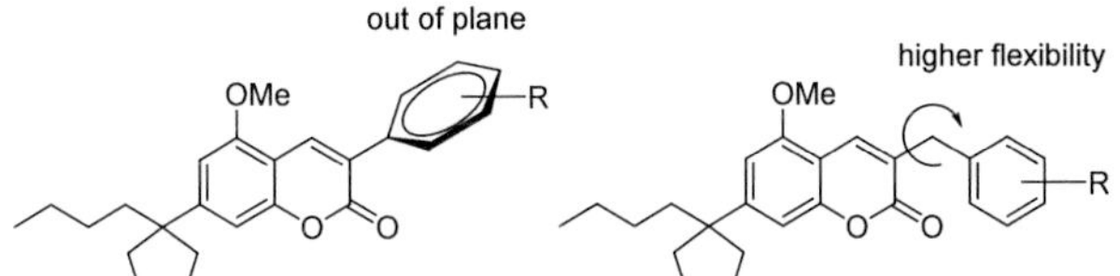

Scheme 35: Structural comparison between 3-arylcoumarins **40** and the established 3-benzylcoumarins **26**.

The out of plane situation was confirmed via X-Ray analysis. The molecular structures of **40d-H, 40e-H** and **40f-H** are shown in Figure 16. It can be assumed that the angle for a substitution in *o*-position is most pronounced. Further crystallographic analysis for *m*- and *p*-substituted regioisomers is pending.

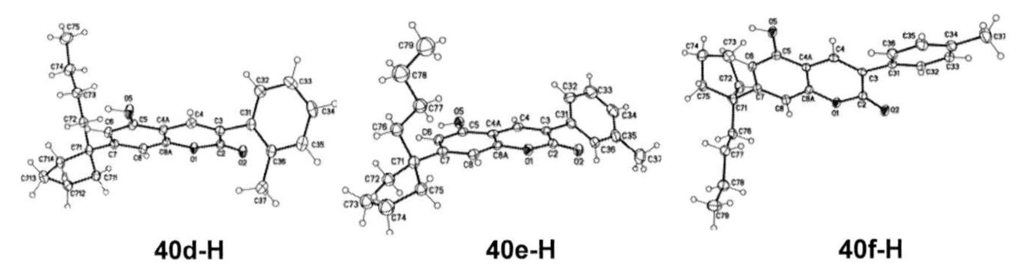

Figure 16: Molecular structures of *o*-methyl-3-arylcoumarin **40d-H**, *m*-methyl-3-arylcoumarin **40e-H** and *p*-methyl-3-arylcoumarin **40f-H**. (X-ray image is depicted with displacement paraeters drawn at 50% probability levels).

3.2.6 3-Styrylcoumarins

The synthesis of 3-styrylcoumarins was performed as described for the 3-arylcoumarins via a SUZUKI coupling. However, when the reaction conditions were transferred without adjustment, the product was obtained in less than 45% yield and contained impurities. An alternative system[186] in a dioxane-water mixture and disodium carbonate gave the product in 74% yield.

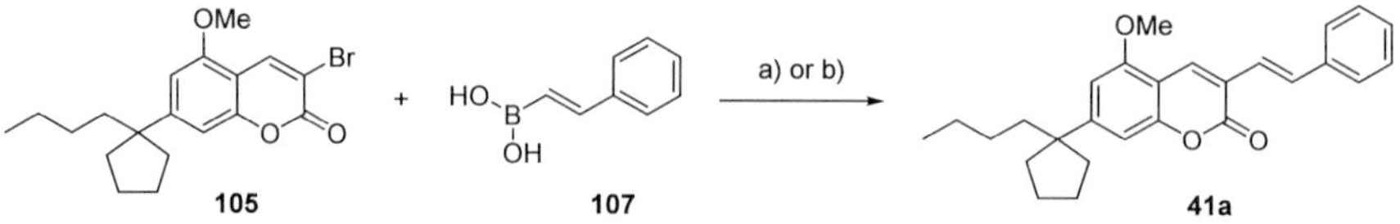

Scheme 36: Synthesis of a 3-styrylcoumarin. *a) 2 equiv. styrylboronic acid **107**, 2 equiv. Cs_2CO_3, 0.05 equiv. $Pd(PPh_3)_4$, degassed dioxane, 90 h, 16 h, < 45%; b) 3 equiv. styrylboronic acid **107**, 8 equiv. Na_2CO_3, 0.05 equiv. $Pd(PPh_3)_4$, degassed dioxane: water 5:1, 90 h, 16 h, 74%.*

In contrast to the multitude of available aryl boronic acids **106**, only a small number of styrylboronic acids **107** are commercially available. Hence an alternative system, with readily accessible building blocks would be preferable. And will be investigated in succeeding work.

An alternative route for the synthesis of 3-styrylcoumarins **41** was published by BRANCO *et al.* while investigating the fluorescent properties of these compounds.[187] The synthesis begins with the dicyclohexylcarbodiimide (**110**, DCC)-mediated esterification of 3-butenoic acid (**108**) with the hydroxy group of a salicylic aldehyde. An ester is formed *in situ* and is then treated with cesium carbonate to yield 3-vinylcoumarin **109**. BRANCO *et al.* were able to isolate the 3-vinylcoumarins **109** and convert them in HECK-reactions to 3-styrylcoumarins. The enormous advantage of the HECK coupling is the commercial availability of halogenated coupling partners.

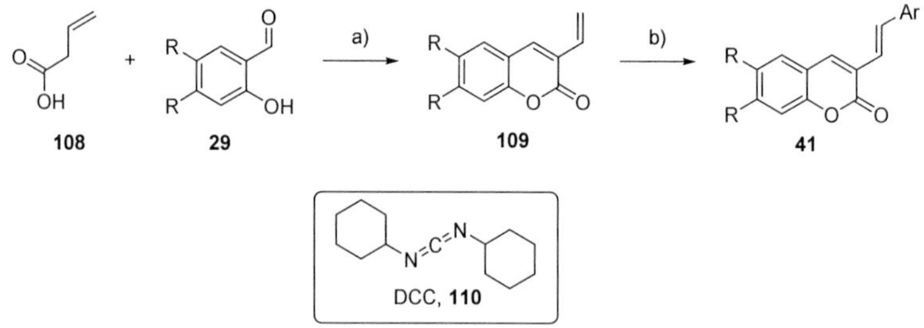

Scheme 37: Schematic synthesis of 3-styrylcoumarins **41** *via* 3-vinylcoumarin **109** synthesis and subsequent HECK-coupling by BRANCO *et al.*[187] *a) 1)butenoic acid (**108**), DCC (**110**), DCM, 2) salicylic aldehyde **29**, DMAP, 3) Cs_2CO_3; b) Ar-X, Pd cat., DMF.*

Biological evaluation of 3-styrylcoumarin **41** and 3-arylcoumarins **40** is still pending. However, if these compound classes should proof affinity towards the cannabinoid receptors, another attribute becomes relevant. 3-Styrylcoumarins[187] show a stronger fluorescence when compared to 3-benzylcoumarins or other cannabinoids. This feature could help to investigate the distribution of cannabinoid receptors as the combination of a fluorescent probe and a selective ligand is of high interest.

3.2.7 2,2-Dimethyl-2*H*-chromenes

While investigating alternative catalyst systems for coumarin cross-couplings, i.e. iron-based systems, KEVIN WAIBEL found in his bachelor project,[188] that in the presence of *N*-methyl-2-pyrolidone (**113**, NMP) coumarins react with methylmagnesium bromide to 2,2-dimethyl-2*H*-chromenes (Scheme 38). The reaction occurred when a KUMADA-coupling reaction on 3-bromocoumarin (**111**) was performed with a catalyst species that later proved to be inactive. The resulting 2,2-dimethyl-2*H*-chromene **112** side product sparked the idea to diversify our existing coumarin libraries further.

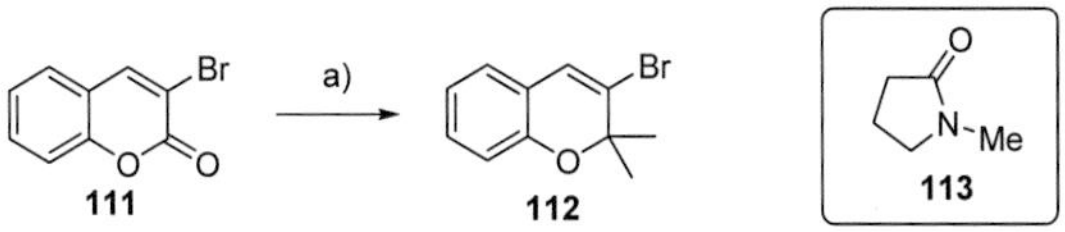

Scheme 38: Formation of 2,2-dimethyl-2*H*-chromenes **112** in the presence of NMP (**113**): *a) THF, NMP 113, 2.20 equiv., MeMgBr, –20 °C, 15 min, 71%.*

Interesting about this reaction is the formation of the product in a single step at low temperature. Established protocols for the synthesis of 2,2-dimethylchromenes from coumarins involve the formation of an intermediate **115** where the lactone ring is opened (Scheme 39). The intermediate **115** then condensates under harsh conditions to the desired 2,2-dimethyl-2*H*-chromene **116**.[189] The presence of NMP (**113**) seemed to allow the transformation in one step and at low temperatures.

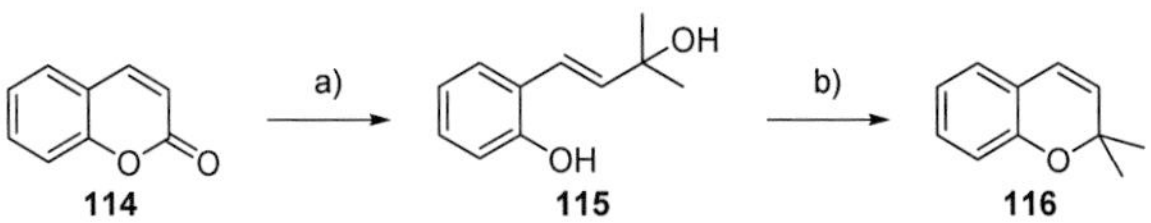

Scheme 39: Example for the synthesis of 2,2-dimethyl chromene involving the formation of an intermediate: *a) 3 equiv. MeMgBr (in diethyl ether), toluene, 0 °C, in situ; b) SiO_2, toluene, reflux, 66%.*[189]

The structural motif of 2,2-dimethyl-2*H*-chromenes in combination with the 3-benzyl moiety led to the idea to synthesize 3-benzyl-2,2-dimehtyl-2*H*-chromenes **117**. Figure 17 shows the respective moieties (red: 2,2-dimethyl; green: 3-benzyl). This combination of structural features from two cannabinoid active compound classes would aspire to constitute a novel active class.

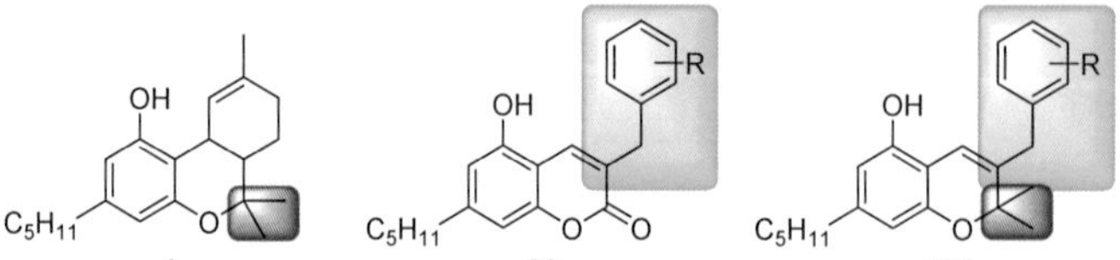

Figure 17: The combination of the structural motifs of THC (1) and 3-benzylcoumarins 26 led to the development of 3-benzyl-2,2-dimehtyl-2*H*-chromenes 117 as target structure.

As a first attempt the reaction was implemented with several synthesized coumarins, where an adequate amount of compound was already available. The results are collected in Table 13.

Table 13: Summary of attempted synthesis of 3-benzyl-2,2-dimehtyl-2*H*-chromenes **117**: *a) THF, NMP (113), 2.20 equiv., MeMgBr, –20 °C, 15 min.*

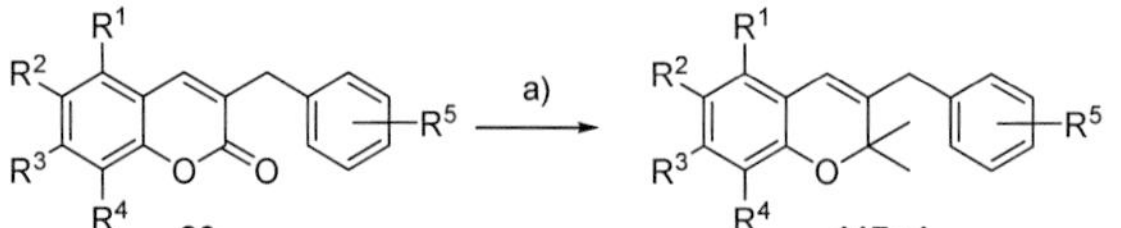

Entry	R^1	R^2	R^3	R^4	R^5	Yield$^{a)}$ [%]	Product
1	OMe	H	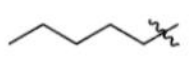 [structure]	H	*o*-Me	77	**117a**
2	OMe	H	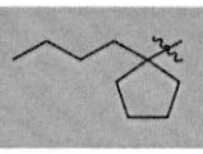 [structure]	H	*o*-OMe	44	**117b**
3	OMe	H	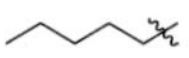 [structure]	H	*o*-OMe	49[a]	**117c**
4	Me	OMe	Me	Me	*o*-OMe	90[b]	**117d**
5	OMe	H	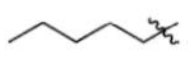 [structure]	H	*p*-OMe	-[c]	**117e**
6	OMe	H	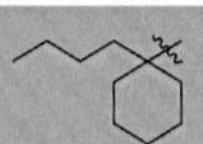 [structure]	H	*p*-F	-[c]	**117f**
7	OH	H	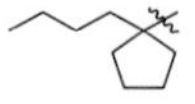 [structure]	H	*o*-OH	-[c]	**117g**
8	 [structure]	H	H	Me	*o*-OMe	45%[a]	**117h**
9	[structure]	H	H	Me	*o*-OH	-[c][d]	**117i**

[a]degraded while being dissolved in $CDCl_3$ for 16 h, [b]contains unknown impurity, [c]no desired product obtained, [d] a side product was identified.

While product formation was observed for compounds with an electron-donating group (EDG) in *ortho*-position of the benzyl group (Table 13, Entry 1-4), no other compounds were isolated. Also, stability of the resulting compounds was shown to be troublesome. While all compounds showed signs of degradation, compounds **117c** and **117d** degraded completely while being dissolved in $CDCl_3$ at room temperature for more than 16 h (e.g. when queued for ^{13}C-NMR measurement). No product formation was observed for **117d** where the EDG is located in *para*-position of the benzyl group as well as for **117f** with an electron-withdrawing group (EWG) e.g. fluorine. In the case of compounds with free hydroxy groups, no product formation was observed. We assume that the protic character of the hydroxy group leads to a reaction with methylmagnesium bromide and thus removes the reagent from the reaction. Interestingly for compound **26c** the formation of a side product **118** involving two condensation reactions was observed (Scheme 40).

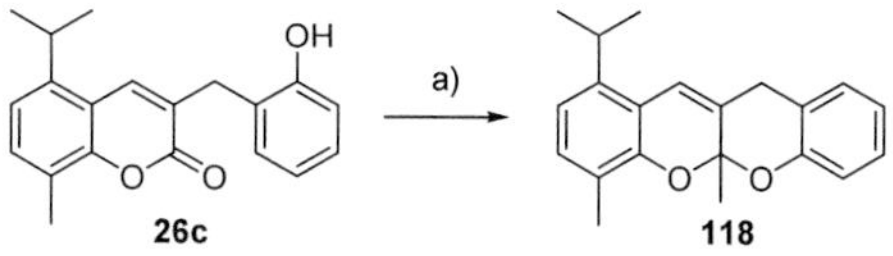

OH a) 26c 118

Scheme 40: Formation of the side product **118**: *a) THF, NMP (113), 2.20 equiv., MeMgBr, –20 °C, 15 min.*

Presumably, since methylmagnesium bromide has been consumed, the NMP (**113**) mediated condensation was faster than a second Grignard reaction, thus enabling the ring closure.

So far, the only described 3-benzyl-2,2-dimethyl-2*H*-chromene in literature bears an nitrile group in the 6-position.[190] The probability that the compounds will not be stable under physiological conditions as well as the synthetic issues led to the discontinuation of the efforts in this field. However, the two successful synthesized and purified compounds **117a** and **117b** were handed over for biological testing and, given potential encouraging results, might lead to the revival of scientific efforts on the 2,2-dichromene framework.

3.3 Controlled Release Systems for Cannabinoids[vi]

The objective of the work package was the design of degradable polymer micro- and nanoparticles that would act as controlled release systems.[154] Encapsulation of some of the most potent cannabinoids (**T1-T3**, Figure 18) was striven for. Controlled release systems are used in medical applications to allow temporal and/or distributional control of a drug and their concentration level. Site specific release allows to achieve higher drug concentrations at the therapeutic site, while temporal control allows to retain drug levels that are stable in the therapeutic window (see Chapter 1.6).[130]

In the context of cannabinoids, temporal control of drug levels over long-time periods would be interesting as, i.e. THC levels, quickly fall shortly after administration.[191-192] Moreover, the distributional control is of interest, as selectivity between the receptors can be hard to achieve via synthetic modification of the ligands. Since the expression of the receptors is distinctly discriminative in different tissues[43] a site specific release could help solve the selective targeting problem between receptors (see Chapter 1.3). Additionally, the ubiquitous expressed receptors could be addressed in defined target organs while forgoing a systemic interference. Some of the most promising cannabinoid candidates **T1-T3** that have been evaluated in preceding work (Figure 18)[169-170] were selected and have been synthesized in adequate amount (100 mg, see Chapter 3.2.2).

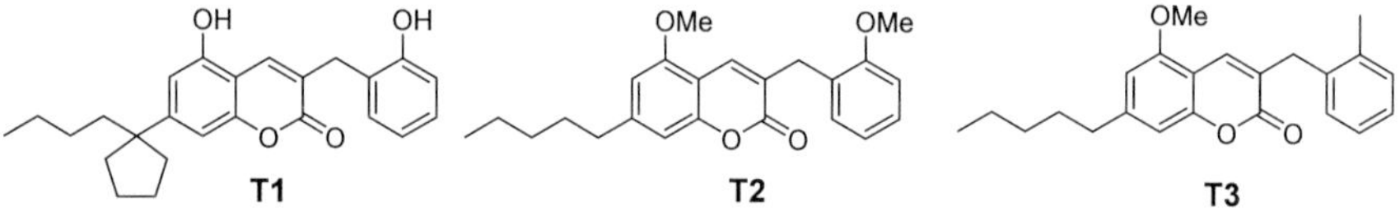

T1
CB_2 agonist
K_i = 68 nM; EC_{50} = 48 nM

T2
CB_1/CB_2 agonist
K_i : CB_1 = 32 nM; CB_2 = 49 nM
EC_{50} : CB_1 = 56 nM; CB_2 = 14 nM

T3
CB_1 antagonist
K_i = 22 nM; EC_{50} = 213 nM

Figure 18: Highly active 7-alkyl-3-benzylcoumarin derivatives: CB_1/CB_2 modulators **T1-T3**.[170,193]

Several physicochemical properties for **T1-T3** are collected in Table 14. Most noteworthy are the free hydroxy groups of **T1**, resulting in an overall higher polarity. The high calculated

[vi] This work was performed at the Helmholtz-Zentrum Geesthacht (HZG), campus Teltow, at the WISCHKE-group as part of the JOINT-project „Coumarin derivatives and microparticulate carriers for immunomodulation", in the context of the BioInterfaces in Technology and Medicine (BIF-TM) program.

octanol-water partition coefficients (log *P*) values show that all three compounds are highly lipophilic to a similar extend.

Table 14: Comparison of the physiochemical properties of **T1-T3**.

Compound	Log *P*[a]	H-bond donor[b]	H-bond acceptor[c]	Number of Atoms	M_w [g/mol]
T1	6.3	2	2	58	393
T2	5.7	0	4	53	366
T3	6.3	0	3	52	350

[a] Octanol-water partition coefficient calculated by ChemDraw® 16 by CRIPPEN'S fragmentation[194], [b] number of functional groups in the molecule that can act as hydrogen bond donor e.g. OH and NH groups, [b] number of functional groups in the molecule that can act as hydrogen bond acceptor e.g. O, N.

As release system, polymer-based micro and nanoparticle carriers were chosen. These particles have several advantages and combine (1) the suitability for application in site specific injection without surgical intervention,[148] (2) a good distribution in tissue and ideally a low foreign body response,[149] (3) feasibility of adjusting surface to volume (A/V) ratio[146] to allow variation of the release rate[195] and (4) a passive targeting function of competent cells that are able to internalize the particles via phagocytosis/endocytosis, eventually leading to intracellular drug release.[154-155]

Particles in three size ranges, i.e. ~ 50 µm, ~ 5 µm and ~ 200 nm, were targeted. Microparticles between 20-100 µm with a polydisperse distribution are interesting as injectable system for long-acting depot applications where the release is controlled by the diffusion barrier of the polymer matrix. Smaller microparticles between 5-10 µm are interesting when intracellular release is targeted. Competent cells can internalize particles in this size range via phagocytosis, ultimately leading to intracellular drug release.[135] Nanoparticles on the other hand are interesting for their biodistribution properties as parenteral drug delivery systems.[145]

The polymers chosen in this study were poly[(D,L-lactide)-co-glycolide]s (Figure 19), i.e. copolymers (50:50 Lactide:Glycolide) commonly used for pharmaceutical applications due to its biocompatible and biodegradable properties.[149-153] The two applied polymers have been purchased from Evonik under the trademark Resomer® RG 503 and RG 503H and are referred to as **PLGA-Et** and **PLGA-COOH** respectively. They bear distinctive end groups on their termini, **PLGA-COOH** carries a free acid group, while **PLGA-Et** is capped as an ethyl ester function. The employed polymer batches were examined prior to use. The molecular weight was determined for **PLGA-Et** to 24 kDa with a PDI of 2.1 and for **PLGA-COOH** to 22 kDa with a PDI of 1.9. The slightly smaller M_w for **PLGA-COOH** can have an influence on the particle formation and drug release.

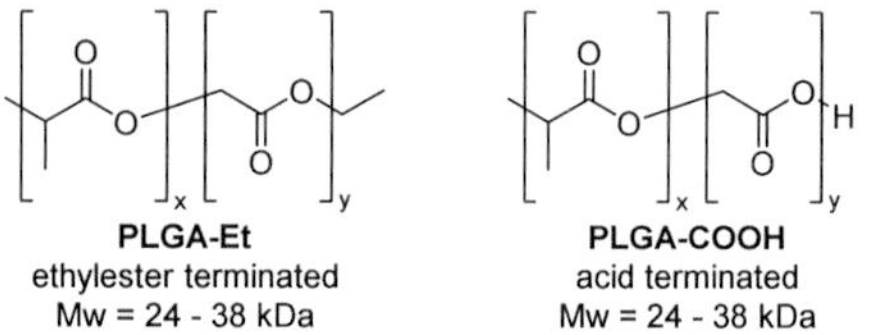

PLGA-Et
ethylester terminated
Mw = 24 - 38 kDa

PLGA-COOH
acid terminated
Mw = 24 - 38 kDa

Figure 19: Structures of the polymers that were employed in the particle preparation: **PLGA-Et** 50:50 poly[(D,L-lactide)-co-glycolide] with an ethyl ester terminus;[196] **PLGA-COOH** 50:50 poly[(D,L-lactide-co-glycolide)] with an acid terminus.[197]

3.3.1 Solubility and Quantification of the Coumarins

In order to determine the quantity of coumarin in a given sample, e.g. in a certain amount of particles or in a solution, a HPLC method was established. The HPLC method was conceptualized with a gradient system of acetonitrile and water (containing 0.1% TFA each) as the mobile phase and a reversed phase silica gel column as stationary phase (RP-18 (5 µm) LiChroCART® 125-4). The acetonitrile content was increased from 70% to 95% over 12 min (Figure 20, A). This technique allowed a high throughput of samples and simultaneous differentiation of all three compounds. With this method, the most polar coumarin **T1** has a retention time of 4.1 min, while **T2** elutes after 6.9 min and **T3** after 7.4 min. The compilation of calibration curves for all three compounds with standard solutions, covering the expected sample range from 0.1 µg/mL to 100 µg/mL (Figure 20, B), allowed the precise determination of sample concentrations and was verified at regular intervals by measuring standard solutions. The UV-absorption determined by the HPLC detector shows a linear relation to the concentration of the compounds. For a given molar concentration, the more electron rich coumarins **T3** and **T2** show a higher UV-absorption than **T1**, meeting the expectations. The UV-channel at 313 nm was employed in regard to the absorption maxima of the coumarins around 310 nm (Figure 38, Additional Data). The lower limit of detection lies at 0.1 µg/mL.

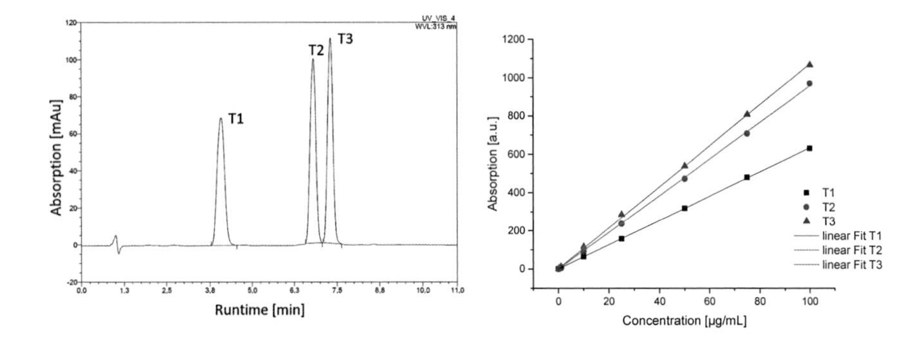

Figure 20: Left: HPLC run for a sample containing a standard concentration of 10 µg/mL of each coumarin **T1-T3**. Right: Calibration curves for Coumarins **T1-T3**.

For the process of encapsulation, the solubility of a drug in organic and aqueous media become important.[135] Accordingly the solubility in process relevant organic solvents and aqueous media has been investigate. The quantity of coumarins soluble in organic solvents was estimated by dissolving weighted amounts of compound in as little as possible solvent as the solubility was generally high. Adequate solubility in acetonitrile (>5 mg/mL), ethanol (>10 mg/mL) and in dichloromethane (> 50 mg/mL) has been found, which are applicable solvents used in preparation and investigation of the particles.

The solubility of the compounds in different aqueous buffer solutions that were relevant for the particle preparation process and release experiments has been determined via HPLC. First, saturated solutions were prepared, remaining solids removed, and the supernatant diluted with acetonitrile to allow the detection with exclusion of compound precipitation or solvent effects. The final solubility was then calculated and can be found in Figure 21 (data is given in Additional Data, Table 20). The solubility for the three compounds in aqueous media differ by a broad margin. They are practically insoluble in water or phosphate-buffered saline solution (PBS) (< 0.05 µg/mL). However, as soon as poly-vinyl alcohol (PVA) is added, coumarin **T1** becomes slightly soluble (60-180 µg/mL), while **T2** and **T3** are still not dissolving. This difference is explained by the structures; **T1** is distinguished from the other compounds by the free hydroxy groups, which leads to an overall more polar structure and therefore a better solubility in polar media. In PBST (PBS + Tween® 80 surfactant) significant higher solubility is found for all coumarins that increases with surfactant content. The solubility for **T1** in PBST is again increased by a factor of 3-5 when compared to **T2** and **T3**.

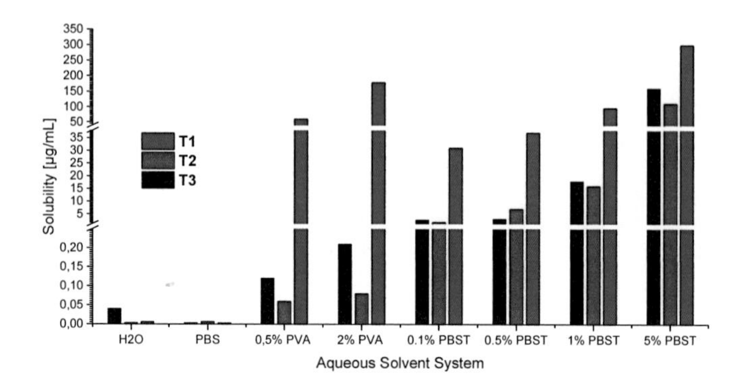

Figure 21: Depiction of the solubility of compounds **T1-T3** at room temperature in different aqueous solutions.

The solubility properties were considered in the later discussed encapsulation efficiency and release of the compounds.

3.3.2 Microparticle Formulation Screening

First, appropriate conditions for the particle synthesis and coumarin encapsulation were systematically explored for three different particle sizes (~ 50 µm, ~ 5 µm and ~ 200 nm). The particles were prepared by a standard oil-in-water (o/w) emulsion technique (Scheme 41).[135,146] For the particle preparation, a polymer is dissolved in an organic solvent, preferably dichloromethane, and mixed with an aqueous solution containing an emulsifier, e.g. 2% polyvinyl alcohol (PVA), to form an emulsion, where the organic oil phase is contained in PVA stabilized droplets (step a). The emulsion is then added to a larger volume of diluted aqueous emulsifier solution (step b), which acts as a hardening bath (step c). The hardening bath is constantly stirred to prevent fusion of the stabilized oil phase droplets. The particles form as the volatile organic solvent extracts from the droplets (step d) and the solidified polymer remains. The coumarins **T1-T3** comply with an important criterion for encapsulation i.e. that they are poorly soluble in the aqueous solution (see above). Different particle sizes from 200 nm to 50 µm should be achieved by varying e.g. the means of emulsification and polymer concentration.[146,198] Higher viscosities, often a consequence of high polymer concentrations or high M_w polymers, lead to bigger oil phase droplets as more agitation is needed to break up droplets in the homogenizing process.[135,199] The emulsification method, e.g. the agitation and energy density, has to be chosen accordingly. Increased polymer concentration often leads to an increased mean diameter since the viscosity of the phase to be dispersed is higher. In this case 'immature' particles can collide and fuse to larger 'final' microparticles.[200]

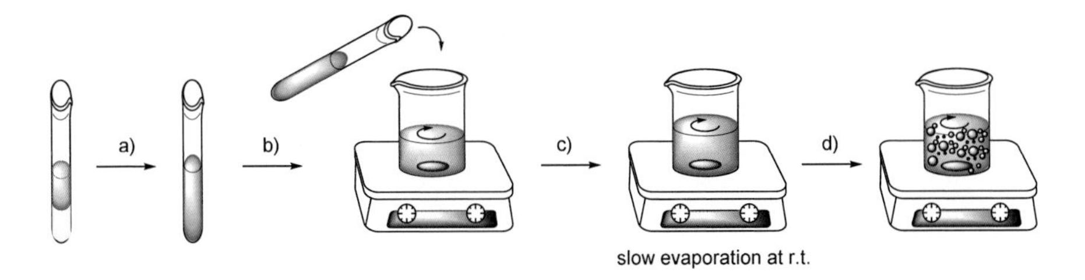

Scheme 41: Schematic depiction of particle preparation with o/w emulsion technique: *a) polymer is dissolved in the oil phase (yellow) and an aqueous phase with emulsifier is added (blue), the mixture is then emulsified, e.g. via vortexing, to result in a turbid emulsion in which PVA stabilized oil phase droplets are formed (green); b) The emulsion is then poured into an larger volume of aqueous emulsifier solution that acts as hardening bath (stir bar indicated in red); c) the solution is stirred at room temperature and volatiles (organic solvent) evaporate; d) particles are formed when the volatiles are completely evaporated, the particles can then be removed from the turbid dispersion by filtration or centrifugation.*

Microparticles in the Range of 50 µm

For particles in the range of ~ 50 µm, the emulsification method of choice was to put the oil phase and an emulsifier solution (2% PVA) into a sealed tube. An emulsion is then generated by vortexing. Concentrations of 10, 15 and 20 wt% polymer in the oil phase were tested and led to various microparticle sizes (Figure 22). The effect of the vessel (shape, material, etc.) in which the emulsions were prepared, the manner of addition of the emulsion to the hardening bath and the mode of stirring of the same had been varied to rule out possible disturbances on the particle formations (Table 15). No significant variance was observed, and the conditions highlighted in bold were chosen for further particle preparation.

Table 15: Summary of the varied conditions for particle preparation with a vortex mixer.

Polymer loading	Vessel[a]	Stirring[b]	Conveyance[c]	Av. Size[d] [µm]
15 wt%	Glass tube	Overhead	poured	53
15 wt%	Glass tube	Overhead	pipetted	51[e]
15 wt%	**Glass tube**	**Magnetic**	**poured**	**51**
15 wt%	Falcon tube	Overhead	poured	53
15 wt%	Plastic tube	Overhead	poured	56
10 wt%	Glass tube	Overhead	poured	45
20 wt%	Glass tube	Overhead	poured	122

[a]Vessel in which the emulsion was prepared, [b]Mode of stirring of the hardening bath, either an overhead stirrer or a magnetic stirrer was used, [c]Means of conveyance of the suspension into the hardening bath, [d]Sizes measured by SLS (volume weighted mean diameter), [e]Distribution of particle sizes appeared broader.

The particles were monitored by static light scattering (SLS) (Figure 22). Particle sizes are evaluated as volume weighted distributions. The different polymer concentrations led to particles with various sizes, 10 wt% polymer (**PLGA-Et**) gave particles with an average diameter of 45 µm (black curve), a polymer loading of 15 wt% gave 51 µm particles (red curve) and 20 wt% resulted in bigger particles averaging 122 µm (green curve). Implementing an alternative polymer, **PLGA-COOH**, resulted in smaller particles with a mean size of 43 µm (blue curve). The narrower distribution when compared to 10 wt% (**PLGA-Et**) leads to the smaller average size. In all samples, smaller particles are formed alongside the desired size. This is explained by the formation of a distribution of oil phase droplet sizes at low stirring rates.[201] For further preparations, 15 wt% polymer in the oil phase was chosen to achieve particles of the desired size around 50 µm.

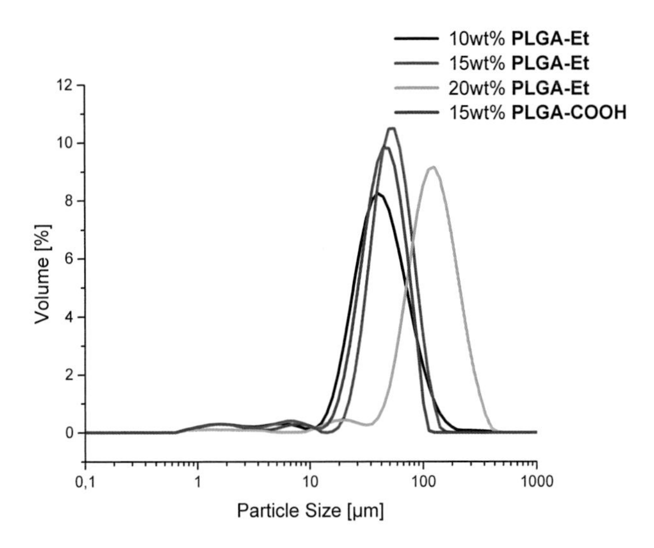

Figure 22: Influence of polymer concentration in the oil phase on the particle size distribution when the emulsion is prepared by vortexing.

Additionally to the SLS measurements, the particles were also studied by light microscopy. Figure 23 shows particles formed with different polymer concentrations. As already evident from the SLS measurements, the particles are not uniformly sized. The majority of particles, however, corresponds to the desired size range and exhibits spherical appearance. While the lower concentration (10 wt% PLGA, A) resulted in smaller smoother particles, higher concentration (15 wt% PLGA, B) led to an apparently cratered surface and distinctly bigger particles. Even larger particles and a more cratered appearance was found for particles prepared with 20 wt% PLGA (C).

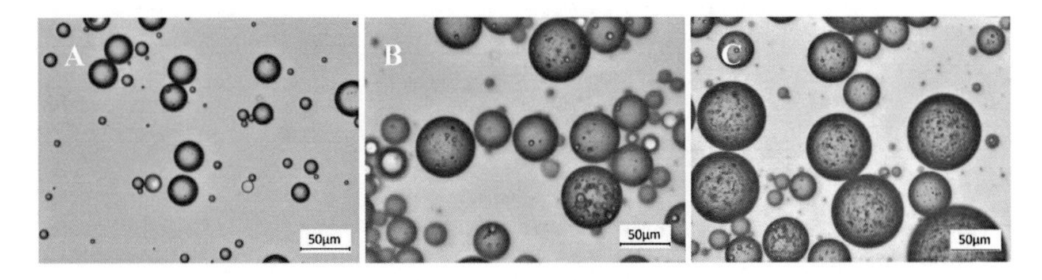

Figure 23: Light microscope pictures at 40-fold magnification of microparticles prepared from a dispersion *via* vortex mixer. **A** 10 wt% **PLGA-Et**, **B** 15 wt% **PLGA-Et**, **C** 20 wt% **PLGA-Et**.

As the particles prepared from 15 wt% polymer in the oil phase met the desired size requirements, this method was chosen for further particle preparation and examined more closely by scanning electron microscopy (SEM, Figure 24). The particles shown in Figure 23 B are from the same batch as Figure 24, A and B. Interestingly, SEM shows a smooth surface with few irregularities. This observation is seemingly contradictory to the light microscopy pictures where a rough appearance was deduced (Figure 23). However, the light microscope can also show the interior structure of the particles. This observation is made possible because the particles are translucent and light that shines through the particles is disturbed by inner irregularities. The observed darker spots in the light microscopy indicates the formation of areas in the particle that have refractive properties that deviate from the bulk of the matrix. Such areas could be trapped aqueous phase from the emulsification process. During homogenization and stabilized oil-droplet formation, smaller aqueous droplets can be trapped inside the oil-phase droplets. When the nascent particles begin to harden, the entrapped aqueous phase remains,

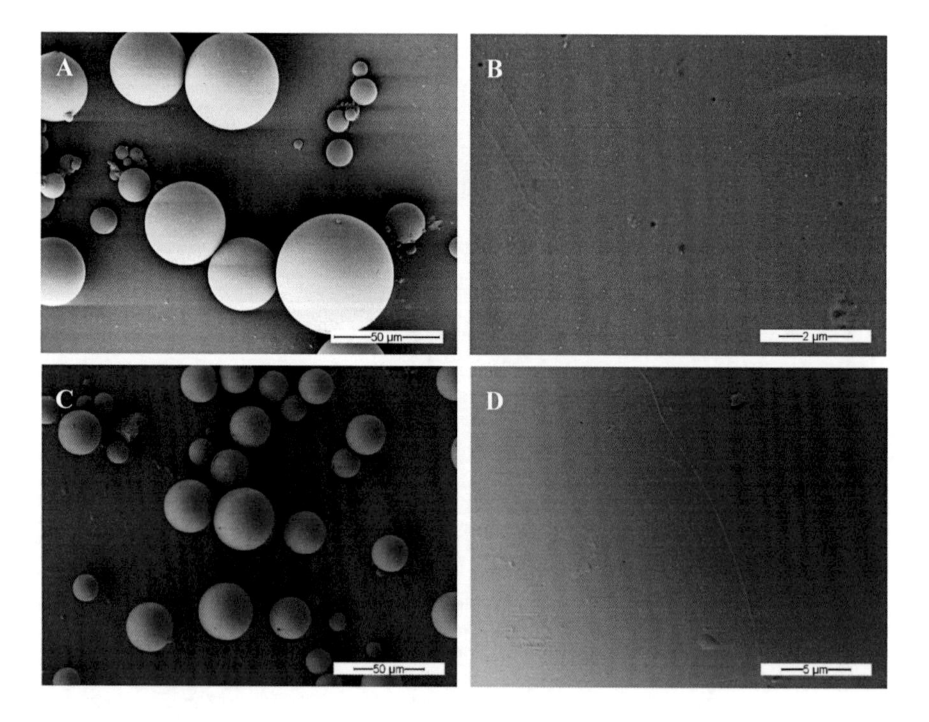

Figure 24: SEM images of particles prepared from a dispersion *via* vortex mixer and 15 wt% polymer in the oil phase. The average particle sizes are in the range of 40-50 µm. **A**: Overview of particle shape and morphology for **PLGA-Et** particles, **B**: Magnification of **PLGA-Et** particle surface, **C**: Overview of particle shape and morphology for **PLGA-COOH** particles, **D**: Magnification of **PLGA-COOH** particle surface.

forming small cavities. Entrapping of aqueous media is observed to a greater extend when s/o/w particle preparation methods are concerned where the entrapment is required for the encapsulation of drugs.[135] With respect to the polymer matrices, particles prepared with 15 wt% **PLGA-COOH** show a smaller average size distribution in accordance with the SLS measurements (Figure 24, C). The exhibited appearance is also spherical and the surface analogical to the **PLGA-Et** particles.

Microparticles in the Range of 2-5 μm

For obtaining particles in the size range of 2-5 μm, the means of emulsification was changed to a high-performance dispersion unit (rotor-stator homogenizer, Ultra Turrax®). At 24,000 rpm, this tool generates considerably smaller oil phase droplets, which results in smaller particles. Again, the obtained particles for 10 wt%, 15 wt% and 20 wt% polymer in the oil phase were compared (Figure 25). For this emulsification method, 15 wt% (5.7 μm) and 20 wt% (14.8 μm) gave particles that were larger than desired, and the size distributions contained shoulders (red and green lines). In contrast, the size distribution for 10 wt% polymer was monomodal (black line) and the average size of 4.3 μm meets the specifications. Implementation of the alternative polymer **PLGA-COOH** resulted in smaller average sizes (2.4 μm, blue line). This may be a result of a smaller inherent viscosity of the resulting oil phase when **PLGA-COOH** is employed and was also observed for the ~ 50 μm particle preparation (see paragraph above).[202]

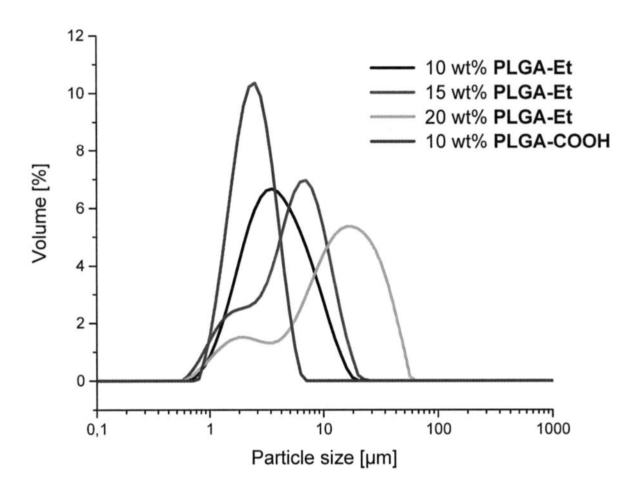

Figure 25: Influence of polymer concentration in the oil phase on the particle sizes for emulsions prepared by a rotor stator homogenizer.

The microscope images (Figure 26) of the particles support the SLS results. Particles prepared with 15% PLGA were broadly distributed (B) while 20% PLGA resulted in particles that were oversized (C). Ultimately the concentration of 10% PLGA was used in the encapsulation experiments for particles in the range of 2-5 μm, as this concentration allowed to form particles in the desired size range (A).

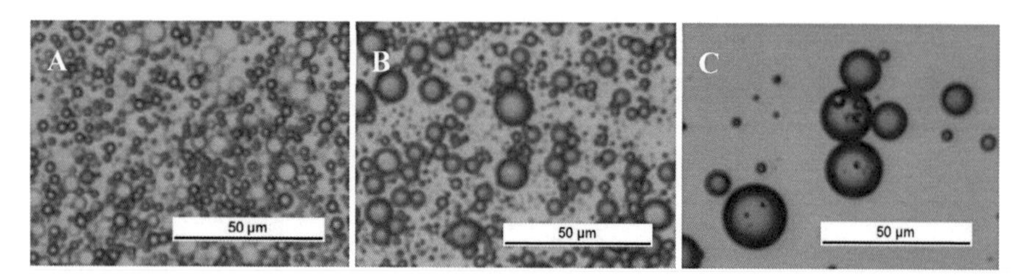

Figure 26: Light microscope pictures of microparticles (40-fold magnification) where the emulsion was prepared by an rotor stator homogenizer with different polymer concentrations: **A**: 10% **PLGA-Et**, **B**: 15% **PLGA-Et**, **C**: 20% **PLGA-Et**.

Closer observation with SEM was also conducted for the particles prepared with 10 wt% polymer in the oil phase for both polymers (Figure 27). Picture A shows that the majority of particles prepared from **PLGA-Et** are spherical and show a smooth surface (B). Aside from the microspheres, several hollow particles have been formed (C) that collapsed during the sample preparation for the SEM.

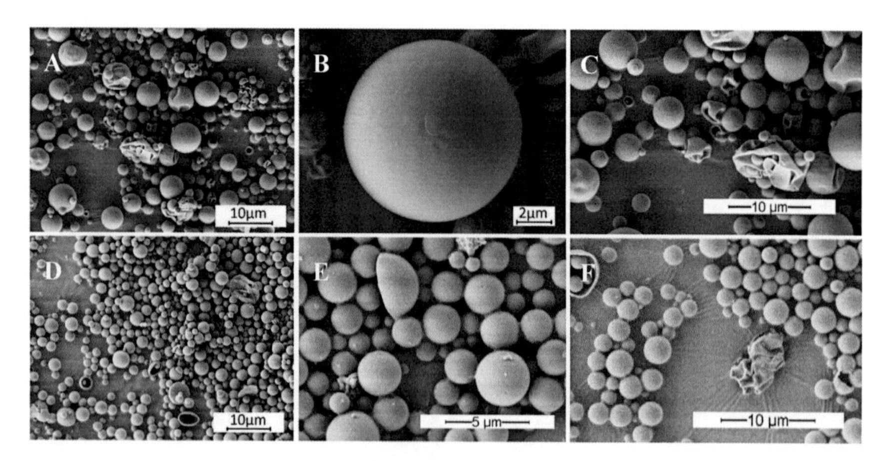

Figure 27: SEM images of particles prepared by rotor stator homogenization and 10 wt% polymer in the oil phase. The average particle sizes are in the range of 4 μm for top row and 2.5 μm for bottom row. **A**: Overview of particle shape for **PLGA-Et** particles, **B**: Magnification of **PLGA-Et** particle surface, **C**: Zoom on defective **PLGA-Et** particles; **D**: Overview of particle shape for **PLGA-COOH** particles, **E**: Magnification of **PLGA-COOH** particle surface, **F**: Zoom on defective **PLGA-COOH** particles.

Hollow particles can form when aqueous medium, possibly assisted by the presence of surfactant, is entrapped in the nascent particles either accidently during emulsification or systematically by diffusion before the completion of the hardening process. Particles prepared from **PLGA-COOH** are smaller but show similar characteristics (D-F). Picture D gives an overview of particles from **PLGA-COOH** and shows mostly spherical particles with smooth surfaces (E) but here, some collapsed hollow particles can be found as well (F).

Nanoparticles

Particles in the 200-300 nm range were synthesized by preparing the o/w emulsion with an ultrasonic homogenizer. Several methods were tested by varying polymer concentration and dispersion conditions (Table 16) and were monitored by dynamic light scattering (DLS, Figure 28). With 10 wt% polymer and 30 seconds sonication, the resulting particles showed a broad distribution (high polydispersity, blue line). Preparation of the suspension from 5 wt% polymer by sonication for 30 seconds led to a narrower distribution (green line); however, a shoulder was observed in the DLS spectrum. Pre-emulsification with a rotor-stator homogenizer (24,000 rpm) for 2 min followed by sonication for 1 min allowed avoiding the formation of the small sized fraction but gave a slightly broader distribution (red line). The best result was achieved when the sonication was applied for 1 min without pre-emulsification (black line) and led to an average diameter of 207 nm and a Polydispersity Index of 0.08 (PDI). This method was therefore, later used for the preparation of coumarin loaded nanoparticles.

Table 16: Summary of the varied conditions for nanoparticle preparation.

Polymer loading	Pre-emulsification[a]	Emulsification[b]	Av. Size[c] [nm]	PDI[c]
10%	-	30 s	308	0.21
5%	-	30 s	242	0.19
5%	120 s	60 s	235	0.12
5%	-	60 s	207	0.08

[a]Pre-emulsification was performed with a rotor-stator homogenizer at 24,000 rpm, [b]Emulsification was performed by sonication with an ultra sound rod at 52% intensity, [c]Sizes and PDI were measured by DLS.

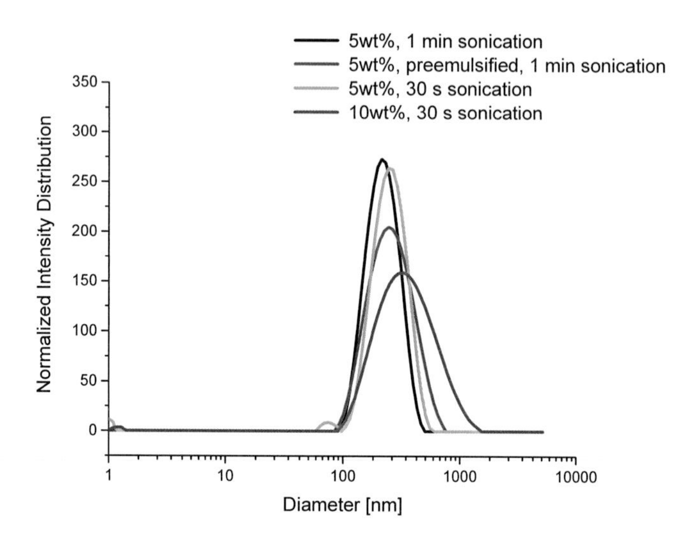

Figure 28: Influence of polymer concentration and sonication conditions on the size distribution of nanoparticles.

The resolution of the light microscope is inadequate to study the nanoparticles Therefore, only SEM images were recorded (Figure 29). Figure A shows an overview of particle distribution. All particles have a spherical appearance. At higher magnification (B) it becomes apparent that the particles form aggregates. Since DLS measurements indicate that the particles are present as individuals, it is likely that the agglomeration is an artefact from the particle preparation for SEM analysis. The particles are prepared by spin-coating and drying during the process, whereby the dehydration can lead to the observed aggregates.[203-204]

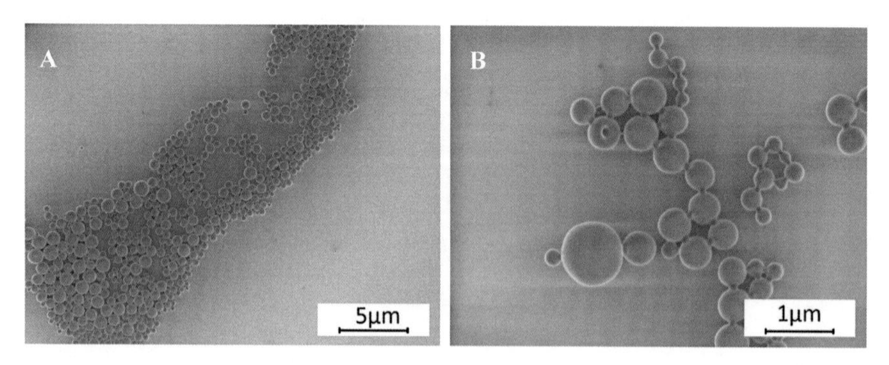

Figure 29: SEM images of nanoparticles from **PLGA-Et**. **A**: Overview; **B**: Zoom on particle conglomerate.

3.3.3 Encapsulation of coumarins

With the developed strategies for particle preparation, coumarins **T1-T3** were added to the oil phase so that a theoretical loading of 4.75 wt% (drug to PLGA ratio) would be obtained in the particles. Microparticles in the size of 50 µm and 2-5 µm were prepared from both PLGA types **PLGA-Et** and **PLGA-COOH**. Nanoparticles were produced from **PLGA-Et** while a preparation of nanoparticles from **PLGA-COOH** was forgone due to an expected too fast release. Encapsulation of the compounds resulted in particle sizes that were slightly smaller than the corresponding blank particles for method A (vortex mixer) und slightly larger for method B (rotor stator homogenizer) as an effect of the encapsulation. The encapsulation efficiency was evaluated in triplicates by dissolving a weighted sample of particles in 1 mL dichloromethane and precipitation of the PLGA by addition of the resulting solution to 9 mL of ethanol, a good solvent for **T1-T3** (compare Chapter 3.3.1). After centrifugation to remove the precipitate, the concentration of coumarin was determined via HPLC and the loading of the particles calculated. Results of the obtained particle sizes and percentages of loaded compound is given in Table 17.

The efficiency of the encapsulation varies for all samples. The differences in the structure and physicochemical properties of the three coumarins leads to deviations in this perspective, especially the higher solubility of coumarin **T1** in the PVA-solution has been considered a probable factor. When testing the supernatant from the preparation of particles containing coumarin **T1**, higher concentrations of **T1** had leaked from the nascent particles when compared to **T2** and **T3**. This effect correlates to a lower encapsulation efficiency (EE) for **T1** in the case of **PLGA-Et**. For **PLGA-COOH** (Table 17, Entry 14 and 18) the EE is higher which might be a result of an increased interaction between the free hydroxy groups in **T1** and the polar end groups of **PLGA-COOH**. Since the variance of the EE is relatively high, other effects have to play a role. Another observation is that the particles prepared by method B (emulsion generated with rotor stator homogenizer) have a slightly lower loading when compared to particles prepared by method A (vortex mixer). This effect may be attributed to the smaller particles resulting from this method. Since smaller particles have a higher surface to volume ratio, it is more likely that part of the encapsulated coumarins is lost during the solvent extraction/evaporation and washing process. Encapsulation efficiencies of more than 100% (Table 17, Entry 3 and 15) have been observed and might be a result of incomplete polymer dissolution in the oil phase. The polymer is then not available for the particle preparation and the total mass of dry particles is lower.

The variance in EE is even more prominent when the nanoparticles prepared by method C (ultrasonic homogenizer) are considered. Here, the loading differs from 3.8 for **T1** to 8.0 **T2** and 13.8 wt% for **T3**. The nanoparticles with **T2** and **T3** exceed the anticipated maximum loading with 170 and 290% (Table 17, Entry 11 and 12). **T1**, the compound most soluble in aqueous media, is to a certain degree lost in by diffusion from the polymer phase into the hardening bath solution and might have been washed out further in the work up process, that involves several cycles of centrifugation and substitution of the supernatant with water. The only plausible cause for the unexpected high EE for **T2** and **T3** is a loss of matrix polymer. No visible polymer film was observed in the particle preparation set-up that could explain the discrepancy. A loss of the polymer to the continuous phase has been observed for low M_w (~ 1 kDa) polymers before,[205] but should not be an issue for the applied **PLGA-Et** (24 kDa). During the homogenization step of the particle preparation, small PVA stabilized PLGA/dichloromethane/payload coumarin droplets are formed that harden out as the solvent is extracted and evaporates. The formation of PLGA/PVA mixed phases on the dichloromethane/water interface have been described.[206-207] If during the homogenization step individual PLGA chains are stabilized as ultrafine particles or individual polymer chains, a significant amount of the polymer matrix could be lost and might result in the high EE values that were observed. Another possible explanation is the hydrolysis of the polymer matrix through the significant energy input of the sonication in the homogenization step and consequent loss of water-soluble matrix polymer fragments.[208]

Table 17: Prepared particles with encapsulated coumarins **T1-T3**. *A: emulsion generated with vortex mixer; B: emulsion generated with rotor stator homogenizer; C: emulsion generated with ultrasonic homogenizer.*

Entry	Preparation Method[a]	Polymer	Coumarin	Av. Size [μm][b]	Loading[wt%]	EE [%]
1	A	**PLGA-Et**	**blank**	51	-	-
2	A	**PLGA-Et**	**T1**	44	4.28 ±0	90 ±0
3	A	**PLGA-Et**	**T2**	41	4.98 ±0.2	105 ±4
4	A	**PLGA-Et**	**T3**	59	4.72 ±0.02	99 ±1
5	B	**PLGA-Et**	**blank**	4.2	-	-
6	B	**PLGA-Et**	**T1**	4.8	4.35 ±0.1	92 ±2
7	B	**PLGA-Et**	**T2**	4.5	4.23 ±0.1	89 ±3
8	B	**PLGA-Et**	**T3**	4.0	3.71 ±0.03	78 ±0.6
9	C	**PLGA-Et**	**blank**	0.242	-	-
10	C	**PLGA-Et**	**T1**	0.207	3.84 ±0.2	81 ±4
11	C	**PLGA-Et**	**T2**	0.220	8.03 ±0.7	169 ±15
12	C	**PLGA-Et**	**T3**	0.226	13.79 ±0.4	290 ±9
13	A	**PLGA-COOH**	**blank**	43	-	-
14	A	**PLGA-COOH**	**T1**	40	4.74 ±0.04	100 ±1
15	A	**PLGA-COOH**	**T2**	40	5.36 ±0.2	113 ±4
16	A	**PLGA-COOH**	**T3**	38	4.54 ±0.1	96 ±1
17	B	**PLGA-COOH**	**blank**	2.4	-	-
18	B	**PLGA-COOH**	**T1**	3.2	3.66 ±0.03	77 ±1
19	B	**PLGA-COOH**	**T2**	3.5	3.41 ±0.1	72 ±2
20	B	**PLGA-COOH**	**T3**	3.9	3.16 ±0.1	66 ±2

[a]For detailed method see experimental part. [b]Sizes measured by SLS for microparticles (volume weighted mean diameter) and DLS for nanoparticles (intensity weighted average diameter).

In Figure 30 (A1-C1), light microscopic pictures for the representative particles in the 40-50 μm range (Table 17, entry 1-3) are shown. When compared to the blank particles in Figure 23, it becomes apparent that the presence of the payload coumarins did not significantly disturb the particle formation. The appearance and shape stayed similar for all encapsulated compounds. A closer inspection of the particles via SEM was performed and is discussed in the following pages.

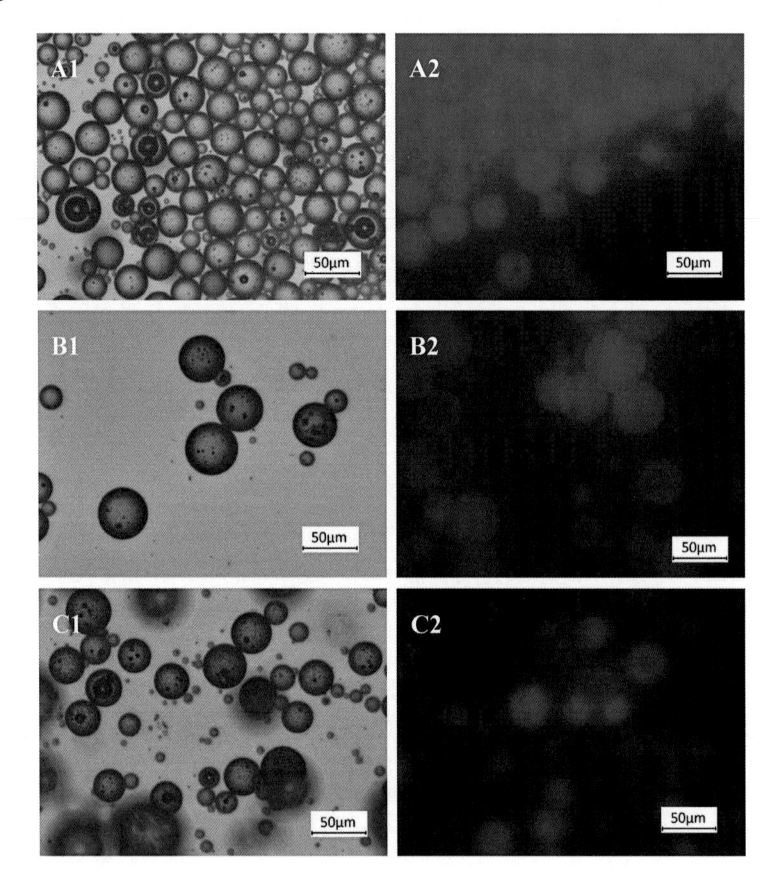

Figure 30: Light and fluorescence microscopy pictures of **PLGA-Et** particles with payload coumarin (40-fold magnification, fluorescence filter A4, excitation with UV light): **A**: Particles with **T1**, **B**: Particles with **T2**, **C**: Particles with **T3**.

The fluorescent properties of the coumarins allow the analysis of the particles with a fluorescence microscope (A2-C2), thereby constituting an opportunity to use fluorescent imaging for tracking of the particles in cells or tissues. Fluorescence spectra for **T1-T3** showed emission maxima at 422 to 433 nm (spectra are shown in Additional data, Figure 30 and Figure 40).

The morphology of the loaded particles was examined via SEM. The analysis of particles in the size range 40-50 μm prepared from **PLGA-Et** and **PLGA-COOH** shows close resemblance and is summarized in Figure 31. As reported in Chapter 3.3.2, blank particles (Figure 31, A and B) showed a smooth surface and no defective particles were observed. Encapsulation of coumarins **T1-T3** (C-H) led to on average slightly smaller particles when compared with the blank particles (A and B). This observation was also emphasized by the SLS measurements (Table 17).

Loading of coumarin **T1** into **PLGA-Et** particles (C) resulted in several defective particles, while the majority exhibited the spherical shape and smoothness of the blank particles. For encapsulation of **T1** into **PLGA-COOH** (D) particles, no defective specimens were observed and the particles showed a similar morphology to the blank particles (B).

In contrast, loading of particles with coumarins **T2** (E and F) and **T3** (G and H) resulted in spherical shapes with pitted surfaces that were more pronounced for **T3** loaded particles. One hypotheses is that the surface roughness might occur due to payload coumarin deposits on the particle surface, developing in the nascent particles during the hardening process, which are subsequently removed in the washing stage. The absence in the case of **T1** might be explained by the solubility in the hardening bath (PVA-solution), resulting in a smooth surface as the surface deposits are already removed during the hardening process. The formation of such deposits is thought to be caused by solvent flux from the nascent particles.[135,201,209]

SEM pictures of particles prepared from **PLGA-Et** and **PLGA-COOH** in the size range of 2-5 μm are shown in Figure 32. As described before (Chapter 3.3.2), some hollow particles that have collapsed can be observed for the blank particles (A and B). The encapsulation of **T1** to **T3** does not alter this behavior and hollow particles can be found in all samples (C-H).

The majority of all prepared particles shows spherical appearance and the blank particles as well as the particles with **T1** loading exhibit smooth surfaces (A-D) for both polymers. As described for the bigger particles, encapsulation of **T2** and **T3** (E-H) leads to small indentations in the particle surfaces and might be caused by coumarins deposits on the particle surfaces that are washed out in the washing process. In the case of the soluble **T1,** the deposits can already diffuse from the nascent particles into the hardening bath solution and thereby do not disturb the particle surfaces (C and D).

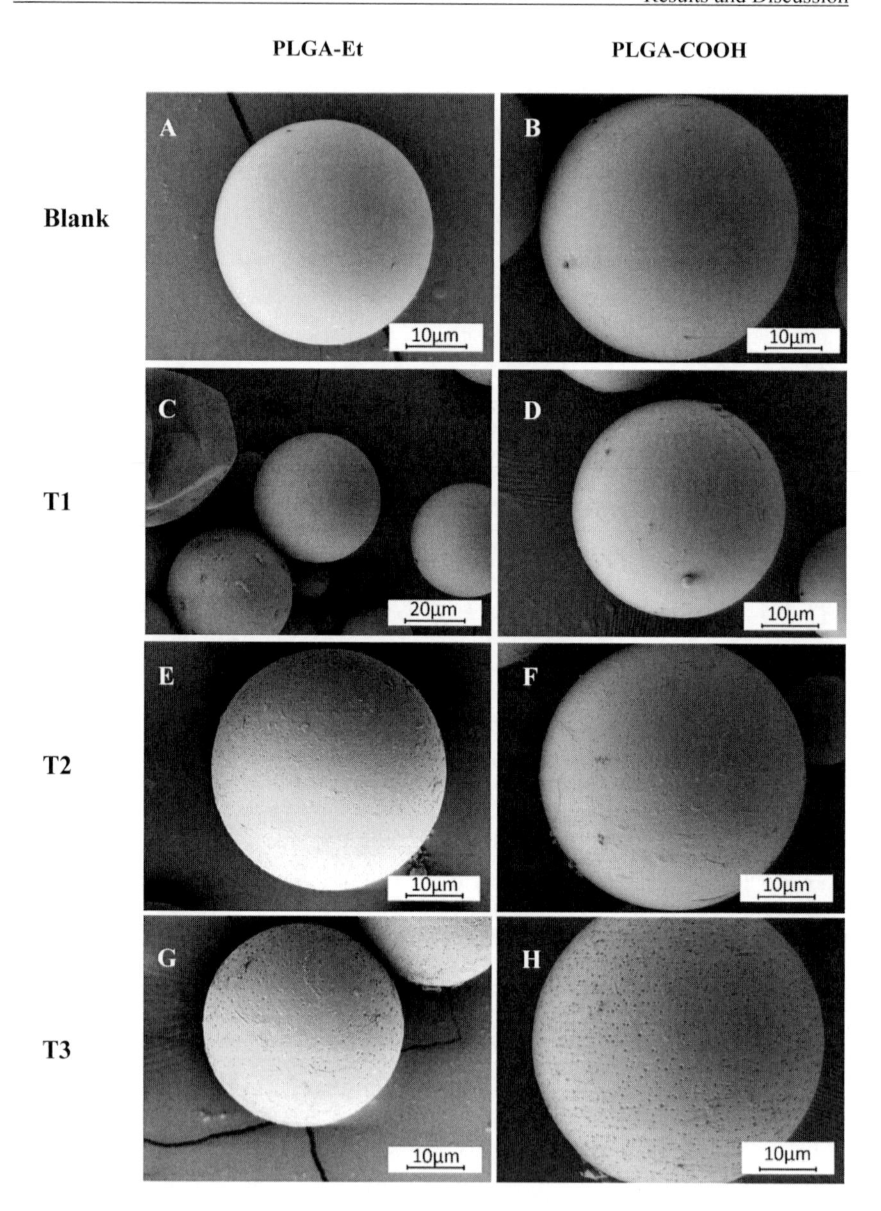

Figure 31: SEM pictures of exemplary **PLGA-Et** and **PLGA-COOH** particles in the 40-50 µm range. **A**: blank **PLGA-Et** particles, **B**: blank **PLGA-COOH** particles, **C**: **PLGA-Et** particles with **T1** payload, **D**: **PLGA-COOH** particles with **T1** payload, **E**: **PLGA-Et** particles with **T2** payload, **F**: **PLGA-COOH** particles with **T2** payload, **G**: **PLGA-Et** particles with **T3** payload, **H**: **PLGA-COOH** particles with **T3** payload.

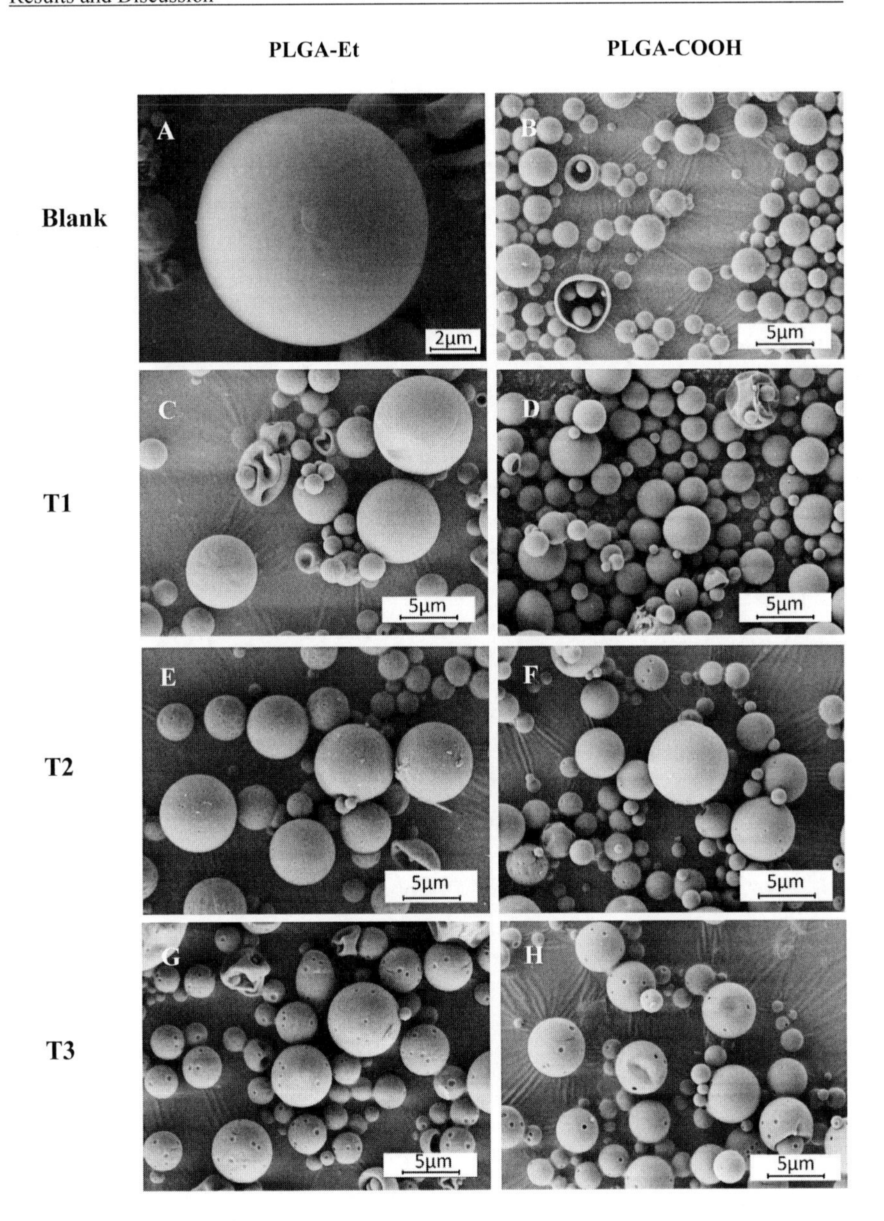

Figure 32: SEM pictures of exemplary **PLGA-Et** and **PLGA-COOH** particles in the 2-5 μm range. **A**: blank **PLGA-Et** particles, **B**: blank **PLGA-COOH** particles, **C**: **PLGA-Et** particles with **T1** payload, **D**: **PLGA-COOH** particles with **T1** payload, **E**: **PLGA-Et** particles with **T2** payload, **F**: **PLGA-COOH** particles with **T2** payload, **G**: **PLGA-Et** particles with **T3** payload, **H**: **PLGA-COOH** particles with **T3** payload.

SEM analysis of nanoparticles prepared from **PLGA-Et** was also performed. The pictures are shown in Figure 33. The sample preparation of the nanoparticles for SEM involves spin coating for better particle separation and mostly led to well isolated particles in case of the loaded specimen (B to D). The loaded particles (B to D) are significantly smaller than the blank particles (A) and show spherical shapes and smooth surfaces.

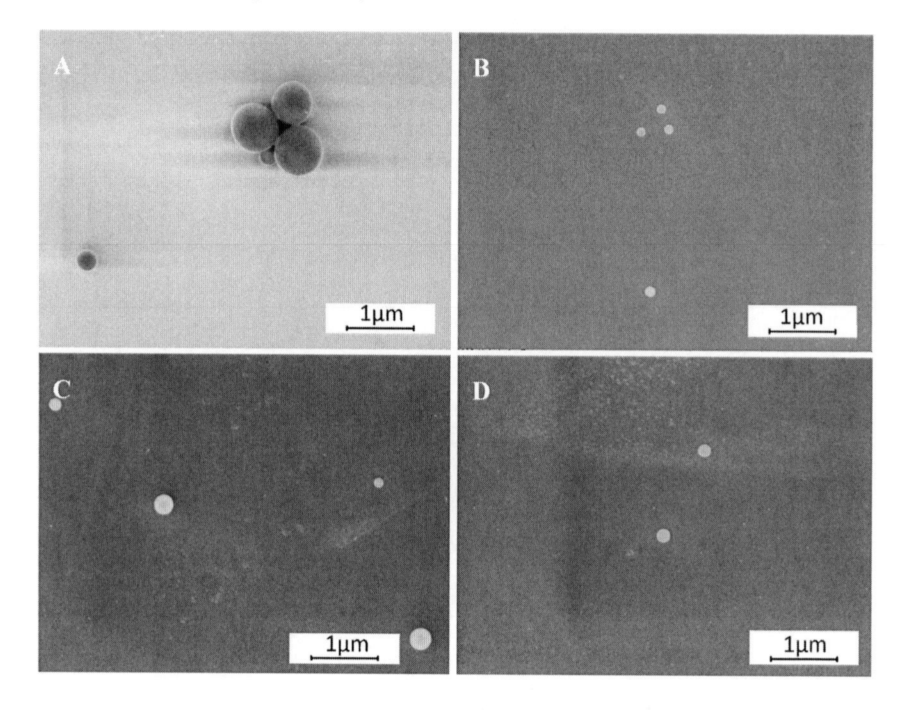

Figure 33: SEM pictures of **PLGA-Et** nanoparticles. **A**: blank particles, **B**: particles with **T1** payload, **C**: particles with **T2** payload, **D**: particles with **T3** payload.

With respect to the planned release experiments at 37.5 °C, the glass transition temperatures (T_g) of the polymer particles become relevant. At this temperature range, a material transforms reversible from a glassy to a viscous or rubbery state.[210] Transition temperatures for PLGA (Resomer® R 503, **PLGA-Et** and R 503H, **PLGA-COOH**) are quoted as 44-48 °C.[196-197] The T_g is generally affected by the symmetry, structural rigidity and secondary forces of the polymer chains as well as by molecular weight and end groups.[210] However, the emulsifier used in particle preparation as well as the encapsulated compounds themselves could act as plasticizer and reduce the transition temperature to a level where it effects the release. To determine if any softening effects take place that might influence the release at the desired temperature, differential scanning calorimetry (DSC) measurements were carried out (Table 18). The DSC measurements confirmed that all three coumarins act as plasticizers and lower the T_g when compared to polymer powder or blank particles (T_g ~ 47-48 °C). The effect for **T1** is, with a T_g reduced by about 1 °C, not as pronounced as for **T2** and **T3**, where a shift by 3 to 4 °C was observed (T_g ~ 44 °C). The bigger impact of **T2** and **T3** is not fully referable to the higher effective loading of the tested particles with deviations of less than 1% (Table 17, Entry 1-4 and Entry 13-16). The plasticizer effect is caused when a small molecule acts as an "internal lubricant" between polymer chains of the matrix. The correlation between the plasticizing effect of drugs and the number of hydroxy groups has been investigated in several studies.[211-215] The plasticizing effect seems to be correlated to the strength of interaction between polymer and plasticizer. The decrease in T_g can be lower than expected if plasticizer and polymer interact strongly.[214-215] In the case of the polar polyester backbone of PLGA, the less polar structures of **T2** and **T3** have a more distinct plasticizer effect. The free hydroxy groups of **T1** on the other hand can form hydrogen bonds with the ester functionalities of the PLGA back bone, thus offsetting the plasticizer effect in part. In extreme cases, antiplasticization can be observed. One example was shown for the interaction of PLGA with Leuprorelin,[212] an extreme polar compound with a high number of hydrogen bond donor and acceptors.

Table 18: Summary of DSC results with T_g determined in second heating cycle.

Entry	Composition	T_{onset} [°C]	T_g [°C][a]	T_{offset} [°C]
1	**PLGA-Et** powder	45	47	47
2	**PLGA-Et** blank particles	44	47	47
3	**PLGA-Et** + **T1** particles	40	46	44
4	**PLGA-Et** + **T2** particles	40	44	44
5	**PLGA-Et** + **T3** particles	43	44	47
6	**PLGA-COOH** powder	44	47	48
7	**PLGA-COOH** blank particles	45	48	48
8	**PLGA-COOH** + **T1** particles	42	47	45
9	**PLGA-COOH** + **T2** particles	41	45	44
10	**PLGA-COOH** + **T3** particles	44	44	47

[a]heating range from -100 °C to 150 °C, heat rate 10.0 K/min, T_g determined in the second heating cycle

The T_g range between onset and offset was found relatively narrow with 3 to 4 °C centered around the T_g. The lowest observed onset of polymer softening was observed at 40 °C. In conclusion, the transition temperatures are still in a range where they should not affect the release of the compounds at 37.5 °C.

3.3.4 Release experiments

The release pattern of a drug from a polymer matrix is affected by several factors. The size of the particles, drug polymer interaction, solubility of both drug or polymer and degradation of the polymer can influence the speed and amount of drug that is released.[135] To observe the release of coumarins **T1-T3**, a release setup was devised in which the particles were put into 2 mL Eppendorf tubes, covered with 1.5 mL PBS at pH 7.4 supplemented with 1% (v/v) polysorbate 80 (Tween®) (PBST) and incubated at 37.5 °C on a rocking platform shaker. Polysorbate 80 acts as an emulsifier and is necessary to assist in the dissolution of lipophilic coumarins **T1-T3** in the aqueous medium (see Chapter 3.3.1, Figure 21). The concentration of 1% was chosen as low as possible while still guaranteeing adequate dissolution of the payload coumarins to maintain sink conditions[vii]. PBST buffer systems have been used for other lipophilic drugs.[147] Furthermore, the mixture mimics the conditions *in vivo*. The bicarbonate buffer of blood with its pH of 7.4 is simulated by the PBS buffer system, while polysorbate 80 mimics the presence of amphiphilic peptides/protein, such as serum albumin, a nonspecific plasma carrier protein that helps to dissolve hydrophobic molecules.[216]

At defined time points, the tubes were centrifuged and 1 ml of the buffer removed and replaced with fresh solution before resuspension and continuation of the incubation. The withdrawn volume was then evaluated with the above described HPLC method to determine the amount of compound that had been released to the medium. HPLC chromatograms can indicate degradation of the payload coumarin, an effect that has been observed for other drugs and can lead to toxic decomposition products.[217] The polymer matrix is slightly acidic, the buffer slightly basic and so decomposed of the compounds could occur. The chromatograms were therefore examined for newly formed peaks that could indicate breakdown of the compounds. No such peaks were observed. This is in line with other findings. The compounds **T1-T3** as

[vii] Sink conditions mean that the concentration of the solution is kept below 10 to 20% of the maximum saturation. This is necessary to allow a constant dissolution rate. The dissolution rate would decrease with approximation to maximum saturation.

well as other 3-benzylcoumarins have been exposed to strong acidic and strong basic conditions and no degradation was observed.

After 30 days, the release experiments were terminated and the particles and solution lyophilized and dissolved to determine the remaining content of the encapsulated coumarin. The release from nanoparticles was significantly faster than from the microparticles (full release in less than 24 h) and will be further discussed later.

Release from Microparticles

Figure 34 shows the release curves for encapsulated compound **T1-T3**, generated from the data of triplicate release studies. The Y-axis is assigned to the cumulative release in percent relative to full quantity of extractable compound from the particles. Release from particles in the 40 μm range are drawn in black for **PLGA-Et** and blue for **PLGA-COOH**. The release from the 4 μm range particles are drawn in red for **PLGA-Et** and green for **PLGA-COOH**.

When the release curves for coumarins **T1-T3** are compared, the first observation that is made is the faster release rate from the smaller particles (~ 4 μm, red and green) when compared with the bigger particles (~ 40 μm, black and blue). Furthermore, the release from ~ 4 μm **PLGA-COOH** particles (green curves) shows a higher initial release compared to **PLGA-Et** particles (red curves). The initial burst release is commonly observed in particles prepared by o/w methods. It is attributed to the presence of drug deposits on or close to the particle surfaces.[135,201] When compared to the bigger particles (black and blue) the observed burst release is significantly higher for small particles (red and green). The higher surface to volume ratio for the small particles should attribute for this phenomenon. The observed burst release for **T1** (A) is higher when compared to **T2** and **T3** (B and C) and results in almost full release for **T1** from ~ 4 μm particles after one week.

This leads to a second observation; **T1** (A), the most polar compound, displays a relatively fast release when compared to **T2-T3** (B and C). This observation is found for both particle size ranges and polymer matrix types. Explanations for the differences in release behavior must be deducible from the difference in properties as a result of the substitution patterns on the coumarin core. **T1**, the compound that shows the fastest release, has two free hydroxy groups

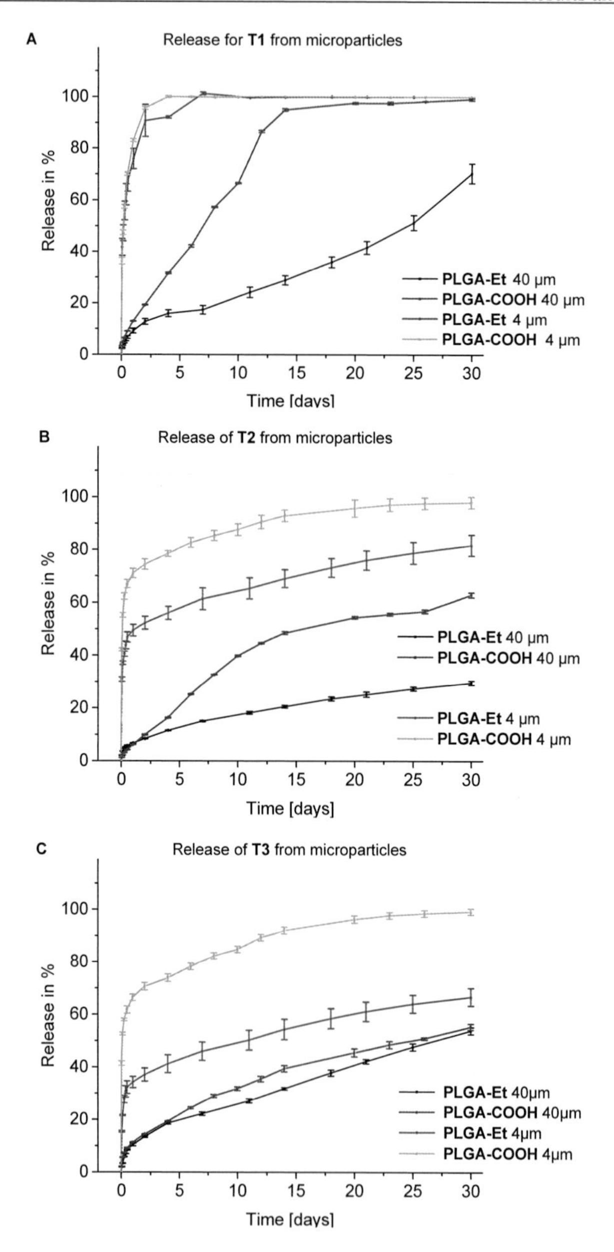

Figure 34: Release curves for compounds **T1-T3** from **PLGA-Et** and **PLGA-COOH** microparticles.

and shows the best solubility in the release medium (compare section 3.3.1, Figure 21). Therefore, the osmotic forces for the formulated compound **T1** should be higher when compared to **T2** and **T3**, accelerating the intrusion rate of water into the polymer. Due to their small molecular weight, the release for the compounds is expected to be mainly influenced by diffusion through the polymer matrix.[147] The strength of interactions between the payload coumarin and the matrix plays therefore a prominent role for the release. The less polar compound **T2** and **T3** should have bigger attractive forces towards the polymer matrix then the polar compound **T1**.

When the release rates for coumarins **T1-T3** from the two different polymer matrixes are compared they show significant differences. The release curves for particles in the range of 40 µm for **PLGA-COOH** particles (A-C, blue graphs) show systematically faster release rates then the **PLGA-Et** particles (A-C, black graphs) of the same size range. The difference is most pronounced for **T1** and **T2**. For ~ 4 µm microparticles, the same systematic faster release from **PLGA-COOH** particles (A-C, green curves) when compared to **PLGA-Et** particles (A-C, red curves) was observed. The polymer matrix influences diffusion controlled release through parameters such as chain length, flexibility and mobility, water uptake and swelling behavior.[135] As the chain length and overall mobility for the used **PLGA-COOH** and **PLGA-Et** should be in the same dimensions, the effect of faster release from **PLGA-COOH** might be caused by faster swelling. The hydrophilic end groups should allow a faster and greater extent of intrusion of aqueous medium into the polymer matrix ultimately leading to the observed faster release rates. The release rate for **T1** from ~ 40 µm **PLGA-COOH** particles (A, blue graph) that showed the fastest constant release after the initial burst period can be explained by a combination of the faster swelling of **PLGA-COOH** and the higher osmotic pressure caused by more hydrophilic coumarin **T1**.

Overall, the observed burst release for ~ 40 µm particles is relatively small and the release almost linear (0. order) over prolonged time spans. A possible explanation is that the solidified coumarins are slowly dissolved in the particle and only then start to diffuse through the polymer matrix. This effect is expected to be more pronounced for larger and more hydrophobic particles and expectations correspond with the observed effects.[135]

Release from Nanoparticles

The release of **T1-T3** from nanoparticles is recapitulated in Figure 35 and show a significantly faster release than the microparticles. The graph for **T1** (black line) clearly shows that more than 90% compound is released after the first measurement at 30 min. In contrast, the less polar substances **T2** and **T3** (red and blue) also show a fast release but the 90% mark is not crossed before 5 h. Working with nanoparticles in the release experiment was also associated with the highest potential for outlier due to handling issues with such samples. Centrifugation of the nanoparticles is not as effective as for the larger particles and the probability of losing particles is therefore higher. Also, aggregation of the particles could have a much higher influence on the effective surface and is hard if not impossible to detect.

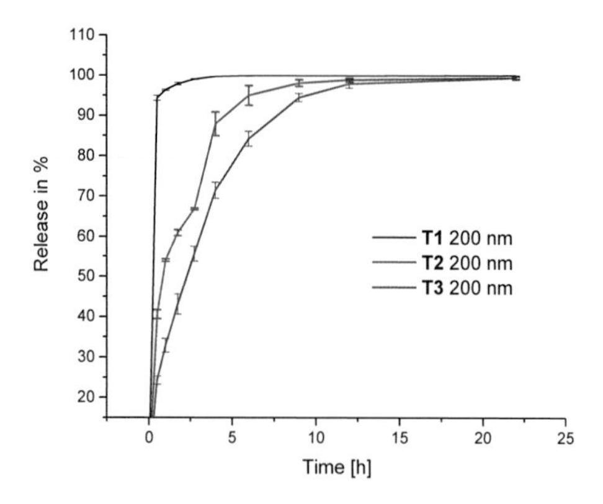

Figure 35: Release curves for **T1-T3** from **PLGA-Et** nanoparticles.

In conclusion, the release experiments show that the liberation of compounds is largely dependent on the size and solubility of the drug as well as the size and composition of the used polymer scaffold. The experimental data corresponds thereby to the expectations and findings of previous research.[130,135] Larger particles tend to release their payload slower and over longer periods of time. This long-time release behavior is highly interesting in long-term exposure settings where the drugs should be administered over a prolonged period. It was shown that depending on the drug and particle system, release over hours, days, weeks and even months can be addressed for all three compounds. The difference in release rates allows the use for a variety of applications and the amount of drug per time can be controlled by the amount of particles.

For all particles, a burst release is observed i.e. a high initial release. The burst release is most likely explained by the adherence of compound on the particle surfaces or close to the surfaces. In the case of **T1,** the most polar compound, this increased initial release is most prominent.

Also, the burst release is more distinct for smaller particles, which relates to the higher surface to volume ratio. Compound that is adsorbed or is loaded close to the surface has a shorter travel distance through the polymer matrix and is thus released faster. In subsequent stages, the release is influenced by diffusion of the encapsulated coumarin through the polymer carrier as well as the degradation of the polymer scaffold that facilitates the release. The burst release for the bigger particles (~ 40 μm) is relative small and almost insignificant.

Countermeasures to remove the significant burst behavior in the small micro- and nanoparticles could be considered. One option is to include washing steps with an alcohol solution that is capable of dissolving the adhering drug deposits from the surface, a technique that has been employed for progesterone.[135,218] Alternatively, the particles could be coated again with a polymer matrix by s/o/w techniques.[135]

The studies have shown that microparticulate carriers are capable of accomplishing a temporal control of release for the cannabinoid analogs **T1-T3**. Fine tuning of the release properties by preparation of particle different particle sizes and choice of the polymer matrices is evidently feasible. Further investigations could include other polymer matrices, especially lower M_w polymers are enticing. The release rates from such particles should be faster and are therefore interesting for faster release of **T2** and **T3**. Slower release of **T1** could be achieved accordingly by the employment of polymers with higher M_w, selection of a PLGA with a lower glycolide content or change to more hydrophobic polymers, e.g. PLA (poly-[lactic acid]) or PCL (polycaprolactone).

4. Conclusion and Outlook

4.1 Work on the THC-Framework

With the successful synthesis of the *α,β*-unsaturated ketone **55**, the long standing goal of our group of a formal total synthesis of THC was concluded. Ketone **55** can be methoxy-deprotected according to HUFFMAN *et al.*[164] via an *in situ* formed thiolate (Scheme 42, a) to generate phenol **21**. Phenol **21** is also an intermediate in the total synthesis route of FAHRENHOLZ *et al.*[40] (Chapter 1.2.1) and thereby presents the conclusion of our formal total synthesis.

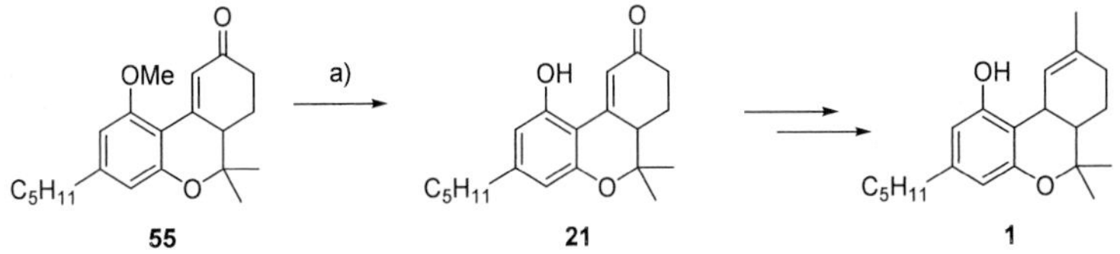

Scheme 42: Further synthetic transformations towards THC (**1**): *a) NaH, n-propane thiol, DMF, reflux, 3 h, 84%.*[164]

More importantly, we were able to show that the DA-approach towards cannabinoids is feasible (Scheme 43). In total six DA-products were isolated, the reaction of diene **39** with maleic anhydride showing quantitative conversion. Further optimization for the conversion of diene **39** with other dienophiles remains necessary. With respect to the findings by MIANMI *et al.*[163] (Chapter 3.1.1), elevation of temperatures needs to be implemented cautiously as isomerization of the double bond can occur. The implementation of catalyst systems would be an alternative approach to higher yields and enantioselective reaction control.[219]

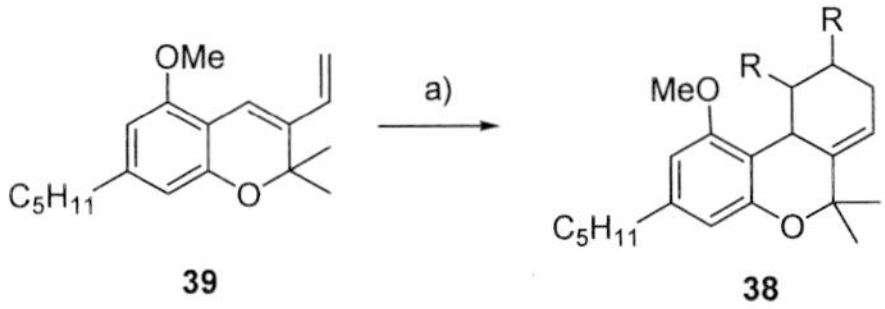

Scheme 43: DIELS-ALDER reactions with diene **38**.

The introduced strategy allows a modular synthesis of the THC-framework. Succeeding work could exploit upon the fact and generate a vast library in a modular fashion and generate even more THC-derivatives by synthetic transformation of the resulting DA-products.

4.2 Cannabinoid Analogs

The synthesis of cannabinoid analogs remains to be an interesting field and will be continued in succeeding work in our group. In this thesis we initiated the expansion of substrate scopes, the inclusion of new motifs and established new synthetic protocols. The obtained libraries are shown in Figure 36 and include 3-benzylcoumarins, 3-pyridylmethylcoumains, 3-alkylcoumarins, 3-arylcoumarins and first examples for 3-styrylcoumarins and 2,2-dimethyl-2*H*-chromenes.

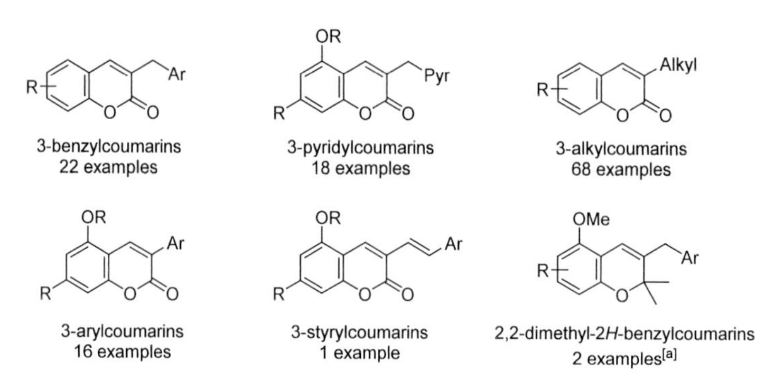

Figure 36: Realized compound libraries. [a] Two stable examples isolated.

The synthesis of 3-benzylcoumarins is well established in our group and the examples presented in this thesis are limited to compounds that have already shown biological activity or are close modifications of active structures. We have been asked repeatedly if the synthesis could be performed in quantities suitable for *in vivo* testing. A major achievement was the successful upscaling of 3-benzylcoumarin **27**. However, the shown synthesis only presents the first step towards large-scale preparation of 3-benzylcoumarins with our method. Optimization in regard of reaction time, equivalent ratio and NHC-loading will be performed. It is also interesting to add molecular sieves (MS) as water absorbent to the reaction mixture as has been shown by JIANG *et al.*[116] The addition of molecular sieves is not beneficiary under microwave conditions but could prove useful in the classical heating setup.

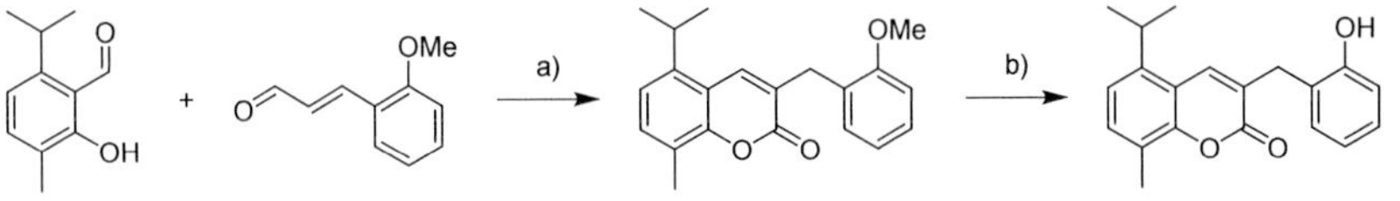

29i 30a 26n 26c

Scheme 44: Upscaling attempt of the coumarin synthesis: *a) IL **31**, K_2CO_3, toluene, reflux,24 h, 62%; b) BBr_3, DCM, 30 min, –78 °C, 16 h, r.t., 96%.*

Furthermore, the substrate scope of the established 3-benzylcoumarin reaction could broadened. Through the incorporation of pyridyl acrylaldehydes **92** in the synthesis protocol we were able to synthesize a small library of 3-pyridylmethylcoumarins **96** that were subsequently deprotected. This library constitutes the first report of 3-pyridylmethylcoumarins **96** where no additional substitution in 4-position is present.

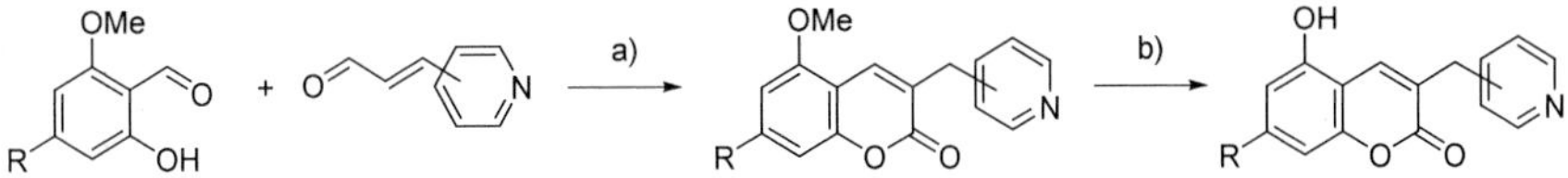

29a-c 92a-c 96a-i 96A-l

Scheme 45: Successful library synthesis of 3-pyridylmethylcoumarins **96**: *a) IL **31**, K_2CO_3, toluene, 110 °C, MWI (max. 200 W, 7 bar), 50 min, 15-21%; b) BBr_3, DCM, 30 min, –78 °C, overnight, r.t., 21-88%.*

Future synthetic effort can include a broader range of hetero aromatic systems. One example was synthesized with the 3-furylmethylcoumarin **98**.

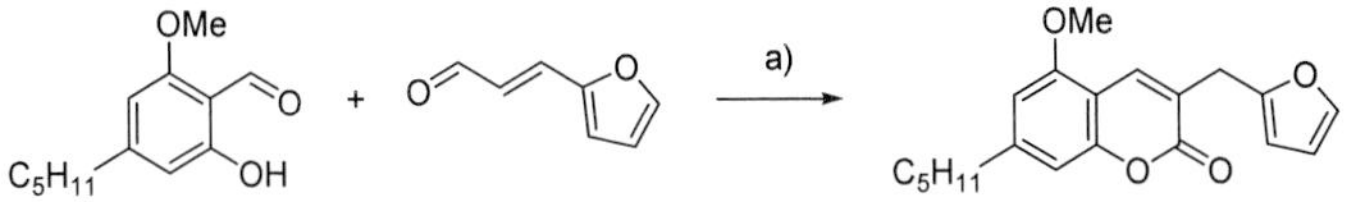

29a 97 98

Scheme 46: Successful synthesis of 3-furylmethylcoumarin **98**: *a) IL **31**, K_2CO_3, toluene, 110 °C, MWI (max. 200 W, 7 bar), 50 min, 30%.*

A large library of 3-alkylcoumarins was synthesized via PERKIN reaction (Scheme 47). Several of the salicylic aldehydes that have been synthesized specifically for this purpose have not been employed in the 3-benzyl synthesis yet. The applied synthetic strategy is simple and allows a fast and modular synthesis. For future endeavors the scope of anhydrides could be expanded. Especially the introduction of polar groups in the side chain is interesting to mimic the pharmacophore NAH position discussed in Chapter 1.4.

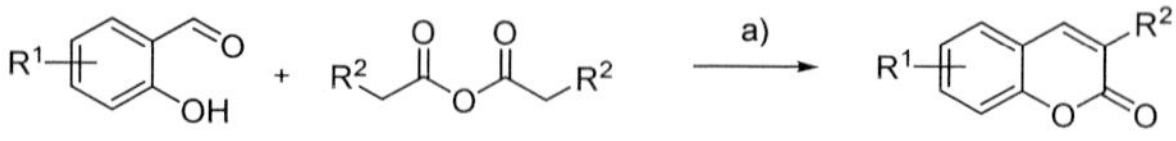

Scheme 47: Synthesis of 3-alkylcoumarins **27**: *a)* K_2CO_3*, 180 °C, MWI (max. 300 W, 15 bar), 1 h.*

For the first time, 3-arylcoumarins **40** have been synthesized in our group. The small library could be rapidly expanded by following the established protocol with alternative salicylic aldehydes.

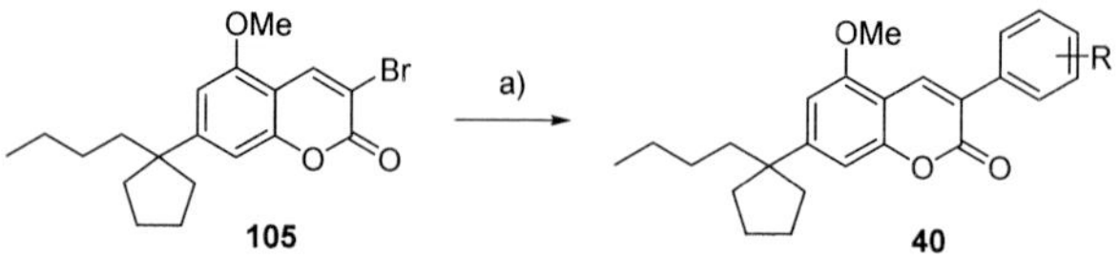

Scheme 48: 3-Arylcoumarins **40** synthesized from 3-bromocoumarins **105**: *a) 2 equiv. boronic acid, 2 equiv.* Cs_2CO_3*, 0.05 equiv.* $Pd(PPh_3)_4$*, degassed dioxane, 90 h, 16 h, 23-90%.*

Synthetic application of the SUZUKI-coupling in this work has been limited so far to aryl- and styrylboronic acids. The SUZUKI-coupling still offers the opportunity to synthesize a broader range of compounds with several more boronic acids. Future plans could also include alkylboronic acids in an aryl-alkyl SUZUKI-coupling.[220] Especially interesting would be structures with a three-dimensional expansion. Substituents such as cyclohexyl **119**, adamantly **120** or propellane **121** moieties (Scheme 49, left) could be included. In addition, we proposed the introduction of the boronic acid moiety on the coumarin **122** (Scheme 49, right), to compare the effect on the coupling yields. However, coumarins with a boronic acid function in 3-position **122** have not been reported in literature so far and the endeavor might prove challenging.

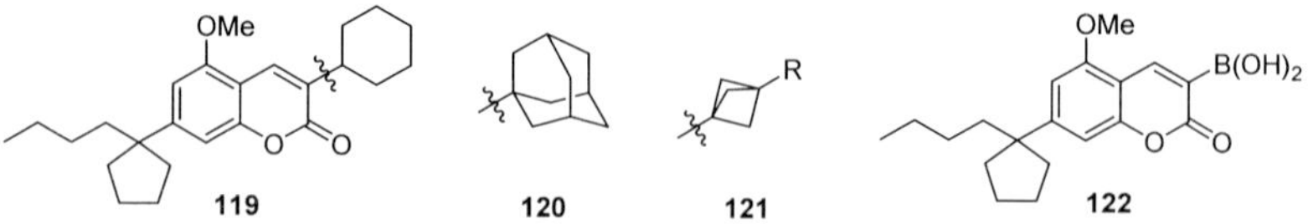

Scheme 49: Ideas for future SUZUKI-coupling products and strategies.

The synthesis of a 3-styrylcoumarin **41a** was also successfully implemented. However, the absence of a broad scope of styrylboronic acids **107**, demanded the focus on alternative synthetic pathways.

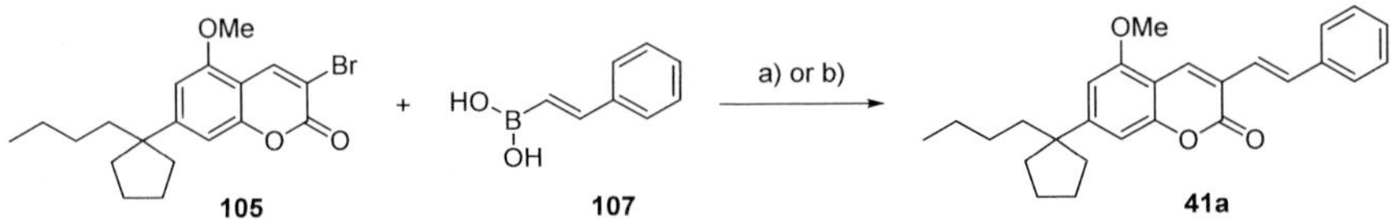

Scheme 50: Synthesis of a 3-styrylcoumarin **41a**. *a) 2 equiv. boronic acid **107**, 2 equiv. Cs_2CO_3, 0.05 equiv. $Pd(PPh_3)_4$, degassed dioxane, 90 h, 16 h, < 45%; b) 3 equiv. boronic acid **107**, 8 equiv. Na_2CO_3, 0.05 equiv. $Pd(PPh_3)_4$, degassed dioxane: water 5:1, 90 h, 16 h, 74%.*

In succeeding work, a new synthetic option will be explored. According to Branco *et al.*,[187] the synthesis of 3-vinylcoumarins **109** can be accomplished by a DCC mediated esterification with subsequent condensation (Scheme 51). The 3-vinylcoumarins **109** could then be converted in Heck-reactions to 3-styrylcoumarins **41**.

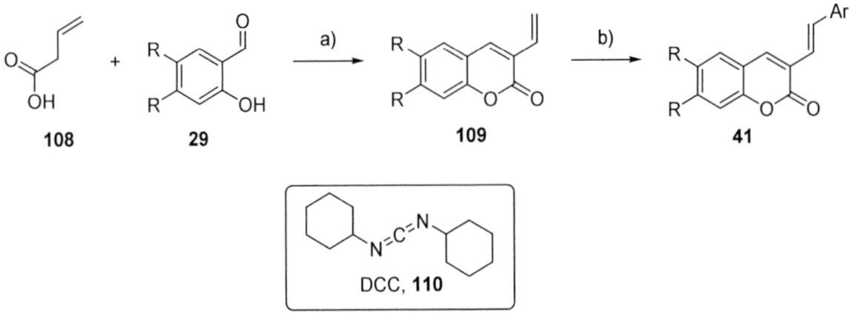

Scheme 51: Schematic synthesis of 3-styrylcoumarins **41** *via* 3-vinylcoumarin **109** synthesis and subsequent Heck-coupling by Branco *et al.*[187] *a) 1)butenoic acid (**108**), DCC (**110**), DCM, 2) salicylic aldehyde **29**, DMAP, 3) Cs_2CO_3; b) Ar-X, Pd cat., DMF.*

A second alternative synthesis pathway towards the intermediate 3-vinylcoumarins **109** would also be the implementation of butenoic anhydride **123** in a Perkin synthesis. However, this reaction has not been reported and it remains open if the compounds would be stable under Perkin-conditions. Especially, the required elevated temperatures, might be inappropriate for the rather labile 3-vinylcoumarins.[163]

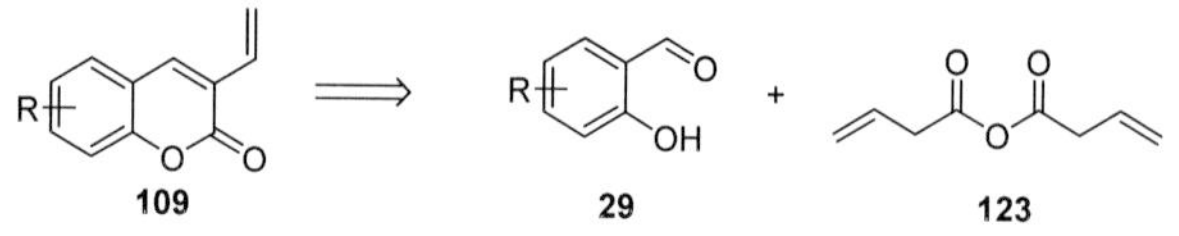

Scheme 52: Synthesis of 3-vinylcoumarins **109** *via* PERKIN reaction.

The synthesis of two 2,2-dimethyl-2*H*-chromenes **117** (Figure 37) was achieved in this work. Thereby we created a novel class of compounds with potential cannabinoid activity that was not reported before. The novel compounds combine structural features of both, the 3-benzylcoumarins and the THC framework.

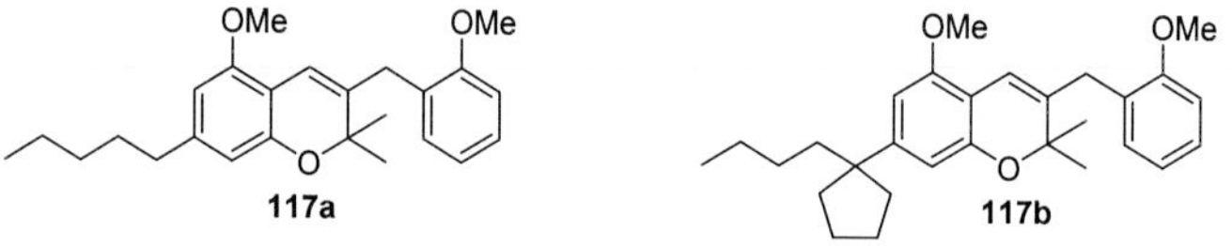

Figure 37: Synthesized 2,2-dimethyl-2*H*-chromenes **117a** and **117b**.

In this work we have been preliminary focused on the synthetic aspect of cannabinoid development. Biological screening for all compounds described in this chapter is planned and we hope that the newly generated data will allow more accurate SAR evaluation and will help to develop new compounds.

4.3 Controlled Release systems for Cannabinoids

In the last part of the present thesis we described the development of controlled release systems for cannabinoids. We could show that the lipophilic 3-benzylcoumarins **T1-T3** could be efficiently encapsulated in micro- and nanoparticulate carriers and subsequently released in a controlled fashion. The liberation is dependent on the properties of the compounds and can be controlled by the size and composition of the particle matrices meeting our expectations.[130,135]

The prepared particle sizes and formulations allow to address different conceptual applications. For instance, the prepared nanoparticles can be used for distributional controlled release by exploitation of their inherent biodistribution properties. The biodistribution could be tuned accordingly by surface modification of the particles.[145] However, *in vivo* studies will be necessary to determine if a selective effect can be achieved in this fashion.

The controlled release from microparticles is highly interesting for depot applications. Depending on the drug and particle system, we were able to achieve controlled release over hours, days, weeks and even months for all three compounds. Other applications include systems where a combination of different particle sizes allows an even more precise control of the drug levels. Additionally, other particle compositions could be employed, e.g. lower M_w polymers for faster release and accordingly higher M_w polymers for slower release.

5. Experimental Part

5.1 Miscellaneous

5.1.1 Analytics and Equipment

Nuclear Magnetic Resonance (NMR):

NMR spectra have been recorded using the following machines:

^{1}H NMR: BRUKER *Avance* 300 (300 MHz), BRUKER *Avance* 400 (400 MHz), BRUKER *Avance DRX 500* (500 MHz). The chemical shift *δ* is expressed in parts per million (ppm) where the residual signal of the solvent has been used as secondary reference: chloroform (^{1}H: δ = 7.26 ppm) acetone (δ = 2.05 ppm), dimethyl sulfoxide (δ = 2.50 ppm) and dichloromethane (δ = 5.32 ppm).[221] The spectra were analyzed according to first order.

^{13}C-NMR: Bruker *Avance* 300 (75 MHz), Bruker *Avance* 400 (101 MHz), Bruker *Avance DRX 500* (126.3 MHz). The chemical shift *δ* is expressed in parts per million (ppm), where the residual signal of the solvent has been used as secondary reference: chloroform (δ = 77.0 ppm), acetone (δ = 30.8 ppm), dimethyl sulfoxide (δ = 39.4 ppm) and dichloromethane (δ = 53.8 ppm).[221] The spectra were 1H-decoupled and characterization of the ^{13}C-NMR-spectra ensued through the DEPT-technique (DEPT = Distortionless Enhancement by Polarization Transfer) and is stated as follows: DEPT: "+" = primary or secondary carbon atoms (positive DEPT-signal), "–" = secondary (negative DEPT-signal), $C_{quart.}$ = quaternary carbon atoms (no DEPT-signal).

All spectra were obtained at room temperature. As solvents, products obtained from EURISOTOP and SIGMA ALDRICH were used: chloroform-d_1, acetone-d_6, dimethylsulfoxide-d_6 and dichloromethane-d_2. For central symmetrical signals the midpoint is given, for multiplets the range of the signal region is given. The multiplicities of the signals were abbreviated as follows: s = singlet, d = doublet, t = triplet, q = quartet, hept = heptet, bs = broad singlet, m = multiplet, b = broad (unresolved) and combinations thereof. All coupling constants *J* are stated as modulus in Hertz [Hz].

Melting Point (MP):

Melting points were measured on a STANFORD RESEARCH SYSTEMS *OptiMelt* at 5 °C/min.

Infrared spectroscopy (IR):

IR-spectra were recorded on a BRUKER *Alpha P* and a BRUKER *IFS 88*. Measurement of the samples was conducted via attenuated total reflection (ATR). The intensity of bands (strength of absorption) was described as follows: vs = very strong (0-9% transmission, T); s = strong (10-39% T); m = middle (40-69% T); w = weak (70-89%); vw = very weak (90-100%). Position of the absorption bands is given as wavenumber $\tilde{\nu}$ with the unit [cm^{-1}].

Ultra violet/visible light absorption spectra (UV/Vis):

UV/Vis spectra have been recorded on an ANALYTIK JENA *Specord 50/plus*.

Fluorescence Emission Spectra

Fluorescence spectra have been recorded on a HORIBA SCIENTIFIC *fluoromax-4*, with a Xenon source used for excitation.

Mass Spectrometry (EI-MS, FAB-MS):

Mass spectra were recorded on a FINNIGAN *MAT 95*. Ionization was achieved through either EI (*Electron Ionization*) or FAB (*Fast Atom Bombardment*). Notation of molecular fragments is given as mass to charge ratio (*m/z*); the intensities of the signals are noted in percent relative to the base signal (100%). As abbreviation for the ionized molecule $[M]^+$ was used. Characteristic fragmentation peaks are given as $[M\text{–fragment}]^+$ and $[\text{fragment}]^+$.

For HRMS (*High Resolution Mass-Spectrometry*) following abbreviations were used: calc. = expected value (calculated); found = value found in analysis.

Gas Chromatography-Mass Spectrometry (GCMS):

GCMS measurements have been recorded with an AGILENT TECHNOLOGIES model *6890N* (electron impact ionization), equipped with a AGILENT *19091S-433* column (5% phenyl methyl siloxane, 30 m, 0.25 µm) and a *5975B VL MSD* detector with turbo pump. As a carrier gas helium was used.

Elemental analysis (EA):

Measurements were conducted on a ELEMENTAR *Vario Micro*. As analytical scale SARTORIUS *M2P* was used. Notation of Carbon (C), Hydrogen (H) and Nitrogen (N) is given in mass percent. Following abbreviations were used: calc. = expected value (calculated); found = value found in analysis.

Thin Layer Chromatography (TLC):

All reactions were monitored by thin layer chromatography (TLC). The TLC plates were purchased from MERK (silica gel 60 on aluminum plate, fluorescence indicator F254, 0.25 mm layer thickness). Detection was carried out under UV-light at $\lambda = 254$ nm and $\lambda = 366$ nm. Alternatively, the TLC plates were stained with a SEEBACH-dip (2.5% phosphor molybdic acid, 1.0% cer(IV)sulfate-tetrahydrate, 6.0% conc. sulfuric acid, 90.5% water; as dip solution) and dried in a hot air stream. Used solvent mixtures were measured volumetrically.

Solvents and reagents:

Solvents of technical quality have been purified by distillation prior to use. Solvents of the grade p.a. (*per analysis*) have been purchased (ACROS, FISHER SCIENTIFIC, SIGMA ALDRICH, ROTH, RIEDEL–DE HAËN) and used without further purification. Absolute solvents have been dried using the methods listed in Table 19, and were stored under argon afterwards or have been purchased from a commercial supplier (abs. acetonitrile (ACROS, < 0.005% water), abs. chloroform (FISCHER, over molecular sieves), abs methanol (FISCHER, < 0.005% water), abs. ethanol (ACROS, < 0.005% water).

Table 19: Methods for the absolutizing of solvents. All distillations were carried out under argon atmosphere

Solvent	Method
Dichloromethane	heating to reflux over calcium hydride, distilled over a packed column
Tetrahydrofuran	heating to reflux over sodium metal (benzophenone as an indicator), distilled over a packed column
Diethyl ether	heating to reflux over sodium metal (benzophenone as an indicator) distilled over a packed column
Toluene	heating to reflux over sodium metal (benzophenone as an indicator) distilled over a packed column

Reagents have been purchased from commercial suppliers (Companies: ABCR, ACROS, ALFA AESAR, APOLLO, CARBOLUTION, CHEMPUR, FLUKA, IRIS, MAYBRIDGE, MERCK, RIEDEL DE

HAËN, TCI, THERMO FISHER SCIENTIFIC, SIGMA ALDRICH). They have been used without further purification unless stated otherwise.

Microwave (MW)

Microwave reactions were carried out in a single mode *CEM* Discover LabMate microwave operated with *CEM*'s Synergy™ software. This instrument works with a constantly focused power source (0–300W). Irradiation can be adjusted *via* power- or temperature control. The temperature was monitored with an infrared sensor and carried out in pressurized closed vials.

Analytical scales:

Used machine: SARTORIUS Basic.

5.1.2 Preparative work

Before the reactions with air or moisture sensitive reagents were carried out, the glass devices have been dried in an oven and under high vaccum. Reactions have been executed according to SCHLENK-techniques using argon as an inert gas. Liquids were added via plastic syringes and V2A-needles. Solids were added in pulverized form. Reactions at 0 °C were cooled with a mixture of ice/water. Reactions at deeper temperatures were tempered with brine/ice mixture (–20 °C), isopropanol/dry ice mixture (–78 °C) or with a cryostat.

All reactions were monitored by TLC or GCMS.

Solvents were removed at 40 °C with a rotary evaporator.

If not stated otherwise, solutions of inorganic salts are saturated aqueous solutions.

If not otherwise specified, the crude products, were purified by flash column chromatography following the concepts of Still *et al.*[222] using silica gel (SIGMA ALDRICH, pore size 60 Å, particle size 40 – 63 μm) and sand (calcined and purified with hydrochloric acid) as stationary phase. Solvents were distilled prior to use or p.a. grade solvents used. Solvent mixtures were prepared individually in terms of volume ratios are given as volumetric. The use of a gradient is indicated in the experimental procedures

Celite® for filtrations was purchased from ALFA AESAR (Celite® 545, treated with Na_2CO_3).

5.2 Synthesis and Characterization

5.2.1 General Procedures

General Procedure for the synthesis of aldehydes from 1,3-dimethoxybenzenes (**GP1**):

A solution of the respective 1.00 equiv. of 1,3-dimethoxybenzene in diethyl ether (10 mL per g phenol) is prepared and 1.50 equiv. of *N*,*N*,*N'*,*N'*-tetramethylethene-1,2-diamine (TMEDA) is added dropwise. The mixture is cooled to 0 °C and 1.50 equiv. of butyl lithium (1.6 M in *n*-hexanes) is added. After stirring for 2 h at room temperature the solution is cooled to 0 °C and 3.00 equiv. of dimethylformamide is added and the mixture was stirred for another 4 h at room temperature. The reaction is quenched by the addition of brine and extracted with diethyl ether. The combined organic layers are dried over sodium sulfate, the volatiles were removed under reduced pressure and the residue is then purified via flash column chromatography.

General Procedure for the synthesis of salicylic aldehydes from phenols (**GP2**):

A solution of the respective 1.00 equiv. of phenol in (dry acetonitrile 30 mL per g phenol) is cooled to 0 °C. Subsequently, 6.75 equiv. of paraformaldehyde and 1.50 equiv. of magnesium chloride are added under argon counterflow. Then 3.75 equiv. of triethylamine is added and the mixture is stirred at reflux for 3 h. The reaction mixture is then allowed to cool to room temperature, quenched with water and the pH is adjusted to 1 by adding concentrated hydrochloric acid. The aqueous layer is extracted with dichloromethane and the combined organic layers are dried over magnesium sulfate and the volatiles are removed under reduced pressure. The crude product is then purified via flash column chromatography.

General Procedure for the selective mono-deprotection towards salicylic aldehydes (**GP3**):

The respective *o*-methoxy aldehyde (1.00 equiv.) is dissolved in a 2:1 mixture of dry acetonitrile (10 mL per g aldehyde) and dichloromethane (5 mL per g aldehyde). The mixture is cooled to 0 °C. Aluminum trichloride (2.50 equiv.) and sodium iodide (2.50 equiv.) are slowly added under argon counterflow. The solution is then stirred for 1 h or until TLC shows complete conversion and quenched with water. The aqueous layer is extracted with dichloromethane, dried over sodium sulfate and the volatiles are removed under reduced pressure. The crude product is then purified via flash column chromatography.

General Procedure for the WILLIAMSON-ether synthesis (**GP4**):

Under argon atmosphere, a flask equipped with condenser is charged with 1.00 equiv. of the respective benzyl alcohol, 8.00 equiv. of bromo propane, which are dissolved in 27.5 mL of tetrahydrofuran per mmol benzyl alcohol. Under argon counterflow 10.0 equiv. sodium hydride is added, and the mixture is heated at reflux for 24 h. The reaction is cooled in an ice bath and quenched by the addition of water. The tetrahydrofuran is removed under reduced pressure and the remaining aqueous layer is extracted with dichloromethane. The combined organic layers are dried over sodium sulfate and volatiles are removed under reduced pressure. The residue is purified via flash chromatography.

General Procedure for the synthesis of 3-benzylcoumarins (**GP5**):

Under argon atmosphere a microwave vial is charged with 1.00 equiv. of the respective salicylic aldehyde, 2.20 equiv. of potassium carbonate, 2.50 equiv. of the respective cinnamic aldehyde or its analog and 1.20 equiv. of 1,3-dimethylimidazolium dimethyl phosphate and suspended in toluene. The reaction mixture is then heated at 110 °C for 50 min at 200 Watt microwave irradiation and a maximum pressure of 7 bars. The mixture is allowed to cool to room temperature and then 5-10 mL of water is added, following by extraction with ethyl acetate and the removal of volatiles under reduced pressure. The residue is purified via flash chromatography.

General Procedure for the synthesis of acrylaldehydes (**GP6**):

To a suspension of 1.00 equiv. of (triphenylphosphoranylidene)-acetaldehyde in 10 mL (2 mmol/mL) of nitrobenzene under argon atmosphere, 1.00 equiv. of pyridine-3-carbaldehyde is added dropwise. The reaction is stirred at room temperature for 24 h under exclusion of light and then taken up in dichloromethane, washed with brine, dried over sodium sulfate and the volatiles are removed under reduced pressure. The crude product is then purified via flash column chromatography.

General Procedure for the PERKIN-Synthesis of coumarins (**GP7**):

The respective salicylic aldehyde, carbonic acid anhydride and potassium carbonate are placed in a microwave vial and heated at 180 °C for 65 min at 300 W microwave irradiation. The resulting mixture is allowed to cool to room temperature, poured onto crushed ice and the pH is adjusted to ~7 with sodium bicarbonate. The mixture is then extracted with ethyl acetate and the organic layer is dried over sodium sulfate. Evaporation of the volatiles and subsequent purification (if necessary) via column chromatography leads to the desired product.

General Procedure for the deprotection of methoxy positions (**GP8**):

Under argon atmosphere, 1.00 equiv. of the respective methoxy ether is dissolved in dichloromethane. The mixture is cooled to –78 °C and 5.00 equiv. of boron tribromide (1 M in dichloromethane) is added dropwise. The mixture is stirred for 30 min at this temperature and is then allowed to warm to room temperature and stirred for 16 h. The reaction is quenched at 0 °C by addition of sodium bicarbonate solution. The aqueous layer is extracted with dichloromethane and the combined organic layers are washed with brine, dried over sodium sulfate and the volatiles are removed under reduced pressure. The crude product is then purified via flash column chromatography.

General Procedure for the synthesis of 3-arylcoumarins (**GP9**):

Under argon atmosphere, 1.00 equiv. of the respective 3-bromo coumarin, 2.00 equiv. of boronic acid, 2.00 equiv. of cesium carbonate and 0.05 equiv. of tetrakis triphenylphosphine palladium (0) are placed in a sealed vial and dioxane (1 mL per 0.1 mmol of bromide) is added. The mixture is degassed with three-freeze-pump-thaw cycles and put under argon atmosphere. After heating the reaction at 90 °C for 16 h the reaction is allowed to cool to room temperature, quenched with 20 mL of water and extracted with ethyl acetate. The combined organic layers are dried over sodium sulfate and the volatiles are removed under reduced pressure. The crude product is then purified via flash column chromatography.

General Procedure for the synthesis 2,2-dimethyl-2*H*-chromenes (**GP10**):

Under argon atmosphere, the respective coumarin is dissolved in 1.5 mL of dry tetrahydrofuran and 9 equiv. N-methyl-pyrrolidone (NMP) are added. The mixture is cooled to –15 °C and 2.2 equiv. of methyl magnesium bromide (3 N solution in diethyl ether) is added dropwise. After stirring for 15 min the mixture is allowed to warm to room temperature and let stirring until full conversion of the starting material is observed. Subsequently, 1 M hydrochloric acid solution is added at –10 °C until pH 1 is achieved to quench the reaction. The aqueous layer is extracted with ethyl acetate and the combined organic layers are washed with sodium carbonate solution, dried over sodium sulfate and the volatiles are removed under reduced pressure. The crude product is then purified via flash column chromatography.

5.2.2 Synthesis and Characterization THC Framework (Chapter 3.1)

5-Methoxy-2,2-dimethyl-7-pentylchroman-4-one (62)

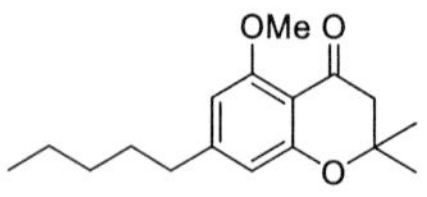

OMe O

O

To a solution of 13.2 g of phenol **63** (50.2 mmol, 1.0 equiv.) in 40 mL of acetone, 20.8 g of potassium carbonate (150.5 mmol, 3.00 equiv.) and 14.7 mL of dimethyl sulfate (150.5 mmol, 3.00 equiv.) were added and the mixture was stirred for 4 h at reflux. After cooling to room temperature, the solids were filtered off, washed with 30 mL of ethyl acetate. The filtrate was washed with 80 mL of 1 M hydrochloric acid and 2 × 80 mL of water, dried over sodium sulfate and the volatiles were removed under reduced pressure. The crude product was then purified via flash column chromatography (CH/EtOAc 3:1) and the product was obtained as 13.4 g of a colorless oil (97%). Analytical data are consistent with the literature.[161]

***R*f** (CH/EtOAc 3:1): 0.24. – **^{1}H NMR** (300 MHz, $CDCl_3$): δ = 6.36 (s, 1H, 8-CH_{Ar}), 6.28 (s, 1H, 6-CH_{Ar}), 3.89 (s, 3H, OCH_3), 2.65 (s, 2H, 3-CH_2), 2.53 (t, *J* = 7.7 Hz 2H, C$H_2C_4H_9$), 1.68 – 1.56 (m, 2H, C$H_2C_3H_7$), 1.42 (s, 6H, 2 × C$H_{3,\,gem.}$), 1.34 – 1.28 (m, 4H, C_2H_4-CH_3), 0.90 (t, *J* = 6.8 Hz, 3H, CH_3) ppm.

3-(Hydroxymethyl)-5-methoxy-2,2-dimethyl-7-pentylchroman-4-on (61)

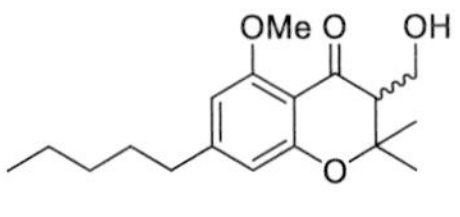

To a solution of 2.22 mL of *n*-BuLi (1.6 M in tetrahydrofuran, 5.54 mmol, 3.0 equiv.) in 5 mL of abs. tetrahydrofuran at –10 °C, 940 µL of tetramethylpiperidine (5.54 mmol, 3.0 equiv.) were added dropwise. The mixture was stirred at 0 °C for 30 min and cooled to –78 °C. Then, a solution of 510 mg of ketone **62** (1.85 mmol, 1.0 equiv.) in 5 mL abs. tetrahydrofuran was added dropwise and stirred for 1 h at –78 °C and afterwards a solution of 650 mg of *N*-hydroxy methyl phthalimide (3.69 mmol, 2.0 equiv.) in 15 mL abs. tetrahydrofuran was added dropwise over 30 min. After complete addition the mixture was stirred for a further 2 h at –78 °C and the reaction was then quenched via the addition of 10 mL water. The mixture was extracted with 3 × 20 mL of diethyl ether and the combined organic layers were washed with 40 mL of a sodium hydroxide-solution (3 M), 40 mL of brine, dried over sodium sulfate and the volatiles were removed. Flash column chromatography (CH/EtOAc 1:1) gave 404 mg (71%) of the product as a pale-yellow oil. Analytical data are consistent with the literature.[161]

***R*f** (CH/EtOAc 1:1): 0.19. – **^{1}H NMR** (300 MHz, $CDCl_3$): δ = 6.35 (s, 1H, 8-CH_{Ar}), 6.29 (s, 1H, 6-CH_{Ar}), 3.98 – 3.94 (m, 1H, 3C-C*H*b), 3.91 (s, 3H, OCH_3), 3.80 (dd, *J* = 11.5 Hz, 3.2 Hz, 1H, 3C-C*H*a), 3.18 (bs, 1H, O*H*), 2.91 (dd, *J* = 7.9 Hz, 3.2 Hz, 1H, 3-C*H*), 2.54 (t, *J* = 7.7 Hz, 2H, C$H_2C_4H_9$), 1.66 – 1.59 (m, 2H, C$H_2C_3H_7$), 1.54 (s, 3H, C$H_{3,\ gem.}$), 1.35 – 1.30 (m, 4H, C_2H_4-CH_3), 1.28 (s, 3H, C$H_{3,\ gem.}$), 0.90 (t, *J* = 6.8 Hz 3H, CH_3) ppm.

3-(Hydroxymethyl)-5-methoxy-2,2-dimethyl-7-pentylchroman-4-ol (60)

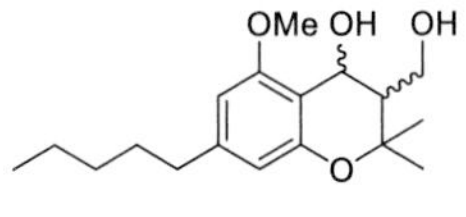

To a solution of 290 mg of ketone **61** (950 µmol, 1.0 equiv.) in 10 mL of abs. tetrahydrofuran at 0 °C and under argon atmosphere, 50.0 mg of $LiAlH_4$ (1.23 mmol, 1.2 equiv.) were added carefully and the mixture was stirred for 2 h at room temperature. The reaction was then quenched at 0 °C by slow addition of 5 mL water and a saturated solution of 5 mL sodium potassium tartrate was added and the reaction mixture stirred overnight at room temperature. The inorganic phase was extracted with 3 × 10 mL diethyl ether, the combined organic layers were dried over sodium sulfate and the volatiles were removed. The resulting 295 mg of white solid (quant.) were directly used in subsequent reactions.

5-Methoxy-2,2-dimethyl-7-pentyl-2*H*-chromene-3-carbaldehyde (59):

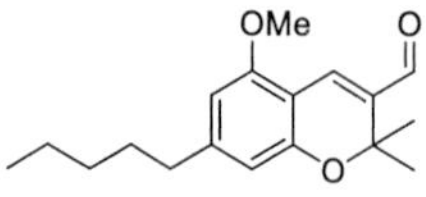

To a solution of 67.0 µL, oxalyl chloride (0.778 mmol, 3.0 equiv.) in 5 mL of dry dichloromethane at –55 °C, a solution of 111 µL dimethyl sulfoxide (1.56 mmol, 6.0 equiv.) in 3 mL of dry dichloromethane was added dropwise. After stirring for 5 min at the same temperature, a solution of 80.0 mg of diol **60** (0.255 mmol, 1.00 equiv.) in 3 mL of dry dichloromethane was added dropwise. After stirring for an additional 30 min at –55 °C, the reaction was quenched via the addition of 0.36 mL of triethylamine (2.60 mmol, 10.0 equiv.). The reaction was then allowed to warm to room temperature and 10 mL of water were added and the mixture was extracted with 3 × 10 mL of dichloromethane. The combined organic layers were then washed with 10 mL of 2 M hydrochloric acid, 10 mL of sodium bicarbonate-solution and 10 mL of brine, dried over sodium sulfate and the volatiles were removed under reduced pressure. Flash column chromatography (CH/EtOAc 20:1) gave the product as a yellow oil, 66.8 mg (89%). Analytical data are consistent with the literature.[161]

R_f (CH/EtOAc 20:1): 0.26. – **^{1}H NMR** (400 MHz, $CDCl_3$): δ = 9.40 (s, 1H, C*H*O), 7.47 (s, 1H, 4-C*H*), 6.29 (s, 1H, H_{Ar}), 6.23 (s, 1H, H_{Ar}) 3.85 (s, 3H, OCH_3), 2.52 (t, J = 8.0 Hz, 2H, $ArCH_2CH_2$), 1.60 (s, 6H, $C(CH_3)_2$), 1.65–1.55 (m, 2H, CH_2), 1.35–1.27 (m, 4H, CH_2), 0.88 (t, J = 7.2 Hz, 3H, CH_2CH_3) ppm.

5-Methoxy-2,2-dimethyl-7-pentyl-3-vinyl-2*H*-chromene (39)

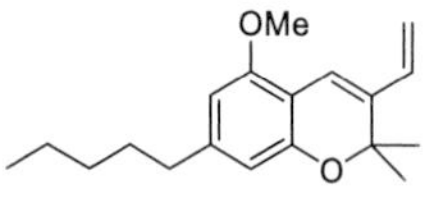

To a solution of 6.69 g of methyl triphenyl phosphonium bromide (19.7 mmol, 1.80 equiv.) in 80 mL of dry tetrahydrofuran, 2.10 g of potassium *tert*-butoxide (18.7 mmol, 1.80 equiv.) were added under argon counterflow and the resulting mixture was stirred for 1.5 h at room temperature. The reaction was then cooled to –78 °C and a solution of 3.00 g of formyl chromene **59** (10.4 mmol, 1.00 equiv.) in 40 mL of dry tetrahydrofuran was added dropwise. The mixture was then stirred for an additional 30 min and quenched by the addition of 20 mL of ammonium chloride-solution. The aqueous layer was extracted with 3 × 20 mL of diethyl ether, the combined organic layers were dried over sodium sulfate and the volatiles were removed carefully at room temperature and under reduced pressure. Flash column chromatography (CH/EtOAc 100:1) gave the product as a colorless oil, 2.65 g (89%). Analytical data are consistent with the literature.[114]

***R*f** (CH/EtOAc, 100:1): 0.40. – **^{1}H NMR** (300 MHz, $CDCl_3$): δ = 6.78 (s, 1H, 6-CH_{Ar}), 6.30 – 6.22 (m, 3H, C*H*=CH_2, 4-C*H*, 8-CH_{Ar}), 5.52 (d, *J* = 14.2, 1.6 Hz, 1H, CH=CH*H*), 5.08 (dd, *J* = 14.2, 2.6 Hz, 1H, CH=C*H*H), 2.51 (t, *J* = 7.7 Hz, 2H, C$H_2C_4H_9$), 3.82 (s, 3H, OCH_3), 1.65–1.55 (m, 2H, C$H_2C_3H_7$), 1.47 (s, 6H, 2 × C$H_{3,\,gem.}$), 1.34–1.27 (m, 4H, $C_2H_4CH_3$), 0.90 (t, *J* = 6.7 Hz, 3H, CH_3) ppm.

(3a*R*,11b*R*,11c*S*)-11-Methoxy-6,6-dimethyl-9-pentyl-4,6,11b,11c-tetrahydro-1*H*-isobenzofuro[5,4-c]chromen-1,3(3a*H*)-dione (38a)

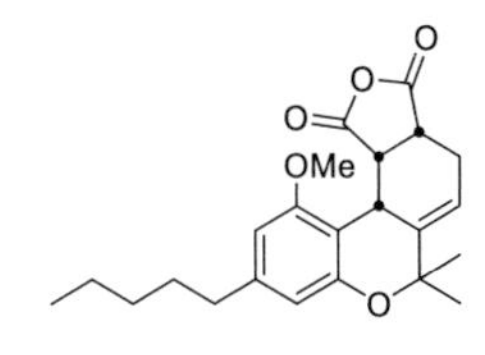

To a solution of 1.49 g of (5.21 mmol, 1.00 equiv.) diene **39** in 30 mL of dry ACN, 510 mg of maleic anhydride (5.21 mmol, 1.00 equiv.) was added and the mixture was stirred overnight at room temperature. The volatiles were removed under reduced pressure which gave the product as a white solid 2.65 g (89%). Analytical data are consistent with the literature.[114]

^{1}H NMR (300 MHz, $CDCl_3$): δ = 6.43 (s, 1H, 10-H_{Ar}) 6.38 (s, 1H, 8-H_{Ar}), 5.94 (dt, *J* = 7.4, 3.2 Hz, 1H, 5-*H*), 4.02 (dd, *J* = 9.7, 6.4 Hz, 1H, 11c-*H*), 3.87 (s, 3H, OCH_3), 3.71 (dt, *J* = 6.0, 2.6 Hz, 1H, 11b-*H*), 3.50 (ddd, *J* = 9.5, 7.3, 1.7 Hz, 1H, 3a-*H*), 2.87 (ddd, *J* = 15.2, 7.4, 1.7 Hz, 1H, 4-H_b), 2.57 (t, *J* = 8.8 Hz, 2H, C$H_2C_4H_9$), 2.35 – 2.25 (dddd, *J* = 15.2, 7.3, 3.3, 2.2 Hz, 1H, 4-H_a), 1.69 – 1.57 (m, 2H, C$H_2C_3H_7$), 1.53 (s, 3H, C$H_{3,\,gem.}$), 1.41 – 1.27 (s, 4H, $C_2H_4CH_3$), 1.22 (s, 3H, C$H_{3,\,gem.}$), 0.90 (t, *J* = 6.9 Hz, 3H, CH_3) ppm.

(3aR)-11-Methoxy-3a,6,6-trimethyl-9-pentyl-4,6,11b,11c-tetrahydro-1H-isobenzofuro[5,4-c]chromene-1,3(3aH)-dione (38b) and (11c*S*)-11-Methoxy-6,6,11c-trimethyl-9-pentyl-4,6,11b,11c-tetrahydro-1H-isobenzofuro[5,4-c]chromene-1,3(3aH)-dione (38c)

To a solution of 27.0 mg of (0.100 mmol, 1.00 equiv.) diene **39** in 2 mL of dry ACN, 11.0 mg of methyl maleic anhydride (0.100 mmol, 1.00 equiv.) was added and the mixture was stirred for 60 h at room temperature. The volatiles were removed under reduced pressure and flash column chromatography (CH/EtOAc 5:1) gave both regio isomers as white solids.

38b 12.5 mg (33%)

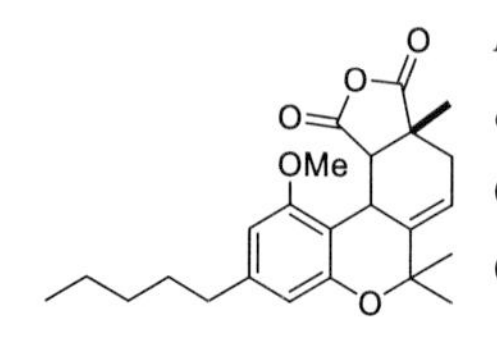

$\boldsymbol{R_f}$ (CH/EtOAc, 5:1): 0.28. – **^{1}H NMR** (400 MHz, $CDCl_3$): δ = 6.46 (d, J = 1.4 Hz, 1H, 8-CH_{Ar}), 6.41 (d, J = 1.4 Hz, 1H, 10-CH_{Ar}), 5.96 (ddd, J = 7.4, 3.8, 2.6 Hz, 1H, 5-C*H*), 3.81 (s, 3H, OCH_3), 3.71 (dd, J = 2.6, 1.4 Hz, 1H, 11b-C*H*), 3.06 (dd, J = 6.4, 1.9 Hz, 1H, 11c-C*H*), 2.85 (ddd, J = 15.0, 7.5, 1.9 Hz, 1H, 4-C*H*H), 2.57 (dd, J = 9.1, 6.6 Hz, 2H, $CH_2C_4H_9$), 2.40 (dddd, J = 14.9, 6.2, 3.9, 1.5 Hz, 1H, 4-CH*H*), 1.68 – 1.60 (m, 2H, $CH_2C_3H_7$), 1.51 (s, 3H, $CH_{3,\,gem.}$), 1.47 (s, 3H, $CH_{3,\,gem.}$), 1.37 – 1.30 (m, 4H, $C_2H_4CH_3$), 1.07 (s, 3H, 3aC-CH_3), 0.93 – 0.87 (m, 3H, $C_4H_8CH_3$) ppm. – **^{13}C NMR** (100 MHz, $CDCl_3$): δ = 174.1 ($C_{quart.}$, *C*OO), 173.4 ($C_{quart.}$, *C*OO), 157.4 ($C_{quart.}$, 11-$C_{Ar}OCH_3$), 156.6 ($C_{quart.}$, 7a-C_{Ar}), 149.0 ($C_{quart.}$, 9-$C_{Ar}C_5H_{11}$), 144.4 ($C_{quart.}$, 5a-*C*), 119.2 (+, 5-*C*H), 112.2 (+, 8-C_{Ar}H), 107.8 ($C_{quart.}$, 11a-C_{Ar}), 104.5 (+, 10-C_{Ar}H), 77.6 ($C_{quart.}$, 6-*C*), 55.1 (+, OCH_3), 52.0 (+, 11c-*C*H), 50.5 (+, 11b-*C*H), 36.6 (–, CH_2), 36.4 (–, CH_2), 31.8 (–, CH_2), 30.8 (–, CH_2), 27.6 (+, $CH_{3,\,gem.}$), 26.9 (+, 3a-CCH_3), 25.2 (–, 4-CH_2), 23.2 (+, $CH_{3,\,gem.}$), 22.7 (–, CH_2), 14.2 (+, $C_4H_8CH_3$) ppm. – **IR** (KBr): $\tilde{\nu}$ = 2929 (w), 2857 (w), 1849 (w), 1777 (s), 1616 (m), 1581 (m), 1451 (m), 1426 (m), 1381 (w), 1346 (w), 1314 (w), 1267 (w), 1203 (m), 1119 (m), 1102 (m), 1085 (m), 1054 (w), 1024 (m), 954 (s), 923 (m), 869 (w), 821 (m), 869 (w), 821 (m), 783 (w), 750 (w), 732 (w), 631 (w), 591 (w), 564 (w) cm^{-1}. – **MS** (70 eV, EI): *m/z* (%) = 398 (14) $[M]^+$, 287 (10), 286 (48), 272 (20), 271 (100), 86 (21), 84 (33). – **HRMS** ($C_{24}H_{30}O_3$): calc. 398.2093, found 398.2093.

38c 3 mg (10%):

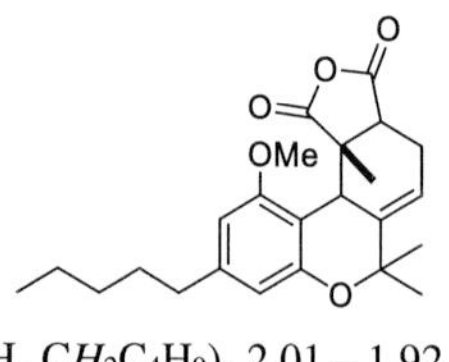

$\boldsymbol{R_f}$ (CH/EtOAc, 5:1: 0.33.**^{1}H NMR** (300 MHz, $CDCl_3$): δ = 6.36 (s, 1H, 8-CH_{Ar}), 6.31 (s, 1H, 10-CH_{Ar}), 5.90 – 5.81 (m, 1H, 5-C*H*), 3.80 (s, 3H, OCH_3), 3.60 (s, 1H, 11b-C*H*), 3.50 (d, J = 5.7 Hz, 1H, 3a-C*H*), 2.75 (dd, J = 15.0, 7.2 Hz, 1H, 4-C*H*H), 2.53 – 2.46 (m, 2H, $CH_2C_4H_9$), 2.01 – 1.92 (m, 1H, 4-C*H*H), 1.53 (s, 2H, $CH_2C_3H_7$), 1.47 (s, 3H, $CH_{3,\,gem.}$), 1.46 (s, 3H, $CH_{3,\,gem.}$), 1.31 – 1.23 (m, 4H, $C_2H_4CH_3$), 1.15 (s, 3H, 11cC-CH_3), 0.87 – 0.79 (m, 3H, $C_4H_8CH_3$) ppm. – **IR** (KBr): $\tilde{\nu}$ = 2927 (w), 2856 (w), 1849 (vw), 1780 (w), 1713 (w), 1615 (w), 1585 (w), 1455 (w), 1423 (w), 1381 (w), 1363 (w), 1260 (w), 1218 (w), 1118 (m), 1086 (w), 1039 (w), 985 (w), 950 (w), 930 (w), 821 (w), 756 (w), 559 (vw), 408 (vw) cm^{-1}. – **MS** (MAT 95, +FAB): *m/z* (%) = 398/ (26) $[M]^+$, 343 (25), 324 (15), 318 (12), 310 (17), 309 (65), 307 (20), 288 (19), 287 (15), 286 (28), 285 (15), 274 (23), 273 (100), 272 (17), 271 (61), 282 (10), 259 (12). – **HRMS** ($C_{24}H_{30}O_5$): calc. 398.2093, found 398.2091.

(3a*S*)-3a-Bromo-11-methoxy-6,6-dimethyl-9-pentyl-4,6,11b,11c-tetrahydro-1H-isobenzofuro[5,4-c]chromene-1,3(3aH)-dione (38d)

To a solution of 22.0 mg (80.0 µmol, 1.00 equiv.) diene **39** in 2 mL of dry ACN, 11 mg of bromo maleic anhydride (80.0 µmol, 1.00 equiv.) was added and the mixture was stirred for 36 h at room temperature. The volatiles were removed under reduced pressure and flash column chromatography (CH/EtOAc 5:1) gave the product as 7.0 mg (17%)of a white solid.

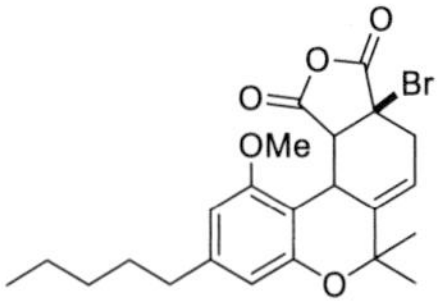

***R*f** (CH/EtOAc, 5:1) = 0.41. – **^{1}H NMR** (400 MHz, $CDCl_3$): δ = 6.46 (d, *J* = 1.4 Hz, 1H, 8-CH_{Ar}), 6.43 (d, *J* = 1.4 Hz, 1H, 10-CH_{Ar}), 6.05 (ddd, *J* = 7.1, 3.8, 2.6 Hz, 1H, 5-C*H*), 4.32 (dd, *J* = 2.6, 1.6 Hz, 1H, 11c-C*H*), 3.86 – 3.83 (m, 4H, 11b-C*H*, OCH_3), 2.86 (ddd, *J* = 15.5, 7.3, 1.8 Hz, 1H, 4-C*H*H), 2.78 – 2.68 (m, 1H, 4-CH*H*), 2.58 (dd, *J* = 9.1, 6.7 Hz, 2H, C$H_2C_4H_9$), 1.68 – 1.61 (m, 2H, C$H_2C_3H_7$), 1.50 (s, 3H, C$H_{3,\ gem.}$), 1.37 – 1.32 (m, 4H, $C_2H_4CH_3$), 1.09 (s, 3H, C$H_{3,\ gem.}$), 0.93 – 0.87 (m, 3H, $C_4H_8CH_3$) ppm. – **^{13}C NMR** (100 MHz, $CDCl_3$): δ = 171.6 ($C_{quart.}$, *C*OO), 166.8 ($C_{quart.}$, *C*OO), 157.8 ($C_{quart.}$, 11-$C_{Ar}OCH_3$), 156.5 ($C_{quart.}$, 7a-C_{Ar}), 148.1 ($C_{quart.}$, 9-$C_{Ar}C_5H_{11}$), 145.4 ($C_{quart.}$, 5a-*C*), 120.6 (+, 5-*C*H), 111.8 (+, 8-C_{Ar}H), 106.0 ($C_{quart.}$, 11a-C_{Ar}), 105.0 (+, 10-C_{Ar}H), 77.5 ($C_{quart.}$, 6-*C*), 56.6 (+, 3a-*C*Br), 55.0 (+, CH_3), 54.8 (+, 11c-*C*H), 37.4 (+, 11b-*C*H), 36.4 (–, CH_2), 31.8 (–, CH_2), 30.7 (–, CH_2), 27.6 (+, $CH_{3,\ gem.}$), 26.8 (+, $CH_{3,\ gem.}$), 25.7 (–, 4-CH_2), 22.7 (–, CH_2), 14.2 (+, $C_4H_8CH_3$) ppm. – **IR** (KBr): ṽ = 2928 (w), 2856 (w), 1857 (w), 1787 (s), 1616 (m), 1581 (m), 1450 (w), 1426 (m), 1382 (w), 1364 (w), 1311 (w), 1256 (w), 1252 (w), 1226 (m), 1200 (m), 1178 (m), 1108 (m), 1009 (w), 942 (s), 872 (m), 822 (m), 768 (w), 741 (w), 632 (m), 607 (w), 591 (w), 559 (w), 511 (vw), 470 (vw) cm^{-1}. – **MS** (MAT 95, +FAB): *m/z* (%) = 463/465 (41/37) $[M]^+$, 383 (15) $[M - Br]^+$, 286 (55) [retro DA], 271 (100). – **HRMS** ($C_{24}H_{28}O_5{}^{79}Br = M^++1$): calc. 463.1120, found 463.1122.

Ethyl 10-formyl-1-methoxy-6, 6-dimethyl-3-pentyl-8, 9, 10, 10a-tetrahydro-6*H*-benzo[c]chromene-9-carboxylate (38f) and ethyl 9-formyl-1-methoxy-6,6-dimethyl-3-pentyl-8,9,10,10a-tetrahydro-6*H*-benzo[c]chromene-10-carboxylate (38g)

To a solution of 99.0 mg of (0.350 mmol, 1.00 equiv.) diene **39** in 4 mL of dry dichloromethane, 44.0 mg of ethyl (E)-4-oxobut-2-enoate (0.350 mmol, 1.00 equiv.) was added and the mixture was stirred for 36 h at room temperature. The volatiles were removed under reduced pressure and flash column chromatography (CH/EtOAc 20:1) gave the two regioisomers as white solids.

38f 22.1 mg (15%)

***R*f** (CH/EtOAc, 10:1): 0.26. – **^{1}H NMR** (400 MHz, $CDCl_3$): δ = 9.14 (s, 1H, C*H*O), 6.34 (d, *J* = 1.4 Hz, 1H, 4-C*H*$_{Ar}$), 6.31 (d, *J* = 1.4 Hz, 1H, 2-C*H*$_{Ar}$), 5.73 (q, *J* = 3.5 Hz, 1H, 7-C*H*), 4.34 – 4.15 (m, 2H, C*H*$_2$CH$_3$), 3.93 (dd, *J* = 4.4, 2.7 Hz, 1H, 10-C*H*), 3.83 (s, 3H, OC*H*$_3$), 3.73 (p, *J* = 3.3 Hz, 1H, 10a-C*H*), 3.37 (ddd, *J* = 8.2, 2.8, 1.3 Hz, 1H, 9-C*H*), 2.64 – 2.56 (m, 1H, 8-C*H*H), 2.55 – 2.49 (m, 2H, C*H*$_2$C$_4$H$_9$), 2.26 (ddt, *J* = 19.1, 7.8, 3.7 Hz, 1H, 8-C*H*H), 1.60 (h, *J* = 7.4, 6.8 Hz, 2H, C*H*$_2$C$_3$H$_7$), 1.52 (s, 3H, C*H*$_{3,\ gem.}$), 1.37 – 1.28 (m, 7H, C$_2$*H*$_4$CH$_3$, CH$_2$C*H*$_3$), 1.25 (s, 3H, C*H*$_{3,\ gem.}$), 0.89 (t, *J* = 6.8 Hz, 3H, C$_4$H$_8$C*H*$_3$) ppm. – **^{13}C NMR** (100 MHz, $CDCl_3$): δ = 203.2 (C$_{quart.}$, *C*HO), 174.9 (C$_{quart.}$, *C*OOEt), 157.4 (C$_{quart.}$, 1-*C*$_{Ar}$OCH$_3$), 155.1 (C$_{quart.}$, 4a-*C*$_{Ar}$), 143.8 (C$_{quart.}$, 3-*C*$_{Ar}$C$_5$H$_{11}$), 138.2 (C$_{quart.}$, 6a-*C*), 120.4 (+, 7-*C*H), 110.8 (+, 4-*C*$_{Ar}$H), 108.1 (C$_{quart.}$, 10c-*C*$_{Ar}$), 103.7 (+, 2-*C*$_{Ar}$H), 76.3 (C$_{quart.}$, 6-*C*), 61.0 (–, COO*C*H$_2$CH$_3$), 55.4 (+, *C*H$_3$), 49.0 (+, 10-*C*H), 38.0 (+, 9-*C*H), 36.2 (–, *C*H$_2$C$_4$H$_9$), 31.7 (–, *C*H$_2$), 30.9 (–, *C*H$_2$), 28.7 (+, 10a-*C*H), 26.0 (+, *C*H$_{3,\ gem.}$), 25.0 (+, *C*H$_{3,\ gem.}$), 22.9 (–, 8-*C*H$_2$), 22.7 (–, *C*H$_2$), 14.5 (+, ethyl-*C*H$_3$), 14.2 (+, C$_4$H$_8$*C*H$_3$) ppm. – **IR** (KBr): $\tilde{\nu}$ = 2929 (w), 2856 (w), 1718 (m), 1615 (w), 1579 (m), 1450 (w), 1423 (m), 1382 (w), 1368 (w), 1342 (w), 1288 (w), 1230 (m), 1189 (m), 1137 (m), 1115 (s), 1085 (m), 1049 (m), 1027 (w), 946 (w), 909 (w), 560 (w), 821 (w), 760 (w), 560 (vw), 529 (w) cm^{-1}. – **MS** (MAT 95, +FAB): *m/z* (%) = 415 (58) [M]$^+$, 399 (100) [M – CH$_3$]$^+$, 286 (55) [retro DA], 271 (100). – **HRMS** (C$_{25}$H$_{35}$O$_5$ = [M + H]$^+$): calc. 415.2484, found 415.2483.

38g 10.3 mg (7%)

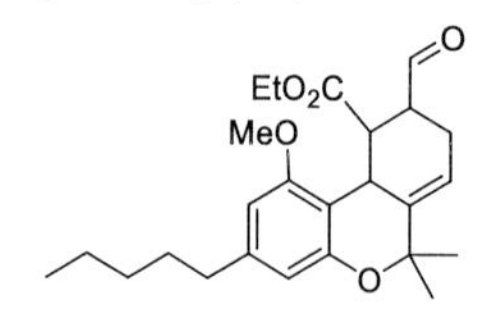

***R*f** (CH/EtOAc, 10:1): 0.12. – **^{1}H NMR** (400 MHz, $CDCl_3$): δ = 9.67 (d, *J* = 1.9 Hz, 1H, C*H*O), 6.33 (d, *J* = 1.5 Hz, 1H, 4-C*H*$_{Ar}$), 6.26 (d, *J* = 1.5 Hz, 1H, 2-C*H*$_{Ar}$), 5.84 (td, *J* = 4.3, 2.8 Hz, 1H, 7-C*H*), 4.04 (q, *J* = 7.1 Hz, 2H, C*H*$_2$CH$_3$), 3.86 – 3.81 (m, 1H, 10a-C*H*),), 3.73 (s, 3H, OC*H*$_3$), 3.08 (td, *J* = 8.2, 7.8, 1.9 Hz, 1H, 10-C*H*), 3.04 – 2.96 (m, 1H, 9-C*H*), 2.53 – 2.47 (m, 2H, C*H*$_2$C$_4$H$_9$), 2.43 – 2.33 (m, 2H, 8-C*H*$_2$), 1.67 – 1.55 (m, 2H, C*H*$_2$C$_3$H$_7$), 1.54 (s, 3H, C*H*$_{3,\ gem.}$), 1.35 – 1.28 (m, 4H, C$_2$*H*$_4$CH$_3$), 1.27 (s, 3H, C*H*$_{3,\ gem.}$), 1.13 (t, *J* = 7.1 Hz, 3H, CH$_2$C*H*$_3$), 0.90 – 0.85 (m, 3H, C$_4$H$_8$C*H*$_3$) ppm – **^{13}C NMR** (100 MHz, $CDCl_3$): δ = 201.7 (C$_{quart.}$, *C*HO), 174.2 (C$_{quart.}$, *C*OOEt), 156.9 (C$_{quart.}$, 1-*C*$_{Ar}$OCH$_3$), 154.7 (C$_{quart.}$, 4a-*C*$_{Ar}$), 143.5 (C$_{quart.}$, 3-*C*$_{Ar}$C$_5$H$_{11}$), 139.2 (C$_{quart.}$, 6a-*C*), 118.9 (+, 7-*C*H), 111.0 (+, 4-*C*$_{Ar}$H), 110.5 (C$_{quart.}$, 10c-*C*$_{Ar}$), 103.5 (+, 2-*C*$_{Ar}$H), 76.6 (C$_{quart.}$, 6-*C*), 61.0 (–, COO*C*H$_2$CH$_3$), 54.8 (+, O*C*H$_3$), 53.0 (+, 10-*C*H), 40.6 (+, 9-*C*H), 36.1 (–, *C*H$_2$), 31.7 (–, *C*H$_2$), 31.3 (+, 10a-

*C*H), 30.9 (–, CH_2), 26.7 (+, $CH_{3, gem.}$), 26.6 (–, 8-CH_2), 25.0 (+, $CH_{3, gem.}$), 22.7 (–, CH_2), 14.2 (+, ethyl-CH_3), 14.1 (+, $C_4H_8CH_3$) ppm. – **IR** (KBr): $\tilde{\nu}$ = 2929 (w), 2855 (w), 1722 (m), 1615 (w), 1577 (m), 1449 (w), 1422 (m), 1367 (w), 1349 (w), 1204 (m), 1179 (m), 1113 (m), 1039 (m), 914 (w), 850 (w), 820 (w), 732 (w), 681 (vw), 564 (vw) cm^{-1}. – **MS** (MAT 95, +FAB): *m/z* (%) = 363/365 (41/37) $[M]^+$, 383 (15) $[M - Br]^+$, 286 (55) [retro DA], 271 (100). – **HRMS** ($C_{25}H_{35}O_5$ = $[M + H]^+$): calc. 415.2484, found 415.2486.

(3a*R*,11b*R*,11c*S*)-11-Methoxy-6,6-dimethyl-9-pentyl-4,6,11b,11c-tetrahydro-1*H*-isobenzofuro[5,4-c]chromen-1,3(3a*H*)-dione (58)

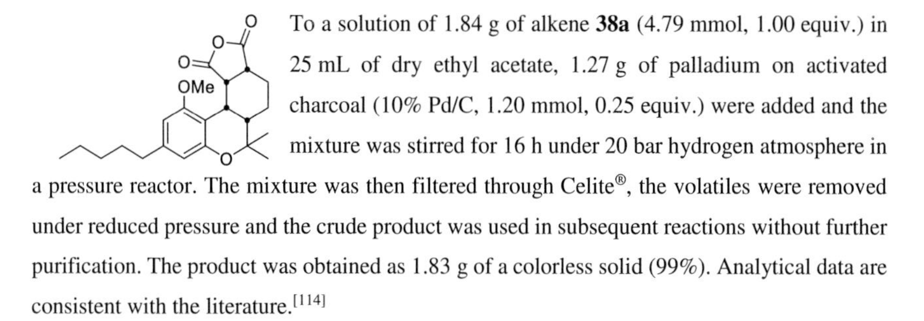

To a solution of 1.84 g of alkene **38a** (4.79 mmol, 1.00 equiv.) in 25 mL of dry ethyl acetate, 1.27 g of palladium on activated charcoal (10% Pd/C, 1.20 mmol, 0.25 equiv.) were added and the mixture was stirred for 16 h under 20 bar hydrogen atmosphere in a pressure reactor. The mixture was then filtered through Celite®, the volatiles were removed under reduced pressure and the crude product was used in subsequent reactions without further purification. The product was obtained as 1.83 g of a colorless solid (99%). Analytical data are consistent with the literature.[114]

^{1}H NMR (300 MHz, $CDCl_3$): δ = 6.43 (s, 1H, 10-CH_{Ar}) 6.38 (s, 1H, 8-H_{Ar}), 5.94 (dt, *J* = 7.4, 3.2 Hz, 1H, 5-C*H*), 4.02 (dd, *J* = 9.7, 6.4 Hz, 1H, 11c-C*H*), 3.87 (s, 3H, OCH_3), 3.71 (dt, *J* = 6.0, 2.6 Hz, 1H, 11b-C*H*), 3.50 (ddd, *J* = 9.5, 7.3, 1.7 Hz, 1H, 3a-C*H*), 2.87 (ddd, *J* = 15.2, 7.4, 1.7 Hz, 1H, 4-CH_b), 2.57 (t, *J* = 8.8 Hz, 2H, $CH_2C_4H_9$,), 2.35 – 2.25 (dddd, *J* = 15.2, 7.3, 3.3, 2.2 Hz, 1H, 4-CH_a), 1.69 – 1.57 (m, 2H, $CH_2C_3H_7$), 1.53 (s, 3H, $CH_{3, gem.}$), 1.41 – 1.27 (s, 4H, $C_2H_4CH_3$), 1.22 (s, 3H, $CH_{3, gem.}$), 0.90 (t, J = 6.9 Hz, 3H, CH_3) ppm.

(6a*S*,9*R*,10*R*,10a*S*)-1-Methoxy-6,6-dimethyl-3-pentyl-6a,7,8,9,10,10a-hexahydro-6*H*-benzo[c]chromene-9,10-dicarboxylic acid (64)

To 1.82 g of the solid anhydride **58** (4.72 mmol, 1.00 equiv.) 100 mL of saturated sodium bicarbonate-solution was added and the mixture was stirred at 90 °C for 3 h (complete dissolution). The reaction was then cooled in an ice bath and concentrated hydrochloric acid added carefully until a pH of 1 was obtained. The aqueous solution was then extracted with 3 × 50 mL of ethyl acetate, the combined organic layers were dried over sodium sulfate and the volatiles were removed under reduced pressure to result in 1.87 g (98%) of the product as a white solid that was used in subsequent reactions without further purification. Analytical data are consistent with the literature.[114]

^{1}H NMR (300 MHz, [D$_6$]-DMSO): δ = 11.81 (bs, 2H, 2 × COO*H*), 6.16 (d, *J* = 14.6 Hz, 2H, C2-*H*, 4-C*H*), 3.52 (s, 3H, OC*H*$_3$), 3.40 – 3.32 (m, 1H, 10a-C*H*), 2.56 – 2.69 (m, 1H, 9-C*H*), 2.45 – 2.40 (m, 3H, 8-C*H*$_b$, C*H*$_2$C$_4$H$_9$), 2.28 – 2.04 (m, 1H, 10-C*H*), 1.67 – 1.46 (m, 4H, 6a-CH, 7-C*H*$_b$, C*H*$_2$C$_3$H$_7$), 1.39 – 1.08 (m, 12H, 2 × C*H*$_{3,\ gem.}$, C$_2$*H*$_4$CH$_3$, 7-C*H*$_a$, 8-C*H*$_b$), 0.89 (t, *J* = 6.6 Hz, 3H, C*H*$_3$) ppm.

(6a*S*,10a*R*)-1-Methoxy-6,6-dimethyl-3-pentyl-6a,7,8,10a-tetrahydro-6*H*-benzo[c]chromene (57)

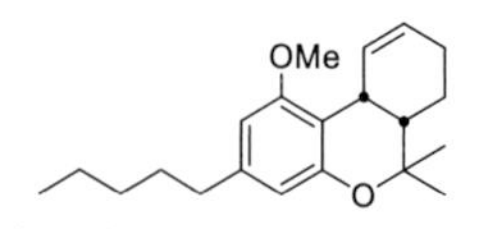

Under argon atmosphere, 1.82 g of dicarboxylic acid **64** (4.50 mmol, 1.00 equiv.) were dissolved in 45 mL of pyridine and 4.20 g of lead tetraacetate (9.00 mmol, 2.00 equiv.) were added. The mixture was stirred for 2 h at 55 °C and then allowed to cool to room temperature. The mixture was poured in 70 mL of 5%-HNO$_3$-solution and was extracted with 3 × 20 mL of diethyl ether. The combined organic layers were washed with 10 mL of 5%-sodium bicarbonate-solution, dried over sodium sulfate and the volatiles were removed under reduced pressure. Flash column chromatography (CH/EtOAc, 100:1) gave the product as a yellow oil, 417 mg (31%). Analytical data are consistent with the literature.[114]

R$_f$ (CH/EtOAc, 80:1): 0.35. – **^{1}H NMR** (300 MHz, CDCl$_3$): δ = 6.50 (ddt, J = 10.1, 4.6, 2.1 Hz, 1H, 9-C*H*), 6.29 (s, 1H, 4-C*H*$_{Ar}$), 6.26 (s, 1H, 2-C*H*$_{Ar}$), 5.75 – 5.66 (m, 1H, 10-C*H*), 3.82 (s, 3H, OC*H*$_3$), 3.61 – 3.50 (m, 1H, 10a-C*H*), 2.51 – 2.47 (m, 2H, C*H*$_2$C$_4$H$_9$), 2.05 (dt, *J* = 5.6, 2.5 Hz, 2H, 8-C*H*$_2$), 1.96 – 1.90 (m, 1H, 7-C*H*$_b$), 1.77 (ddd, *J* = 10.9, 6.1, 2.9 Hz, 1H, 6a-C*H*), 1.65 – 1.53 (m, 2H, C*H*$_2$C$_3$H$_7$), 1.49 – 1.43 (m, 1H, 7-C*H*$_a$), 1.41 (s, 3H, C*H*$_{3,gem.}$), 1.37 – 1.30 (m, 4H, C$_2$*H*$_4$CH$_3$), 1.27 (s, 3H, C*H*$_{3,gem.}$), 0.89 (t, *J* = 6.7 Hz, 3H, C*H*$_3$) ppm.

(1a*R*,3a*S*,9b*R*,9c*S*)-9-methoxy-4,4-dimethyl-7-pentyl-1a,2,3a,4,9b,9c-hexahydro-3H-oxireno[2',3':3,4]benzo[1,2-c]chromene (56)

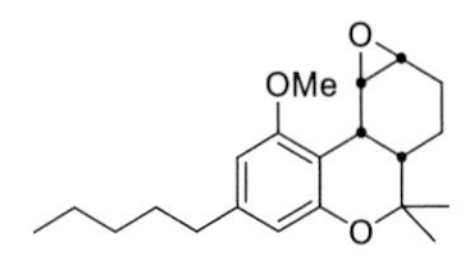

To a solution of 160 mg of alkene **57** (509 µmol, 1.00 equiv.) in 10 mL of acetone at 0 °C, 9.56 mL of dimethyl dioxirane (79.8 µmol/mL in acetone, 763 µmol, 1.50 equiv.) were added dropwise. After stirring for 2 h at 0 °C, the volatiles were removed by condensing them into a cool trap under high vacuum. Subsequent flash column chromatography (CH/EtOAc, 10:1) gave the product as a yellow oil, 47.4 mg (28%). Analytical data are consistent with the literature.[1]

R_f (CH/EtOAc, 10:1): 0.32. – **^{1}H NMR** (400 MHz, $CDCl_3$): δ = 6.30 (s, 2H, 6-H_{Ar}, 8-CH_{Ar}), 3.86 – 3.84 (m, 1H, 9c-C*H*), 3.84 (s, 3H, OCH_3), 3.42 (d, *J* = 6.1 Hz, 1H, 9b-C*H*), 3.16 (t, *J* = 3.8 Hz, 1H, 1a-C*H*), 2.55 – 2.46 (m, 2H, $CH_2C_4H_9$), 1.98 – 1.89 (m, 2H, 2-CH_2), 1.85 (ddd, *J* = 9.0, 4.3, 2.3 Hz, 1H, 3a-C*H*), 1.73 – 1.64 (m, 1H, 3-H_b), 1.63 – 1.57 (m, 2H, $CH_2C_3H_7$), 1.45 – 1.40 (m, 1H, 3-H_a), 1.25 – 1.37 (m, 10H, $C_2H_4CH_3$, 2 × $CH_{3,\ gem.}$), 0.89 (t, *J* = 6.4 Hz, 3H, $C_4H_8CH_3$) ppm. – **^{13}C-NMR** (100 MHz, $CDCl_3$): δ = 158.8 ($C_{quart.}$ 9-$C_{Ar}OCH_3$) 154.6 ($C_{quart.}$, 5a-C_{Ar}O), 143.5 ($C_{quart.}$, 7-$C_{Ar}C_5H_{11}$), 110.2 (+, 6-CH_{Ar}), 107.9 ($C_{quart.}$, 9a-C), 103.0 (+, 8-CH_{Ar}), 76.4 ($C_{quart.}$, $C(CH_3)_2$), 55.4 (+, OCH_3), 55.2 (+, 9c-CH), 53.5 (+, 1a-CH), 37.1 (+, 3a-CH), 36.2 (–, $CH_2C_4H_9$,), 31.8 (–, CH_2, $CH_2C_2H_5$), 31.7 (+, 9b-CH), 30.9 (–, CH_2, $CH_2C_3H_7$), 26.0 (+, 1 × 4-$CH_{3,\ gem.}$), 25.2 (+, CH_3, 1 × 4-$CH_{3,\ gem.}$), 22.7 (–, CH_2CH_3), 21.9 (–, 2-CH_2), 18.8 (–, 3-CH_2), 14.2 (+, $C_4H_8CH_3$) ppm.– **IR** (KBr): ṽ = 2928 (m), 2848 (m), 1614 (m), 1580 (s), 1451 (m), 1421 (s), 1348 (m), 1296 (w), 1208 (m), 1173 (m), 1140 (m), 1105 (vs), 1070 (m), 1014 (w), 973 (m), 914 (m), 885 (w), 869 (w), 852 (w), 824 (m), 812 (m), 779 (w), 732 (m), 654 (w), 568 (m), 533 (w), 460 (w), 408 (w) cm^{-1}. – **MS** (EI, 70 eV): *m/z* (%) = 330 (88) $[M]^+$, 288 (19), 287 (31), 275 (25), 259 (11), 245 (11), 231 (19), 207 (14), 161 (12), 119 (11), 86 (67), 84 (100). – **HRMS** ($C_{21}H_{30}O_3$): calc. 330.2189 found. 330.2187.

(6a*S*,9*R*)-1-Methoxy-6,6-dimethyl-3-pentyl-6a,7,8,9-tetrahydro-6*H*-benzo[c]chromen-9-ol (71)

Under argon atmosphere, a solution of 39.0 mg of epoxide **56** (118 µmol, 1.00 equiv.) in 5 mL of abs. tetrahydrofuran was cooled to –70 ° and 0.22 mL of methyl lithium (1.6 M in diethyl ether, 354 µmol, 3.00 equiv.) were added and the mixture was stirred for 12 h at room temperature. The reaction was quenched by addition of 10 mL of brine and the aqueous solution was extracted with 3 × 10 mL of diethyl ether, dried over sodium sulfate and the volatiles were removed under reduced pressure. Subsequent flash column chromatography (CH/EtOAc, 10:1) gave the product as a white solid 25 mg (64%). Analytical data are consistent with the literature.[1]

***R**$_f$* (CH/EtOAc, 3:2): 0.53. – **^{1}H NMR** (400 MHz, $CDCl_3$): δ = 6.98 – 6.96 (m, 1H, 10-C*H*) 6.30 (s, 1H, 4-CH_{Ar}), 6.27 (s, 1H, 2-CH_{Ar}), 4.40 – 4.35 (m, 1H, 9-C*H*), 3.85 (s, 3H, OCH_3), 2.51 – 2.44 (m, 3H, C$H_2$$C_4H_9$, 6a-C*H*), 2.23 – 2.17 (m, 1H, 8-CH_b), 1.99 – 1.88 (m, 1H, 7-H_b), 1.66 (bs, 1H, O*H*), 1.61 – 1.55 (m, 2H, C$H_2$$C_3H_5$), 1.50 – 1.40 (m, 1H, 8-H_a) 1.38 (s, 3H, 1 × 6-C$H_{3,\text{ gem.}}$), 1.35 – 1.16 (m, 5H, $C_2H_4CH_3$, 7-H_a), 1.08 (s, 3H, 1 × 6-C$H_{3,\text{ gem.}}$), 0.88 (dt, *J* = 7.5, 4.0 Hz, 3H, C_4H_8CH_3) ppm. – **^{13}C-NMR** (100 MHz, CDC13): δ = 158.7 ($C_{quart.}$, 1-$C_{Ar}$$OCH_3$) 154.5 ($C_{quart.}$, 4a-$C_{Ar}$O), 144.2 ($C_{quart.}$, 3-$C_{Ar}$$C_5H_{11}$), 129.7 (+, 10-*C*H), 129.6 (10a-*C*), 110.3 (+, 4-C_{Ar}H), 107.6 ($C_{quart.}$, 10b-C_{Ar}), 103.4 (+, 2-C_{Ar}H), 77.5 ($C_{quart.}$, 6-*C*$(CH_3)_2$), 68.6 ($C_{quart.}$, 9-*C*OH), 55.4 (+, OCH_3), 43.8 (+, 6a-*C*H), 36.2 (–, $CH_2C_4H_9$), 32.2 (–, 8-CH_2), 31.7 (–, CH_2CH_3), 30.8 (–, $CH_2C_3H_7$), 27.3 (+, 1 × 6-$CH_{3,\text{ gem.}}$), 23.0 (–, 7-CH_2), 22.7 (–, $CH_2C_2H_5$), 19.3 (+, 1 × 6-$CH_{3,\text{ gem.}}$), 14.2 (+, $C_4H_8CH_3$) ppm. – **IR** (KBr): $\tilde{\nu}$ = 3259 (w), 2952 (m), 2855 (m), 1610 (m), 1563 (m), 1450 (m), 1420 (m), 1352 (m), 1226 (m), 1190 (m), 1173 (m), 1129 (m), 1104 (s), 1084 (s), 1044 (s), 967 (m), 918 (m), 889 (m), 858 (m), 819 (m), 730 (m), 561 (m), 486 (w) cm^{-1}. – **MS** (EI, 70 eV): *m/z* (%) = 330 (18) $[M]^+$, 313 (11) $[M - OH]^+$, 312 (35), 298 (100), 297 (23), 259 (11). – **HRMS** ($C_{21}H_{30}O_3$): calc. 330.2189 found 330.2187.

(*S*)-1-Methoxy-6,6-dimethyl-3-pentyl-6,6a,7,8-tetrahydro-9H-benzo[c]chromen-9-one (55)

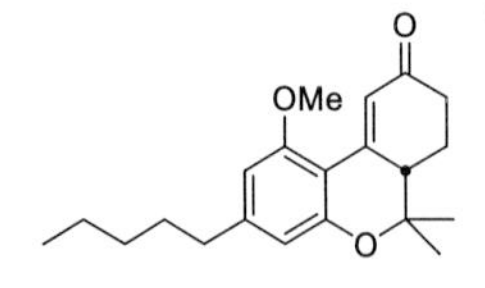

To a solution of 30 mg alcohol **71** under argon atmosphere in 3 mL of dichloromethane, 0.42 mL of DESS-MARTIN-periodinane solution (0.3 M in dichloromethane, 127 µmol, 1.40 equiv.) were added dropwise at room temperature and the mixture was stirred for 4 h. The reaction was quenched with 0.5 mL of sodium thiosulfate-solution and stirred for an additional 10 min at room temperature. The mixture was then diluted with further 5 mL of sodium thiosulfate-solution and was extracted with 2 × 5 mL of dichloromethane. The combined organic layers were washed with sodium bicarbonate-solution and brine, dried over sodium sulfate and the volatiles were removed under reduced pressure. Subsequent flash column chromatography (CH/EtOAc, 5:1) gave the product as a white solid 12.2 mg (49%).

R_f (CH/EtOAc, 5:1): 0.17. – **^{1}H NMR** (400 MHz, CD_2Cl_2): δ = 7.28 (d, *J* = 1.8 Hz, 1H, 10-C*H*), 6.34 (s, 2H, 4-H_{Ar}, 2-H_{Ar}), 3.88 (s, 3H, OCH_3), 2.77 (ddd, *J* = 11.7, 4.5, 2.3 Hz, 1H, 6a-*H*), 2.56 – 2.50 (m, 2H, C$H_2$$C_4H_9$), 2.50 – 2.44 (m, 1H, 8-H_a), 2.37 (ddd, *J* = 16.1, 14.6, 5.0 Hz, 1H, 8-H_b), 2.21 – 2.12 (m, 1H, 7-H_a), 1.72 – 1.55 (m, 3H, 7-H_b, C$H_2$$C_3H_5$), 1.47 (s, 3H, 1 × 6-C$H_{3,\,gem.}$), 1.39 – 1.28 (m, 4H, $C_2H_4CH_3$), 1.12 (s, 3H, 1 × 6-C$H_{3,\,gem.}$), 0.90 (t, *J* = 6.9 Hz, 3H, C_4H_8CH_3) ppm. – **^{13}C-NMR** (100 MHz, CD_2Cl_2): δ = 200.3 ($C_{quart.}$, 9-*C*=O), 160.9 ($C_{quart.}$, 1-$C_{Ar}$$OCH_3$), 156.7 ($C_{quart.}$, 4a-$C_{Ar}$O), 149.5 (10a-*C*), 148.8 ($C_{quart.}$, 3-$C_{Ar}$$C_5H_{11}$), 125.2 (+, 10-*C*H), 111.0 (+, 4-C_{Ar}H), 107.5 ($C_{quart.}$, 10b-C_{Ar}), 104.3 (+, 2-C_{Ar}H), 78.1 ($C_{quart.}$, 6-*C*(CH_3)$_2$), 56.0 (+, O*C*H_3), 45.2 (+, 6a-*C*H), 37.4 (–, 8-CH_2), 36.7 (–, *C*$H_2C_4H_9$), 32.1 (–, *C*$H_2C_3H_7$), 31.0 (–, *C*$H_2C_2H_5$), 27.8 (+, 1 × 6-*C*$H_{3,\,gem.}$), 24.9 (–, 7-CH_2), 23.1 (–, *C*H_2CH_3), 19.4 (+, 1 × 6-*C*$H_{3,\,gem.}$), 14.4 (+, C_4H_8*C*H_3) ppm. – **IR** (KBr): ṽ = 2927 (m), 2855 (w), 1656 (m), 1612 (m), 1578 (m), 1557 (m), 1451 (m), 1422 (m), 1386 (w), 1356 (m), 1261 (w), 1230 (m), 1209 (m), 1177 (m), 1168 (m), 1110 (m), 1089 (m), 1016 (w), 961 (w), 889 (m), 853 (w), 822 (w), 798 (w), 755 (w), 732 (w), 665 (vw), 569 (w), 509 (w) cm^{-1}. – **MS** (EI, 70 eV): *m/z* (%) = 328 (39) $[M]^+$, 313 (11) $[M-CH_3]^+$, 295 (19), 285 (11) $[M-C_3H_7]^+$, 272 (11) $[M-C_4H_9]^+$, 57 (100) $[C_4H_9]^+$. – **HRMS** ($C_{21}H_{28}O_3$): calc. 328.2038 found 328.2040.

5.2.3 Synthesis and Characterization Salicylic Aldehydes (Chapter 3.2.1)

1,3-Dimethoxy-5-pentylbenzene (73)

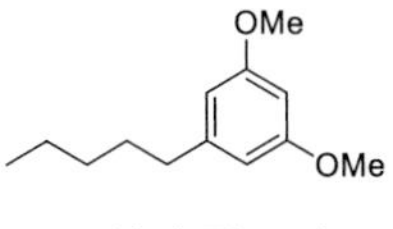

To a suspension of 20 g of Olivetol (**72**) (111 mmol, 1.00 equiv.) and 46.0 g of potassium carbonate (333 mmol, 3.00 equiv.) in 200 mL of acetone, 32.6 mL of dimethyl sulfate (43.3 g, 333 mmol, 3.00 equiv.) were added. The mixture was heated at reflux for 4H, allowed to cool to room temperature, then filtered off and the solid residue washed with 30 mL of diethyl ether. The filtrate was washed with 200 mL of 1 M hydrochloric acid, 2 × 200 mL of water and dried over sodium sulfate. Removal of the volatiles under reduced pressure and purification via flash column chromatography (CH/EtOAc 5:1) resulted in 18.9 g (82%) of the pure product as colorless oil. Analytical data are consistent with the literature.[54,114]

R_f (CH/EtOAc 5:1): 0.20. – **^{1}H NMR** (300 MHz, $CDCl_3$): δ = 6.35 (d, *J* = 2.2 Hz, 2H, 2 × H_{Ar}), 6.30 (t, *J* = 2.3 Hz, 1H, H_{Ar}), 3.78 (s, 6H, 2 × OC*H*3), 2.61 – 2.49 (m, 2H, CH_2), 1.68 – 1.56 (m, 2H, CH_2), 1.36 – 1.25 (m, 3H, 2 × CH_2), 0.94 – 0.84 (m, 3H, CH_3) ppm.

2,6-dimethoxy-4-pentylbenzaldehyde (74)

According to **GP1**, 19.0 g of 1,3-dimethoxy-5-pentylbenzol (**73**, 91.1 mmol, 1.00 equiv.) were dissolved in 150 mL of diethyl ether and 20.6 mL of TMEDA (15.9 g, 136.6 mmol, 1.50 equiv.) were added dropwise. The solution was cooled to 0 °C and 54.6 mL of *n*-butyl lithium (2.5 M in *n*-hexanes, 136 mmol, 1.50 equiv.) was added slowly. After stirring for 2 h at room temperature, the solution was cooled to 0 °C, 21.0 mL of dimethylformamide (20.0 g, 273.2 mmol, 3.00 equiv.) were added and the mixture was stirred for another 4 h at room temperature. The reaction was quenched by the addition of 200 mL of brine and extracted with 3 × 100 mL of diethyl ether. The combined organic layers were dried over sodium sulfate, the volatiles were removed under reduced pressure and the residue was then purified via flash column chromatography (CH/EtOAc 10:1) to result in 20.3 g (95%) of the pure product as a yellow oil. Analytical data are consistent with the literature.[54,114]

***R*f** (CH/ EtOAc 5:1): 0.51. – **^{1}H NMR** (300 MHz, $CDCl_3$): δ = 10.45 (s, 1H, C*H*O), 6.38 (s, 2H, 2 × H_{Ar}), 3.89 (s, 6H, 2 × OCH_3), 2.62 – 2.57 (m, 2H, CH_2), 1.66 – 1.61 (m, 2H, CH_2), 1.36 – 1.31 (m, 4H, 2 × CH_2), 0.93 – 0.88 (m, 3H, CH_3) ppm.

2-Hydroxy-6-methoxy-4-pentylbenzaldehyde (29a)

According to **GP3** 20.33 g of **74** (1.00 eq, 86.03 mmol) was dissolved in a mixture of 100 mL of dry acetonitrile and 200 mL of dry dichloromethane, cooled to 0 °C and 28.7 g of aluminum trichloride (2.50 eq, 215.1 mmol) and 32.2 g of sodium iodide (2.50 eq, 215.1 mmol) were added slowly under argon counterflow. The reaction mixture was stirred for 1.5 h at room temperature, quenched with water, extracted with 3 × 100 mL of dichloromethane and the combined organic layers were dried over sodium sulfate. After removal of volatiles under reduced pressure, the crude product was purified via flash column chromatography (CH/EtOAc 100:1) to result in 15.3 g (80%) of the product as a yellow oil. Analytical data are consistent with the literature.[114,159]

***R*f** (CH/EtOAc 10:1): 0.69. – **^{1}H NMR** (300 MHz, $CDCl_3$): δ = 12.00 (s, 3H, 2-CO*H*), 10.25 (s, 1H, C*H*O), 6.36 (s, 1H, H_{Ar}), 6.19 (s, 1H, H_{Ar}), 3.88 (s, 3H, OCH_3), 2.61 – 2.49 (m, 2H, CH_2), 1.70 – 1.53 (m, 2H, CH_2), 1.38 – 1.27 (m, 4H, 2 × CH_2), 0.92 – 0.87 (m, 3H, CH_3) ppm.

1-(3,5-Dimethoxyphenyl)cyclopentane-1-carbonitrile (76a)

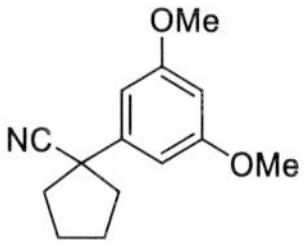

To a solution of 7.50 g of 2-(3,5-dimethoxyphenyl)acetonitrile (**75**) (42.3 mmol, 1.00 equiv.) in 200 mL of abs. tetrahydrofuran under argon counterflow at –16 °C, 25.3 g of potassium bis(trimethylsilyl)amide (KHMDS) (127 mmol, 3.00 equiv.) were added. The mixture was stirred for 3 min at the same temperature and then 5.56 mL of 1,4-dibromobutane (10.1 g, 46.6 mmol, 1.10 equiv.), diluted in 50 mL of abs. tetrahydrofuran, were added dropwise. The mixture was allowed to warm to room temperature and stirred overnight. The reaction was quenched via the addition of ammonium chloride solution (150 mL) and diluted with 100 mL of diethyl ether. The organic layers were extracted with 3 × 200 mL of diethyl ether and the combined organic layers were dried over sodium sulfate. Removal of the volatiles under reduced pressure and purification via flash column chromatography (CH/EtOAc 5:1) resulted in 8.63 g (88%) of the pure product as a colorless oil. Analytical data are consistent with the literature.[111,113]

***R*f** (CH/EtOAc 5:1): 0.31. – **^{1}H NMR** (300 MHz, $CDCl_3$): δ = 6.59 (d, *J* = 2.3 Hz, 2H, 2 × H_{Ar}), 6.39 (t, *J* = 2.2 Hz, 1H, H_{Ar}), 3.81 (s, 6H, 2 × OCH_3), 2.53 – 2.38 (m, 2H, CH_2), 2.17 – 1.86 (m, 6H, CH_2) ppm.

1-(3,5-Dimethoxyphenyl)cyclohexane-1-carbonitrile (76b)

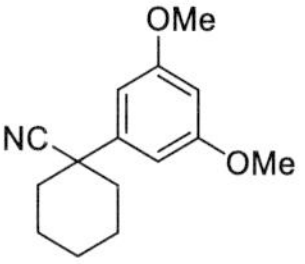

To a solution of 7.50 g of 2-(3,5-dimethoxyphenyl)acetonitrile (**75**) (42.3 mmol, 1.00 equiv.) in 200 mL of abs. tetrahydrofuran under argon counterflow at –16 °C, 25.3 g of KHMDS (127 mmol, 3.00 equiv.) were added. The mixture was stirred for 3 min at the same temperature and then 6.24 mL of 1,4-dibromopentane (10.6 g, 46.6 mmol, 1.10 equiv.), diluted in 50 mL of abs. tetrahydrofuran, were added dropwise. The mixture was allowed to warm to room temperature and stirred overnight. The reaction was quenched via the addition of ammonium chloride solution (150 mL) and diluted with 100 mL of diethyl ether. The organic layers were extracted with 3 × 200 mL of diethyl ether and the combined organic layers were dried over sodium sulfate. Removal of the volatiles under reduced pressure and purification via flash column chromatography (CH/EtOAc 5:1) resulted in 8.09 g (78%) of the pure product as a colorless oil. Analytical data are consistent with the literature.[111-112]

***R*f** (CH/EtOAc 5:1): 0.43. – **^{1}H NMR** (300 MHz, $CDCl_3$): δ = 6.63 (d, *J* = 2.2 Hz, 2H, 2 × H_{Ar}), 6.40 (t, *J* = 2.2 Hz, 1H, H_{Ar}), 3.81 (s, 6H, 2 × OCH_3), 2.21 – 2.16 (m, 2H, CH_2), 1.93 – 1.65 (m, 6H, CH_2), 1.46 – 1.02 (m, 2H, CH_2) ppm.

1-(3,5-Dimethoxyphenyl)cyclopentane-1-carbaldehyde (77a)

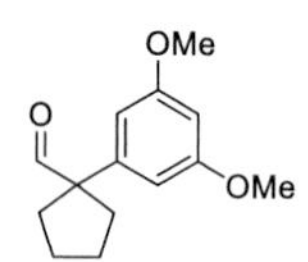

A solution of 9.30 g of 1-(3,5-dimethoxyphenyl)cyclopentane-1-carbonitrile (**76a**) (40.2 mmol, 1.00 equiv.) in 300 mL of abs. dichloromethane under argon atmosphere was cooled to –78 °C and 100 mL of diisobutylaluminium hydride (DIBAL-H) (1 M in dichloromethane, 100 mmol, 2.50 equiv.) was added dropwise. The mixture was stirred for an additional 1 h at the same temperature and then the reaction was quenched by dropwise addition of 150 mL of 10% aqueous sodium potassium tartrate. After thawing up to room temperature the mixture was stirred for another 40 min and the aqueous layer was extracted with 3 × 200 mL of ethyl acetate. The combined organic layers were washed with 300 mL of brine and dried over sodium sulfate. Removal of the volatiles under reduced pressure and purification via flash column chromatography (CH/EtOAc 10:1) resulted in 8.14 g (86%) of the pure product as a colorless oil. Analytical data are consistent with the literature.[111,113]

***R*f** (CH/EtOAc 5:1): 0.40. – **^{1}H NMR** (300 MHz, $CDCl_3$): δ = 9.37 (s, 1H, C*H*O), 6.42 – 6.35 (m, 3H, 3 × H_{Ar}), 3.78 (s, 6H, 2 × OCH_3), 2.53 – 2.39 (m, 2H, CH_2), 1.95 – 1.60 (m, 6H, CH_2) ppm.

1-(3,5-Dimethoxyphenyl)cyclohexane-1-carbaldehyde (77b)

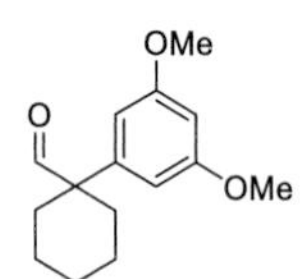

A solution of 7.97 g of 1-(3,5-dimethoxyphenyl)cyclohexane-1-carbonitrile (**76b**) (32.5 mmol, 1.00 equiv.) in 250 mL of abs. dichloromethane under argon atmosphere was cooled to –78 °C and 81.2 mL of DIBAL-H (1 M in dichloromethane, 81.2 mmol, 2.50 equiv.) were added dropwise. The mixture was stirred for an additional 1 h at the same temperature and then the reaction was quenched by dropwise addition of 120 mL of 10% aqueous sodium potassium-tartrate. After thawing up to room temperature the mixture was stirred for another 40 min and the aqueous layer extracted with 3 × 200 mL of ethyl acetate. The combined organic layers were washed with 300 mL of brine and dried over sodium sulfate. Removal of the volatiles under reduced pressure and purification via flash column chromatography (CH/EtOAc 10:1) resulted in 7.39 g (92%) of the pure product as a colorless oil. Analytical data are consistent with the literature.[111-112]

***R*f** (CH/EtOAc 20:1): 0.20. – **^{1}H NMR** (300 MHz, $CDCl_3$): $\delta = 9.34$ (s, 1H, C*HO*), 6.46 (d, $J = 2.2$ Hz, 2H, 2 × H_{Ar}), 6.37 (t, $J = 2.2$ Hz, 1H, H_{Ar}), 3.78 (s, 6H, 2 × OCH_3), 2.27 – 2.22 (m, 2H, CH_2), 1.85 – 1.78 (m, 2H, CH_2), 1.69 – 1.57 (m, 3H, CH_2), 1.52 – 1.25 (m, 3H, CH_2) ppm.

(Z)-1-(1-(But-1-en-1-yl)cyclopentyl)-3,5-dimethoxybenzene (78a)

To a suspension of 38.9 g of *n*-propyltriphenylphosphonium bromide (101.1 mmol, 3.00 equiv.) in 300 mL of abs. tetrahydrofuran at 0 °C, 20.2 g of KHMDS (101.1 mmol, 3.00 equiv.) were added under argon counterflow. The mixture was stirred for 30 min at 10 °C and a solution of 7.90 g of 1-(3,5-dimethoxyphenyl)cyclopentane-1-carbaldehyde (**77a**) (33.7 mmol, 1.00 equiv.) in 50 mL of abs. tetrahydrofuran were added dropwise. After stirring for another 60 min the reaction was quenched by the addition of 200 mL of ammonium chloride solution. The aqueous layer was extracted with 3 × 200 mL of diethyl ether and the combined organic layers were dried over sodium sulfate. Removal of the volatiles under reduced pressure and purification via flash column chromatography (CH/EtOAc 10:1) resulted in 8.70 g (99%) of the pure product as a colorless oil. Analytical data are consistent with the literature.[111]

***R*f** (CH/EtOAc 5:1): 0.65. – **^{1}H NMR** (300 MHz, $CDCl_3$): $\delta = 6.53$ (d, $J = 2.3$ Hz, 2H, 2 × H_{Ar}), 6.28 (t, $J = 2.3$ Hz, 1H, H_{Ar}), 5.67 (dt, $J_{cis} = 11.2$, 1.7 Hz, 1H, H_{DB}), 5.27 (dt, $J_{cis} = 11.2$, 7.4 Hz, 1H, H_{DB}), 3.78 (s, 6H, 2 × OCH_3), 2.08 – 1.85 (m, 4H, CH_2), 1.84 – 1.61 (m, 6H, CH_2), 0.74 (t, $J = 7.5$ Hz, 3H, CH_3) ppm.

(Z)-1-(1-(But-1-en-1-yl)cyclohexyl)-3,5-dimethoxybenzene (78b)

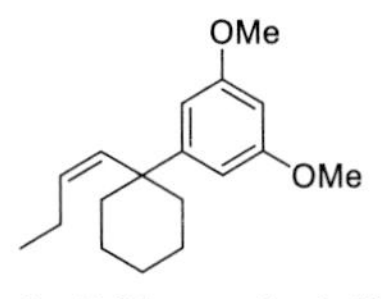

To a suspension of 36.3 g of *n*-propyl triphenylphosphonium bromide (94.3 mmol, 3.00 equiv.) in 300 mL of abs. tetrahydrofuran at 0 °C, 18.8 g of KHMDS (94.3 mmol, 3.00 equiv.) were added under argon counterflow. The mixture was stirred for 30 min at 10 °C and a solution of 7.81 g of 1-(3,5-dimethoxyphenyl)cyclohexane-1-carbaldehyde (**77b**) (33.7 mmol,

1.00 equiv.) in 50 mL of abs. tetrahydrofuran was added dropwise. After stirring for another 60 min the reaction was quenched by the addition of 200 mL of ammonium chloride solution. The aqueous layer was extracted with 3 × 200 mL of diethyl ether and the combined organic layers were dried over sodium sulfate. Removal of the volatiles under reduced pressure and purification via flash column chromatography (CH/EtOAc 10:1) resulted in 8.62 g (99%) of the pure product as a colorless oil. Analytical data are consistent with the literature.[111]

***R*f** (CH/EtOAc 5:1): 0.65. – **^{1}H NMR** (300 MHz, $CDCl_3$): δ = 6.58 (d, J = 2.3 Hz, 2H, 2 × H_{Ar}), 6.28 (t, J = 2.3 Hz, 1H, H_{Ar}), 5.63 (dt, J_{cis} = 11.2, 1.7 Hz, 1H, H_{DB}), 5.34 (dt, J_{cis} = 11.2, 7.4 Hz, 1H, H_{DB}), 3.78 (s, 6H, 2 × OCH_3), 1.95 – 1.90 (m, 2H, CH_2), 1.72 – 1.56 (m, 9H, CH_2), 1.31 – 1.24 (m, 1H, CH_2), 0.72 (t, J = 7.5 Hz, 3H, CH_3) ppm.

1-(1-Butylcyclopentyl)-3,5-dimethoxybenzene (79a)

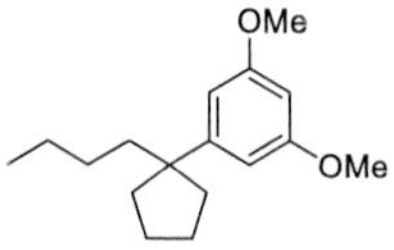

OMe

OMe

To a solution of 8.50 g of (Z)-1-(1-(but-1-en-1-yl)cyclopentyl)-3,5-dimethoxybenzene (**78a**) in ethyl acetate 1.73 g of palladium on activated charcoal (10% Pd/C) were added. Hydrogen gas was bubbled through the solution for several hours and subsequently kept under hydrogen atmosphere for 24 h. Filtration through Celite®, rinsing with ethyl acetate and removal of the volatiles resulted in 8.55 g (99%) of the pure product as a colorless oil. Analytical data are consistent with the literature.[111]

***R*f** (CH/EtOAc 5:1): 0.68. – **^{1}H NMR** (300 MHz, $CDCl_3$): δ = 6.44 (d, J = 2.3 Hz, 2H, 2 × H_{Ar}), 6.29 (t, J = 2.3 Hz, 1H, H_{Ar}), 3.79 (s, 6H, 2 × OCH_3), 1.96 – 1.49 (m, 10H, 5 × CH_2), 1.23 – 1.09 (m, 2H, CH_2), 1.04 – 0.90 (m, 2H, CH_2), 0.79 (t, J = 7.3 Hz, 3H, CH_3) ppm.

1-(1-Butylcyclohexyl)-3,5-dimethoxybenzene (79b)

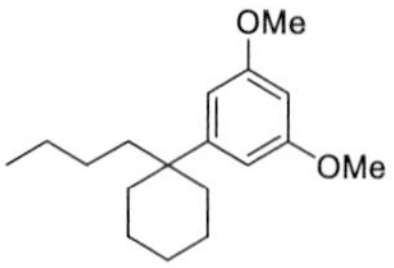

OMe

OMe

To a solution of 8.50 g of ((Z)-1-(1-(but-1-en-1-yl)cyclohexyl)-3,5-dimethoxybenzene (**78b**) in ethyl acetate 1.73 g of palladium on activated charcoal (10% Pd/C) were added. Hydrogen gas was bubbled through the solution for several hours and subsequently kept under hydrogen atmosphere for 24 h. Filtration through Celite®, rinsing with ethyl acetate and removal of the volatiles resulted in 8.31 g (97%) of the pure product as a colorless oil. Analytical data are consistent with the literature.[111]

***R*f** (CH/EtOAc 5:1): 0.68. – **^{1}H NMR** (300 MHz, $CDCl_3$): δ = 6.48 (d, *J* = 2.3 Hz, 2H, 2 × H_{Ar}), 6.3 (t, *J* = 2.3 Hz, 1H, H_{Ar}), 3.80 (s, 6H, 2 × OCH_3), 2.02 – 1.98 (m, 2H, CH_2), 1.57 – 1.36 (m, 10H, 5 × CH_2), 1.18 – 1.09 (m, 2H, CH_2), 0.96 – 0.88 (m, 2H, CH_2), 0.78 (t, *J* = 7.3 Hz, 3H, CH_3) ppm.

4-(1-Butylcyclopentyl)-2,6-dimethoxybenzaldehyde (80a)

OMe O

OMe

According to **GP1**, 8.50 g of 1-(1-butylcyclopentyl)-3,5-dimethoxybenzene (**79a**, 32.4 mmol, 1.00 equiv.) was dissolved in 60 mL of diethyl ether and 7.29 mL of TMEDA (5.65 g, 48.6 mmol, 1.50 equiv.) were added dropwise. The solution was cooled to 0 °C and 19.5 mL of *n*-butyl lithium (2.5 M in *n*-hexanes, 48.6 mmol, 1.50 equiv.) were added slowly. After stirring for 4 h at room temperature, the solution was cooled to 0 °C, 7.48 mL of dimethylformamide (7.11 g, 97.3 mmol, 3.00 equiv.) were added and the mixture was stirred for another 4 h at room temperature. The reaction was quenched by the addition of 60 mL of brine and extracted with 3 × 15 mL of diethyl ether. The combined organic layers were dried over sodium sulfate, the volatiles were removed under reduced pressure and the residue was then purified via flash column chromatography (CH/EtOAc 15:1) to result in 9.4 g (with traces of dimethylformamide) of the product as a yellow oil that was used directly in the next step. Analytical data are consistent with the literature.[111]

***R*f** (CH/ EtOAc 10:1): 0.18. – **^{1}H NMR** (300 MHz, $CDCl_3$): δ = 10.45 (s, 1H, C*H*O), 6.47 (s, 2H, 2 × H_{Ar}), 3.89 (s, 6H, 2 × OCH_3), 2.00 – 1.51 (m, 10H, 5 × CH_2), 1.28 – 1.08 (m, 2H, CH_2), 1.04 – 0.89 (m, 2H, CH_2), 0.80 (t, *J* = 7.3 Hz, 3H, CH_3) ppm.

4-(1-Butylcyclohexyl)-2,6-dimethoxybenzaldehyde (80b)

OMe O

OMe

According to **GP1**, 8.26 g of 1-(1-butylcyclohexyl)-3,5-dimethoxybenzene (**79b**, 29.9 mmol, 1.00 equiv.) were dissolved in 60 mL of diethyl ether and 6.72 mL of TMEDA (5.21 g, 44.8 mmol, 1.50 equiv.) were added dropwise. The solution was cooled to 0 °C and 17.9 mL of *n*-butyl lithium (2.5 M in *n*-hexanes, 44.8 mmol, 1.50 equiv.) were added slowly. After stirring for 4 h at room temperature, the solution was cooled to 0 °C, 6.89 mL of dimethylformamide (6.551 g, 89.7 mmol, 3.00 equiv.) were added and the mixture was stirred for another 4 h at room temperature. The reaction was quenched by the addition of 60 mL of brine and extracted with 3 × 15 mL of diethyl ether. The combined organic layers were dried over sodium sulfate, the volatiles were removed under reduced pressure and the residue was then purified via flash column chromatography (CH/EtOAc 10:1) to result in 8.31 g (91%) of the product as yellow oil that was used directly in the next step. Analytical data are consistent with the literature.[111]

$\boldsymbol{R_f}$ (CH/ EtOAc 10:1): 0.19. – **^{1}H NMR** (300 MHz, $CDCl_3$): δ = 10.46 (s, 1H, C*H*O), 6.52 (s, 2H, 2 × H_{Ar}), 3.89 (s, 6H, 2 × OCH_3), 2.07 – 1.94 (m, 2H, CH_2), 1.66 – 1.33 (m, 10H, 5 × CH_2), 1.29 – 1.04 (m, 2H, CH_2), 1.01 – 0.85 (m, 2H, CH_2), 0.79 (t, *J* = 7.3 Hz, 3H, CH_3) ppm.

4-(1-Butylcyclopentyl)-2-hydroxy-6-methoxybenzaldehyde (29b)

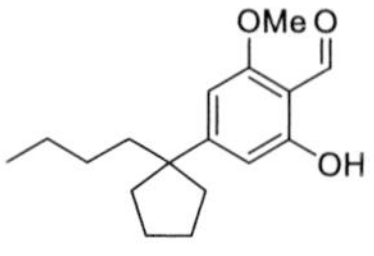

According to **GP3** 10.3 g of **80a** (1.00 eq, 35.3 mmol) were dissolved in a mixture of 200 mL of dry acetonitrile and 100 mL of dry dichloromethane, cooled to 0 °C and 11.8 g of aluminum trichloride (88.24 mmol, 2.50 eq,) and 13.2 g of sodium iodide (88.24 mmol, 2.50 eq,) were added slowly under argon counterflow. The reaction mixture was stirred for 1.5 h at room temperature, quenched with water, extracted with 3 × 50 mL of dichloromethane, the combined organic layers were washed with saturated sodium thiosulfate solution, dried over sodium sulfate and after removal of the volatiles the crude product was purified via flash column chromatography (CH/EtOAc 100:1) to result in 6.69 g (69%) of a yellow oil. Analytical data are consistent with the literature.[111]

$\boldsymbol{R_f}$ (CH/EtOAc 40:1): 0.39. – **^{1}H NMR** (300 MHz, $CDCl_3$): δ = 11.95 (s, 1H, C2-O*H*), 10.25 (s, 1H, C*H*O), 6.45 (d, *J* = 1.5 Hz, 1H, H_{Ar}), 6.28 (d, *J* = 1.5 Hz, 1H H_{Ar}), 3.88 (s, 3H, OCH_3), 1.95 – 1.50 (m, 10H, 5 × CH_2), 1.26 – 1.10 (m, 2H, CH_2), 1.05 – 0.89 (m, 2H, CH_2), 0.80 (t, *J* = 7.3 Hz, 3H, CH_3) ppm.

4-(1-Butylcyclohexyl)-2-hydroxy-6-methoxybenzaldehyde (29c)

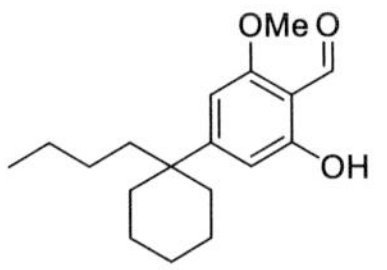

According to **GP3** 8.00 g of **80b** (1.00 eq, 26.3 mmol) were dissolved in a mixture of 100 mL of dry acetonitrile and 50 mL of dry dichloromethane, cooled to 0 °C and 8.76 g of aluminum trichloride (65.7 mmol, 2.50 eq,) and 9.85 g of sodium iodide (65.7 mmol, 2.50 eq,) were added slowly under argon counterflow. The reaction mixture was stirred for 1.5 h at room temperature, quenched with water, extracted with 3 × 30 mL of dichloromethane, the combined organic layers were washed with sodium thiosulfate solution, dried over sodium sulfate and after removal of volatiles, the crude product was purified via flash column chromatography (CH/EtOAc 40:1) to result in 6.11 g (80%) of a yellow oil. Analytical data are consistent with the literature.[111]

***R*f** (CH/EtOAc 40:1): 0.26. – **^{1}H NMR** (300 MHz, $CDCl_3$): δ = 11.92 (s, 1H, C2-O*H*), 10.26 (s, 1H, C*H*O), 6.51 (d, *J* = 1.3 Hz, 1H, H_{Ar}), 6.34 (d, *J* = 1.3 Hz, 1H H_{Ar}), 3.88 (s, 3H, OCH_3), 2.00 – 1.32 (m, 12H, 6 × CH_2), 1.20 – 1.10 (m, 2H, CH_2), 0.96 – 0.88 (m, 2H, CH_2), 0.79 (t, *J* = 7.3 Hz, 3H, CH_3) ppm.

1,3-Dimethoxy-5-(propoxymethyl)benzene (82a)[172]

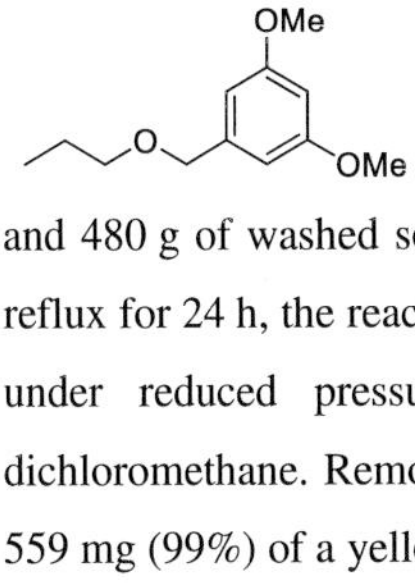

According to **GP4**, 450 mg of (3,5-dimethoxyphenyl)methanol (**81a**, 2.68 mmol, 1.00 equiv.) and 1.46 mL of bromo propane (1.97 g, 16.0 mmol, 5.98 equiv.) were dissolved in 55 mL of tetrahydrofuran and 480 g of washed sodium hydride (20.0 mmol, 7.48 equiv.) were added. After heating at reflux for 24 h, the reaction was quenched with 50 mL of water, tetrahydrofuran was removed under reduced pressure and the aqueous layer was extracted with 2 × 30 mL of dichloromethane. Removal of the volatiles under reduced pressure resulted in the product as 559 mg (99%) of a yellow oil.

^{1}H NMR (400 MHz, $CDCl_3$): δ = 6,51 (d, *J* = 2.3 Hz, 2H, 2 × H_{Ar}), 6.38 (t, *J* = 2.3 Hz, 1H, H_{Ar}), 4.46 (s, 2H, $C_3H_7OCH_2$-Ar), 3.79 (s, 6H, 2 × OCH_3), 3.43 (t, *J* = 6.7 Hz, 2H, $CH_3CH_2CH_2O$), 1.71 – 1.58 (m, 2H, $CH_3CH_2CH_2O$), 0.95 (t, *J* = 7.4 Hz, 3H, $CH_3CH_2CH_2O$) ppm. – **^{13}C NMR** (100 MHz, $CDCl_3$): δ = 161.0 ($C_{quart.}$, 2 × $C_{Ar}OCH_3$), 141.4 ($C_{quart.}$, C_{Ar}), 105.4 (+, 2 × C_{Ar}H), 99.7 (+, C_{Ar}H), 72.9 (–, CH_2), 72.2 (–, CH_2), 55.4 (+, 2 × OCH_3), 23.1 (–, $CH_3CH_2CH_2O$), 10.8 (+, CH_3) ppm. – **IR** (KBr): ṽ = 2932 (w), 2838 (w), 1595 (m), 1458 (m),

1428 (m), 1358 (m), 1318 (w), 1294 (w), 1203 (m), 1151 (s), 1100 (m), 1065 (m), 961 (w), 921 (w), 831 (m), 690 (w), 592 (w), 538 (w) cm^{-1}. – **MS** (70 eV, EI): *m/z* (%) = 210 (18) $[M]^+$, 153 (9), 152 (100) $[M - C_3H_6O]^+$, 151 (20) $[M - C_3H_7O]^+$. – **HRMS** ($C_{12}H_{18}O_3$): calc. 210.1250, found 210.1249.

2,6-Dimethoxy-4-(propoxymethyl)benzaldehyde (83a)[172]

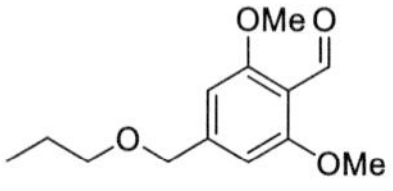

According to **GP1**, 12.2 g of 1,3-dimethoxy-5-(propoxymethyl)benzene (**82a**, 58.1 mmol, 1.00 equiv.) were dissolved in 100 mL of diethyl ether and 13.0 mL of TMEDA (10.1 g, 87.1 mmol, 1.50 equiv.) were added dropwise. The solution was cooled to 0 °C and 34.9 mL of *n*-butyl lithium (2.5 M in *n*-hexanes, 87.13 mmol, 1.50 equiv.) were added slowly. After stirring for 2 h at room temperature, the solution was cooled to 0 °C, then 21.0 mL of dimethylformamide (20.0 g, 273.2 mmol, 3.00 equiv.) were added and the mixture was stirred for another 4 h at room temperature. The reaction was quenched by the addition of 200 mL of brine and extracted with 3 × 100 mL of diethyl ether. The combined organic layers were dried over sodium sulfate, the volatiles were removed under reduced pressure and the residue was then purified via flash column chromatography (CH/EtOAc 2:1) to result in 6.00 g (43%) of the pure product as a yellow oil.

$\boldsymbol{R_f}$ (CH/EtOAc 2:1): 0.27. – **^{1}H NMR** (400 MHz, $CDCl_3$): δ = 10.47 (s, 1H, C*H*O), 6.56 (s, 2H, 2 × H_{Ar}), 4.50 (s, 2H, $C_3H_7OCH_2$-Ar), 3.90 (s, 6H, 2 × OCH_3), 3.47 (t, J = 6.6 Hz, 2H, $CH_3CH_2CH_2O$), 1.73 – 1.62 (m, 2H, $CH_3CH_2CH_2O$), 0.97 (t, J = 7.4 Hz, 3H, $CH_3CH_2CH_2O$) ppm. – **^{13}C NMR** (100 MHz, $CDCl_3$): δ = 189.2 (+, *C*HO), 162.5 ($C_{quart.}$, 2 × $C_{Ar}OCH_3$), 148.2 ($C_{quart.}$, C_{Ar}) 113.5 ($C_{quart.}$, C_{Ar}), 102.3 (+, 2 × $C_{Ar}H$), 72.8(–, CH_2), 72.6 (–, CH_2), 56.2 (+, 2 × OCH_3), 23.1 (–, $CH_3CH_2CH_2O$), 10.8 (+, CH_3) ppm. – **IR** (KBr): $\tilde{\nu}$ = 2937 (w), 2873 (w), 1730 (w), 1683 (m), 1608 (m), 1574 (w), 1459 (m), 1405 (w), 1358 (w), 1325 (w), 1229 (m), 1195 (w), 1124 (m), 968 (w), 924 (vw), 827 (w), 622 (vw), 579 (vw) cm^{-1}. – **MS** (70 eV, EI): *m/z* (%) = 238 (52) $[M]^+$, 226 (25), 194 (34), 182 (32), 181 (30), 180 (100), 168 (75), 167 (25), 166 (12), 151 (15), 139 (28). – **HRMS** ($C_{13}H_{18}O_4$): calc. 238.1200, found 238.1198.

2-Hydroxy-6-methoxy-4-(propoxymethyl)benzaldehyde (29d) [172]

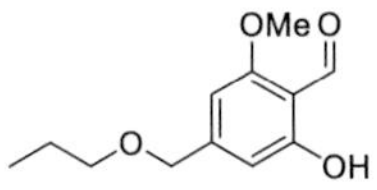

According to **GP3** 66 mg of **83a** (1.00 equiv., 0.28 mmol) was dissolved in a mixture of 2 mL of dry acetonitrile and 1 mL of dry dichloromethane, cooled to 0 °C, 92 mg of aluminum trichloride (0.69 mmol, 2.50 equiv.) and 103 mg of sodium iodide (0.69 mmol, 2.50 equiv.) were added slowly under argon counterflow. The reaction mixture was stirred for 1.5 h at room temperature, quenched with water, extracted with 3 × 5 mL of dichloromethane, the combined organic layers were washed with saturated sodium thiosulfate solution, dried over sodium sulfate and after removal of the volatiles the crude product was purified via flash column chromatography (CH/EtOAc 5:1) to result in 50.9 mg (82%) of an orange oil.

***R*f** (CH/EtOAc 5:1): 0.39. – **^{1}H NMR** (400 MHz, $CDCl_3$): δ = 12.01 (s, 1H, O*H*), 10.28 (s, 1H, C*HO*), 6.48 (d, *J* = 0.6 Hz, 1H, H_{Ar}), 6.40 (d, *J* = 0.8 Hz, 1H, H_{Ar}), 4.45 (s, 2H, $C_3H_7OCH_2$-Ar), 3.89 (s, 3H, OCH_3), 3.45 (t, *J* = 6.6 Hz, 3H, $CH_3CH_2CH_2O$), 1.71 – 1.59 (m, 2H, $CH_3CH_2CH_2O$), 0.96 (t, *J* = 7.4 Hz, 3H, $CH_3CH_2CH_2O$) ppm. – **^{13}C NMR** (100 MHz, $CDCl_3$): δ = 193.9 (+, *C*HO), 163.8 ($C_{quart.}$, C_{Ar}O), 162.7 ($C_{quart.}$, C_{Ar}O), 151.1 ($C_{quart.}$, $C_{Ar}C_{quart.}$), 110.1 ($C_{quart.}$, C_{Ar}CHO), 108.0 (+, C_{Ar}H), 99.4 (+, C_{Ar}H), 72.7(–, CH_2), 72.4 (–, CH_2), 55.9 (+, OCH_3), 23.1 (–, $CH_3CH_2CH_2O$), 10.8 (+, $CH_3CH_2CH_2O$) ppm. – **IR** (KBr): ṽ = 2961 (w), 2935 (w), 2874 (w), 1642 (s), 1573 (m), 1455 (m), 1425 (m), 1395 (m), 1306 (m), 1216 (s), 1200 (m), 1181 (m), 1150 (m), 1100 (s), 1003 (w), 962 (w), 806 (m), 730 (m), 510 (w) cm^{-1}. – **MS** (70 eV, EI): *m/z* (%) = 224 (16) $[M]^+$, 181 (12) $[M - C_3H_7]^+$, 137 (14), 131 (10), 69 (20). – **HRMS** ($C_{12}H_{16}O_4$): calc. 224.1043, found 224.1044.

1,3-Dimethoxy-5-(2-propoxyprop-2-yl)benzene (82b)[172]

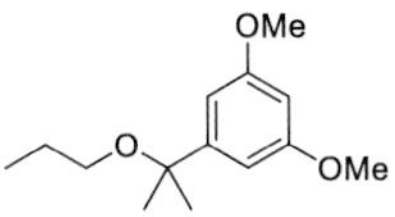

According to **GP4**, 390 mg of 2-(3,5-dimethoxyphenyl)propan-2-ol (**81b**, 2.00 mmol, 1.00 equiv.) and 1.46 mL of bromo propane (1.97 g, 16.0 mmol, 5.98 equiv.) were dissolved in 55 mL of tetrahydrofuran and 480 g of washed sodium hydride (20.0 mmol, 7.48 equiv.) were added. After heating at reflux for 24 h, the reaction was quenched with 50 mL of water, tetrahydrofuran was removed under reduced pressure and the aqueous layer was extracted with 2 × 30 mL of dichloromethane. Removal of the volatiles under reduced pressure and purification via flash column chromatography (CH/EtOAc 20:1) resulted in 260 mg (54%) of the pure product as a yellow oil.

***R*f** (CH/EtOAc 20:1): 0.42. – **^{1}H NMR** (400 MHz, $CDCl_3$): δ = 6.58 (d, *J* = 2.3 Hz, 2H, 2 × H_{Ar}), 6.35 (t, *J* = 2.3 Hz, 1H, H_{Ar}), 3.80 (s, 6H, 2 × OCH_3), 3.12 (t, *J* = 6.9 Hz, 2H, $CH_3CH_2CH_2O$), 1.61 – 1.52 (m, 2H, $CH_3CH_2CH_2O$), 1.50 (s, 6H, 2 x CH_3), 0.89 (t, *J* = 7.4 Hz, 3H, $CH_3CH_2CH_2O$) ppm. – **^{13}C NMR** (100 MHz, $CDCl_3$): δ = 160.8 ($C_{quart.}$, 2 × $C_{Ar}OCH_3$), 149.8 ($C_{quart.}$, C_{Ar}), 104.3 (+, 2 × $C_{Ar}H$), 98.5 (+, $C_{Ar}H$), 76.5 ($C_{quart.}$, $C(CH_3)_2$), 64.7 (–, $CH_3CH_2CH_2O$), 55.4 (+, 2 × OCH_3), 28.6 (+, 2 × CH_3), 23.8 (–, $CH_3CH_2CH_2O$), 10.9 (+, $CH_3CH_2CH_2O$) ppm. – **IR** (KBr): ṽ = 2959 (w), 2933 (w), 2873 (w), 2835 (w), 1594 (m), 1455 (m), 1422 (m), 1377 (w), 1341 (w), 1312 (w), 1291 (m), 1252 (vw), 1203 (m), 1151 (s), 1107 (w), 1063 (m), 1010 (m), 994 (w), 922 (w), 889 (w), 843 (m), 752 (vw), 700 (m), 631 (vw), 591 (vw), 543 (w), 460 (vw) cm^{-1}. – **MS** (70 eV, EI): *m/z* (%) = = 238 (14) $[M]^+$, 181 (19) $[M-C_3H_5O]^+$, 180 (100) $[M-C_3H_6O]^+$, 179 (21) $[M-C_3H_7O]^+$, 139 (13).– **HRMS** ($C_{14}H_{22}O_3$): calc. 238.1563, found 238.1562.

2,6-Dimethoxy-4-(2-propoxypropan-2-yl)benzaldehyde (83b)[172]

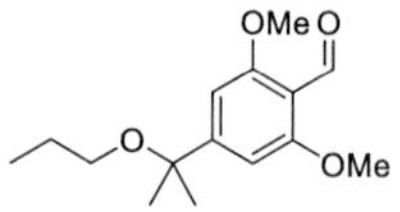

According to **GP1**, 12.2 g of 1,3-dimethoxy-5-(2-propoxypropan-2-yl)benzene (**82b**, 58.1 mmol, 1.00 equiv.) were dissolved in 5 mL of diethyl ether and 0.37 mL of TMEDA (285 Mg, 2.45 mmol, 1.50 equiv.) was added dropwise. The solution was cooled to 0 °C and 0.98 mL of *n*-butyl lithium (2.5 m in *n*-hexanes, 2.45 mmol, 1.50 equiv.) were added slowly. After stirring for 2 h at room temperature, the solution was cooled to 0 °C, 0.38 mL of dimethylformamide (359 mg, 4.91 mmol, 3.00 equiv.) was added and the mixture was stirred for another 4 h at room temperature. The reaction was quenched by the addition of 5 mL of brine and extracted with 3 × 5 mL of diethyl ether. The combined organic layers were dried over sodium sulfate, the volatiles were removed under reduced pressure and the residue was then purified via flash column chromatography (CH/EtOAc 5:1) to result in 374 mg (86%) of the pure product as a yellow oil.

***R*f** (CH/EtOAc 5:1): 0.15. – **^{1}H NMR** (400 MHz, $CDCl_3$): δ = 10.46 (s, 1H, CHO), 6.63 (s, 2H, 2 × H_{Ar}), 3.89 (s, 6H, 2 × OCH_3), 3.18 – 3.09 (m, 2H, $CH_3CH_2CH_2O$), 1.64 – 1.55 (m, 2H, $CH_3CH_2CH_2O$), 1.52 (s, 6H, 2 × CH_3), 0.91 (t, *J* = 7.4 Hz, 3H, $CH_3CH_2CH_2O$) ppm. – **^{13}C NMR** (100 MHz, $CDCl_3$): δ = 189.2 (+, CHO), 162.3 ($C_{quart.}$, 2 × $C_{Ar}OCH_3$), 156.6 ($C_{quart.}$, C_{Ar}), 113.1 ($C_{quart.}$, $C_{Ar}CHO$), 101.5 (+, 2 × $C_{Ar}H$), 76.8 ($C_{quart.}$, $C(CH_3)_2$), 64.9 (–, $CH_3CH_2CH_2O$), 56.1 (+, 2 × OCH_3), 28.3 (+, 2 × CH_3), 23.7 (–, $CH_3CH_2CH_2O$), 10.9 (+,

C$H_3CH_2CH_2O$) ppm. – **IR** (KBr): ṽ = 2965 (w), 2933 (w), 2871 (w), 1728 (w), 1683 (m), 1603 (m), 1567 (m), 1453 (m), 1401 (m), 1360 (w), 1316 (w), 1233 (m), 1197 (m), 1164 (w), 1125 (s), 1070 (m), 1010 (m), 994 (w), 955 (w), 926 (w), 899 (w), 831 (m), 791 (w), 674 (w), 624 (w), 592 (w), 510 (w) cm^{-1}. – **MS** (70 eV, EI): *m/z* (%) = 266 (3) $[M]^+$, 208 (20) $[M - C_3H_6O]^+$, 180 (10), 179 (15), 178 (100), 149 (8), 91 (17). – **HRMS** ($C_{15}H_{22}O_4$): calc. 266.1513, found 266.1512.

2-Hydroxy-6-methoxy-4-(2-propoxypropan-2-yl)benzaldehyde (29e)[172]

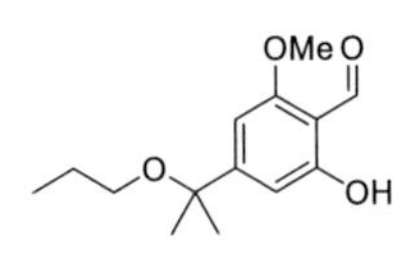

According to **GP3** 220 mg of 2,6-dimethoxy-4-(2-propoxypropan-2-yl)benzaldehyde (**83b**, 1.00 equiv., 0.830 mmol) were dissolved in a mixture of 2 mL of dry acetonitrile and 1 mL of dry dichloromethane, cooled to 0 °C and 280 mg of aluminum trichloride (2.08 mmol, 2.50 equiv.) and 310 mg of sodium iodide (2.08 mmol, 2.50 equiv.) were added slowly under argon counterflow. The reaction mixture was stirred for 1.5 h at room temperature, quenched with water, extracted with 3 × 5 mL of dichloromethane, the combined organic layers were washed with sodium thiosulfate solution, dried over sodium sulfate and after removal of volatiles the crude product was purified via flash column chromatography (CH/EtOAc 5:1) to result in 138.7 mg (66%) of a yellow oil.

$\boldsymbol{R_f}$ (CH/EtOAc 5:1): 0.5. – **^{1}H NMR** (400 MHz, $CDCl_3$): δ = 11.97 (s, 1H, O*H*), 10.28 (s, 1H, C*HO*), 6.54 (d, *J* = 1.4 Hz, 1H, H_{Ar}), 6.52 (d, *J* = 0.5 Hz, 1H, H_{Ar}), 3.89 (s, 3H, OCH_3), 3.15 (t, *J* = 6.8 Hz, 2H, $CH_3CH_2CH_2O$), 1.69 – 1.51 (m, 2H, $CH_3CH_2CH_2O$), 1.49 (s, 6H, 2 × CH_3), 0.91 (t, *J* = 7.4 Hz, 3H, $CH_3CH_2CH_2O$) ppm. – **^{13}C NMR** (100 MHz, $CDCl_3$): δ = 193.8 (+, *C*HO), 163.7 ($C_{quart.}$, $C_{Ar}O$), 162.5 ($C_{quart.}$, $C_{Ar}O$), 159.5 ($C_{quart.}$, C_{Ar}), 109.7 ($C_{quart.}$, $C_{Ar}CHO$), 107.4 (+, $C_{Ar}H$), 98.8 (+, $C_{Ar}H$), 76.7 ($C_{quart.}$, $C(CH_3)_2$), 64.9 (–, $CH_3CH_2CH_2O$), 55.9 (+, OCH_3), 28.1 (+, 2 × CH_3), 23.7 (–, $CH_3CH_2CH_2O$), 10.9 (+, C_3H_7O) ppm. – **IR** (KBr): ṽ = 2973 (w), 2935 (w), 2873 (w), 1638 (s), 1569 (m), 1461 (m), 1414 (m), 1394 (m), 1345 (m), 1310 (m), 1219 (s), 1198 (m), 1183 (m), 1166 (m), 1110 (s), 1071 (m), 1011 (m), 995 (w), 952 (w), 901 (w), 844 (m), 792 (m), 731 (m), 677 (w), 590 (w), 555 (w), 515 (m) cm^{-1}. – **MS** (70 eV, EI): *m/z* (%) = 252 (8) $[M]^+$, 195 (27), 194 (100), 193 (15). – **HRMS** ($C_{14}H_{20}O_4$): calc. 252.1356, found 252.1355.

1-(3,5-Dimethoxyphenyl)cyclopentan-1-ol (81c)[172]

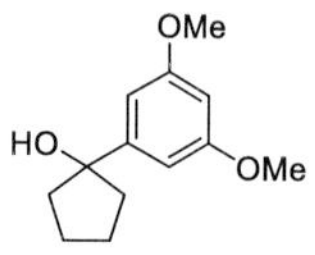

Under argon atmosphere, 100 mL of tetrahydrofuran were added to 0.94 g magnesium shavings (39.0 mmol, 1.10 equiv.) and 1/20 of a solution of 7.70 g of 1-bromo-3,5-dimethoxybenzene (**84**, 35.5 mmol, 1.00 equiv.) in 200 mL of tetrahydrofuran were added slowly. A small amount of I_2 was added to start up the reaction and the rest of the solution was added, so that the reaction was boiling. After complete addition the mixture was heated at reflux for 2 h. The reaction was allowed to cool to room temperature and 18.9 mL of cyclopentanone (17.9 g, 213 mmol, 6.00 equiv.) was added. After stirring for 1 h at room temperature the reaction was quenched with 50 mL of 1 M hydrochloric acid solution. The aqueous layer was extracted with 3 × 50 mL of diethyl ether. The combined organic layers were washed with 50 mL of brine, dried over sodium sulfate, the volatiles were removed under reduced pressure and the residue was then purified via flash column chromatography (CH/EtOAc 10:1) to result in 6.02 g (76%) of the pure product as a yellow oil.

***R*f** (CH/EtOAc 10:1): 0.21. – **^{1}H NMR** (400 MHz, $CDCl_3$): $\delta = 6.66$ (d, $J = 2.3$ Hz, 2H, 2 × H_{Ar}), 6.36 (t, $J = 2.3$ Hz, 1H, H_{Ar}), 3.80 (s, 6H, 2 × OCH_3), 2.06 – 1.90 (m, 6H, 6 × H_{Cp}), 1.89 – 1.77 (m, 2H, 2 × H_{Cp}) ppm. – **^{13}C NMR** (100 MHz, $CDCl_3$): $\delta = 160.8$ ($C_{quart.}$, 2 × $C_{Ar}OCH_3$) 150.0 ($C_{quart.}$, C_{Ar}), 103.6 (+, 2 × $C_{Ar}H$), 98.6 (+, $C_{Ar}H$), 83.7 ($C_{quart.}$, COH), 55.5 (+, 2 × OCH_3), 42.0 (–, 2 × CH_2), 24.0 (–, 2 × CH_2) ppm. – **IR** (KBr): $\tilde{\nu} = 3428$ (w), 2940 (w), 2870 (w), 2835 (w), 1592 (m), 1454 (m), 1421 (m), 1336 (w), 1295 (w), 1201 (m), 1150 (s), 1061 (m), 1039 (m), 1004 (m), 921 (w), 837 (m), 696 (m), 544 (w), 449 (vw), 389 (vw) cm^{-1}. – **MS** (70 eV, EI): *m/z* (%) = 222 (10) $[M]^+$, 181 (23), 179 (14), 178 (100), 152 (10), 131 (24), 109 (10), 91 (16), 77 (10), 69 (41). – **HRMS** ($C_{13}H_{18}O_3$): calc. 222.1250, found 222.1249.

1,3-Dimethoxy-5-(1-propoxycyclopentyl)benzene (82c)[172]

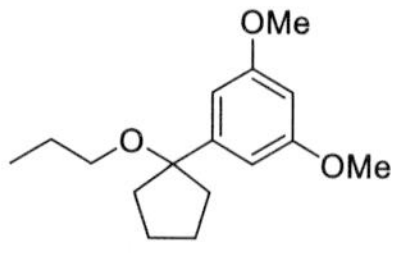

According to **GP4**, 7.34 mg of 2-(3,5-dimethoxyphenyl)propan-2-ol (**81c**, 33.0 mmol, 1.00 equiv.) and 24.1 mL of 1-bromopropane (32.5 g, 264.2 mmol, 8.00 equiv.) were dissolved in 600 mL of tetrahydrofuran and 21.1 g of sodium hydride (60% suspension in mineral oil, 528.3 mmol, 16.00 equiv.) were added. After heating at reflux for 24 h, the reaction was quenched with 500 mL of water, tetrahydrofuran was removed under reduced pressure and the aqueous layer was extracted with 2 × 300 mL of dichloromethane. Removal of the volatiles under reduced pressure and purification via flash column chromatography (CH/EtOAc 20:1) resulted in 7.40 mg (85%) of the pure product as a yellow oil.

***R*f** (CH/EtOAc 10:1): 0.58. – **^{1}H NMR** (400 MHz, $CDCl_3$): δ = 6.59 (d, *J* = 2.3 Hz, 2H, 2 × H_{Ar}), 6.36 (t, *J* = 2.3 Hz, 1H, H_{Ar}), 3.79 (s, 6H, 2 × OCH_3), 3.02 (t, *J* = 6.8 Hz, 2H, $CH_3CH_2CH_2O$), 2.18 – 2.04 (m, 2H, 2 × H_{Cp}), 1.93 – 1.63 (m, 6H, 6 x H_{Cp}), 1.56 – 1.44 (m, 2H, $CH_3CH_2CH_2O$), 0.85 (t, *J* = 7.4 Hz, 3H, $CH_3CH_2CH_2O$) ppm. – **^{13}C NMR** (100 MHz, $CDCl_3$): δ = 160.7 ($C_{quart.}$, 2 × $C_{Ar}OCH_3$), 147.2 ($C_{quart.}$, C_{Ar}), 105.0 (+, 2 × $C_{Ar}H$), 98.6 (+, $C_{Ar}H$), 87.9 ($C_{quart.}$, $CH_3CH_2CH_2OC$), 64,8 (–, $CH_3CH_2CH_2O$), 55.4 (+, 2 × OCH_3), 37.2 (–, 2 × $C_{Cp}H_2$), 23.7 (–, $CH_3CH_2CH_2O$), 23.2 (–, 2 × $C_{Cp}H_2$), 11.0 (+, $CH_3CH_2CH_2O$) ppm. – **IR** (KBr): ṽ = 2957 (w), 2870 (w), 2835 (w), 1593 (m), 1454 (m), 1422 (m), 1337 (w), 1310 (w), 1291 (w), 1251 (w), 1202 (m), 1151 (s), 1064 (m), 1008 (m), 925 (w), 835 (m), 697 (m), 543 (w) cm^{-1}. – **MS** (70 eV, EI): *m/z* (%) = 264 (10) $[M]^+$, 207 (16), 206 (100), 205 (11) $[M - C_3H_7O]^+$, 181 (20), 179 (13), 178 (90), 177 (14), 165 (13), 154 (12), 151 (16), 131 (22), 109 (10), 91 (16), 69 (38).– **HRMS** ($C_{16}H_{24}O_3$): calc. 264.1720, found 264.1721.

2,6-Dimethoxy-4-(1-propoxycyclopentyl)benzaldehyde (83c)[172]

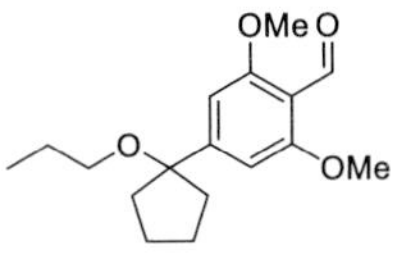

According to **GP1**, 141 mg of 1,3-dimethoxy-5-(2-propoxypropan-2-yl)benzene (**82c**, 0.53 mmol, 1.00 equiv.) was dissolved in 5 mL of diethyl ether and 0.12 mL of TMEDA (93 mg, 0.80 mmol, 1.50 equiv.) were added dropwise. The solution was cooled to 0 °C and 0.32 mL of *n*-butyl lithium (2.5 M in *n*-hexanes, 0.80 mmol, 1.50 equiv.) were added slowly. After stirring for 2 h at room temperature, the solution was cooled to 0 °C, 0.12 mL of dimethylformamide (117 mg, 1.60 mmol, 3.00 equiv.) were added and the mixture was stirred for another 4 h at room temperature. The reaction was quenched by the addition of 5 mL of brine and extracted with 3 × 5 mL of diethyl ether. The combined organic layers were dried over sodium sulfate, the volatiles were removed under reduced pressure and the residue was then purified via flash column chromatography (CH/EtOAc 5:1) to result in 81.8 mg (53%) of the pure product as a yellow oil.

***R*f** (CH/EtOAc 5:1): 0.23. – **^{1}H NMR** (400 MHz, $CDCl_3$): δ = 10.47 (s, 1H, CHO), 6.64 (s, 2H, 2 × H_{Ar}), 3.89 (s, 6H, 2 × OCH_3), 3.05 (t, *J* = 6.7 Hz, 2H, $CH_3CH_2CH_2O$), 2.19 – 2.07 (m, 2H, 2 × H_{Cp}), 1.96 – 1.69 (m, 6H, 6 × H_{Cp}), 1.59 – 1.48 (m, 2H, $CH_3CH_2CH_2O$), 0.89 (t, *J* = 7.4 Hz, 3H, $CH_3CH_2CH_2O$) ppm. – **^{13}C NMR** (100 MHz, $CDCl_3$): δ = 189.2 (+, CHO), 162.2 ($C_{quart.}$,

2 × $C_{Ar}OCH_3$), 154.1 ($C_{quart.}$, $C_{Ar}C_{quart.z}$), 113.2 ($C_{quart.}$, $C_{Ar}CHO$), 102.2 (+, 2 × $C_{Ar}H$), 88.2 ($C_{quart.}$, $CH_3CH_2CH_2OC$), 65.0 (–, $CH_3CH_2CH_2O$), 56.1 (+, 2 × OCH_3), 37.3 (–, 2 × $C_{Cp}H_2$), 23.7 (–, $CH_3CH_2CH_2O$), 23.3 (–, 2 × $C_{Cp}H_2$), 11.0 (+, $CH_3CH_2CH_2O$) ppm. – **IR** (KBr): $\tilde{\nu}$ = 2959 (m), 2936 (m), 2866 (w), 1683 (m), 1645 (w), 1603 (m), 1565 (m), 1452 (m), 1396 (m), 1316 (w), 1232 (m), 1192 (m), 1123 (s), 1075 (s), 1032 (m), 1009 (m), 976 (w), 948 (w), 929 (w), 903 (w), 884 (w), 844 (m), 821 (m), 796 (m), 729 (w), 658 (w), 576 (m), 509 (w), 454 (w) cm^{-1}. – **MS** (70 eV, EI): *m/z* (%) = 292 (9) $[M]^+$, 235 (16), 234 (100), 233 (6). – **HRMS** ($C_{17}H_{24}O_4$): calc. 292.1670, found 292.1669.

4-(Cyclopent-1-en-1-yl)-2-hydroxy-6-methoxybenzaldehyde (29g)[172]

According to **GP3** 1.14 g of 2,6-dimethoxy-4-(1-propoxycyclopentyl)benzaldehyde **83c** (1.00 equiv., 3.89 mmol) were dissolved in a mixture of 50 mL of dry acetonitrile and 25 mL of dry dichloromethane, cooled to 0 °C and 1.30 g of aluminum trichloride (9.72 mmol, 2.50 equiv.) and 1.46 g of sodium iodide (9.72 mmol, 2.50 equiv.) were added slowly under argon counterflow. The reaction mixture was stirred for 1.5 h at room temperature, quenched with water and extracted with 3 × 5 mL of dichloromethane. The combined organic layers were washed with sodium thiosulfate solution, dried over sodium sulfate and after removal of volatiles the crude product was purified via flash column chromatography (CH/EtOAc 5:1) to result in 216.4 mg (24%) of a yellow solid.

***R*f** (CH/EtOAc 5:1): 0.43. – **^{1}H NMR** (400 MHz, $CDCl_3$): δ = 12.03 (s, 1H, *OH*), 10.25 (s, 1H, C*H*O), 6.54 (s, 1H, H_{Ar}), 6.45 (d, *J* = 1.4 Hz, 1H, H_{Ar}), 6.39 – 6.35 (m, 1H, C*H*), 3.90 (s, 3H, OCH_3), 2.72 – 2.63 (m, 2H, CH_2), 2.60 – 2.52 (m, 2H, CH_2), 2.09 – 1.97 (m, 2H, $CH_2CH_2CH_2$) ppm. – **^{13}C NMR** (100 MHz, $CDCl_3$): δ = 193.5 (+, *C*HO), 163.8 ($C_{quart.}$, $C_{Ar}OH$), 162.2 ($C_{quart.}$, $C_{Ar}OCH_3$), 147.0 ($C_{quart.}$, C_{Ar}), 142.4 ($C_{quart.}$, $C_{Ar}CHO$), 131.88 (+, $C_{Ar}H$), 109.95 ($C_{quart.}$, C_{Ar}), 107.1 (+, $C_{Ar}H$), 98.4 (+, *C*H), 55.8 (+, OCH_3), 33.7 (–, CH_2), 33.1 (–, CH_2), 23.4 (–, CH_2) ppm. – **IR** (KBr): $\tilde{\nu}$ = 2951 (w), 2901 (w), 2841 (w), 1612 (m), 1559 (m), 1465 (w), 1446 (w), 1411 (m), 1359 (m), 1305 (m), 1206 (m), 1110 (m), 1038 (w), 974 (w), 959 (w), 849 (w), 820 (m), 786 (m), 725 (m), 683 (w), 622 (w), 542 (w), 510 (w), 445 (w), 410 (w) cm^{-1}. – **MS** (70 eV, EI): *m/z* (%) = 218 (100) $[M]^+$, 200 (13), 179 (15). – **HRMS** ($C_{13}H_{14}O_3$): calc. 218.0937, found 218.0936.

2,5-Dihydroxy-3,4,6-trimethylbenzaldehyde (87)

A solution of 4.00 g of 2,3,5-trimethylbenzene-1,4-diol (**86**, 1.00 equiv., 26.3 mmol) in 200 mL of absolute dichloromethane was cooled to 0 °C and subsequently 5.76 mL of titanium tetrachloride (2.00 equiv., 9.97 g, 52.6 mmol) and 2.45 mL of 1,1-dichlordimethylether (1.05 equiv. 3.17 g, 27.6 mmol) were added. The reaction mixture was then allowed to warm to room temperature, stirred for additional 15 h and then poured onto ice and stirred for 2 h. The organic layer was separated and the aqueous layer was extracted two more times with dichloromethane. The combined organic layers were dried over magnesium sulfate and the volatiles were removed under reduced pressure. The crude product was then purified twice via flash column chromatography using the solvent mixtures of 10:1 and 5:1 (CH/EtOAc) respectively. Subsequent recrystallization from ethyl acetate led to 1.97 g (42%) of yellow crystals.

R_f (CH/EtOAc 5:1): 0.24. – **MP**: 152.1 °C – **^{1}H NMR** (400 MHz, $CDCl_3$): δ = 12.08 (s, 1H, 2-CO*H*), 10.23 (s, 1H, 5-C*H*O), 4.44 (s, 1H, 5-CO*H*), 2.45 (s, 3H, CH_3), 2.24 (s, 3H, CH_3), 2.16 (s, 3H, CH_3) ppm. – **^{13}C NMR** (100 MHz, $CDCl_3$): δ = 195.0 (+, CHO), 156.2 ($C_{quart.}$, C_{Ar}), 144.2 ($C_{quart.}$, C_{Ar}), 135.7 ($C_{quart.}$, C_{Ar}), 123.4 ($C_{quart.}$, C_{Ar}), 122.3 ($C_{quart.}$, C_{Ar}), 115.9 ($C_{quart.}$, C_{Ar}), 13.8 (+, CH_3), 11.3 (+, CH_3), 10.1 (+, CH_3) ppm. – **IR** (KBr): $\tilde{\nu}$ = 3440 (vw), 2902 (vw), 1620 (w), 1568 (vw), 1475 (vw), 1401 (vw), 1381 (vw), 1299 (w), 1269 (w), 1202 (w), 1131 (vw), 1086 (w), 1053 (w), 1032 (vw), 905 (vw), 844 (vw), 714 (w), 665 (vw), 608 (vw), 578 (vw), 514 (vw), 455 (vw), 420 (vw), 391 (vw) cm^{-1}. – **MS** (70 eV, EI): *m/z* (%) = 180 (100) $[M]^+$, 179 (37) $[M-H]^+$, 165 (4) $[M-CH_3]^+$, 152 (20) $[M-CO]^+$, 151 (13) $[M-CHO]^+$, 69 (18). – **HRMS** ($C_{10}H_{12}O_3$): calc. 180.0781, found 180.0782. – **Elemental analysis**: $C_{10}H_{12}O_3$ (316.39): calc. C 66.65, H 6.71, found C 66.65, H 6.76.

2,5-Dimethoxy-3,4,6-trimethylbenzaldehyde (88)

A mixture of 1.37 g of **87** (1.00 equiv., 7.57 mmol), 9.42 g of powdered anhydrous potassium carbonate (9.00 equiv., 68.2 mmol) and 2.87 mL of dimethyl sulfate (4.00 eq, 3.82 g, 30.3 mmol) in 60 mL of dry acetone were stirred at reflux for 12 h. The reaction mixture was allowed to cool to room temperature and the inorganic material filtered off. Water (60 mL) was added to the filtrate and the mixture was stirred for 1 h to hydrolyze the excess of dimethyl sulfate. The acetone was then removed under reduced pressure and the remaining aqueous layer was extracted with 3 × 20 mL of diethyl ether, the combined organic layers were washed with 20 mL of water and 20 mL of brine and dried over sodium sulfate. The volatiles were removed under reduced pressure and the residue was purified via flash column chromatography CH/EtOAc 10:1 resulted in a yellow solid 1.18 g (75%). Analytical data are consistent with the literature.[114,223]

R_f (CH/EtOAc 10:1): 0.22. – **^{1}H NMR** (300 MHz, $CDCl_3$): δ = 10.48 (s, 1H, C*H*O), 3.77 (s, 3H, 5-COCH_3), 3.65 (s, 3H, 2-COCH_3), 2.49 (s, 3H, 3-CCH_3), 2.27 (s, 3H, 6-CCH_3), 2.21 (s, 3H, 4-CCH_3) ppm.

2-Hydroxy-5-methoxy-3,4,6-trimethylbenzaldehyde (29h)

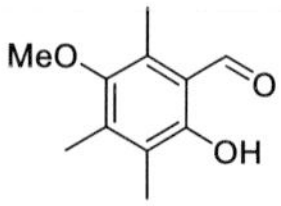

According to **GP3** 1.10 g of **88** (1.00 eq, 5.29 mmol) was dissolved in a mixture of 40 mL of dry acetonitrile and 20 mL of dry dichloromethane, cooled to room temperature, 1.76 g of aluminum trichloride (2.50 eq, 13.2 mmol) and 1.98 g of sodium iodide (2.50 eq, 13.2 mmol) were added slowly under argon counterflow. The reaction mixture was stirred for 1 h at room temperature, quenched with water, extracted with 3 × 20 mL of dichloromethane, the combined organic layers were dried over sodium sulfate and the crude product was purified via flash column chromatography (CH/EtOAc 10:1) to result in 899 mg (89%) of a yellow solid. Analytical data are consistent with the literature.[114]

R_f (CH/EtOAc 10:1): 0.22. – **^{1}H NMR** (300 MHz, $CDCl_3$): δ = 12.2 (s, 3H, 2-CO*H*), 10.2 (s, 1H, C*H*O), 3.64 (s, 3H, 5-COCH_3), 2.51 (s, 3H, 3-CCH_3), 2.27 (s, 3H, 6-CCH_3), 2.15 (s, 3H, 4-CCH_3) ppm.

5-Isopropyl-4-methoxy-2-methylphenol (89c)[177]

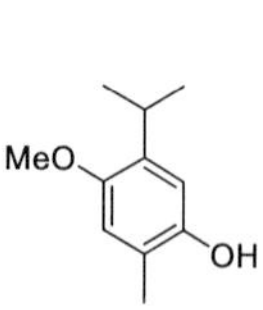

To a mixture of 8.33 g of copper iodide (43.7 mmol, 2.00 equiv.), 5.00 g of 4-bromo carvacrol (**89b**) (21.8 mmol, 1.00 equiv.) and 33.9 mL of sodium methoxide (183 mmol, 30% in methanol, 6.60 equiv.), 45 mL of abs. dimethylformamide were added slowly. The mixture was heated at 110 °C for 48 h, allowed to cool to room temperature, diluted with 215 mL of diethyl ether and 215 mL of saturated ammonium chloride solution was added. The aqueous layer was extracted with 3 × 100 mL of diethyl ether and the combined organic layers washed with 100 mL of sodium bicarbonate-solution and 100 mL of brine and dried over magnesium sulfate. The volatiles were removed under reduced pressure and the residue was purified via flash column chromatography (CH/EtOAc 20:1) that resulted in red crystals 2.76 g (70%).[22]

***R*f** (CH/EtOAc 20:1): 0.10. – **^{1}H NMR** (400 MHz, $CDCl_3$): δ = 6.66 (s, 1H, H_{Ar}), 6.65 (s, 1H, H_{Ar}), 4.54 (s, 1H, O*H*), 3.79 (s, 3H, OCH_3), 3.26 (hept, *J* = 7.28 Hz, 1H, C*H*(CH_3)$_2$), 2.24 (s, 3H, CH_3), 1.18 (d, *J* = 6.99 Hz, 6H, CH(CH_3)$_2$) ppm. – **^{13}C-NMR** (100 MHz, $CDCl_3$): *δ* = 150.8 ($C_{quart.}$, $C_{Ar}OCH_3$), 147.6 ($C_{quart.}$, C_{Ar}OH), 136.1 ($C_{quart.}$, $C_{Ar}{}^iPr$), 121.0 ($C_{quart.}$, $C_{Ar}CH_3$), 114.0 (+, C_{Ar}H), 113.3 (+, C_{Ar}H), 56.5 (+, OCH_3), 26.5 (+, *C*H(CH_3)$_2$), 23.0 (+, (CH_3)$_2$CH), 15.9 (+, CH_3) ppm. – **IR** (KBr): ṽ = 3369 (s), 3266 (vs), 2953 (s), 2831 (vs), 1620 (vs), 1514 (s), 1457 (s), 1405 (w), 1361 (s), 1271 (s), 1234 (vs), 1198 (vw), 1172 (s), 1106 (s), 1060 (vs), 1048 (m), 1005 (s), 985 (m), 899 (s), 877 (m), 868 (vs), 851 (vs), 814 (m), 708 (s), 663 (m), 595 (vs), 474 (s), 430 (vs) cm^{-1}. – **MS** (EI, 70 eV): m/z (%) = 180 (59) $[M]^+$, 166 (12), 165 (100) $[M – CH_3]^+$, 150 (20), 149 (3) $[M – OCH_3]^+$, 137 (5) $[M – C_3H_7]^+$, 121 (5), 107 (4), 91 (9), 77 (6). – **HRMS** ($C_{11}H_{16}O_3$): calc. 180.1145, found 180.1146.

3-Bromo-6-hydroxy-2-isopropyl-5-methylbenzaldehyde (29k)[177]

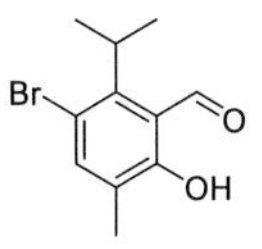

According to **GP2**, 5.0 g of 4-bromo carvacrol **89b** (21.8 mmol, 1.00 equiv.) was dissolved in 120 mL of dry acetonitrile, 4.44 g of paraformaldehyde (148 mmol, 6.79 equiv.) and 3.13 g of magnesium chloride (32.8 mmol, 1.50 equiv.) were added under argon counterflow. Then 11.5 mL of dry triethylamine (83.0 mmol, 3.81 equiv.) were added and the mixture was heated at reflux for 12 h. The reaction mixture was then allowed to cool to room temperature, quenched with 60 mL of water and concentrated hydrochloric acid was added until pH was adjusted to 1. The mixture was extracted with 3 × 100 mL of diethyl ether and dried over sodium sulfate. The volatiles were removed under reduced pressure and the residue was purified via flash column chromatography (CH/EtOAc 20:1) resulted in yellow crystals 572 mg (10%).

***R*f** (CH/EtOAc 20:1): 0.72. – **^{1}H NMR** (400 MHz, $CDCl_3$): δ = 12.59 (s, 1H, OH) 10.55 (s, 1H, CHO), 7.56 (s, 1H, H_{Ar}), 3.89 (hept, *J* = 6.91 Hz, 1H, C*H*(CH_3)$_2$), 2.21 (s, 3H, CH_3), 1.49 (d, *J* = 7.48 Hz, 6H, CH(CH_3)$_2$) ppm. – **^{13}C-NMR** (100 MHz, $CDCl_3$): *δ* = 195.6 (+, *C*HO) 162.4 ($C_{quart.}$, C_{Ar}OH), 147.4 ($C_{quart.}$, C_{Ar}*i*Pr), 141.9 (+, C_{Ar}H), 127.7 ($C_{quart.}$, $C_{Ar}CH_3$), 119.0 ($C_{quart.}$, C_{Ar}CHO), 113.6 ($C_{quart.}$, C_{Ar}Br), 33.6 (+, *C*H(CH_3)$_2$), 23.7 (+, CH(CH_3)$_2$), 14.9 (+, CH_3) ppm. – **IR** (KBr): ṽ = 2973 (s), 2928 (m), 1612 (w), 1441 (s), 1415 (m), 1379 (s), 1292 (m),

1264 (s), 1235 (vw), 1201 (s), 1140 (vs), 1039 (m), 970 (m), 907 (s), 874 (vs), 787 (s), 764 (w), 718 (m), 670 (vs), 613 (m), 573 (m), 614 (vs), 497 (s), 453 (s). – **MS** (EI, 70 eV): m/z (%) = 258/256 (60/61) [M^+], 243/241 (7/9) [$M - CH_3$] $^+$, 240/238 (16/16) [$M - H_2O$]$^+$, 225/223 (28/29) [$M - CH_3, - H_2O$]$^+$, 162 (12) [$M - CH_3, - Br$]$^+$, 160 (13) [M – OH, – Br]$^+$, 159 (100) [$M - H_2O - Br$]$^+$, 158 (9), 147 (4), 144 (13) [$M - H_2O - Br - CH_3$]$^+$, 141 (6), 134 (6), 133 (8), 132 (11), 131 (11), 105 (8), 103 (8), 91 (18), 77 (18). – **HRMS** ($C_{11}H_{13}Br\ O_2$): calc. 256.0093, found 256.0094.

2-Hydroxy-6-isopropyl-3-methylbenzaldehyde (29i)

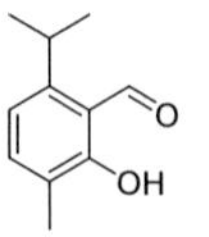

According to **GP2**, 15.0 g of carvacrol (**89a**, 99.9 mmol, 1.00 equiv.) is dissolved in 500 mL of dry acetonitrile, 20.2 g of paraformaldehyde (674 mmol, 6.75 equiv.) and 14.3 g magnesium chloride (150 mmol, 1.50 equiv.) were added under argon counterflow. Then 52.3 mL of dry triethylamine (37.9 g, 374 mmol, 3.75 equiv.) were added and the mixture was heated at reflux for 4 h. The reaction mixture was then allowed to cool to room temperature, quenched with 350 mL of water and concentrated hydrochloric acid was added until the pH was adjusted to 1. The mixture was extracted with 3 × 250 mL of diethyl ether and dried over sodium sulfate. The volatiles were removed under reduced pressure and the residue was purified via flash column chromatography CH/EtOAc 40:1 resulted in a yellow oil 9.83 g (55%). Analytical data are consistent with the literature.[111,224]

***R*f** (CH/EtOAc 20:1): 0.50. – **^{1}H NMR** (300 MHz, $CDCl_3$): δ = 12.42 (s, 1H, 2-CO*H*), 10.40 (s, 1H, C*H*O), 7.32 (d, *J* = 7.8 Hz, 1H, H_{Ar}), 6.77 (d, *J* = 7.8 Hz, 1H H_{Ar}), 3.61 (hept, *J* = 6.8 Hz, 1H, C*H*(CH_3)$_2$), 2.21 (s, 3H, CH_3), 1.32 (d, *J* = 6.9 Hz, 6H, CH(C*H*$_3$)$_2$) ppm.

2-Hydroxy-6-isopropyl-5-methoxy-3-methylbenzaldehyde (29j)

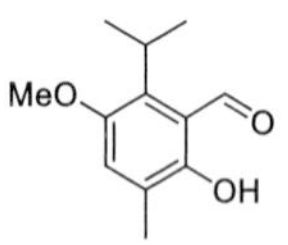

According to **GP2**, 2.37 g of 4-methoxy carvacrol (**89c**) (13.1 mmol, 1.00 equiv.) was dissolved in 80 mL of dry acetonitrile, 1.97 g of paraformaldehyde (65.7 mmol, 6.75 equiv.) and 3.75 g of magnesium chloride (39.4 mmol, 1.50 equiv.) were added under argon counterflow. Then 5.50 mL of dry triethylamine (3.99 g, 39.4 mmol, 3.75 equiv.) were added and the mixture was heated at reflux for 5 h. The reaction mixture was then allowed to cool to room temperature, quenched with 25 mL of water and concentrated hydrochloric acid was added until pH was adjusted to 1. The mixture was extracted with 3 × 25 mL of diethyl ether and dried over sodium sulfate. The volatiles were removed under reduced pressure and the residue was twice purified via flash column chromatography CH/EtOAc 10:1 and 20:1 resulted in 481 mg of a yellow oil that contained 6% of a side product. The amount of product was calculated as 461 mg (24%, calculated by ^{1}H NMR) and was used in subsequent reactions without further purification.

***R*f** (CH/EtOAc 10:1): 0.55. – **^{1}H NMR** (400 MHz, $CDCl_3$): δ = 12.08 (s, 1H, O*H*), 10.47 (s, 1H, C*H*O), 7.04 (s, 1H, C*H*$_{Ar}$), 3.82 – 3.72 (m, 4H, OC*H*$_3$, C*H*(CH_3)$_2$), 2.23 (s, 3H, C*H*$_3$), 1.40 (d, *J* = 7.1 Hz, 6H, CH(C*H*$_3$)$_2$) ppm. – **^{13}C NMR** (100 MHz, $CDCl_3$): δ = 196.0 (+, *C*HO), 156.3 ($C_{quart.}$, *C*$_{Ar}$OH), 149.9 ($C_{quart.}$, *C*$_{Ar}$OCH$_3$), 137.1 ($C_{quart.}$, *C*$_{Ar}$), 125.1 (+, C_{Ar}H), 125.0 ($C_{quart.}$, *C*$_{Ar}$), 117.7 ($C_{quart.}$, *C*$_{Ar}$), 57.2 (+, O*C*H$_3$), 25.6 (+, *C*(CH_3)$_2$), 22.6 (+, 2 × *C*H$_3$), 15.5 (+, *C*H$_3$) ppm.

2-Hydroxy-3-isopropyl-6-methylbenzaldehyde (29l)

According to **GP2**, 5.00 g of thymol (**89d**) (33.3 mmol, 1.00 equiv.) was dissolved in 165 mL of dry acetonitrile, 6.75 g of paraformaldehyde (225 mmol, 6.75 equiv.) and 4.75 g of magnesium chloride (49.9 mmol, 1.50 equiv.) were added under argon counterflow. Then 17.4 mL of dry triethylamine (12.9 g, 124.8 mmol, 3.75 equiv.) were added and the mixture was heated at reflux for 4 h. The reaction mixture was then allowed to cool to room temperature, quenched with 75 mL of water and concentrated hydrochloric acid was added until the pH was adjusted to 1. The mixture was extracted with 3 × 75 mL of diethyl ether and dried over sodium sulfate. The volatiles were removed under reduced pressure and the residue was purified twice via flash column chromatography CH/EtOAc 5:1 and 10:1 resulted in 1.14 g of a yellow oil (19%).

***R*f** (CH/EtOAc 20:1): 0.35. – **^{1}H NMR** (400 MHz, $CDCl_3$): δ = 12.29 (s, 1H, O*H*), 10.31 (s, 1H, C*H*O), 7.32 (d, *J* = 7.7 Hz, 1H, *H*$_{Ar}$), 6.68 (d, *J* = 7.6 Hz, 1H, *H*$_{Ar}$), 3.32 (hept, *J* = 6.9 Hz, 1H, C*H(*CH_3)$_2$), 2.57 (s, 3H, CH_3), 1.22 (d, *J* = 6.9 Hz, 6H, CH(C*H*$_3$)$_2$) ppm.– **^{13}C NMR** (100 MHz, $CDCl_3$): δ = 195.8 (+, *C*HO), 161.0 ($C_{quart.}$, *C*$_{Ar}$OH), 139.4 ($C_{quart.}$, *C*$_{Ar}$), 135.5 $C_{quart.}$, *C*$_{Ar}$), 134.1 (+, C_{Ar}H), 121.5 (+, C_{Ar}H), 118.1 ($C_{quart.}$, *C*$_{Ar}$), 26.1 (+, *C*(CH_3)$_2$), 22.4 (+, 2 × *C*H$_3$), 18.1 (+, *C*H$_3$) ppm. – **IR** (KBr): ṽ = 2959 (m), 2872 (w), 1633 (s), 1422 (m), 1380 (w), 1324 (m), 1299 (m), 1271 (m), 1246 (m), 1233 (m), 1149 (m), 1095 (w), 1066 (w), 1048 (w), 964 (w), 865 (w), 819 (w), 772 (m), 733 (m), 683 (w), 594 (m), 547 (m), 489 (w) cm^{-1}. – **MS** (70 eV, EI): *m/z* (%) = 178 (78) $[M]^+$, 169 (6), 163 (100) $[M - CH_3]^+$, 162 (32), 69 (24). – **HRMS**

($C_{11}H_{14}O_2$): calc. 178.0988, found 178.0986. – **Elemental analysis**: $C_{11}H_{14}O_2$ (178.10): calc. C 74.13, H 7.92, found C 74.35 H 7.65.

4-Bromo-2-hydroxy-6-isopropyl-3-methylbenzaldehyde (29m)[177]

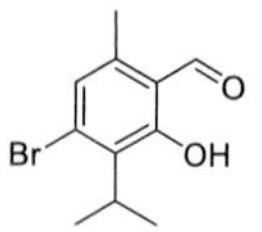

According to **GP2**, 8.36 g of 4-bromothymol (**89e**, 21.8 mmol, 1.00 equiv.) was dissolved in 140 mL of dry acetonitrile, 7.40 g of paraformaldehyde (246 mmol, 6.75 equiv.) and 5.21 g of magnesium chloride (54.7 mmol, 1.50 equiv.) were added under argon counterflow. Then 19.1 mL of dry triethylamine (137 mmol, 3.75 equiv.) were added and the mixture was heated at reflux for 5 h. The reaction mixture was then allowed to cool to room temperature, quenched with 60 mL of water and concentrated hydrochloric acid added until the pH was adjusted to 1. The mixture was extracted with 3 × 100 mL of diethyl ether and dried over sodium sulfate. The volatiles were removed under reduced pressure and the residue was purified via flash column chromatography (CH/EtOAc 100:1) resulted in a yellow solid 601 mg that contains 13% dibromothymol originating from the starting material (measured by NMR), the calculated product yield was 520 mg (6%). The mixture was used in subsequent reactions without further purification.

***R*f** (CH/EtOAc 20:1): 0.72. – **^{1}H NMR** (300 MHz, $CDCl_3$): δ = 12.47 (s, 1H, CO*H*), 10.34 (s, 1H, C*H*O), 7.55 (s, 1H, CH_{Ar}), 3.30 (hept, J = 6.7 Hz, 1H, C*H*(CH_3)$_2$), 2.65 (s, 3H, CH_3), 1.21 (d, J = 6.9 Hz, 6H, CH(CH_3)$_2$) ppm. – **MS** (EI, 70 eV): m/z (%) = 256/258 (70/70) [M^+/M^+ +2], 243/241 (100/99) [$M^+ - CH_3$]. – **HRMS** ($C_{11}H_{13}Br O_2$): calc. 256.0093, found 256.0093.

5.2.4 Synthesis and Characterization 3-Benzylcoumarins (Chapter 3.2.2)

3-Benzyl-5-methoxy-7-pentyl-2*H*-chromen-2-one (26d)

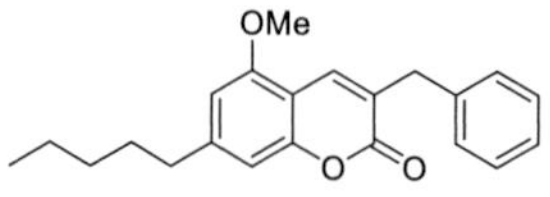

According to GP5, 200 mg of salicylic aldehyde **29a** (0.900 mmol, 1.00 equiv.), 274 mg of potassium carbonate (1.98 mmol, 2.20 equiv.), 297 mg of cinnamic aldehyde (2.25 mmol, 2.50 equiv.) and 240 mg of 1,3-dimethylimidazolium dimethyl phosphate (**31**) (1.08 mmol, 1.20 equiv.) were suspended in 3 mL of toluene. The reaction mixture was heated at 110 °C for 50 min via microwave irradiation. After cooling to room temperature, 5 mL of water was added and the mixture was extracted with 3 ×15 mL of ethyl acetate. Removal of the volatiles under reduced pressure and purification via flash column chromatography (CH/EtOAc 20:1) resulted in 71.0 mg (23%) of the pure product as a white solid. Analytical data are consistent with the literature.[111]

***R*f** (CH/EtOAc 5:1): 0.66. – **^{1}H NMR** (400 MHz, $CDCl_3$): δ = 7.65 (s, 1H, 4-C*H*), 7.27 – 7.22 (m, 3H, 3 × H_{Ar}), 7.18 – 7.13 (m, 2H, 2 × H_{Ar}), 6.65 (s, 1H, H_{Ar}), 6.42 (s, 1H, H_{Ar}), 3.79 (s, 2H, CH_2), 3.79 (s, 3H, OCH_3), 2.55 (t, *J* = 7.7 Hz, 2H, CH_2), 1.58 – 1.51 (m, 2H, CH_2), 1.27 – 1.22 (m, 2H, 2 × CH_2), 0.82 (t, *J* = 6.9 Hz, 3H, CH_3) ppm.

5-Methoxy-3-(2-methoxybenzyl)-7-pentyl-2*H*-chromen-2-one (26e)

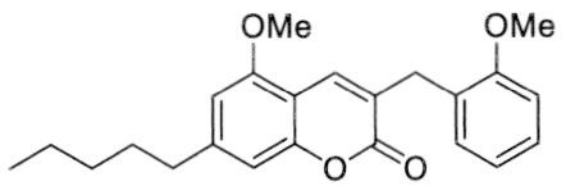

According to GP5, 150 mg of salicylic aldehyde **29a** (0.675 mmol, 1.00 equiv.), 205 mg of potassium carbonate (1.48 mmol, 2.20 equiv.), 274 mg of cinnamic aldehyde (1.68 mmol, 2.50 equiv.) and 180 mg of 1,3-dimethylimidazolium dimethyl phosphate (**31**) (0.810 mmol, 1.20 equiv.) were suspended in 3 mL of toluene. The reaction mixture was heated at 110 °C for 50 min via microwave irradiation. After cooling to room temperature, 5 mL of water was added and the mixture was extracted with 3 ×15 mL of ethyl acetate. Removal of the volatiles under reduced pressure and purification via flash column chromatography (CH/EtOAc 20:1) resulted in 140 mg (57%) of the pure product as a white solid. Analytical data are consistent with the literature.[111]

***R*f** (CH/EtOAc 5:1): 0.35. – **^{1}H NMR** (300 MHz, $CDCl_3$): δ = 7.64 (s, 1H, 4-C*H*), 7.27 – 7.23 (m, 2H, 2 × H_{Ar}), 6.95 – 6.88 (m, 2H, 2 × H_{Ar}), 6.73 (s, 1H, H_{Ar}), 6.47 (s, 1H, H_{Ar}), 3.86 (s, 2H, CH_2), 3.84 (s, 3H, OCH_3), 3.81 (s, 3H, OCH_3), 2.62 (t, *J* = 7.7 Hz, 2H, CH_2), 1.69 – 1.56 (m, 2H, CH_2), 1.42 – 1.20 (m, 2H, 2 × CH_2), 0.89 (t, *J* = 6.8 Hz, 3H, CH_3) ppm.

5-Methoxy-3-(2-methylbenzyl)-7-pentyl-2*H*-chromen-2-one (26f)

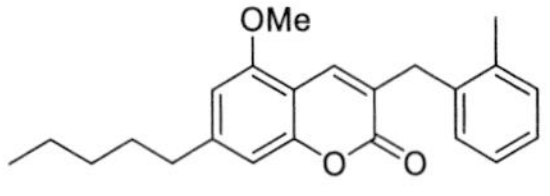

According to GP5, 150 mg of salicylic aldehyde **29a** (0.675 mmol, 1.00 equiv.), 205 mg of potassium carbonate (1.48 mmol, 2.20 equiv.), 247 mg of cinnamic aldehyde (1.68 mmol, 2.50 equiv.) and 180 mg of 1,3-dimethylimidazolium dimethyl phosphate (**31**)

(0.810 mmol, 1.20 equiv.) were suspended in 4 mL of toluene. The reaction mixture was heated at 110 °C for 50 min via microwave irradiation. After cooling to room temperature, 5 mL of water was added and the mixture was extracted with 3 ×15 mL of ethyl acetate. Removal of the volatiles under reduced pressure and purification via flash column chromatography (CH/EtOAc 40:1) resulted in 159 mg (67%) of the pure product as a white solid. Analytical data are consistent with the literature.[111]

$\boldsymbol{R_f}$ (CH/EtOAc 20:1): 0.25. – **^{1}H NMR** (300 MHz, $CDCl_3$): δ = 7.43 (s, 1H, 4-C*H*), 7.22 – 7.19 (m, 4H, 4 × H_{Ar}), 6.74 (s, 1H, H_{Ar}), 6.47 (s, 1H, H_{Ar}), 3.86 (s, 2H, CH_2), 3.81 (s, 3H, OCH_3), 2.66 (t, *J* = 7.7 Hz, 2H, CH_2), 2.28 (s, 3H, CH_3), 1.67 – 1.60 (m, 2H, CH_2), 1.37 – 1.28 (m, 4H, 2 × CH_2), 0.89 (t, *J* = 6.9 Hz, 3H, CH_3) ppm.

3-(4-(Dimethylamino)benzyl)-5-methoxy-7-pentyl-2*H*-chromen-2-one (26g)

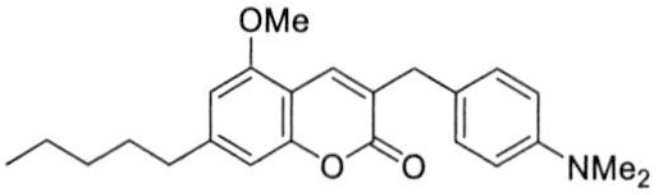

According to GP5, 150 mg of salicylic aldehyde **29a** (670 µmol, 1.00 equiv.), 112 mg of potassium carbonate (810 µmol, 1.20 equiv.), 295 mg of cinnamic aldehyde (1.69 mmol, 2.50 equiv.) and 180 mg of 1,3-dimethylimidazolium dimethyl phosphate (**31**) (810 µmol, 1.20 equiv.) were suspended in 4 mL of toluene. The reaction mixture was heated at 110 °C for 50 min via microwave irradiation. After cooling to room temperature, 10 mL of water was added and the mixture was extracted with 3 × 15 mL of ethyl acetate. Removal of the volatiles under reduced pressure and purification via flash column chromatography (CH/EtOAc 5:1) resulted in 118 mg (46%) of the pure product as a red oil. Analytical data are consistent with the literature.[114]

$\boldsymbol{R_f}$ (CH/EtOAc 5:1): 0.30. – **^{1}H NMR** (400 MHz, $CDCl_3$): δ = 7.71 (s, 1H, 4-C*H*), 7.18 (d, *J* = 8.7 Hz, 2H, 2 × CH_{Ar}), 6.75 – 6.70 (m, 3H, 3 × CH_{Ar}), 6.48 (d, *J* = 1.5 Hz, 1H, CH_{Ar}), 3.86 (s, 3H, OCH_3), 3.77 (d, *J* = 1.2 Hz, 2H, BnCH_2), 2.93 (s, 6H, N(CH_3)$_2$), 2.67 – 2.59 (m, 2H, CH_2), 1.68 – 1.56 (m, 2H, CH_2), 1.38 – 1.26 (m, 4H, 2 × CH_2), 0.94 – 0.85 (m, 3H, CH_3) ppm.

3-(4-(Dimethylamino)benzyl)-5-hydroxy-7-pentyl-2*H*-chromen-2-one (26k)

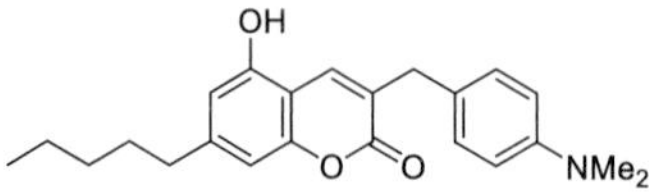

According to **GP8**, 155 mg of 5-methoxycoumarin **26g** (407 µmol, 1.00 equiv.) was dissolved in 3 mL of dry dichloromethane. The solution was cooled to

–78 °C and 1.83 mL of boron tribromide (1 M in dichloromethane, 1.83 mmol, 4.50 equiv.) was added dropwise. The mixture was stirred for 30 min at this temperature and then allowed to warm to room temperature. The reaction was quenched after 16 h at 0 °C by addition of sodium bicarbonate. The aqueous layer was extracted with 3 × 15 mL of dichloromethane and the combined organic layers were washed with brine, dried over sodium sulfate and the volatiles were removed under reduced pressure. The crude product was then purified via flash column chromatography (CH/EtOAc 5:1 to 2:1) to give the product as 94.0 mg (63%) of a red solid.

$\boldsymbol{R_f}$ (CH/EtOAc 5:1): 0.11. – **^{1}H NMR** (400 MHz, $CDCl_3$): δ = 7.77 (s, 1H, 4-C*H*), 7.18 (d, J = 8.6 Hz, 2H, 2 × H_{Ar}), 6.73 (d, J = 8.7 Hz, 2H, 2 × H_{Ar}), 6.66 (d, J = 1.3 Hz, 1H, H_{Ar}), 6.48 (d, J = 1.3 Hz, 1H, H_{Ar}), 3.78 (d, J = 1.1 Hz, 2H, BnCH_2), 2.91 (s, 6H, N(CH_3)$_2$), 2.57 – 2.49 (m, 2H, CH_2), 1.62 – 1.50 (m, 2H, CH_2), 1.36 – 1.20 (m, 4H, 2 × CH_2), 0.86 (t, J = 6.9 Hz, 3H, CH_3) ppm. – **^{13}C NMR** (100 MHz, $CDCl_3$): δ = 163.0 ($C_{quart.}$, *C*OO), 154.3 ($C_{quart.}$, C_{Ar}), 152.5 ($C_{quart.}$, C_{Ar}), 149.6 ($C_{quart.}$, C_{Ar}), 147.6 ($C_{quart.}$, C_{Ar}), 134.9 (+, C_{Ar}H), 130.0 (+, 2 × C_{Ar}H), 126.6 ($C_{quart.}$, C_{Ar}), 126.5 ($C_{quart.}$, C_{Ar}), 113.5 (+, 2 × C_{Ar}H), 110.5 (+, C_{Ar}H), 108.3 (+, C_{Ar}H), 107.5 ($C_{quart.}$, C_{Ar}), 41.1 (+, N(*C*H_3)$_2$), 36.2 (–, CH_2), 35.9 (–, CH_2), 31.4 (–, CH_2), 30.7 (–, CH_2), 22.6 (–, *C*H_2), 14.1 (+, *C*H_3) ppm. – **IR** (KBr): $\tilde{\nu}$ = 3291 (m), 2925 (m), 1676 (s), 1624 (s), 1581 (m), 1519 (m), 1438 (m), 1342 (m), 1267 (m), 1181 (m), 1134 (m), 1082 (m), 1059 (m), 946 (w), 897 (w), 874 (m), 819 (m), 807 (m), 726 (m), 670 (m), 560 (w), 525 (m), 455 (w), 426 (w) cm^{-1}. – **MS** (70 eV, EI): *m/z* (%) = 365 (100) $[M]^+$, 364 (14). – **HRMS** ($C_{23}H_{27}O_3N$): calc. 365.1985, found 365.1983.

7-(1-Butylcyclopentyl)-5-methoxy-3-(2-methoxybenzyl)-2*H*-chromen-2-one (26h)

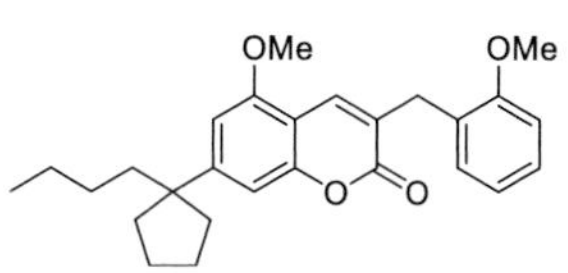

According to GP5, 200 mg of salicylic aldehyde **29b** (0.724 mmol, 1.00 equiv.), 220 mg of potassium carbonate (1.59 mmol, 2.20 equiv.), 294 mg of cinnamic aldehyde (1.81 mmol, 2.50 equiv.) and 193 mg of 1,3-dimethylimidazolium dimethyl phosphate (**31**) (868 µmol, 1.20 equiv.) were suspended in 2.5 mL of toluene. The reaction mixture was heated at 110 °C for 50 min via microwave irradiation. After cooling to room temperature, 10 mL of water was added and the mixture was extracted with 3 × 15 mL of ethyl acetate. Removal of the volatiles under reduced pressure and purification via flash column chromatography (CH/EtOAc 40:1) resulted in 155 mg (51%) of the pure product as a yellow oil. Analytical data are consistent with the literature.[111]

***R*f** (CH/EtOAc 5:1): 0.35. – **^{1}H NMR** (300 MHz, $CDCl_3$): δ = 7.65 (s, 1H, 4-C*H*), 7.27 – 7.22 (m, 2H, 2 × H_{Ar}), 6.95 – 6.82 (m, 3H, 3 × H_{Ar}), 6.57 (d, *J* = 1.3 Hz, 1H, H_{Ar}), 3.87 (s, 2H, CH_2), 3.85 (s, 3H, OCH_3), 3.81 (s, 3H, OCH_3), 1.62 – 1.92 (m, 8H, 4 × CH_2), 1.59 – 1.55 (m, 2H, CH_2), 1.20 – 1.11 (m, 2H, CH_2), 0.99 – 0.89 (m, 2H, CH_2), 0.77 (t, *J* = 7.3 Hz, 3H, CH_3) ppm.

7-(1-Butylcyclopentyl)-5-hydroxy-3-(2-hydroxybenzyl)-2*H*-chromen-2-one (26l)

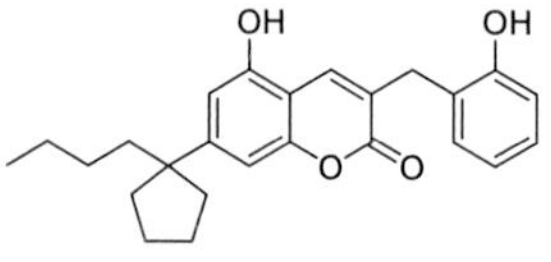

According to **GP8**, 325 mg of 5-methoxycoumarin **26h** (0.770 mmol, 1.00 equiv.) was dissolved in 5 mL of dry dichloromethane. The solution was cooled to –78 °C and 4.27 mL of boron tribromide (1 M in dichloromethane, 4.27 mmol, 5.00 equiv.) was added dropwise. The mixture was stirred for 30 min at this temperature and then allowed to warm to room temperature. The reaction was quenched after 16 h at 0 °C by addition of sodium bicarbonate. The aqueous layer was extracted with 3 × 15 mL of dichloromethane and the combined organic layers were washed with brine, dried over sodium sulfate and the volatiles were removed under reduced pressure. The crude product was then purified via flash column chromatography (CH/EtOAc 5:1) to give the product as 239 mg (79%) of an off-white solid. Analytical data of the product are consistent with the literature.[111]

***R*f** (CH/EtOAc 5:1): 0.18. – **^{1}H NMR** (300 MHz, $CDCl_3$): δ = 8.15 (s, 1H, 4-C*H*), 8.09 (s, 1H, O*H*), 7.26 – 7.20 (m, 1H, H_{Ar}), 7.20 – 7.08 (m, 1H, H_{Ar}), 7.00 – 6.90 (m, 1H, H_{Ar}), 6.90 – 6.82 (m, 1H, H_{Ar}), 6.81 (d, J = 1.4 Hz, 1H, H_{Ar}), 6.61 (d, J = 1.5 Hz, 1H, H_{Ar}), 5.96 (s, 1H, O*H*), 3.85 (s, 2H, CH_2), 1.91 – 1.47 (m, 10H, 5 × CH_2), 1.12 (p, J = 7.3, 6.8 Hz, 2H, CH_2), 1.20 – 1.07 (m, 2H, CH_2), 0.76 (t, J = 7.3 Hz, 3H, CH_3).ppm.

7-(1-Butylcyclopentyl)-3-(4-(dimethylamino)benzyl)-5-methoxy-2*H*-chromen-2-one (26i)

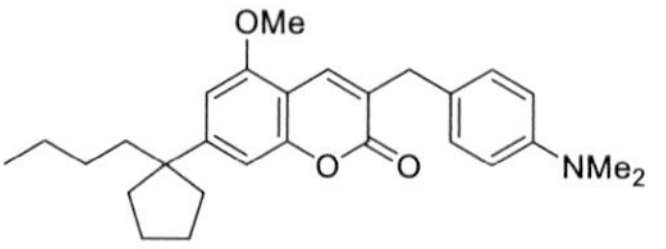

According to GP5, 150 mg of salicylic aldehyde **29b** (0.54 mmol, 1.00 equiv.), 90 mg of potassium carbonate (0.65 mmol, 1.20 equiv.), 238 mg of cinnamic aldehyde (1.36 mmol, 2.50 equiv.) and 145 mg of 1,3-dimethylimidazolium dimethyl phosphate (**31**) (0.65 mmol, 1.20 equiv.) were suspended in 4 mL of toluene. The reaction mixture was heated at 110 °C for 50 min via microwave irradiation. After cooling to room temperature, 10 mL of water was added and the mixture was extracted with 3 × 15 mL of ethyl acetate. Removal of the volatiles under reduced pressure and purification via flash column chromatography (CH/EtOAc 5:1) resulted in 101 mg (43%) of the pure product as a red oil.

R**f** (CH/EtOAc 5:1): 0.40. – **^{1}H NMR** (400 MHz, $CDCl_3$): δ = 7.71 (s, 1H, 4-C*H*), 7.21 – 7.15 (m, 2H, 2 × H_{Ar}), 6.81 (d, *J* = 1.4 Hz, 1H, H_{Ar}), 6.73 (d, *J* = 8.1 Hz, 2H, 2 × H_{Ar}), 6.58 (d, *J* = 1.5 Hz, 1H, H_{Ar}), 3.87 (s, 3H, OCH_3), 3.78 (d, *J* = 1.3 Hz, 2H, $BnCH_2$), 2.93 (s, 6H, $N(CH_3)_2$), 1.94 – 1.54 (m, 10H, 5 × CH_2), 1.21 – 1.08 (m, 2H, CH_2), 0.99 – 0.86 (m, 2H, CH_2), 0.77 (t, *J* = 7.3 Hz, 3H, CH_3) ppm. – **^{13}C NMR** (100 MHz, $CDCl_3$): δ = 162.4 ($C_{quart.}$, C_{Ar}), 155.2 ($C_{quart.}$, C_{Ar}), 154.0 ($C_{quart.}$, C_{Ar}), 153.9 ($C_{quart.}$, C_{Ar}), 149.6 ($C_{quart.}$, C_{Ar}), 134.2 (+, $C_{Ar}H$), 130.0 (+, 2 × $C_{Ar}H$), 127.1 ($C_{quart.}$, C_{Ar}), 126.5 ($C_{quart.}$, C_{Ar}), 113.2 (+, 2 × $C_{Ar}H$), 108.0 ($C_{quart.}$, C_{Ar}), 107.6 (+, $C_{Ar}H$), 104.0 (+, $C_{Ar}H$), 55.9 (+, OCH_3), 52.0 ($C_{quart.}$, C_{CP}), 41.7 (–, CH_2), 41.0 (+, 2 × CH_3, $N(CH_3)_2$), 37.8 (–, 2 × CH_2), 35.9 (–, CH_2), 27.6 (–, CH_2), 23.4 (–, CH_2), 23.3 (–, 2 × CH_2), 14.1 (+, CH_3) ppm. – **IR** (KBr): ṽ = 2926 (m), 2868 (w), 1713 (s), 1611 (s), 1567 (m), 1519 (m), 1492 (m), 1456 (m), 1415 (m), 1347 (m), 1247 (m), 1161 (m), 1109 (s), 1043 (m), 946 (w), 840 (m), 803 (m), 783 (w), 726 (w), 681 (w), 619 (vw), 576 (w), 551 (w) cm^{-1}. – **MS** (70 eV, EI): *m/z* (%) = 443 (24) $[M]^+$, 69 (100). – **HRMS** ($C_{28}H_{35}O_3N$): calc. 443.2611, found 443.2610.

7-(1-Butylcyclopentyl)-3-(4-(dimethylamino)benzyl)-5-hydroxy-2*H*-chromen-2-one (26m)

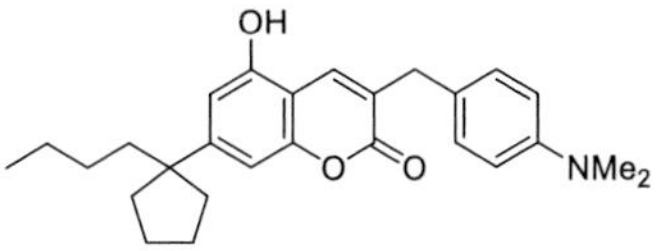

According to **GP8**, 124 mg of 5-methoxycoumarin **26i** (285 µmol, 1.00 equiv.) was dissolved in 3 mL of dry dichloromethane. The solution was cooled to –78 °C and 1.28 mL of boron tribromide (1 M in dichloromethane, 1.28 mmol, 4.50 equiv.) was added dropwise. The mixture was stirred for 30 min at this temperature and then allowed to warm to room temperature. The reaction was quenched after 16 h at 0 °C by addition of sodium bicarbonate. The aqueous layer was extracted with 3 × 15 mL of dichloromethane and the combined organic layers were washed with brine, dried over sodium sulfate and the volatiles were removed under reduced pressure. The crude product was then twice purified via flash column chromatography (CH/EtOAc 5:1; 2:1) to give the product as 58.4 mg (49%) of a red solid.

***R*f** (CH/EtOAc 5:1): 0.15. – **^{1}H NMR** (400 MHz, $CDCl_3$): δ = 7.70 (s, 1H, 4-C*H*), 7.18 (d, *J* = 8.6 Hz, 2H, 2 × H_{Ar}), 6.79 – 6.76 (m, 1H, H_{Ar}), 6.73 (d, *J* = 8.3 Hz, 2H, 2 × H_{Ar}), 6.55 (d, *J* = 1.6 Hz, 1H, H_{Ar}), 5.81 (s, 1H, O*H*), 3.78 (d, *J* = 1.1 Hz, 2H, BnCH_2), 2.92 (s, 6H, $N(CH_3)_2$), 1.90 – 1.45 (m, 10H, 5 × CH_2), 1.19 – 1.06 (m, 2H, CH_2), 0.97 – 0.85 (m, 2H, CH_2), 0.76 (t, *J* = 7.3 Hz, 3H, CH_3) ppm. – **^{13}C NMR** (100 MHz, $CDCl_3$): δ = 162.6 ($C_{quart.}$, *C*OO), 154.1 ($C_{quart.}$, 2 × C_{Ar}), 151.7 ($C_{quart.}$, C_{Ar}), 149.5 ($C_{quart.}$, C_{Ar}), 134.1 (+, C_{Ar}H), 130.1 (+, 2 × C_{Ar}H, $C_{quart.}$, C_{Ar}), 127.1 ($C_{quart.}$, C_{Ar}), 113.5 (+, 2 × C_{Ar}H), 109.0 (+, C_{Ar}H), 107.6 (+, C_{Ar}H), 107.1 ($C_{quart.}$, C_{Ar}), 51.7 ($C_{quart.}$, C_{CP}), 41.7 (–, CH_2), 41.1 (+, $N(CH_3)_2$), 37.8 (–, 2 × CH_2), 35.9 (–, CH_2), 27.6 (–, CH_2), 23.4 (–, CH_2), 23.3 (–, 2 × CH_2), 14.1 (+, CH_3) ppm. – **IR** (KBr): $\tilde{\nu}$ = 3327 (vw), 2923 (w), 1673 (w), 1617 (m), 1572 (w), 1521 (w), 1422 (w), 1349 (w), 1280 (w), 1230 (w), 1185 (w), 1164 (w), 1083 (w), 947 (vw), 876 (vw), 847 (w), 800 (w), 784 (w), 749 (vw), 723 (vw), 668 (w), 576 (vw), 545 (vw), 422 (vw) cm^{-1}. – **MS** (70 eV, EI): *m/z* (%) = 419 (72) $[M]^+$, 69 (100). – **HRMS** ($C_{27}H_{33}O_3N$): calc. 419.2460, found 419.2462.

7-(1-Butylcyclohexyl)-3-(4-fluorobenzyl)-5-methoxy-2*H*-chromen-2-one (26j)

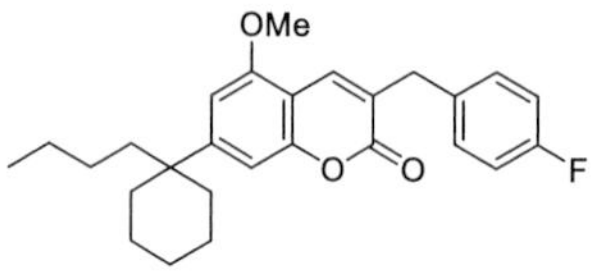

According to GP5, 300 mg of salicylic aldehyde **29c** (1.03 mmol, 1.00 equiv.), 314 mg of potassium carbonate (2.27 mmol, 2.20 equiv.), 388 mg of cinnamic aldehyde (2.58 mmol, 2.50 equiv.) and 275 mg of 1,3-dimethylimidazolium dimethyl phosphate (**31**) (1.24 mmol, 1.20 equiv.) were suspended in 4 mL of toluene. The reaction mixture was heated at 110 °C for 50 min via microwave irradiation. After cooling to room temperature, 10 mL of water was added and the mixture was extracted with 3 × 15 mL of ethyl acetate. Removal of the volatiles under reduced pressure and purification via flash column chromatography (CH/EtOAc 20:1) resulted in 108 mg (25%) of the pure product as an orange oil. Analytical data are consistent with the literature.[114]

***R*f** (CH/EtOAc 20:1): 0.25. – **^{1}H NMR** (300 MHz, $CDCl_3$): δ = 7.65 (s, 1H, 4-C*H*), 7.27 – 7.22 (m, 2H, 2 × H_{Ar}), 6.95 – 6.82 (m, 3H, 3 × H_{Ar}), 6.57 (d, *J* = 1.3 Hz, 1H, H_{Ar}), 3.87 (s, 2H, BnCH_2), 3.85 (s, 3H, OCH_3), 3.81 (s, 3H, OCH_3), 1.62 – 1.92 (m, 10H, 5 × CH_2), 1.59 – 1.55 (m, 2H, CH_2), 1.20 – 1.11 (m, 2H, CH_2), 0.99 – 0.89 (m, 2H, CH_2), 0.77 (t, *J* = 7.3 Hz, 3H, CH_3) ppm.

5-Isopropyl-3-(2-methoxybenzyl)-8-methyl-2*H*-chromen-2-one (26n)

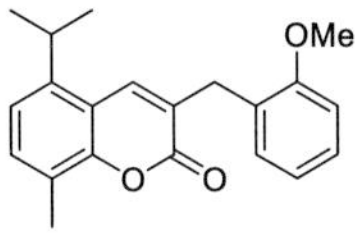

Variant A: According to GP5, 130 mg of salicylic aldehyde **29i** (0.729 mmol, 1.00 equiv.), 121 mg of potassium carbonate (0.875 mmol, 1.20 equiv.), 308 mg of cinnamic aldehyde **30a** (1.82 mmol, 2.50 equiv.) and 195 mg of 1,3-dimethylimidazolium dimethyl phosphate (**31**) (0.875 mmol, 1.20 equiv.) were suspended in 2.50 mL of toluene. The reaction mixture was heated at 110 °C for 50 min via microwave irradiation. After cooling to room temperature, 5 mL of water was added and the mixture was extracted with 3 × 15 mL of ethyl acetate. Removal of the volatiles under reduced pressure and purification via flash column chromatography (CH/EtOAc 20:1) resulted in a mixture of the product and a methyl ester impurity as a solid. The mixture was stirred at room temperature for 30 min in 80 mL of 1 M sodium hydroxide-solution resulting in a dispersion. The remaining solid was dissolved in ethyl acetate and the organic layer was separated and washed again with 40 mL of 1 M sodium hydroxide-solution and then dried over sodium sulfate. Removal of the volatiles gave 130 mg (55%) of the pure product as a pale yellow solid.[111]

Variant B: In a 500 mL of flask with reflux condenser, 9.80 g of salicylic aldehyde **29i** (54.9 mmol, 1.00 equiv.), 9.12 g of potassium carbonate (65.9 mmol, 1.20 equiv.), 23.2 g of cinnamic aldehyde **30a** (138 mmol, 2.50 equiv.) and 14.6 g of 1,3-dimethylimidazolium dimethyl phosphate (**31**) (65.9 mmol, 1.20 equiv.) were suspended in 180 mL of toluene and the atmosphere replaced with argon. The reaction mixture was heated at reflux for 24 h in an oil bath. After cooling to room temperature, 100 mL of water was added and the mixture was extracted with 3 × 100 mL of ethyl acetate (or until no further product was noticeable via TLC), and the organic layer was washed with 3 × 100 mL of 1 M sodium hydroxide-solution. After removal of the volatiles a short filter column (CH/EtOAc 2:1) and a second flash column (CH/EtOAc 80:1 to 40:1) gave 11.8 g of (62%) of the product as a pale-yellow powder. (Purity 98%, an aliquot of 300 mg of was further purified by washing several times with 1 M sodium hydroxide-solution, the rest was used directly without further purification). Analytical data of the product are consistent with the literature.[111]

***R*f** (CH/EtOAc 10:1): 0.46. – **^{1}H NMR** (300 MHz, $CDCl_3$): δ = 7.64 (s, 1H, 4-C*H*), 7.33 – 7.23 (m, 3H, 3 × H_{Ar}), 7.04 – 7.02 (m, 1H, H_{Ar}), 6.97 – 6.89 (m, 2H, 2 × H_{Ar}), 3.91 (s, 2H, BnCH_2), 3.81 (s, 3H, OCH_3), 3.17 (hept, *J* = 6.9 Hz, 1H, C*H*(CH_3)$_2$), 2.41 (s, 3H, CH_3), 1.21 (d, *J* = 6.8 Hz, 6H, CH(CH_3)$_2$) ppm.

3-(2-Hydroxybenzyl)-5-isopropyl-8-methyl-2*H*-chromen-2-one (26c)

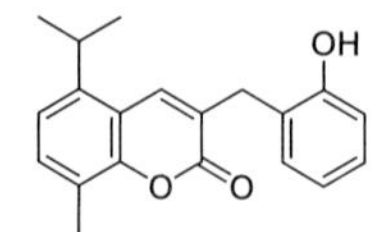

According to **GP8**, 11 g of 5-methoxycoumarin **26n** (34.1 mmol, 1.00 equiv.) was dissolved in 200 mL of dry dichloromethane. The solution was cooled to –78 °C and 154 mL of boron tribromide (1 M in dichloromethane, 154 mmol, 5.00 equiv.) was added dropwise. The mixture was stirred for 30 min at this temperature and then allowed to warm to room temperature. The reaction was quenched after 16 h at 0 °C by addition of sodium bicarbonate. The aqueous layer was extracted with 3 × 150 mL of dichloromethane and the combined organic layers were washed with brine, dried over sodium sulfate and the volatiles were removed under reduced pressure. The crude product was then purified via flash column chromatography (CH/EtOAc 15:1) to give the product as 10.0 g (96%) of an off-white solid. Analytical data of the product are consistent with the literature.[111]

***R*f** (CH/EtOAc 10:1): 0.28. – **^{1}H NMR** (400 MHz, $CDCl_3$): δ = 8.11 (s, 1H, 4-C*H*), 8.08 – 8.03 (m, 1H, O*H*), 7.31 (d, *J* = 7.9 Hz, 1H, H_{Ar}), 7.23 – 7.08 (m, 3H, 3 × H_{Ar}), 6.97 (dd, *J* = 8.1, 1.3 Hz, 1H, H_{Ar}), 6.88 (td, *J* = 7.4, 1.3 Hz, 1H, H_{Ar}), 3.89 (s, 2H, CH_2), 3.41 (hept, *J* = 6.7 Hz, 1H, C*H*(CH_3)$_2$), 2.40 (d, *J* = 0.8 Hz, 3H, CH_3), 1.32 (d, *J* = 6.9 Hz, 6H, CH(CH_3)$_2$) ppm.

3-Benzyl-8-isopropyl-5-methyl-2*H*-chromen-2-one (26o)

According to GP5, 150 mg of salicylic aldehyde **29l** (0.842 mmol, 1.00 equiv.), 140 mg of potassium carbonate (1.01 mmol, 1.20 equiv.), 281 mg of cinnamic aldehyde (2.01 mmol, 2.50 equiv.) and 224 mg of 1,3-dimethylimidazolium dimethyl phosphate (**31**) (1.01 mmol, 1.20 equiv.) were suspended in 2.50 mL of toluene. The reaction mixture was heated at 110 °C for 50 min via microwave irradiation. After cooling to room temperature, 5 mL of water was added and the mixture was extracted with 3 × 15 mL of ethyl acetate. Removal of the volatiles under reduced pressure and purification via flash column chromatography (CH/EtOAc 20:1) resulted in a mixture of the product and a methyl ester impurity as a solid. The mixture was stirred at room temperature for 30 min in 80 mL of 1 M sodium hydroxide-solution resulting in a dispersion. The remaining solid was dissolved in ethyl acetate and the organic layer was separated and washed again with 40 mL of 1 M sodium hydroxide-solution and then dried over sodium sulfate. Removal of the volatiles gave 100 mg (41%) of the pure product as a white solid.

***R*f** (CH/EtOAc 20:1): 0.45. – **MP**: 146.4 °C – **^{1}H NMR** (400 MHz, $CDCl_3$): δ = 7.54 (t, *J* = 1.2 Hz, 1H, 4-C*H*), 7.38 – 7.24 (m, 6H, 6 × H_{Ar}), 7.02 (d, *J* = 7.9 Hz, 1H, H_{Ar}), 3.92 (s, 2H, BnCH_2), 3.59 (hept, *J* = 6.9 Hz, 1H, C*H*$(CH_3)_2$), 2.37 (s, 3H; CH_3), 1.27 (d, *J* = 6.9 Hz, 6H, $CH(CH_3)_2$) ppm. – **^{13}C NMR** (100 MHz, $CDCl_3$): δ = 161.8 ($C_{quart.}$, C_{Ar}), 151.0 ($C_{quart.}$, C_{Ar}), 138.2 ($C_{quart.}$, C_{Ar}), 137.0 (+, C_{Ar}H), 134.2 ($C_{quart.}$, C_{Ar}), 132.8 ($C_{quart.}$, C_{Ar}), 129.3 (+, 2 × C_{Ar}H), 128.8 (+, 2 × C_{Ar}H), 128.1 ($C_{quart.}$, C_{Ar}), 127.7 (+, C_{Ar}H), 126.8 (+, C_{Ar}H), 125.5 (+, C_{Ar}H), 118.0 ($C_{quart.}$, C_{Ar}), 36.9 (–, CH_2), 26.4 (+, *C*H), 22.9 (+, 2 × CH_3, $CH(CH_3)_2$), 18.2 (+, CH_3) ppm. – **IR** (KBr): $\tilde{\nu}$ = 3026 (vw), 2959 (w), 1699 (m), 1636 (w), 1598 (w), 1489 (w), 1453 (w), 1304 (w), 1258 (w), 1207 (w), 1170 (m), 1068 (m), 1033 (m), 997 (w), 918 (w), 892 (w), 846 (w), 828 (w), 771 (w), 754 (w), 722 (w), 698 (m), 664 (w), 623 (w), 610 (w), 568 (w), 498 (w) cm^{-1}. – **MS** (70 eV, EI): *m/z* (%) = 292 (100) $[M]^+$, 278 (16), 277 (70) $[M - CH_3]^+$, 171 (13), 128 (6), 115 (8), 91 (48). – **HRMS** ($C_{20}H_{20}O_2$): calc. 292.1458, found 292.1457. – **Elemental analysis**: $C_{20}H_{20}O_2$ (292.15): calc. C 82.16, H 6.90, found C 81.82, H 6.98.

8-Isopropyl-3-(2-methoxybenzyl)-5-methyl-2*H*-chromen-2-one (26p)

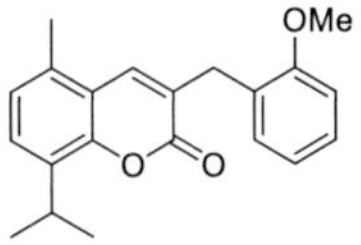

According to GP5, 150 mg of salicylic aldehyde **29l** (0.842 mmol, 1.00 equiv.), 140 mg of potassium carbonate (1.01 mmol, 1.20 equiv.), 356 mg of *o*-methoxy cinnamic aldehyde (2.10 mmol, 2.50 equiv.) and 224 mg of 1,3-dimethylimidazolium dimethyl phosphate (**31**) (1.01 mmol, 1.20 equiv.) were suspended in 2.50 mL of toluene. The reaction mixture was heated at 110 °C for 50 min via microwave irradiation. After cooling to room temperature, 5 mL of water was added and the mixture was extracted with 3 × 15 mL of ethyl acetate. Removal of the volatiles under reduced pressure and purification via flash column chromatography (CH/EtOAc 20:1) resulted in a mixture of the product and a methyl ester impurity as a solid. The mixture was stirred at room temperature for 30 min in 80 mL of 1 M sodium hydroxide-solution resulting in a dispersion. The remaining solid was dissolved in ethyl acetate and the organic layer was separated and washed again with 40 mL of 1 M sodium hydroxide-solution and then dried over sodium sulfate. Removal of the volatiles gave 72.7 mg (28%) of the pure product as a yellow solid.

***R*f** (CH/EtOAc 20:1): 0.36. – **MP**: 95.8 °C – **^{1}H NMR** (400 MHz, $CDCl_3$): δ = 7.53 (t, *J* = 1.3 Hz, 1H, 4-C*H*), 7.34 – 7.25 (m, 2H, 2 × H_{Ar}), 7.25 (d, *J* = 7.7 Hz, 2H, H_{Ar}), 7.00 (d, *J* = 7.8 Hz, 1H, H_{Ar}), 6.95 (td, *J* = 7.4, 1.1 Hz, 1H, H_{Ar}), 6.90 (dd, *J* = 8.2, 1.1 Hz, 1H, H_{Ar}),

3.91 (s, 2H, BnCH_2), 3.82 (s, 3H, OCH_3), 3.60 (hept, J = 6.8 Hz, 1H, $CH(CH_3)_2$), 2.35 (s, 3H, CH_3), 1.27 (d, J = 6.9 Hz, 6H, CH(CH_3)$_2$) ppm. – **^{13}C NMR** (100 MHz, $CDCl_3$): δ = 162.0 ($C_{quart.}$, C_{Ar}), 157.7 ($C_{quart.}$, C_{Ar}), 150.9 ($C_{quart.}$, C_{Ar}), 136.7 (+, C_{Ar}H), 134.1($C_{quart.}$, C_{Ar}), 132.6 ($C_{quart.}$, C_{Ar}), 131.3 (+, C_{Ar}H), 128.3 (+, C_{Ar}H), 127.4 (+, C_{Ar}H), 127.3 ($C_{quart.}$, C_{Ar}), 126.4 ($C_{quart.}$, C_{Ar}), 125.3 (+, C_{Ar}H), 120.8 (+, C_{Ar}H), 118.3 ($C_{quart.}$, C_{Ar}), 110.6 (+, C_{Ar}H), 55.4(+, OCH_3), 31.3 (–, CH_2), 26.4 (+, $CH(CH_3)_2$), 22.9 (+, 2 × CH_3, CH(CH_3)$_2$), 18.2 (+, CH_3) ppm. – **IR** (KBr): $\tilde{\nu}$ = 2957 (w), 1706 (vw), 1597 (w), 1488 (w), 1382 (vw), 1296 (vw), 1241 (m), 1171 (w), 1108 (w), 1069 (w), 1029 (w), 996 (w), 923 (w), 892 (vw), 832 (w), 814 (w), 747 (w), 664 (w), 611 (w), 532 (vw), 476 (w) cm^{-1}. – **MS** (70 eV, EI): *m/z* (%) = 322 (100) $[M]^+$, 307 (10) $[M - CH_3]^+$, 216 (10), 215 (40), 199 (11), 121 (6), 107 (6). – **HRMS** ($C_{21}H_{22}O_3$): calc. 322.1563, found 322.1563.

3-(4-Fluorobenzyl)-8-isopropyl-5-methyl-2*H*-chromen-2-one (26q)

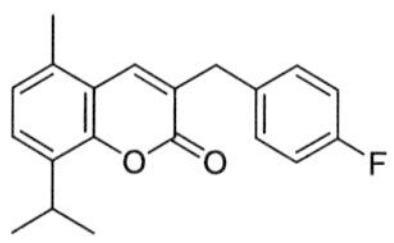

According to GP5, 150 mg of salicylic aldehyde **29l** (0.842 mmol, 1.00 equiv.), 140 mg of potassium carbonate (1.01 mmol, 1.20 equiv.), 319 mg of *p*-fluoro cinnamic aldehyde (2.11 mmol, 2.50 equiv.) and 224 mg of 1,3-dimethylimidazolium dimethyl phosphate (**31**) (1.01 mmol, 1.20 equiv.) were suspended in 2.50 mL of toluene. The reaction mixture was heated at 110 °C for 50 min via microwave irradiation. After cooling to room temperature, 5 mL of water was added and the mixture was extracted with 3 × 15 mL of ethyl acetate. Removal of the volatiles under reduced pressure and purification via flash column chromatography (CH/EtOAc 10:1) resulted in a mixture of the product and a methyl ester impurity as a solid. The mixture was stirred at room temperature for 30 min in 80 mL of 1 M sodium hydroxide-solution resulting in a dispersion. The remaining solid was dissolved in ethyl acetate and the organic layer was separated and washed again with 40 mL of 1 M sodium hydroxide-solution and then dried over sodium sulfate. Removal of the volatiles gave 104 mg (40%) of the pure product as a yellow solid.

***R*f** (CH/EtOAc 10:1): 0.32. – **MP**: 167.8 °C – **^{1}H NMR** (400 MHz, $CDCl_3$): δ =7.53 (t, J = 1.2 Hz, 1H, 4-CH), 7.29 – 7.24 (m, 3H, 3 × H_{Ar}), 7.04 – 6.97 (m, 3H, 3 × H_{Ar}), 3.87 (s, 2H, BnCH_2), 3.57 (hept, J = 6.9 Hz, 1H, $CH(CH_3)_2$), 2.38 (s, 3H, CH_3), 1.25 (d, J = 6.9 Hz, 6H, CH(CH_3)$_2$) ppm. – **^{13}C NMR** (100 MHz, $CDCl_3$): δ = 161.9 (+, d, $C_{quart.}$, CF), 161.7 ($C_{quart.}$, C_{Ar}), 151.0 ($C_{quart.}$, C_{Ar}), 137.0 (+, C_{Ar}H), 134.2 ($C_{quart.}$, C_{Ar}), 133.9 +, (+, d, J = 3.3 Hz, C_{Ar}),

132.8 ($C_{quart.}$, C_{Ar}), 130.8 (d, J = 7.9 Hz, 2 × C_{Ar}H), 127.9 ($C_{quart.}$, C_{Ar}), 127.8 (+, C_{Ar}H), 125.6 (+, C_{Ar}H), 118.0 ($C_{quart.}$, C_{Ar}), 115.6 (+, d, J = 21.4 Hz, 2 × C_{Ar}H) 36.2 (–, CH_2), 26.4 (+, *C*H(CH_3)$_2$), 22.9 (+, 2 × CH_3, CH(CH_3)$_2$), 18.2 (+, CH_3) ppm. – **IR** (KBr): ṽ = 2960 (w), 1700 (m), 1600 (m), 1510 (m),1488 (m),1381 (w), 1305 (vw), 1258 (w), 1226 (m), 1207 (w), 1176 (m), 1156 (m), 1068 (m), 1043 (m), 998 (w), 893 (w), 878 (w), 859 (w), 822 (m), 771 (m), 741 (w), 716 (w), 663 (w), 563 (vw), 535 (w), 489 (m), 450 (w), 420 (w) cm^{-1}. – **MS** (70 eV, EI): *m/z* (%) = 310 (15) $[M]^+$, 295 (18) $[M - CH_3]^+$, 281 (9), 269 (10), 262 (8), 255 (6), 243 (9), 231 (16), 181 (56), 169 (7), 162 (9), 131 (55), 119 (14), 109 (18), 100 (12), 69 (100). – **HRMS** ($C_{20}H_{19}O_2F_1$): calc. 310.1364, found 310.1362.

3-(4-Chlorobenzyl)-8-isopropyl-5-methyl-2*H*-chromen-2-one (26r)

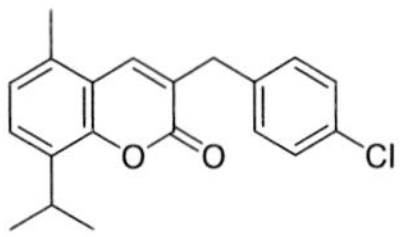

According to **GP5**, 85.0 mg of salicylic aldehyde **29l** (0.477 mmol, 1.00 equiv.), 79.0 mg of potassium carbonate (0.572 mmol, 1.20 equiv.), 201 mg of *p*-chloro cinnamic aldehyde (1.19 mmol, 2.50 equiv.) and 127 mg of 1,3-dimethylimidazolium dimethyl phosphate (**31**) (0.572 mmol, 1.20 equiv.) were suspended in 2.50 mL of toluene. The reaction mixture was heated at 110 °C for 50 min via microwave irradiation. After cooling to room temperature, 5 mL of water was added and the mixture was extracted with 3 × 15 mL of ethyl acetate. Removal of the volatiles under reduced pressure and purification via flash column chromatography (CH/EtOAc 20:1) resulted in a mixture of the product and a methyl ester impurity as a solid. The mixture was stirred at room temperature for 30 min in 80 mL of 1 M sodium hydroxide-solution resulting in a dispersion. The remaining solid was dissolved in ethyl acetate and the organic layer was separated and washed again with 40 mL of 1 M sodium hydroxide-solution and then dried over sodium sulfate. Removal of the volatiles gave 66 mg (24%) of the pure product as a white solid.

***R*f** (CH/EtOAc 10:1): 0.17. – **MP**: 131.6 °C – **^{1}H NMR** (400 MHz, $CDCl_3$): δ = 7.57 (t, J = 1.2 Hz, 1H, 4-C*H*), 7.31 – 7.27 (m, 3H, 3 × H_{Ar}), 7.27 – 7.23 (m, 2H, 2 × H_{Ar}), 7.03 (dd, J = 7.8, 0.9 Hz, 1H, H_{Ar}), 3.88 (s, 2H, BnCH_2), 3.58 (hept, J = 6.9 Hz, 1H, C*H*(CH_3)$_2$), 2.40 (s, 3H, CH_3), 1.27 (d, J = 6.9 Hz, 6H, CH(CH_3)$_2$) ppm. – **^{13}C NMR** (100 MHz, $CDCl_3$): δ = 161.6 ($C_{quart.}$, C_{Ar}), 151.1 ($C_{quart.}$, C_{Ar}), 137.2(+, C_{Ar}H), 136.8 ($C_{quart.}$, C_{Ar}), 134.2 ($C_{quart.}$, C_{Ar}), 132.8 ($C_{quart.}$, C_{Ar}), 132.7 ($C_{quart.}$, C_{Ar}), 130.6(+, 2 × C_{Ar}H), 128.9 (+, 2 × C_{Ar}H), 127.9 (+, C_{Ar}H), 127.6 ($C_{quart.}$, C_{Ar}), 125.6 (+, C_{Ar}H), 117.9 ($C_{quart.}$, C_{Ar}), 36.4 (–, CH_2), 26.4 (+, *C*H(CH_3)$_2$), 22.9 (+, 2 × CH_3, CH(CH_3)$_2$), 18.3 (+, CH_3) ppm. – **IR** (KBr): ṽ = 2962 (w), 1702 (m), 1633 (w), 1599 (w), 1489 (m), 1382 (w), 1303 (w), 1258 (w), 1207 (w), 1174 (m), 1093 (w), 1067 (m), 1041 (m), 1020 (m), 998 (w), 894 (w), 877 (w), 858 (w), 832 (m), 809 (m), 770 (m), 733 (w), 684 (w), 659 (w), 610 (w), 519 (w), 477 (w), 437 (w), 402 (w) cm^{-1}. – **MS** (70 eV, EI): *m/z* (%) = 326/328 (6/2) $[M]^+$, 311 (6) $[M-CH_3]^+$, 169 (5), 162 (9), 119 (13), 100 (13). – **HRMS** ($C_{20}H_{19}O_2{}^{35}Cl_1$): calc. 326.1068, found 326.1067.

3-Benzyl-5-methoxy-7-(propoxymethyl)-2*H*-chromen-2-one (26s) [172]

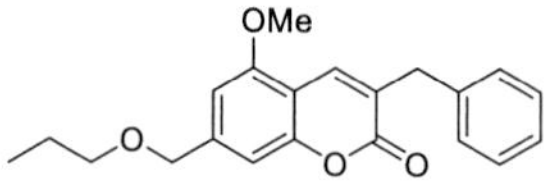

According to **GP5**, 100 mg of salicylic aldehyde **29d** (0.446 mmol, 1.00 equiv.), 73.9 mg of potassium carbonate (0.535 mmol, 1.20 equiv.), 147 mg of cinnamic aldehyde (1.12 mmol, 2.50 equiv.) and 119 mg of 1,3-dimethylimidazolium dimethyl phosphate (**31**) (0.535 mmol, 1.20 equiv.) were suspended in 2.50 mL of toluene. The reaction mixture was heated at 110 °C for 50 min via microwave irradiation. After cooling to room temperature, 5 mL of water was added and the mixture was extracted with 3 × 15 mL of ethyl acetate. Removal of the volatiles under reduced pressure and purification via flash column chromatography (CH/EtOAc 5:1) gave 60 mg (40%) of the pure product as a yellow solid.

R_f (CH/EtOAc 5:1): 0.30. – **^{1}H NMR** (400 MHz, $CDCl_3$): δ = 7.68 (s, 1H, 4-C*H*), 7.14 – 7.35 (m, 5H, 5 × H_{Ar}), 6.81 (t, J = 1.0 Hz, 1H, H_{Ar}), 6.64 (d, J = 1.2 Hz, 1H, H_{Ar}), 4.47 (s, 2H, CH_2OCH_2-Ar), 3.82 (s, 5H, OCH_3, $BnCH_2$), 3.39 (t, J = 6.6 Hz, 2H, $CH_3CH_2CH_2O$), 1.66 – 1.54 (m, 2H, $CH_3CH_2CH_2O$), 0.90 (t, J = 7.4 Hz, 3H, $CH_3CH_2CH_2O$) ppm. – **^{13}C NMR** (100 MHz, $CDCl_3$): δ = 162.0 ($C_{quart.}$, COO), 155.9 ($C_{quart.}$, $C_{Ar}OCH_3$), 154.3 ($C_{quart.}$, C_{Ar}-O-CO), 143.5 ($C_{quart.}$, $C_{Ar}CH_2OCH_2CH_2CH_3$), 138.4 ($C_{quart.}$, C_{Ar}), 134.6 (+, $C_{Ar}H$), 129.3 (+, 2 × $C_{Ar}H$), 128.8 (+, 2 × $C_{Ar}H$), 126.9 ($C_{quart.}$, C_{Ar}), 126.8 (+, $C_{Ar}H$), 109.3 ($C_{quart.}$, C_{Ar}), 107.5 (+, $C_{Ar}H$), 103.8 (+, $C_{Ar}H$), 72.6 (–, CH_2), 72.4 (–, CH_2), 56.0 (+, OCH_3), 36.9 (–, CH_2), 23.1 (–, $CH_3CH_2CH_2O$), 10.8 (+, $CH_3CH_2CH_2O$) ppm. – **IR** (KBr): $\tilde{\nu}$ = 2967 (w), 2872 (w), 1718 (s), 1616 (m), 1574 (m), 1494 (m), 1453 (m), 1421 (m), 1366 (m), 1287 (w), 1268 (w), 1240 (m), 1200 (w), 1162 (m), 1111 (s), 1082 (s), 1046 (m), 1015 (m), 948 (m), 892 (m), 876 (w), 841 (m), 788 (w), 758 (m), 727 (m), 704 (s), 640 (w), 593 (w), 565 (m) cm^{-1}. – **MS** (70 eV, EI): *m/z* (%) = 338 (51) $[M]^+$, 308 (13), 292 (10), 281 (15), 280 (65), 264 (19), 235 (16), 234 (100), 219 (13), 181 (21), 177 (13). – **HRMS** ($C_{21}H_{22}O_4$): calc. 338.1513, found 338.1511.

5-Methoxy-3-(2-methoxybenzyl)-7-(propoxymethyl)-2*H*-chromen-2-one (26t) [172]

According to **GP5**, 100 mg of salicylic aldehyde **29d** (0.446 mmol, 1.00 equiv.), 73.9 mg of potassium carbonate (0.535 mmol, 1.20 equiv.), 182 mg of cinnamic aldehyde (1.12 mmol, 2.50 equiv.) and 119 mg of 1,3-dimethylimidazolium dimethyl phosphate (**31**) (0.535 mmol, 1.20 equiv.) were suspended in 2.50 mL of toluene. The reaction mixture was heated at 110 °C for 50 min via microwave irradiation. After cooling to room temperature, 5 mL of water was added and the mixture was extracted with 3 × 15 mL of ethyl acetate. Removal of the volatiles under reduced pressure and purification via flash column chromatography (CH/EtOAc 5:1) gave 60.9 mg (37%) of the pure product as an off-white solid.

***R*f** (CH/EtOAc 5:1): 0.17. – **^{1}H NMR** (400 MHz, $CDCl_3$): δ = 7.65 (s, 1H, 4-C*H*) 7.30 – 7.21 (m, 2H, 2 × H_{Ar}), 6.97 – 6.84 (m, 3H, 3 × H_{Ar}), 6.68 (d, *J* = 1.2 Hz, 1H, H_{Ar}), 4.52 (s, 2H, CH_2OCH_2-Ar), 3.87 (s, 2H, BnCH_2), 3.86 (s, 3H, OCH_3), 3.81 (s, 3H, OCH_3), 3.44 (t, *J* = 6.6 Hz, 2H, $CH_3CH_2CH_2O$), 1.71 – 1.61 (m, 2H, $CH_3CH_2CH_2O$), 0.95 (t, *J* = 7.4 Hz, 3H, $CH_3CH_2CH_2O$) ppm. – **^{13}C NMR** (100 MHz, $CDCl_3$): δ = 162.2 ($C_{quart.}$, *C*OO), 157.8 ($C_{quart.}$, $C_{Ar}OCH_3$), 155.8 ($C_{quart.}$, $C_{Ar}OCH_3$), 154.2 ($C_{quart.}$, C_{Ar}-O-CO), 143.0 ($C_{quart.}$, $C_{Ar}CH_2OCH_2CH_2CH_3$), 134.2 (+, C_{Ar}H), 131.2 (+, C_{Ar}H), 128.2 (+, C_{Ar}H), 126.6 ($C_{quart.}$, C_{Ar}), 126.3 ($C_{quart.}$, C_{Ar}), 120.8(+, C_{Ar}H), 110.7 (+, C_{Ar}H), 109.6 ($C_{quart.}$, C_{Ar}), 107.5 (+, C_{Ar}H), 103.8 (+, C_{Ar}H), 72.6 (–, CH_2), 72.4 (–, CH_2), 56.0 (+, OCH_3), 55.5 (+, OCH_3), 31.1 (–, CH_2), 23.1 (–, $CH_3CH_2CH_2O$), 10.8 (+, $CH_3CH_2CH_2O$) ppm. – **IR** (KBr): ṽ = 2964 (w), 2872 (w), 1720 (m), 1619 (m), 1573 (w), 1494 (m), 1457 (m), 1409 (m), 1362 (w), 1315 (w), 1243 (m), 1188 (w), 1162 (m), 1104 (s), 1049 (w), 1033 (m), 1019 (m), 947 (w), 900 (w), 821 (m), 771 (w), 748 (m), 725 (m), 675 (w), 618 (vw), 587 (w), 560 (w), 528 (w), 465 (w) cm^{-1}. – **MS** (70 eV, EI): *m/z* (%) = 368 (98) $[M]^+$, 311 (21), 310 (100), 309 (17), 308 (24), 273 (68), 261 (58), 181 (88). – **HRMS** ($C_{22}H_{24}O_5$): calc. 368.1618, found 368.1619.

5-Methoxy-3-(4-methoxybenzyl)-7-(propoxymethyl)-2*H*-chromen-2-one (26u)[172]

According to **GP5**, 130 mg of salicylic aldehyde **29d** (0.580 mmol, 1.00 equiv.), 96.2 mg of potassium carbonate (0.696 mmol, 1.20 equiv.), 235 mg of cinnamic aldehyde (1.45 mmol, 2.50 equiv.) and 155 mg of 1,3-dimethylimidazolium dimethyl phosphate (**31**) (0.696 mmol, 1.20 equiv.) were suspended in 2.50 mL of toluene. The reaction mixture was heated at 110 °C for 50 min via microwave irradiation. After cooling to room temperature, 5 mL of water was added and the mixture was extracted with 3 × 15 mL of ethyl acetate. Removal of the volatiles under reduced pressure and purification via flash column chromatography (CH/EtOAc 5:1) gave 87.4 mg (41%) of the pure product as a red solid.

***R*f** (CH/EtOAc 5:1): 0.21. – **^{1}H NMR** (400 MHz, $CDCl_3$): δ = 7.71 (s, 1H, 4-C*H*) 7.25 – 7.18 (m, 2H, 2 × H_{Ar}), 6.90 – 6.84 (m, 3H, 3 × H_{Ar}), 6.69 (d, *J* = 1.2 Hz, 1H, H_{Ar}), 4.52 (s, 2H, CH_2OCH_2-Ar), 3.88 (s, 3H, OCH_3), 3.81 (s, 2H, CH_2), 3.80 (s, 3H, OCH_3), 3.44 (t, *J* = 6.6 Hz, 2H, $CH_3CH_2CH_2O$), 1.71 – 1.61 (m, 2H, $CH_3CH_2CH_2O$), 0.95 (t, *J* = 7.4 Hz, 3H, CH_3) ppm. – **^{13}C NMR** (100 MHz, $CDCl_3$): δ = 162.0 ($C_{quart.}$, *C*OO), 158.5 ($C_{quart.}$, $C_{Ar}OCH_3$), 155.9 ($C_{quart.}$, $C_{Ar}OCH_3$), 154.2 ($C_{quart.}$, C_{Ar}), 143.4 ($C_{quart.}$, $C_{Ar}CH_2OCH_2CH_2CH_3$), 134.3 (+, C_{Ar}H), 130.4 (+, 2 × C_{Ar}H), 130.3 ($C_{quart.}$, C_{Ar}), 127.3 ($C_{quart.}$, C_{Ar}), 114.2 (+, 2 × C_{Ar}H), 109.4 ($C_{quart.}$, C_{Ar}), 107.5 (+, C_{Ar}H), 103.8 (+, C_{Ar}H), 72.6 (–, CH_2), 72.4 (–, CH_2), 56.0 (+, OCH_3), 55.4 (+, OCH_3), 36.1 (–, BnCH_2), 23.1 (–, $CH_3CH_2CH_2O$), 10.8 (+, $CH_3CH_2CH_2O$) ppm. – **IR** (KBr): $\tilde{\nu}$ = 2957 (w), 1713 (s), 1617 (m), 1573 (w), 1514 (m), 1495 (w), 1452 (m), 1410 (m), 1366 (w), 1301 (w), 1252 (m), 1205 (w), 1164 (m), 1103 (s), 1030 (m), 950 (w), 897 (w), 877 (w), 846 (m), 808 (m), 791 (m), 762 (m), 725 (m), 672 (w), 608 (w), 582 (w), 555 (w), 533 (w), 505 (w) cm^{-1}. – **MS** (70 eV, EI): *m/z* (%) = 368 (100) $[M]^+$, 311 (15), 310 (70), 309 (13), 296 (18), 267 (11), 181 (21), 134 (44). – **HRMS** ($C_{22}H_{24}O_5$): calc. 368.1618, found 368.1618.

3-(4-Chlorobenzyl)-5-methoxy-7-(propoxymethyl)-2*H*-chromen-2-one (26v)[172]

According to GP5, 100 mg of salicylic aldehyde **29d** (0.446 mmol, 1.00 equiv.), 73.9 mg of potassium carbonate (0.535 mmol, 1.20 equiv.), 188 mg of cinnamic aldehyde (1.12 mmol, 2.50 equiv.) and 119 mg of 1,3-dimethylimidazolium dimethyl phosphate (**31**) (0.535 mmol, 1.20 equiv.) were suspended in 2.50 mL of toluene. The reaction mixture was heated at 110 °C for 50 min via microwave irradiation. After cooling to room temperature, 5 mL of water was added and the mixture was extracted with 3 × 15 mL of ethyl acetate. Removal of the volatiles under reduced pressure and purification via flash column chromatography (CH/EtOAc 5:1) resulted in a mixture of the product and a methyl ester impurity as a solid. The mixture was stirred at room temperature overnight in 80 mL of 1 M sodium hydroxide-solution resulting in a dispersion. The remaining solid was dissolved in ethyl acetate and the organic layer was separated and washed again with 40 mL of 1 M sodium hydroxide and then dried over sodium sulfate. Removal of the volatiles gave 57.1 mg (34%) of the pure product as an off-white solid.

***R*f** (CH/EtOAc 5:1): 0.29. – **^{1}H NMR** (400 MHz, $CDCl_3$): δ = 7.73 (s, 1H, 4-C*H*), 7.33 – 7.20 (m, 4H, 4 × H_{Ar}), 6.87 (s, 1H, H_{Ar}), 6.71 (s, 1H, H_{Ar}), 4.53 (s, 2H, CH_2OCH_2-Ar), 3.90 (s, 3H, OCH_3), 3.83 (s, 2H, CH_2), 3.45 (t, *J* = 6.6 Hz, 2H, $CH_3CH_2CH_2O$), 1.71 – 1.60 (m, 2H, $CH_3CH_2CH_2O$), 0.96 (t, *J* = 7.4 Hz, 3H, $CH_3CH_2CH_2O$) ppm. – **^{13}C NMR** (100 MHz, $CDCl_3$): δ = 161.9 ($C_{quart.}$, *C*OO), 155.9 ($C_{quart.}$, $C_{Ar}OCH_3$), 154.3 ($C_{quart.}$, C_{Ar}), 143.7 ($C_{quart.}$, $C_{Ar}CH_2OCH_2CH_2CH_3$), 136.9 ($C_{quart.}$, C_{Ar}), 134.7 (+, $C_{Ar}H$), 132.6 ($C_{quart.}$, C_{Ar}), 130.7 (+, 2 × $C_{Ar}H$), 128.9 (+, 2 × $C_{Ar}H$), 126.4 ($C_{quart.}$, C_{Ar}), 109.2 ($C_{quart.}$, C_{Ar}), 107.5 (+, $C_{Ar}H$), 103.9 (+, $C_{Ar}H$), 72.6 (–, CH_2), 72.4 (–, CH_2), 56.1 (+, OCH_3), 36.4 (–, CH_2), 23.1 (–, $CH_3CH_2CH_2O$), 10.8 (+, $CH_3CH_2CH_2O$) ppm. – **IR** (KBr): ṽ = 2920 (w), 1703 (s), 1617 (m), 1489 (m), 1463 (m), 1426 (m), 1359 (w), 1300 (w), 1256 (m), 1195 (w), 1091 (s), 1063 (m), 1041 (m), 1013 (m), 952 (w), 876 (w), 842 (m), 798 (m), 778 (w), 744 (w), 715 (w), 647 (w), 585 (w), 554 (w), 500 (w), 465 (w), 399 (w) cm^{-1}. – **MS** (70 eV, EI): *m/z* (%) = 372/374 (41/17) $[M]^+$, 316 (36), 315 (24), 314 (100), 313 (12), 308 (14), 289 (15). – **HRMS** ($C_{21}H_{21}O_4{}^{35}Cl$): calc. 372.1123, found 372.1124.

5-Methoxy-3-(2-methylbenzyl)-7-(propoxymethyl)-2*H*-chromen-2-one (26w)[172]

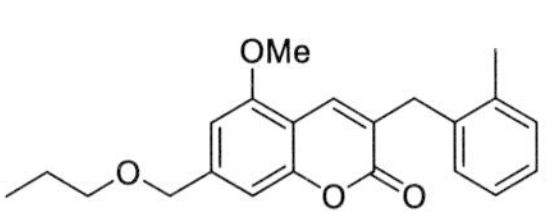

According to GP5, 130 mg of salicylic aldehyde **29d** (0.580 mmol, 1.00 equiv.), 96.2 mg of potassium carbonate (0.696 mmol, 1.20 equiv.), 212 mg of cinnamic aldehyde (1.45 mmol, 2.50 equiv.) and 155 mg of 1,3-dimethylimidazolium dimethyl phosphate (**31**) (0.696 mmol, 1.20 equiv.) were suspended in 2.50 mL of toluene. The reaction mixture was heated at 110 °C for 50 min via microwave irradiation. After cooling to room temperature, 5 mL of water was added and the mixture was extracted with 3 × 15 mL of ethyl acetate. Removal of the volatiles under reduced pressure and purification via flash column chromatography (CH/EtOAc 5:1) gave 165 mg (81%) of the pure product as a yellow solid.

***R*f** (CH/EtOAc 5:1): 0.39. – **^{1}H NMR** (400 MHz, $CDCl_3$): δ = 7.45 (s, 1H, 4-C*H*), 7.24 – 7.16 (m, 4H, 4 × H_{Ar}), 6.89 (s, 1H, H_{Ar}), 6.68 (s, 1H, H_{Ar}), 4.53 (s, 2H, CH_2OCH_2-Ar), 3.87 (d, *J* = 1.5 Hz, 2H, Lacton-CH_2-Ar), 3.83 (s, 3H, OCH_3), 3.45 (t, *J* = 6.6 Hz, 2H, $CH_3CH_2CH_2O$), 2.28 (s, 3H, CH_3), 1.71 – 1.60 (m, 2H, $CH_3CH_2CH_2O$), 0.96 (t, *J* = 7.4 Hz, 3H, $CH_3CH_2CH_2O$) ppm. – **^{13}C NMR** (100 MHz, $CDCl_3$): δ = 162.1 ($C_{quart.}$, *C*=O), 155.8 ($C_{quart.}$, $C_{Ar}OCH_3$), 154.1

($C_{quart.}$, C_{Ar}), 143.4 ($C_{quart.}$, $C_{Ar}CH_2OCH_2CH_2CH_3$), 137.0 ($C_{quart.}$, C_{Ar}), 136.1 ($C_{quart.}$, C_{Ar}), 134.1 (+, $C_{Ar}H$), 130.7 (+, $C_{Ar}H$), 130.4 (+, $C_{Ar}H$), 127.2 (+, $C_{Ar}H$), 126.4 (+, $C_{Ar}H$), 126.2 ($C_{quart.}$, C_{Ar}), 109.3 ($C_{quart.}$, C_{Ar}), 107.5 (+, $C_{Ar}H$), 103.9 (+, $C_{Ar}H$), 72. (–, CH_2), 72.4 (–, CH_2), 56.0 (+, OCH_3), 34.2 (–, CH_2), 23.1 (–, $CH_3CH_2CH_2O$), 19.7 (+,CH_3), 10.8 (+, $CH_3CH_2CH_2O$) ppm. – **IR** (KBr): ṽ = 2960 (w), 2875 (w), 1722 (m), 1616 (m), 1574 (w), 1494 (w), 1463 (m), 1424 (m), 1365 (w), 1294 (w), 1267 (w), 1241 (m), 1162 (m), 1111 (m), 1079 (m), 1038 (m), 1013 (m), 953 (w), 898 (w), 875 (w), 843 (m), 768 (m), 749 (m), 728 (m), 703 (w), 642 (w), 577 (w), 556 (w), 539 (w) cm^{-1}. – **MS** (70 eV, EI): *m/z* (%) = 352 (100) $[M]^+$, 335 (8), 295 (13), 294 (64), 261 (10), 181 (9), 118 (12). – **HRMS** ($C_{22}H_{24}O_4$): calc. 352.1669, found 352.1670.

5-Methoxy-3-(2-methoxybenzyl)-7-(2-propoxypropan-2-yl)-2*H*-chromen-2-one (26x)[172]

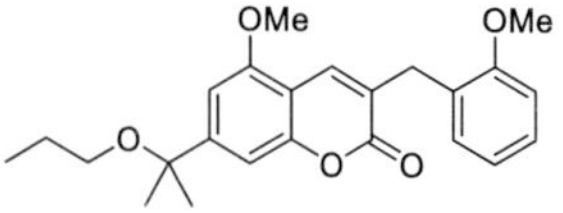

OMe OMe O O O

According to GP5, 100 mg of salicylic aldehyde **29e** (0.396 mmol, 1.00 equiv.), 65.7 mg of potassium carbonate (0.476 mmol, 1.20 equiv.), 161 mg of cinnamic aldehyde (0.991 mmol, 2.50 equiv.) and 106 mg of 1,3-dimethylimidazolium dimethyl phosphate (**31**) (0.476 mmol, 1.20 equiv.) were suspended in 2.50 mL of toluene. The reaction mixture was heated at 110 °C for 50 min via microwave irradiation. After cooling to room temperature, 5 mL of water was added and the mixture was extracted with 3 × 15 mL of ethyl acetate. Removal of the volatiles under reduced pressure and purification via flash column chromatography (CH/EtOAc 20:1) gave 94.9 mg (60%) of the pure brown solidifying oil.

***R*f** (CH/EtOAc 20:1): 0.27. – **^{1}H NMR** (400 MHz, $CDCl_3$): δ = 7.67 (s, 1H, 4-CH), 7.33 – 7.20 (m, 2H, 2 × H_{Ar}), 7.03 – 6.86 (m, 3H, 3 × H_{Ar}), 6.82 (d, *J* = 1.5 Hz, 1H, H_{Ar}), 3.88 (s, 2H, CH_2), 3.87 (s, 3H, OCH_3), 3.82 (s, 3H, OCH_3), 3.12 (t, *J* = 6.8 Hz, 2H, $CH_3CH_2CH_2O$), 1.65 – 1.54 (m, 2H, $CH_3CH_2CH_2O$), 1.53 (s, 6H, 2 × CH_3), 0.90 (t, *J* = 7.4 Hz, 3H, $CH_3CH_2CH_2O$) ppm. – **^{13}C NMR** (100 MHz, $CDCl_3$): δ = 162.2 ($C_{quart.}$, COO), 157.7 ($C_{quart.}$, $C_{Ar}OCH_3$), 155.7 ($C_{quart.}$, $C_{Ar}OCH_3$), 154.1 ($C_{quart.}$, C_{Ar}-O-CO), 151.5 ($C_{quart.}$, C_{Ar}), 134.1 (+, $C_{Ar}H$), 131.1 (+, $C_{Ar}H$), 128.2 (+, $C_{Ar}H$), 126.6 ($C_{quart.}$, C_{Ar}), 126.3 ($C_{quart.}$, C_{Ar}), 120.7 (+, $C_{Ar}H$), 110.7 (+, $C_{Ar}H$), 109.1 ($C_{quart.}$, C_{Ar}), 106.4 (+, $C_{Ar}H$), 102.7 (+, $C_{Ar}H$), 76.5 ($C_{quart.}$, $CH_3CH_2CH_2OC(CH_3)_2$), 64.8 (–, CH_2), 55.9 (+, OCH_3), 55.5 (+, OCH_3), 31.1 (–, CH_2), 28.4 (+, 2 × CH_3), 23.7 (–, $CH_3CH_2CH_2O$), 10.9 (+, $CH_3CH_2CH_2O$) ppm. – **IR** (KBr): ṽ = 2932 (w), 1717 (s), 1613 (s), 1571 (w), 1492 (m), 1460 (m), 1415 (m), 1359 (w), 1294 (w), 1242 (s), 1160 (m), 1109 (s), 1029 (m), 949 (w), 838 (m), 751 (s), 671 (w), 586 (w), 565 (w), 477 (w) cm^{-1}. – **MS** (70 eV,

EI): *m/z* (%) = 396 (6) $[M]^+$, 338 (16), 296 (10), 210 (24), 195 (11), 194 (44), 192 (12), 181 (13), 166 (11), 161 (19), 153 (12), 152 (100), 151 (25). – **HRMS** ($C_{24}H_{28}O_5$): calc. 396.1931, found 396.1930.

5-Methoxy-3-(2-methylbenzyl)-7-(2-propoxypropan-2-yl)-2*H*-chromen-2-one (26y)[172]

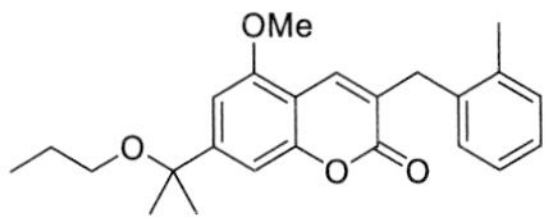

According to GP5, 130 mg of salicylic aldehyde **29e** (0.515 mmol, 1.00 equiv.), 85.4 mg of potassium carbonate (0.618 mmol, 1.20 equiv.), 188 mg of cinnamic aldehyde (1.29 mmol, 2.50 equiv.) and 155 mg of 1,3-dimethylimidazolium dimethyl phosphate (**31**) (0.618 mmol, 1.20 equiv.) were suspended in 2.50 mL of toluene. The reaction mixture was heated at 110 °C for 50 min via microwave irradiation. After cooling to room temperature, 5 mL of water was added and the mixture was extracted with 3 × 15 mL of ethyl acetate. Removal of the volatiles under reduced pressure and purification via flash column chromatography (CH/EtOAc 10:1) gave 90.5 mg (46%) of the pure product as an off-white solid.

$\boldsymbol{R}_f$ (CH/EtOAc 10:1): 0.42. – **^{1}H NMR** (400 MHz, $CDCl_3$): δ = 7.45 (s, 1H, 4-C*H*), 7.24 – 7.17 (m, 4H, 4 × H_{Ar}), 6.93 (d, *J* = 1.3 Hz, 1H, H_{Ar}), 6.81 (d, *J* = 1.4 Hz, 1H, H_{Ar}), 3.87 (d, *J* = 1.5 Hz, 2H, CH_2), 3.83 (s, 3H, OCH_3), 3.12 (t, *J* = 6.8 Hz, 2H, $CH_3CH_2CH_2O$), 2.28 (s, 3H, $C_{Ar}CH_3$), 1.63 – 1.54 (m, 2H, $CH_3CH_2CH_2O$), 1.53 (s, 6H, 2 × CH_3), 0.90 (t, *J* = 7.4 Hz, 3H, $CH_3CH_2CH_2O$) ppm. – **^{13}C NMR** (100 MHz, $CDCl_3$): δ = 162.2 ($C_{quart.}$, *C*=O), 155.7 ($C_{quart.}$, $C_{Ar}OCH_3$), 154.0 ($C_{quart.}$, C_{Ar}), 151.9 ($C_{quart.}$, C_{Ar}), 137.0 ($C_{quart.}$, C_{Ar}), 136.2 ($C_{quart.}$, C_{Ar}), 134.0 (+, $C_{Ar}H$), 130.7 (+, $C_{Ar}H$), 130.3 (+, $C_{Ar}H$), 127.1 (+, $C_{Ar}H$), 126.4 (+, $C_{Ar}H$), 126.2 ($C_{quart.}$, C_{Ar}), 108.9 ($C_{quart.}$, C_{Ar}), 106.4 (+, $C_{Ar}H$), 102.8 (+, $C_{Ar}H$), 76.4 ($C_{quart.}$, $CH_3CH_2CH_2OC(CH_3)_2$), 64.9 (–, $CH_3CH_2CH_2O$), 55.9 (+, OCH_3), 34.2 (–, CH_2), 28.4 (+, 2 × CH_3), 23.7 (–, $CH_3CH_2CH_2O$), 19.7 (+, $C_{Ar}CH_3$), 10.9 (+, $CH_3CH_2CH_2O$) ppm. – **IR** (KBr): ṽ = 2957 (w), 1722 (m), 1612 (m), 1568 (w), 1492 (w), 1462 (w), 1417 (m), 1346 (w), 1306 (w), 1247 (w), 1213 (w), 1159 (w), 1110 (m), 1075 (m), 1041 (m), 998 (w), 952 (w), 899 (w), 868 (w), 848 (w), 761 (w), 747 (w), 726 (m), 670 (w), 589 (w), 559 (w), 499 (vw), 470 (w), 437 (w) cm^{-1}. – **MS** (70 eV, EI): *m/z* (%) = 380 (34) $[M]^+$, 365 (12), 323 (45), 322 (100), 321 (16), 264 (10), 231 (24), 189 (23). – **HRMS** ($C_{22}H_{28}O_4$): calc. 380.1982, found 380.1983.

7-(Cyclopent-1-en-1-yl)-5-methoxy-3-(2-methoxybenzyl)-2*H*-chromen-2-one (26z)[172]

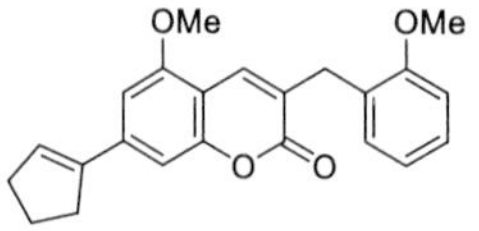

According to GP5, 130 mg of salicylic aldehyde **29g** (0.580 mmol, 1.00 equiv.), 96.2 mg of potassium carbonate (0.696 mmol, 1.20 equiv.), 235 mg of cinnamic aldehyde (1.45 mmol, 1.20 equiv.) and 155 mg of 1,3-dimethylimidazolium dimethyl phosphate (**31**) (0.696 mmol, 1.20 equiv.) were suspended in 2.50 mL of toluene. The reaction mixture was heated at 110 °C for 50 min via microwave irradiation. After cooling to room temperature, 5 mL of water was added and the mixture was extracted with 3 × 15 mL of ethyl acetate. Removal of the volatiles under reduced pressure and purification via flash column chromatography (CH/EtOAc 5:1) gave 87.4 mg (41%) of the pure product as a red solid.

***R*f** (CH/EtOAc 5:1): 0.21. – **^{1}H NMR** (400 MHz, $CDCl_3$): δ = 7.71 (s, 1H, 4-C*H*) 7.18 – 7.25 (m, 2H, 2 × H_{Ar}), 6.84 – 6.90 (m, 3H, 3 × H_{Ar}), 6.69 (d, *J* = 1.2 Hz, 1H, H_{Ar}), 4.52 (s, 2H, CH_2OCH_2-Ar), 3.88 (s, 3H, OCH_3), 3.81 (s, 2H, CH_2), 3.80 (s, 3H, OCH_3), 3.44 (t, *J* = 6.6 Hz, 2H, $CH_3CH_2CH_2O$), 1.61 – 1.71 (m, 2H, $CH_3CH_2CH_2O$), 0.95 (t, *J* = 7.4 Hz, 3H, CH_3) ppm. – **^{13}C NMR** (100 MHz, $CDCl_3$): δ = 162.0 ($C_{quart.}$, *C*=O), 158.5 ($C_{quart.}$, $C_{Ar}OCH_3$), 155.9 ($C_{quart.}$, $C_{Ar}OCH_3$), 154.2 ($C_{quart.}$, C_{Ar}), 143.4 ($C_{quart.}$, $C_{Ar}CH_2OCH_2CH_2CH_3$), 134.3 (+, C_{Ar}H), 130.4 (+, 2 × C_{Ar}H), 130.3 ($C_{quart.}$, C_{Ar}), 127.3 ($C_{quart.}$, C_{Ar}), 114.2 (+, 2 × C_{Ar}H), 109.4 ($C_{quart.}$, C_{Ar}), 107.5 (+, C_{Ar}H), 103.8 (+, C_{Ar}H), 72.6 (–, CH_2), 72.4 (–, CH_2), 56.0 (+, OCH_3), 55.4 (+, OCH_3), 36.1 (–, Lacton-CH_2-Ar), 23.1 (–, $CH_3CH_2CH_2O$), 10.8 (+, $CH_3CH_2CH_2O$) ppm. – **IR** (KBr): ṽ = 2957 (w), 1713 (s), 1617 (m), 1573 (w), 1514 (m), 1495 (w), 1452 (m), 1410 (m), 1366 (w), 1301 (w), 1252 (m), 1205 (w), 1164 (m), 1103 (s), 1030 (m), 950 (w), 897 (w), 877 (w), 846 (m), 808 (m), 791 (m), 762 (m), 725 (m), 672 (w), 608 (w), 582 (w), 555 (w), 533 (w), 505 (w) cm^{-1}. – **MS** (70 eV, EI): *m/z* (%) = 368 (100) $[M]^+$, 311 (15), 310 (70), 309 (13), 296 (18), 267 (11), 181 (21), 134 (44). – **HRMS** ($C_{22}H_{24}O_5$): calc. 368.1618, found 368.1618.

5.2.5 Synthesis and Characterization 3-Pyridylmethylcoumarins (Chapter 3.2.3)

(*E*)-3-(Pyridin-2-yl)acrylaldehyde (92a)

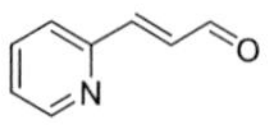

According to **GP6**, 6.65 g of (triphenylphosphoranylidene)-acetaldehyde (**91**, 21.9 mmol, 1.00 equiv.) were suspended in 10 mL of nitrobenzene under argon atmosphere and 2.06 mL of pyridine-3-carbaldehyde (2.34 g, 21.9 mmol, 1.00 equiv.) were added dropwise. After stirring for 24 h under exclusion of light, the mixture was taken up in 150 mL of dry dichloromethane, washed with 3 × 50 mL of brine, dried over sodium sulfate and the volatiles were removed under reduced pressure. The crude product was then filtered twice through a silica column (CH/EtOAc 1:1) and the pure product was obtained as an orange solid (1.07 g, 46% yield).

R_f (CH/EtOAc 1:1): 0.19. – **^{1}H NMR** (300 MHz, $CDCl_3$): δ = 9.80 (d, *J* = 7.8 Hz, 1H, C*H*O), 8.70 (d, *J* = 4.0 Hz, 1H, 4-CH_{Ar}), 7.77 (td, *J* = 7.7, 1.8 Hz, 1H, H_{Ar}), 7.60 – 7.47 (m, 2H, =C*H*, H_{Ar}), 7.32 (ddd, *J* = 7.6, 4.8, 1.2 Hz, 1H, H_{Ar}), 7.08 (dd, *J* = 15.8, 7.8 Hz, 1H, =C*H*CHO) ppm.

(*E*)-3-(Pyridin-3-yl)acrylaldehyde (92b)

Variant A: According to **GP6**, 5.98 g of (triphenylphosphoranylidene)-acetaldehyde (**91**, 19.7 mmol, 1.00 equiv.) were suspended in 10 mL of nitrobenzene under argon atmosphere and 1.85 mL of pyridine-3-carbaldehyde (2.11 g, 19.7 mmol, 1.00 equiv.) were added dropwise. After stirring for 24 h under exclusion of light, the mixture was washed with 5 × 10 mL of 2 N hydrochloric acid, potassium carbonate was added to the aqueous layer and extracted with 3 × 50 mL of diethyl ether. The combined ether layers were dried over sodium sulfate and the volatiles were removed under reduced pressure. The crude product was then filtered through a silica column (CH/EtOAc 1:1) and the product was obtained as a mixture with triphenylphosphine (40%), that was used in subsequent reactions without further purification (1.22 g, 42% yield, determined by ^{1}H NMR spectroscopy).

Variant B: A suspension of 1.00 g of 2-iodopyridine (**94b**, 4.88 mmol, 1.00 equiv.), 0.51 mL of acrolein (432 mg, 7.70 mmol, 1.00 equiv.), 1.80 g of tetrabutylammonium iodide (4.88 mmol, 1.00 equiv.), 1.03 g of sodium bicarbonate (12.2 mmol, 2.50 equiv.) and 27.4 mg of palladium acetate (0.122 mmol, 0.05 equiv.) in 45 mL DMF was heated at 60 °C for 24 h. After the reaction mixture cooled to room temperature and water added. The aqueous layer was extracted with 3 × 50 mL of ethyl acetate, washed with 50 mL brine and dried over sodium sulfate. The volatiles were removed under reduced pressure and the crude product purified via flash chromatography (CH/EtOAc 1:1) and the product was obtained as 460 mg of an orange solid (71%). Analytical data are consistent with the literature.[225]

***R*f** (CH/EtOAc 1:1): 0.19. – **^{1}H NMR** (300 MHz, $CDCl_3$): δ = 9.75 (d, *J* = 7.5 Hz, 1H, C*H*O), 8.79 (d, *J* = 2.3 Hz, 1H, 2-CH_{Ar}), 8.67 (dd, *J* = 4.9, 1.7 Hz, 1H, 4-CH_{Ar}), 7.90 (dt, *J* = 8.0, 2.0 Hz, 1H, 6-CH_{Ar}), 7.48 (d, *J* = 2.8 Hz, 1H, =C*H*), 7.39 (dd, *J* = 8.2, 5.0 Hz, 1H, 5C-H_{Ar}), 6.78 (ddd, *J* = 16.0, 7.5, 1.2 Hz, 1H, =C*H*CHO) ppm.

(*E*)-3-(Pyridin-4-yl)acrylaldehyde (92c)

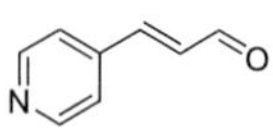

According to **GP6** 5.32 g of (triphenylphosphoranylidene)-acetaldehyde (**91**, 14.5 mmol, 1.00 equiv.) were suspended in 10 mL of nitrobenzene under argon atmosphere and 1.66 mL of pyridine-3-carbaldehyde (1.87 g, 17.5 mmol, 1.00 equiv.) were added dropwise. After stirring for 24 h under exclusion of light, washed with 5 × 10 mL of 2 N hydrochloric acid, potassium carbonate added to the aqueous layer and extracted with 3 × 50 mL of diethyl ether. The combined organic layers were dried over sodium sulfate and the volatiles were removed under reduced pressure. The crude product was then filtered through a silica column (CH/EtOAc 1:1) and the product obtained as a mixture with triphenylphosphine (36%) that was used in subsequent reactions without further purification (1.13 g, 43% yield, determined by ^{1}H NMR spectroscopy). Analytical data are consistent with the literature.[226]

***R*f** (CH/EtOAc 1:1): 0.44. – **^{1}H NMR** (300 MHz, $CDCl_3$): 9.78 (d, *J* = 7.4 Hz, 1H, C*H*O), 8.72 (d, *J* = 5.6 Hz, 2H, 2 × H_{Ar}, 3,5-CH), 7.44 – 7.42 (m, 1H, =C*H*), 7.42 – 7.40 (m, 2H, 2 × H_{Ar}, 2,6-CH), 6.85 (dd, *J* = 16.1, 7.5 Hz, 1H, =C*H*CHO) ppm.

5-Methoxy-7-pentyl-3-(pyridin-2-ylmethyl)-2*H*-chromen-2-one (96a-Me)

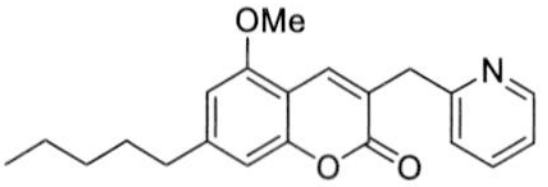

According to GP5, 150 mg of salicylic aldehyde **29a** (0.675 mmol, 1.00 equiv.), 205 mg of potassium carbonate (1.49 mmol, 2.20 equiv.), 225 mg of acrylaldehyde **92a** (1.69 mmol, 2.20 equiv.) and 180 mg of 1,3-dimethylimidazolium dimethyl phosphate (**31**) (0.810 mmol, 1.20 equiv.) were suspended in 3.00 mL of toluene. The reaction mixture was heated at 110 °C for 50 min via microwave irradiation. After cooling to room temperature, 5 mL of water was added and the mixture was extracted with 3 × 15 mL of ethyl acetate. Removal of the volatiles under reduced pressure and purification via flash column chromatography (CH/EtOAc 1:1) resulted in a mixture of the product and a methyl ester impurity as a solid. The mixture was stirred at room temperature for 30 min in 80 mL of 1 M sodium hydroxide-solution resulting in a dispersion.

The remaining solid was dissolved in ethyl acetate and the organic layer was separated and washed again with 40 mL of 1 M sodium hydroxide-solution and then dried over sodium sulfate. Removal of the volatiles and purification via flash column chromatography (CH/EtOAc 1:1) gave 41.8 mg (18%) of the pure product as an off-white solid.

***R*f** (CH/EtOAc 1:1): 0.19. – **^{1}H NMR** (400 MHz, $CDCl_3$): δ = 8.53 (ddd, *J* = 4.9, 1.9, 0.9 Hz, 1H, H_{Ar}), 7.97 (s, 1H, 4-C*H*), 7.61 (td, *J* = 7.7, 1.8 Hz, 1H, H_{Ar}), 7.37 (dt, *J* = 7.8, 1.1 Hz, 1H, H_{Ar}), 7.13 (ddd, *J* = 7.6, 4.9, 1.2 Hz, 1H, H_{Ar}), 6.71 (d, *J* = 1.2 Hz, 1H, H_{Ar}), 6.49 (d, *J* = 1.3 Hz, 1H, H_{Ar}), 4.04 (d, *J* = 1.0 Hz, 2H, BnCH_2), 3.87 (s, 3H, OCH_3), 2.66 – 2.58 (m, 2H, CH_2), 1.69 – 1.55 (m, 2H, CH_2), 1.42 – 1.22 (m, 4H, 2 × CH_2), 0.88 (t, *J* = 6.8 Hz, 3H, CH_3 ppm. – **^{13}C NMR** (100 MHz, $CDCl_3$): δ = 162.1 ($C_{quart.}$, *C*OO), 158.6 ($C_{quart.}$, C_{Ar}), 155.7 ($C_{quart.}$, C_{Ar}), 154.5 ($C_{quart.}$, C_{Ar}), 149.6 (+, C_{Ar}H), 147.9 ($C_{quart.}$, C_{Ar}), 136.8 (+, C_{Ar}H), 136.0 (+, C_{Ar}H), 124.1 ($C_{quart.}$, C_{Ar}), 123.9 (+, C_{Ar}H), 121.8 (+, C_{Ar}H), 108.5 (+, C_{Ar}H), 108.2 ($C_{quart.}$, C_{Ar}), 105.7 (+, C_{Ar}H), 55.9 (+, OCH_3), 39.7 (–, CH_2), 36.7 (–, CH_2), 31.5 (–, CH_2), 30.8 (–, CH_2), 22.6 (–, CH_2), 14.1 (+, CH_3) ppm. – **IR** (KBr): ṽ = 2927 (m), 2855 (m), 1701 (s), 1613 (s), 1568 (m), 1495 (m), 1426 (s), 1297 (m), 1255 (m), 1182 (m), 1139 (m), 1111 (s), 1055 (s), 995 (m), 832 (m), 766 (m), 745 (m), 688 (m), 628 (w), 601 (m), 573 (m), 490 (w), 403 (m) cm^{-1}. – **MS** (70 eV, EI): *m/z* (%) = 337/338 (100/25) $[M]^+$, 310 (9), 309 (40), 294 (8), 253 (10), 252 (31), 238 (7), 237 (7), 209 (9). – **HRMS** ($C_{21}H_{23}O_3N$): calc. 337.1672, found 337.1672.

5-Methoxy-7-pentyl-3-(pyridin-3-ylmethyl)-2*H*-chromen-2-one (96b-Me)

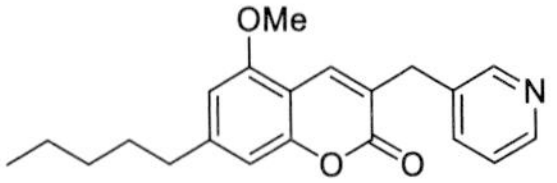

According to GP5, 100 mg of salicylic aldehyde **29a** (0.450 mmol, 1.00 equiv.), 137 mg of potassium carbonate (0.990 mmol, 2.20 equiv.), 150 mg of acrylaldehyde **92b** (1.13 mmol, 2.50 equiv.) and 120 mg of 1,3-dimethylimidazolium dimethyl phosphate (**31**) (0.540 mmol, 1.20 equiv.) were suspended in 3.00 mL of toluene. The reaction mixture was heated at 110 °C for 50 min via microwave irradiation. After cooling to room temperature, 5 mL of water was added and the mixture was extracted with 3 × 15 mL of ethyl acetate. Removal of the volatiles under reduced pressure and purification via flash column chromatography (CH/EtOAc 1:1) gave 27.4 mg (18%) of the pure product as an off-white solid.

***R*f** (CH/EtOAc 1:1): 0.16. – **^{1}H NMR** (400 MHz, $CDCl_3$): δ = 8.61 – 8.55 (m, 1H, H_{Ar}), 8.50 (dd, *J* = 4.9, 1.6 Hz, 1H, H_{Ar}), 7.78 (q, *J* = 1.0 Hz, 1H, 4-C*H*), 7.69 (dt, *J* = 7.8, 2.0 Hz, 1H, H_{Ar}), 7.31 – 7.22 (m, 1H, H_{Ar}), 6.73 (d, *J* = 1.3 Hz, 1H, H_{Ar}), 6.51 (d, *J* = 1.3 Hz, 1H, H_{Ar}), 3.88

(s, 3H, OCH_3), 3.87 (s, 2H, BnCH_2), 2.68 – 2.60 (m, 2H, CH_2), 1.71 – 1.56 (m, 2H, CH_2), 1.39 – 1.27 (m, 4H, 2 × CH_2), 0.95 – 0.85 (m, 3H, CH_3) ppm. – ^{13}**C NMR** (100 MHz, $CDCl_3$): δ = 162.3 ($C_{quart.}$, *C*OO), 156.0 ($C_{quart.}$, C_{Ar}), 154.8 ($C_{quart.}$, C_{Ar}), 150.5 ($C_{quart.}$, C_{Ar}), 148.7 (+, C_{Ar}H), 148.2 (+, C_{Ar}H), 137.5 (+, C_{Ar}H), 135.6 (+, C_{Ar}H), 134.7 ($C_{quart.}$, C_{Ar}), 125.1 ($C_{quart.}$, C_{Ar}), 124.1 (+, C_{Ar}H), 108.9 (+, C_{Ar}H), 108.2 ($C_{quart.}$, C_{Ar}), 106.1 (+, C_{Ar}H), 56.3 (+, O*C*H_3), 37.1 (–, *C*H_2), 34.6 (–, *C*H_2), 31.8 (–, *C*H_2), 31.2 (–, *C*H_2), 22.9 (–, *C*H_2), 14.5 (+, *C*H_3) ppm. – **IR** (KBr): $\tilde{\nu}$ = 2925 (m), 2855 (w), 1702 (s), 1614 (s), 1572 (m), 1496 (m), 1478 (m), 1462 (m), 1422 (m), 1376 (w), 1351 (w), 1300 (w), 1255 (m), 1181 (m), 1141 (m), 1112 (s), 1059 (m), 1027 (m), 1004 (m), 870 (w), 832 (m), 790 (m), 775 (m), 741 (w), 714 (m), 630 (w), 605 (w), 571 (w), 549 (w) cm^{-1}. – **MS** (70 eV, EI): *m/z (%)* = 337 (100) $[M]^+$, 336 (9), 282 (9), 281 (44), 280 (14), 52 (6). – **HRMS** ($C_{21}H_{23}O_3N$): calc. 337.1672, found 337.1672.

5-Methoxy-7-pentyl-3-(pyridin-4-ylmethyl)-2*H*-chromen-2-one (96c-Me)

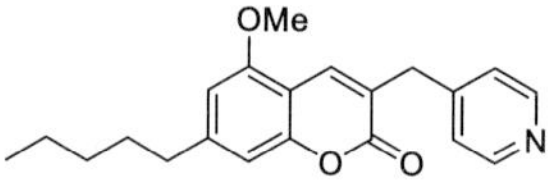

According to GP5, 150 mg of salicylic aldehyde **29a** (0.675 mmol, 1.00 equiv.), 205 mg of potassium carbonate (1.49 mmol, 2.20 equiv.), 225 mg of acrylaldehyde **92c** (1.69 mmol, 2.50 equiv.) and 180 mg of 1,3-dimethylimidazolium dimethyl phosphate (**31**) (0.810 mmol, 1.20 equiv.) were suspended in 3.00 mL of toluene. The reaction mixture was heated at 110 °C for 50 min via microwave irradiation. After cooling to room temperature, 5 mL of water was added and the mixture was extracted with 3 × 15 mL of ethyl acetate. The mixture was stirred at room temperature for 30 min in 80 mL of 1 M sodium hydroxide-solution resulting in a dispersion. The remaining solid was dissolved in ethyl acetate and the organic layer was separated and washed again with 40 mL of 1 M sodium hydroxide-solution and then dried over sodium sulfate. Removal of the volatiles and purification via flash column chromatography (CH/EtOAc 1:1) gave 53.2 mg (23%) of the pure product as an off-white solid.

R_f (CH/EtOAc 1:2): 0.16. – ^{1}**H NMR** (400 MHz, $CDCl_3$): δ = 8.54 (b, 2H, 2 × H_{Ar}), 7.79 (s, 1H, 4-C*H*), 7.24 (d, *J* = 5.1 Hz, 2H, 2 × H_{Ar}), 6.74 (d, *J* = 1.2 Hz, 1H, H_{Ar}), 6.51 (d, *J* = 1.4 Hz, 1H, H_{Ar}), 3.89 (s, 3H, OCH_3), 3.85 (s, 2H, BnCH_2), 2.73 – 2.60 (m, 2H, CH_2), 1.71 – 1.56 (m, 2H, CH_2), 1.42 – 1.23 (m, 4H, 2 × CH_2), 0.92 – 0.84 (m, 3H, CH_3) ppm. – ^{13}**C NMR** (100 MHz, $CDCl_3$): δ = 161.9 ($C_{quart.}$, *C*OO), 155.7 ($C_{quart.}$, C_{Ar}), 154.5 ($C_{quart.}$, C_{Ar}), 150.1 (+, 2 × C_{Ar}H), 148.5 ($C_{quart.}$, C_{Ar}), 147.7 ($C_{quart.}$, C_{Ar}), 135.6 (+, C_{Ar}H), 124.5 (+, 2 × C_{Ar}H), 124.0 ($C_{quart.}$, C_{Ar}), 108.7 (+, C_{Ar}H), 107.8 ($C_{quart.}$, C_{Ar}), 105.8 (+, C_{Ar}H), 56.0 (+, O*C*H_3), 36.8 (–,

CH_2), 36.4 (–, CH_2), 31.5 (–, CH_2), 30.8 (–, CH_2), 22.6 (–, CH_2), 14.1 (+, CH_3) ppm. – **IR** (KBr): $\tilde{\nu}$ = 2925 (m), 2854 (w), 1703 (s), 1614 (s), 1598 (m), 1557 (w), 1495 (m), 1462 (m), 1414 (m), 1351 (w), 1299 (w), 1256 (m), 1219 (w), 1183 (m), 1141 (m), 1112 (m), 1056 (m), 993 (w), 945 (w), 871 (w), 834 (m), 801 (m), 776 (m), 734 (w), 690 (w), 636 (w), 600 (m), 567 (w), 549 (w) cm^{-1}. – **MS** (70 eV, EI): *m/z* (%) = 337/338 (100/25) $[M]^+$, 311 (13), 295 (9), 294 (11), 282 (16), 281 (89), 280 (19), 255 (13), 238 (9), 209 (6), 180 (9). – **HRMS** ($C_{21}H_{23}O_3N$): calc. 337.1672, found 337.1672.

7-(1-Butylcyclopentyl)-5-methoxy-3-(pyridin-2-ylmethyl)-2*H*-chromen-2-one (96d-Me)

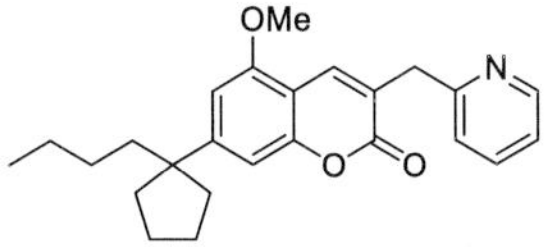

According to GP5, 150 mg of salicylic aldehyde **29b** (0.543 mmol, 1.00 equiv.), 165 mg of potassium carbonate (1.19 mmol, 2.20 equiv.), 181 mg of acrylaldehyde **92a** (1.36 mmol, 2.50 equiv.) and 145 mg of 1,3-dimethylimidazolium dimethyl phosphate (**31**) (0.651 mmol, 1.20 equiv.) were suspended in 3.00 mL of toluene. The reaction mixture was heated at 110 °C for 50 min via microwave irradiation. After cooling to room temperature, 5 mL of water was added and the mixture was extracted with 3 × 15 mL of ethyl acetate. Removal of the volatiles under reduced pressure and purification via flash column chromatography (CH/EtOAc 3:2) gave 51.8 mg (25%) of the pure product as an off-white solid.

***R*f** (CH/EtOAc 1:1): 0.36. – **^{1}H NMR** (400 MHz, $CDCl_3$): δ = 8.53 (ddd, *J* = 4.9, 1.9, 0.9 Hz, 1H, H_{Ar}), 7.98 (s, 1H, 4-C*H*), 7.61 (td, *J* = 7.7, 1.9 Hz, 1H, H_{Ar}), 7.37 (dt, *J* = 7.8, 1.1 Hz, 1H, H_{Ar}), 7.13 (ddd, *J* = 7.6, 4.9, 1.1 Hz, 1H, H_{Ar}), 6.81 (d, *J* = 1.3 Hz, 1H, H_{Ar}), 6.59 (d, *J* = 1.4 Hz, 1H, H_{Ar}), 4.04 (s, 2H, BnCH_2), 3.88 (s, 3H, OCH_3), 1.94 – 1.76 (m, 4H, 2 × CH_2), 1.75 – 1.50 (m, 6H, 3 × CH_2), 1.21 – 1.05 (m, 2H, CH_2), 0.97 – 0.85 (m, 2H, CH_2), 0.77 (t, *J* = 7.3 Hz, 3H, CH_3) ppm. – **^{13}C NMR** (100 MHz, $CDCl_3$): δ = 162.2 ($C_{quart.}$, *C*OO), 158.7 ($C_{quart.}$, C_{Ar}), 155.4 ($C_{quart.}$, C_{Ar}), 154.4 ($C_{quart.}$, C_{Ar}), 154.2 ($C_{quart.}$, C_{Ar}), 149.5 (+, C_{Ar}H), 136.7 (+, C_{Ar}H), 135.9 (+, C_{Ar}H), 124.2 ($C_{quart.}$, C_{Ar}), 123.9 (+, C_{Ar}H), 121.7 (+, C_{Ar}H), 108.0 ($C_{quart.}$, C_{Ar}), 107.6 (+, C_{Ar}H), 104.1 (+, C_{Ar}H), 55.9 (+, OCH_3), 52.1 ($C_{quart.}$, C_{CP}), 41.7 (–, CH_2), 39.7 (–, CH_2), 37.8 (–, 2 × CH_2), 27.6 (–, CH_2), 23.3 (–, CH_2), 23.3 (–, 2 × CH_2), 14.1 (+, CH_3). ppm. – **IR** (KBr): $\tilde{\nu}$ = 2927 (m), 2869 (w), 1715 (s), 1613 (s), 1568 (m), 1493 (w), 1460 (m), 1433 (m), 1415 (m), 1351 (w), 1291 (w), 1251 (m), 1181 (m), 1110 (s), 1051 (m), 995 (w), 901 (w), 836 (m), 768 (m), 748 (m), 691 (w), 603 (w), 558 (w), 502 (w), 403 (w) cm^{-1}. – **MS** (70 eV, EI):

m/z (%) = 391 (15) $[M]^+$, 289 (20), 276 (31), 234 (11), 233 (65), 232 (11), 221 (18), 220 (100), 219 (21), 218 (11), 207 (52), 194 (19), 193 (20), 192 (11), 180 (38), 179 (15), 165 (13), 153 (14), 137 (14), 131 (11). – **HRMS** ($C_{21}H_{23}O_3N$): calc. 391.2142, found 391.2140.

7-(1-Butylcyclopentyl)-5-methoxy-3-(pyridin-3-ylmethyl)-2*H*-chromen-2-one (96e-Me)

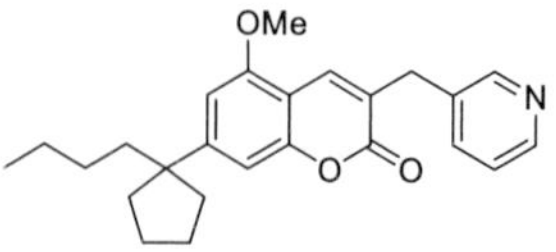

According to GP5, 150 mg of salicylic aldehyde **29b** (0.543 mmol, 1.00 equiv.), 165 mg of potassium carbonate (1.19 mmol, 2.20 equiv.), 181 mg of acrylaldehyde **92b** (1.36 mmol, 2.50 equiv.) and 145 mg of 1,3-dimethylimidazolium dimethyl phosphate (**31**) (0.651 mmol, 1.20 equiv.) were suspended in 3.00 mL of toluene. The reaction mixture was heated at 110 °C for 50 min via microwave irradiation. After cooling to room temperature, 5 mL of water was added and the mixture was extracted with 3 × 15 mL of ethyl acetate. Removal of the volatiles under reduced pressure and purification via flash column chromatography (CH/EtOAc 3:2) gave 38.8 mg (18%) of the pure product as an off-white solid.

R_f (CH/EtOAc 1:1): 0.18. – **^{1}H NMR** (400 MHz, $CDCl_3$): δ = 8.57 (bs, 1H, H_{Ar}), 8.53 – 8.45 (m, 1H, H_{Ar}), 7.78 (s, 1H, 4-C*H*), 7.66 (dt, *J* = 7.9, 2.0 Hz, 1H, H_{Ar}), 7.23 (dd, *J* = 7.9, 4.8 Hz, 1H, H_{Ar}), 6.82 (d, *J* = 1.4 Hz, 1H, H_{Ar}), 6.60 (d, *J* = 1.4 Hz, 1H, H_{Ar}), 3.88 (s, 3H, OCH_3), 3.86 (s, 2H, BnCH_2), 1.92 – 1.77 (m, 4H, 2 × CH_2), 1.76 – 1.52 (m, 6H, 3 × CH_2), 1.21 – 1.07 (m, 2H, CH_2), 1.00 – 0.87 (m, 2H, CH_2), 0.76 (t, *J* = 7.3 Hz, 3H, CH_3) ppm. – **^{13}C NMR** (100 MHz, $CDCl_3$): δ = 162.0 ($C_{quart.}$, *C*OO), 155.3 ($C_{quart.}$, C_{Ar}), 154.8 ($C_{quart.}$, C_{Ar}), 154.1 ($C_{quart.}$, C_{Ar}), 150.5 (+, C_{Ar}H), 148.2 (+, C_{Ar}H), 136.8 (+, C_{Ar}H), 135.2 (+, C_{Ar}H), 134.2 ($C_{quart.}$, C_{Ar}), 125.0 ($C_{quart.}$, C_{Ar}), 123.6 (+, C_{Ar}H), 107.7 ($C_{quart.}$, C_{Ar}), 107.6 (+, C_{Ar}H), 104.2 (+, C_{Ar}H), 55.9 (+, OCH_3), 52.1 ($C_{quart.}$, C_{CP}), 41.6 (–, CH_2), 37.8 (–, 2 × CH_2), 34.3 (–, CH_2), 27.6 (–, CH_2), 23.3 (–, CH_2), 23.3 (–, 2 × CH_2), 14.1 (+,CH_3) ppm. – **IR** (KBr): $\tilde{\nu}$ = 2926 (w), 2869 (w), 1708 (m), 1612 (m), 1568 (w), 1492 (w), 1442 (m), 1413 (m), 1377 (w), 1287 (w), 1245 (m), 1169 (w), 1108 (m), 1051 (m), 1028 (m), 948 (w), 897 (w), 838 (m), 796 (w), 762 (w), 717 (m), 627 (w), 594 (w), 555 (w), 434 (vw), 403 (vw) cm^{-1}. – **MS** (70 eV, EI): *m/z* (%) = 391/392 (72/16) $[M]^+$, 336 (9), 335 (48), 334 (100) $[M – C_4H_9]^+$, 281 (12), 280 (42), 268 (14), 243 (10), 231 (11), 207 (16), 181 (28), 131 (37), 119 (11), 93 (11), 69 (55), 67 (12), 57 (26), 55 (14). – **HRMS** ($C_{21}H_{23}O_3N$): calc. 391.2142, found 391.2144.

7-(1-Butylcyclopentyl)-5-methoxy-3-(pyridin-4-ylmethyl)-2*H*-chromen-2-one (96f-Me)

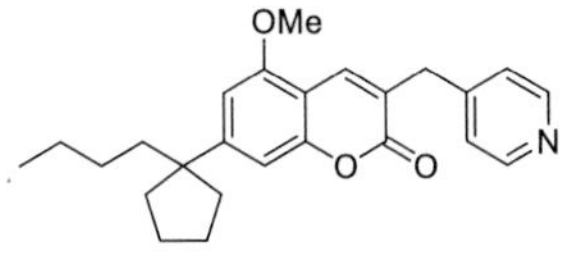

According to GP5, 150 mg of salicylic aldehyde **29b** (0.543 mmol, 1.00 equiv.), 165 mg of potassium carbonate (1.19 mmol, 2.50 equiv.), 181 mg of acrylaldehyde **92c** (1.36 mmol, 1.20 equiv.) and 145 mg of 1,3-dimethylimidazolium dimethyl phosphate (**31**) (0.651 mmol, 1.20 equiv.) were suspended in 3.00 mL of toluene. The reaction mixture was heated at 110 °C for 50 min via microwave irradiation. After cooling to room temperature, 5 mL of water was added and the mixture was extracted with 3 × 15 mL of ethyl acetate. Removal of the volatiles under reduced pressure and purification via flash column chromatography (CH/EtOAc 3:2) gave 50.0 mg (24%) of the pure product as an off-white solid.

***R*f** (CH/EtOAc 1:1): 0.26. – **^{1}H NMR** (400 MHz, $CDCl_3$): δ = 8.53 (d, *J* = 5.2 Hz, 2H, 2 × H_{Ar}), 7.80 (s, 1H, 4-C*H*), 7.26 – 7.23 (m, 2H, 2 × H_{Ar}), 6.83 (d, *J* = 1.3 Hz, 1H, H_{Ar}), 6.61 (d, *J* = 1.4 Hz, 1H, H_{Ar}), 3.89 (s, 3H, OCH_3), 3.86 (s, 2H, BnCH_2), 1.94 – 1.78 (m, 4H, 2 × CH_2), 1.78 – 1.50 (m, 6H, 3 × CH_2), 1.22 – 1.07 (m, 2H, CH_2), 0.98 – 0.84 (m, 2H, CH_2), 0.77 (t, *J* = 7.3 Hz, 3H, CH_3) ppm. – **^{13}C NMR** (100 MHz, $CDCl_3$): δ = 162.7 ($C_{quart.}$, *C*OO), 156.1 ($C_{quart.}$, C_{Ar}), 155.8 ($C_{quart.}$, C_{Ar}), 154.9 ($C_{quart.}$, C_{Ar}), 150.9 (+, 2 × C_{Ar}H), 148.5 ($C_{quart.}$, C_{Ar}), 136.3 (+, C_{Ar}H), 125.2 (+, 2 × C_{Ar}H), 124.9 ($C_{quart.}$, C_{Ar}), 108.4 (+, C_{Ar}H), 108.4 ($C_{quart.}$, C_{Ar}), 105.0 (+, C_{Ar}H), 56.7 (+, OCH_3), 52.9 ($C_{quart.}$, C_{CP}), 42.4 (–, CH_2), 38.6(–, 2 × CH_2), 37.2 (–, CH_2), 28.4 (–, CH_2), 24.1 (–, CH_2), 24.1(–, 2 × CH_2), 14.9 (+, CH_3) ppm. – **IR** (KBr): $\tilde{\nu}$ = 2926 (m), 2862 (w), 1715 (s), 1612 (s), 1567 (m), 1493 (m), 1458 (m), 1413 (m), 1351 (w), 1290 (w), 1253 (m), 1179 (m), 1110 (s), 1050 (m), 993 (w), 927 (w), 842 (m), 795 (w), 778 (m), 730 (w), 678 (w), 600 (w), 556 (w), 475 (w) cm^{-1}. – **MS** (70 eV, EI): *m/z* (%) = 391/392 (71/16) $[M]^+$, 365 (12), 337 (21), 336 (11), 335 (50), 334 (100) $[M – C_4H_9]^+$, 281 (28), 280 (60), 233 (50), 220 (33), 213 (22), 193 (23), 181 (44), 134 (26), 131 (60). – **HRMS** ($C_{21}H_{23}O_3N$): calc. 391.2142, found 391.2139.

7-(1-Butylcyclohexyl)-5-methoxy-3-(pyridin-2-ylmethyl)-2*H*-chromen-2-one (96g-Me)

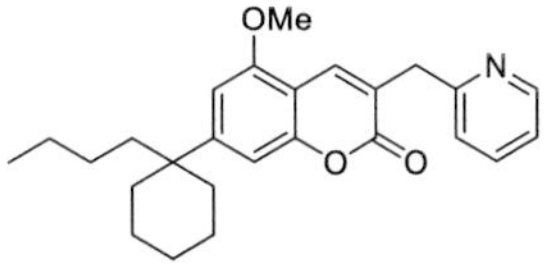

According to GP5, 150 mg of salicylic aldehyde **29c** (0.517 mmol, 1.00 equiv.), 157 mg of potassium carbonate (1.14 mmol, 2.20 equiv.), 172 mg of acrylaldehyde **92a** (1.29 mmol, 2.50 equiv.) and 138 mg of

1,3-dimethylimidazolium dimethyl phosphate (**31**) (0.620 mmol, 1.20 equiv.) were suspended in 3.00 mL of toluene. The reaction mixture was heated at 110 °C for 50 min via microwave irradiation. After cooling to room temperature, 5 mL of water was added and the mixture was extracted with 3 × 15 mL of ethyl acetate. Removal of the volatiles under reduced pressure and purification via flash column chromatography (CH/EtOAc 3:2) gave 38.9 mg (19%) of the pure product as an off-white solid.

***R*f** (CH/EtOAc 1:1): 0.34. – **^{1}H NMR** (400 MHz, $CDCl_3$): δ = 8.57 – 8.48 (m, 1H, H_{Ar}), 8.00 (s, 1H, 4C-H_{Ar}), 7.62 (td, J = 7.7, 1.9 Hz, 1H, H_{Ar}), 7.38 (d, J = 7.8 Hz, 1H, H_{Ar}), 7.13 (dd, J = 7.5, 5.0 Hz, 1H, H_{Ar}), 6.86 (s, 1H, H_{Ar}), 6.63 (d, J = 1.5 Hz, 1H, H_{Ar}), 4.05 (s, 2H, BnCH_2), 3.88 (s, 3H, OCH_3), 2.05 – 1.95 (m, 2H, CH_2), 1.64 – 1.27 (m, 10H, 5 × CH_2), 1.18 – 1.06 (m, 2H, CH_2), 0.95 – 0.81 (m, 2H, CH_2), 0.75 (t, J = 7.3 Hz, 3H, CH_3) ppm. – **^{13}C NMR** (100 MHz, $CDCl_3$): δ = 162.2 ($C_{quart.}$, COO), 158.6 ($C_{quart.}$, C_{Ar}), 155.7 ($C_{quart.}$, C_{Ar}), 154.5 ($C_{quart.}$, C_{Ar}), 153.0 ($C_{quart.}$, C_{Ar}), 149.5 (+, C_{Ar}H), 136.8 (+, C_{Ar}H), 136.0 (+, C_{Ar}H), 124.2 ($C_{quart.}$, C_{Ar}), 123.9 (+, C_{Ar}H), 121.8 (+, C_{Ar}H), 107.9 (+, C_{Ar}H), 107.9 ($C_{quart.}$, C_{Ar}), 103.9 (+, C_{Ar}H), 55.9 (+, OCH_3), 43.6 (–, CH_2), 42.3 ($C_{quart.}$, C_{CH}), 39.6 (–, CH_2), 36.4 (–, 2 × CH_2), 26.5 (–, CH_2), 25.7 (–, CH_2), 23.4 (–, CH_2), 22.5 (–, 2 × CH_2), 14.1 (+, CH_3) ppm. – **IR** (KBr): $\tilde{\nu}$ = 2926 (m), 2855 (m), 1715 (s), 1612 (s), 1589 (m), 1568 (m), 1495 (w), 143 (m), 1433 (m), 1414 (m), 1343 (w), 1288 (w), 1253 (m), 1182 (m), 1111 (s), 1051 (m), 994 (w), 916 (w), 841 (m), 747 (m), 685 (w), 628 (vw), 605 (w), 560 (w), 487 (w), 404 (w) cm^{-1}. – **MS** (70 eV, EI): *m/z* (%) = 405 (22) $[M]^+$, 290 (28), 247 (13), 235 (23), 234 (82), 233 (16), 194 (20), 181 (25), 180 (24), 179 (13), 137 (20), 131 (31), 86 (63), 84 (100), 69 (45). – **HRMS** ($C_{26}H_{31}O_3N$): calc. 405.2298, found 405.2298.

7-(1-Butylcyclohexyl)-5-methoxy-3-(pyridin-3-ylmethyl)-2*H*-chromen-2-one (96h-Me)

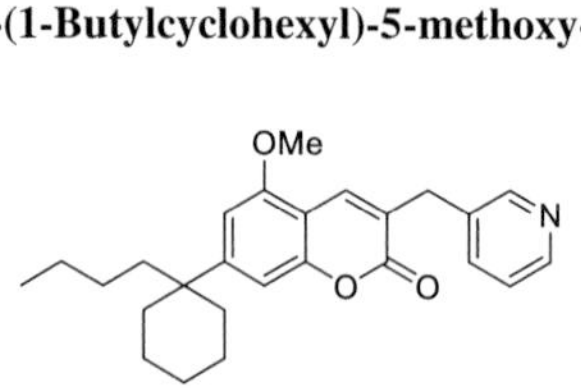

According to GP5, 150 mg of salicylic aldehyde **29c** (0.517 mmol, 1.00 equiv.), 157 mg of potassium carbonate (1.14 mmol, 2.20 equiv.), 172 mg of acrylaldehyde **92b** (1.29 mmol, 2.50 equiv.) and 138 mg of 1,3-dimethylimidazolium dimethyl phosphate (**31**) (0.620 mmol, 1.20 equiv.) were suspended in 3.00 mL of toluene. The reaction mixture was heated at 110 °C for 50 min via microwave irradiation. After cooling to room temperature, 5 mL of water was added and the mixture was extracted with 3 × 15 mL of ethyl acetate. Removal of the volatiles under reduced pressure and purification via flash column chromatography (CH/EtOAc 3:2) gave 31.9 mg (15%) of the pure product as an off-white solid.

***R*f** (CH/EtOAc 1:1): 0.16. – **^{1}H NMR** (400 MHz, $CDCl_3$): δ = 8.58 (d, *J* = 2.2 Hz, 1H, H_{Ar}), 8.49 (dd, *J* = 4.8, 1.6 Hz, 1H, H_{Ar}), 7.79 (s, 1H, 4-C*H*), 7.67 (dt, *J* = 7.8, 2.0 Hz, 1H, H_{Ar}), 7.27 – 7.20 (m, 1H, H_{Ar}), 6.88 (d, *J* = 1.3 Hz, 1H, H_{Ar}), 6.65 (d, *J* = 1.5 Hz, 1H, H_{Ar}), 3.88 (s, 3H, OCH_3), 3.87 (s, 2H, BnCH_2), 2.05 – 1.91 (m, 2H, CH_2), 1.64 – 1.28 (m, 10H, 5 × CH_2), 1.18 – 1.05 (m, 2H, CH_2), 0.94 – 0.82 (m, 2H, CH_2), 0.76 (t, *J* = 7.3 Hz, 3H, CH_3) ppm. – **^{13}C NMR** (100 MHz, $CDCl_3$): δ = 162.0 ($C_{quart.}$, *C*OO), 155.5 ($C_{quart.}$, C_{Ar}), 154.4 ($C_{quart.}$, C_{Ar}), 153.4 ($C_{quart.}$, C_{Ar}), 150.5 (+, C_{Ar}H), 148.2 (+, C_{Ar}H), 136.8 (+, C_{Ar}H), 135.1 (+, C_{Ar}H), 134.2 ($C_{quart.}$, C_{Ar}), 125.1 ($C_{quart.}$, C_{Ar}), 123.6 (+, C_{Ar}H), 107.9 (+, C_{Ar}H), 107.6 ($C_{quart.}$, C_{Ar}), 104.0 (+, C_{Ar}H), 55.9 (+, OCH_3), 43.6 (–, CH_2), 42.4 ($C_{quart.}$, C_{CH}), 36.4 (–, 2 × CH_2), 34.3 (–, CH_2), 26.5 (–, CH_2), 25.8 (–, CH_2), 23.4 (–, CH_2), 22.5 (–, 2 × CH_2), 14.1 (+, CH_3) ppm. – **IR** (KBr): $\tilde{\nu}$ = 2926 (m), 2855 (w), 1715 (m), 1612 (m), 1568 (w), 1496 (w), 1453 (m), 1415 (m), 1343 (w), 1288 (w), 1253 (m), 1180 (w), 1111 (m), 1051 (m), 1026 (w), 915 (w), 841 (w), 778 (w), 712 (m), 630 (w), 607 (w), 559 (w), 479 (w), 401 (w) cm^{-1}. – **MS** (70 eV, EI): *m/z* (%) = 405 (17) $[M]^+$, 349 (9), 348 (17) $[M - C_4H_9]^+$, 280 (17), 247 (23), 234 (26), 221 (28), 191 (19), 153 (17), 132 (56), 130 (29), 118 (20), 117 (43), 97 (13), 86 (62), 85 (16), 84 (100). – **HRMS** ($C_{26}H_{31}O_3N$): calc. 405.2298, found 405.2297.

7-(1-Butylcyclohexyl)-5-methoxy-3-(pyridin-4-ylmethyl)-2*H*-chromen-2-one (96i-Me)

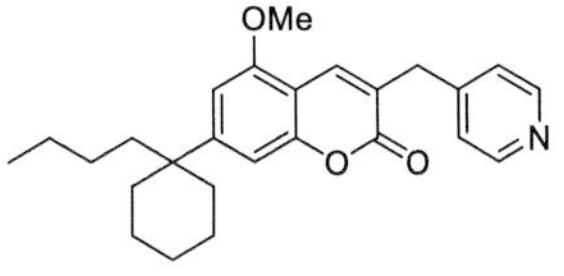

According to GP5, 150 mg of salicylic aldehyde **29c** (0.517 mmol, 1.00 equiv.), 157 mg of potassium carbonate (1.14 mmol, 2.20 equiv.), 172 mg of acrylaldehyde **92c** (1.29 mmol, 2.50 equiv.) and 138 mg of 1,3-dimethylimidazolium dimethyl phosphate (**31**) (0.620 mmol, 1.20 equiv.) were suspended in 3.00 mL of toluene. The reaction mixture was heated at 110 °C for 50 min via microwave irradiation. After cooling to room temperature, 5 mL of water was added and the mixture was extracted with 3 × 15 mL of ethyl acetate. Removal of the volatiles under reduced pressure and purification via flash column chromatography (CH/EtOAc 1:1) gave 53.6 mg (26%) of the pure product as an off-white solid.

***R*f** (CH/EtOAc 1:2): 0.30. – **^{1}H NMR** (400 MHz, $CDCl_3$): δ = 8.58 – 8.46 (m, 2H, 2 × H_{Ar}), 7.81 (s, 1H, 4-C*H*), 7.28 – 7.22 (m, 2H, 2 × H_{Ar}), 6.89 (d, *J* = 1.4 Hz, 1H, H_{Ar}), 6.66 (d,

J = 1.5 Hz, 1H, H_{Ar}), 3.90 (s, 3H, OCH_3), 3.87 (s, 2H, $BnCH_2$), 2.07 – 1.94 (m, 2H, CH_2), 1.65 – 1.22 (m, 10H, 5 × CH_2), 1.19 – 1.06 (m, 2H, CH_2), 0.96 – 0.83 (m, 2H, CH_2), 0.77 (t, J = 7.3 Hz, 3H, CH_3) ppm. – **^{13}C NMR** (100 MHz, $CDCl_3$): δ = 162.0 ($C_{quart.}$, COO), 155.6 ($C_{quart.}$, C_{Ar}), 154.4 ($C_{quart.}$, C_{Ar}), 153.6 ($C_{quart.}$, C_{Ar}), 150.1 (+, 2 × $C_{Ar}H$), 147.8 ($C_{quart.}$, C_{Ar}), 135.6 (+, $C_{Ar}H$), 124.5 (+, 2 × $C_{Ar}H$), 124.2 ($C_{quart.}$, C_{Ar}), 108.0 (+, $C_{Ar}H$), 107.5 ($C_{quart.}$, C_{Ar}), 104.0 (+, $C_{Ar}H$), 55.9 (+, OCH_3), 43.6 (–, CH_2), 42.4 ($C_{quart.}$, C_{CH}), 36.4 (–, CH_2), 36.4 (–, 2 × CH_2), 26.5 (–, CH_2), 25.8 (–, CH_2), 23.4 (–, CH_2), 22.5 (–, 2 × CH_2), 14.1 (+, CH_3) ppm. – **IR** (KBr): $\tilde{\nu}$ = 2926 (m), 2855 (m), 1716 (s), 1612 (s), 1567 (m), 1495 (m), 1453 (m), 1413 (m), 1342 (w), 1288 (w), 1254 (m), 1188 (w), 1182 (m), 1111 (s), 1048 (m), 993 (w), 915 (w), 842 (m), 796 (w), 778 (m), 729 (m), 685 (w), 602 (w), 556 (w), 487 (w) cm^{-1}. – **MS** (70 eV, EI): m/z (%) = 405 (7) $[M]^+$, 379 (7), 348 (5) $[M - C_4H_9]^+$, 303 (11), 247 (19), 234 (20), 181 (24), 166 (18), 131 (28), 119 (11), 88 (11), 86 (64), 84 (100), 82 (23). – **HRMS** ($C_{26}H_{31}O_3N$): calc. 405.2298, found 405.2298.

5-Hydroxy-7-pentyl-3-(pyridin-2-ylmethyl)-2*H*-chromen-2-one (96a-H)

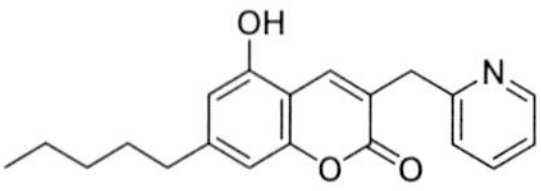

According to **GP8**, 27.0 mg of 5-methoxycoumarin **96a-Me** (80.0 µmol, 1.00 equiv.) was dissolved in 3 mL of dry dichloromethane. The solution was cooled to –78 °C and 0.36 mL of boron tribromide (1 M in dichloromethane, 360 µmol, 4.50 equiv.) was added dropwise. The mixture was stirred for 30 min at this temperature and then allowed to warm to room temperature. The reaction was quenched after 16 h at 0 °C by addition of sodium bicarbonate. The aqueous layer was extracted with 3 × 15 mL of dichloromethane and the combined organic layers were washed with brine, dried over sodium sulfate and the volatiles were removed under reduced pressure. The crude product was then purified twice via flash column chromatography (CH/EtOAc 1:1, 2:1) to give the product as 5.5 mg (21%) of a white solid.

***R*f** (CH/EtOAc 1:1): 0.14. – **^{1}H NMR** (400 MHz, $CDCl_3$): δ = 8.57 – 8.52 (m, 2H, 2 × H_{Ar}), 7.96 (td, J = 7.7, 1.7 Hz, 1H, H_{Ar}), 7.83 (d, J = 7.9 Hz, 1H, H_{Ar}), 7.45 (t, J = 6.4 Hz, 1H, H_{Ar}), 6.45 (s, 1H, H_{Ar}), 6.41 (s, 1H, H_{Ar}), 4.16 (s, 2H, $BnCH_2$), 2.52 – 2.42 (m, 2H, CH_2), 1.57 – 1.45 (m, 2H, CH_2), 1.38 – 1.19 (m, 4H, 2 × CH_2), 0.86 (t, J = 6.9 Hz, 3H, CH_3) ppm. – **^{13}C NMR** (100 MHz, $CDCl_3$): δ = 162.4 ($C_{quart.}$, COO), 157.5 ($C_{quart.}$, C_{Ar}), 154.9 ($C_{quart.}$, C_{Ar}), 154.8 ($C_{quart.}$, C_{Ar}), 148.7 ($C_{quart.}$, C_{Ar}), 139.5 (+, $C_{Ar}H$), 126.8 (+, 2 × $C_{Ar}H$), 123.4 (+, 2 × $C_{Ar}H$),

120.0 ($C_{quart.}$, C_{Ar}), 110.3 (+, $C_{Ar}H$), 107.6 ($C_{quart.}$, C_{Ar}), 107.0 (+, $C_{Ar}H$), 38.4 (–, CH_2), 36.3 (–, CH_2), 31.5 (–, CH_2), 30.6 (–, CH_2), 22.6 (–, CH_2), 14.1 (+, CH_3) ppm. – **IR** (KBr): ṽ = 2924 (m), 2853 (w), 1713 (m), 1616 (m), 1598 (m), 1570 (m), 1476 (w), 1435 (m), 1393 (w), 1358 (m), 1289 (m), 1252 (w), 1185 (m), 1137 (w), 1098 (m), 1078 (m), 1047 (m), 1012 (m), 922 (vw), 834 (m), 766 (m), 749 (m), 735 (w), 665 (w), 636 (w), 606 (w), 569 (w), 522 (w), 504 (w) cm^{-1}. – **MS** (70 eV, EI): *m/z* (%) = 323 (100) $[M]^+$, 57 (100). – **HRMS** ($C_{20}H_{21}O_3N$): calc. 323.1521, found 323.1521.

5-Hydroxy-7-pentyl-3-(pyridin-3-ylmethyl)-2*H*-chromen-2-one (96b-H)

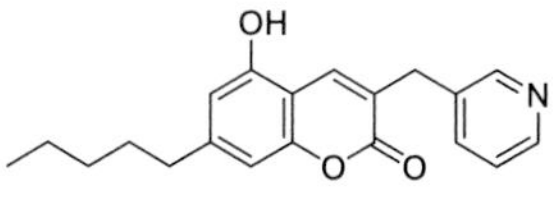

According to **GP8**, 19.0 mg of 5-methoxycoumarin **96b-Me** (56.0 µmol, 1.00 equiv.) was dissolved in 3 mL of dry dichloromethane. The solution was cooled to –78 °C and 0.25 mL of boron tribromide (1 M in dichloromethane, 253 µmol, 4.50 equiv.) was added dropwise. The mixture was stirred for 30 min at this temperature and then allowed to warm to room temperature. The reaction was quenched after 16 h at 0 °C by addition of sodium bicarbonate. The aqueous layer was extracted with 3 × 15 mL of dichloromethane and the combined organic layers were washed with brine, dried over sodium sulfate and the volatiles were removed under reduced pressure. The crude product was then purified via flash column chromatography (CH/EtOAc 1:1) to give the product as 9.4 mg (52%) of a white solid.

***R*f** (CH/EtOAc 1:1): 0.13. – **^{1}H NMR** (400 MHz, $CDCl_3$): δ = 8.56 (s, 1H, H_{Ar}), 8.49 (d, *J* = 5.1 Hz, 1H, H_{Ar}), 8.03 (d, *J* = 7.9 Hz, 1H, H_{Ar}), 8.00 (s, 1H, H_{Ar}), 7.47 (dd, *J* = 7.9, 5.0 Hz, 1H, H_{Ar}), 6.62 (s, 1H, H_{Ar}), 6.60 (s, 1H, H_{Ar}), 3.91 (s, 2H, BnCH_2), 2.57 – 2.47 (m, 2H, CH_2), 1.60 – 1.50 (m, 2H, CH_2), 1.32 – 1.17 (m, 4H, 2 × CH_2), 0.85 (t, *J* = 6.8 Hz, 3H, CH_3) ppm. – **^{13}C NMR** (100 MHz, $CDCl_3$): δ = 162.3 ($C_{quart.}$, *C*OO), 154.7 ($C_{quart.}$, C_{Ar}), 154.4 ($C_{quart.}$, C_{Ar}), 148.8 ($C_{quart.}$, C_{Ar}), 147.3 (+, $C_{Ar}H$), 144.9 (+, $C_{Ar}H$), 140.5 (+, $C_{Ar}H$), 136.8 (+, $C_{Ar}H$), 136.6 ($C_{quart.}$, C_{Ar}), 124.9 (+, $C_{Ar}H$), 122.9 ($C_{quart.}$, C_{Ar}), 110.9 (+, $C_{Ar}H$), 107.4 (+, $C_{Ar}H$), 107.4 ($C_{quart.}$, C_{Ar}), 36.4 (–, CH_2), 34.5 (–, CH_2), 31.5 (–, CH_2), 30.7 (–, CH_2), 22.6 (–, CH_2), 14.1 (+, CH_3) ppm. – **IR** (KBr): ṽ = 2922 (m), 2851 (m), 1706 (s), 1619 (s), 1431 (m), 1395 (m), 1369 (m), 1346 (m), 1278 (m), 1246 (m), 1168 (m), 1137 (m), 1079 (m), 1043 (m), 883 (w), 831 (m), 800 (m), 761 (m), 726 (w), 712 (m), 642 (m), 610 (w), 562 (w), 521 (m), 411 (m) cm^{-1}. – **MS** (70 eV, EI): *m/z* (%) = 323 (19) $[M]^+$, 57 (100). – **HRMS** ($C_{20}H_{21}O_3N$): calc. 323.1521, found 323.1520.

5-Hydroxy-7-pentyl-3-(pyridin-4-ylmethyl)-2*H*-chromen-2-one (96c-H)

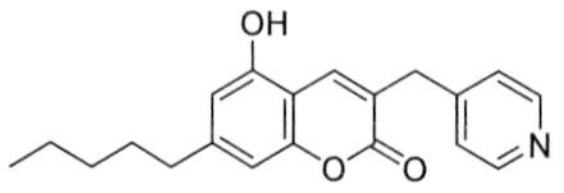

According to **GP8**, 21.1 mg of 5-methoxycoumarin **96c-Me** (63.0 µmol, 1.00 equiv.) was dissolved in 2 mL of dry dichloromethane. The solution was cooled to –78 °C and 0.28 mL of boron tribromide (1 M in dichloromethane, 281 µmol, 4.50 equiv.) was added dropwise. The mixture was stirred for 30 min at this temperature and then allowed to warm to room temperature. The reaction was quenched after 16 h at 0 °C by addition of sodium bicarbonate. The aqueous layer was extracted with 3 × 15 mL of dichloromethane and the combined organic layers were washed with brine, dried over sodium sulfate and the volatiles were removed under reduced pressure. The crude product was then purified via flash column chromatography (CH/EtOAc 1:1) to give the product as 9.0 mg (45%) of a yellow solid.

***R*f** (CH/EtOAc 1:2): 0.13. – **^{1}H NMR** (400 MHz, $CDCl_3$): δ = 8.50 (d, *J* = 5.3 Hz, 2H, 2 × H_{Ar}), 7.82 (s, 1H, H_{Ar}), 7.39 – 7.33 (m, 2H, 2 × H_{Ar}), 6.66 (d, *J* = 1.3 Hz, 1H, H_{Ar}), 6.53 (d, *J* = 1.3 Hz, 1H), 3.91 (s, 2H, BnCH_2), 2.61 – 2.51 (m, 2H, CH_2), 1.63 – 1.51 (m, 2H, CH_2), 1.35 – 1.23 (m, 4H, 2 × CH_2), 0.86 (t, *J* = 6.8 Hz, 3H, CH_3) ppm. – **^{13}C NMR** (100 MHz, $CDCl_3$): δ = 162.3 ($C_{quart.}$, COO), 154.7 ($C_{quart.}$, C_{Ar}), 154.3 ($C_{quart.}$, C_{Ar}), 149.6 ($C_{quart.}$, C_{Ar}), 148.7 (+, 2 × C_{Ar}H), 148.7 ($C_{quart.}$, C_{Ar}), 136.4 (+, C_{Ar}H), 125.3 (+, 2 × C_{Ar}H), 122.9 ($C_{quart.}$, C_{Ar}), 110.6 (+, C_{Ar}H), 107.6 (+, C_{Ar}H), 107.4 ($C_{quart.}$, C_{Ar}), 36.3 (–, CH_2), 36.3 (–, CH_2), 31.4 (–, CH_2), 30.7 (–, CH_2), 22.6 (–, CH_2), 14.1 (+, CH_3) ppm. – **IR** (KBr): ṽ = 2922 (m), 2852 (w), 1720 (m), 1613 (m), 1562 (w), 1428 (m), 1351 (w), 1293 (w), 1249 (w), 1220 (w), 1176 (w), 1139 (w), 1082 (m), 1051 (m), 1012 (m), 921 (m), 836 (m), 803 (m), 778 (m), 747 (w), 732 (w), 659 (w), 599 (w), 568 (w), 501 (w), 408 (w) cm^{-1}. – **MS** (70 eV, EI): *m/z* (%) = 323 (17) $[M]^+$, 69 (100). – **HRMS** ($C_{20}H_{21}O_3N$): calc. 323.1516, found 323.1517.

7-(1-Butylcyclopentyl)-5-hydroxy-3-(pyridin-2-ylmethyl)-2*H*-chromen-2-one (96d-H)

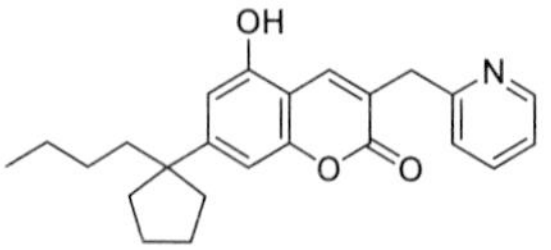

According to **GP8**, 9.1 mg of 5-methoxycoumarin **96d-Me** (23 µmol, 1.00 equiv.) was dissolved in 2 mL of dry dichloromethane. The solution was cooled to –78 °C and 0.10 mL of boron tribromide (1 M in dichloromethane, 105 µmol, 4.50 equiv.) was added dropwise. The mixture was stirred for 30 min at this temperature and then allowed to warm to room temperature. The reaction was quenched after 16 h at 0 °C by addition of sodium bicarbonate. The aqueous layer was extracted with

3 × 15 mL of dichloromethane and the combined organic layers were washed with brine, dried over sodium sulfate and the volatiles were removed under reduced pressure. The crude product was then purified via flash column chromatography (CH/EtOAc 1:1) to give the product as 5.3 mg (60%) of a white solid.

***R*f** (CH/EtOAc 1:1): 0.39. – **^{1}H NMR** (400 MHz, $CDCl_3$): δ = 8.69 (s, 1H, H_{Ar}), 8.55 (dd, J = 5.2, 1.4 Hz, 1H, H_{Ar}), 7.97 (td, J = 7.8, 1.7 Hz, 1H, H_{Ar}), 7.89 (d, J = 7.9 Hz, 1H, H_{Ar}), 7.46 (t, J = 6.2 Hz, 1H, H_{Ar}), 6.65 (d, J = 1.5 Hz, 1H, H_{Ar}), 6.63 (d, J = 1.4 Hz, 1H, H_{Ar}), 4.19 (s, 2H, BnCH_2), 1.82 – 1.54 (m, 8H, 4 × CH_2), 1.53 – 1.46 (m, 2H, CH_2), 1.17 – 1.07 (m, 2H, CH_2), 0.95 – 0.86 (m, 2H, CH_2), 0.76 (t, J = 7.3 Hz, 3H, CH_3) ppm. – **^{13}C NMR** (100 MHz, $CDCl_3$): δ = 162.4 ($C_{quart.}$, COO), 157.3 ($C_{quart.}$, C_{Ar}), 155.3 ($C_{quart.}$, C_{Ar}), 154.6 ($C_{quart.}$, C_{Ar}), 154.5 ($C_{quart.}$, C_{Ar}), 145.4 (+, C_{Ar}H), 140.7 (+, C_{Ar}H), 139.5 (+, C_{Ar}H), 127.1 (+, C_{Ar}H), 123.4 (+, C_{Ar}H), 120.2 ($C_{quart.}$, C_{Ar}), 109.4 (+, C_{Ar}H), 107.4 ($C_{quart.}$, C_{Ar}), 106.1 (+, C_{Ar}H), 51.7 ($C_{quart.}$, C_{CP}), 41.6 (–, CH_2), 37.7 (–, 2 × CH_2), 29.9 (–, CH_2), 27.6 (–, CH_2), 23.4 (–, CH_2), 23.3 (–, 2 × CH_2), 14.2 (+, CH_3) ppm. – **IR** (KBr): ṽ = 2923 (m), 2854 (w), 1711 (m), 1616 (m), 1570 (w), 1421 (m), 1345 (w), 1288 (w), 1256 (w), 1181 (w), 1052 (w), 1009 (w), 920 (vw), 839 (w), 767 (w), 731 (w), 671 (w), 626 (w), 522 (w), 408 (vw) cm^{-1}. – **MS** (70 eV, EI): *m/z* (%) = 377 (12) $[M]^+$, 385 (9) $[M - C_6H_6N]^+$, 284 (31), 279 (33), 57 (100) $[C_4H_9]^+$. – **HRMS** ($C_{24}H_{27}O_3N$): calc. 377.1991, found 377.1990.

7-(1-Butylcyclopentyl)-5-hydroxy-3-(pyridin-3-ylmethyl)-2*H*-chromen-2-one (96e-H)

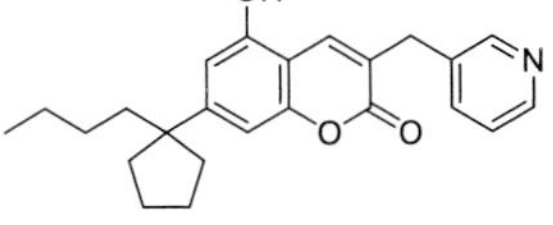

According to **GP8**, 20.5 mg of 5-methoxycoumarin **96e-Me** (52.0 µmol, 1.00 equiv.) was dissolved in 2 mL of dry dichloromethane. The solution was cooled to –78 °C and 0.24 mL of boron tribromide (1 M in dichloromethane, 236 µmol, 4.50 equiv.) was added dropwise. The mixture was stirred for 30 min at this temperature and then allowed to warm to room temperature. The reaction was quenched after 16 h at 0 °C by addition of sodium bicarbonate. The aqueous layer was extracted with 3 × 15 mL of dichloromethane and the combined organic layers were washed with brine, dried over sodium sulfate and the volatiles were removed under reduced pressure. The crude product was then purified via flash column chromatography (CH/EtOAc 1:1) to give the product as 10.3 mg (52%) of a yellow solid.

***R*f** (CH/EtOAc 1:2): 0.18. – **^{1}H NMR** (500 MHz, $CDCl_3$): δ = 8.51 (d, *J* = 2.1 Hz, 1H, H_{Ar}), 8.47 (dd, *J* = 5.0, 1.5 Hz, 1H, H_{Ar}), 7.93 (s, 1H), 7.91 (dt, *J* = 7.8, 1.9 Hz, 1H, H_{Ar}), 7.38 (dd, *J* = 7.9, 5.0 Hz, 1H, H_{Ar}), 6.74 (d, *J* = 1.4 Hz, 1H, H_{Ar}), 6.67 (d, *J* = 1.5 Hz, 1H, H_{Ar}), 3.89 (s, 2H, BnCH_2), 1.85 – 1.55 (m, 8H, 4 × CH_2), 1.54 – 1.48 (m, 2H, CH_2), 1.16 – 1.09 (m, *J* = 7.2 Hz, 2H, CH_2), 0.91 (dp, *J* = 15.4, 7.0, 5.7 Hz, 2H, CH_2), 0.74 (t, *J* = 7.3 Hz, 3H, CH_3) ppm. – **^{13}C NMR** (125 MHz, $CDCl_3$): δ = 162.3 ($C_{quart.}$, *C*OO), 155.0 ($C_{quart.}$, C_{Ar}), 154.4 ($C_{quart.}$, C_{Ar}), 153.8 ($C_{quart.}$, C_{Ar}), 148.6 (+, C_{Ar}H), 146.2 (+, C_{Ar}H), 139.1 (+, C_{Ar}H), 136.2 (+, C_{Ar}H), 135.8 ($C_{quart.}$, C_{Ar}), 124.5 (+, C_{Ar}H), 123.7 ($C_{quart.}$, C_{Ar}), 109.5 (+, C_{Ar}H), 107.3 ($C_{quart.}$, C_{Ar}), 106.5 (+, C_{Ar}H), 51.8 ($C_{quart.}$, C_{CP}), 41.6 (–, CH_2), 37.8 (–, 2 × CH_2), 34.5 (–, CH_2), 27.6 (–, CH_2), 23.4 (–, CH_2), 23.3 (–, 2 × CH_2), 14.1 (+, CH_3) ppm. – **IR** (KBr): ṽ = 2922 (w), 2852 (w), 1712 (w), 1616 (w), 1574 (w), 1421 (w), 1345 (w), 1294 (vw), 1252 (w), 1176 (w), 1049 (w), 934 (vw), 839 (vw), 778 (vw), 708 (w), 642 (vw), 528 (vw), 415 (vw) cm^{-1}. – **MS** (70 eV, EI): *m/z* (%) = 377 (37) $[M]^+$, 369 (9), 331 (12), 321 (44), 320 (100) $[M - C_4H_9]^+$, 319 (11), 292 (14), 281 (18), 269 (21), 266 (14), 238 (19), 231 (20), 219 (25), 181 (44), 169 (31), 131 (41). – **HRMS** ($C_{24}H_{27}O_3N$): calc. 377.1985, found 377.1983.

7-(1-Butylcyclopentyl)-5-hydroxy-3-(pyridin-4-ylmethyl)-2*H*-chromen-2-one (96f-H)

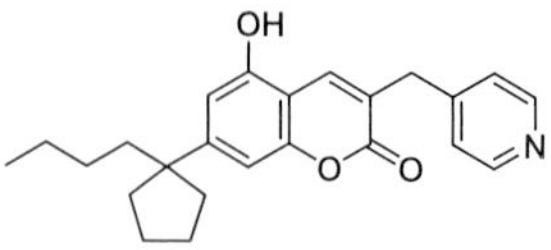

According to **GP8**, 36.0 mg of 5-methoxycoumarin **96f-Me** (92.0 µmol, 1.00 equiv.) was dissolved in 2 mL of dry dichloromethane. The solution was cooled to –78 °C and 0.41 mL of boron tribromide (1 M in dichloromethane, 414 µmol, 4.50 equiv.) was added dropwise. The mixture was stirred for 30 min at this temperature and then allowed to warm to room temperature. The reaction was quenched after 16 h at 0 °C by addition of sodium bicarbonate. The aqueous layer was extracted with 3 × 15 mL of dichloromethane and the combined organic layers were washed with brine, dried over sodium sulfate and the volatiles were removed under reduced pressure. The crude product was then purified via flash column chromatography (CH/EtOAc 1:1) to give the product as 14.6 mg (42%) of a yellow oil.

***R*f** (CH/EtOAc 1:1): 0.14. – **^{1}H NMR** (400 MHz, $CDCl_3$): δ = 8.57 – 8.44 (m, 2H, 2 × H_{Ar}), 7.93 (s, 1H, H_{Ar}), 7.49 – 7.40 (m, 2H, 2 × H_{Ar}), 6.75 (d, *J* = 1.4 Hz, 1H, H_{Ar}), 6.72 (d, *J* = 1.5 Hz, 1H, H_{Ar}), 3.94 (s, 2H, BnCH_2), 1.90 – 1.55 (m, 8H, 4 × CH_2), 1.54 – 1.44 (m, 2H, CH_2), 1.12 (h, *J* = 7.4 Hz, 2H, CH_2), 0.96 – 0.84 (m, 2H, CH_2), 0.74 (t, *J* = 7.3 Hz, 3H, CH_3)

ppm. – ^{13}C **NMR** (100 MHz, $CDCl_3$): δ = 162.3 ($C_{quart.}$, *C*OO), 155.4 ($C_{quart.}$, C_{Ar}), 154.4 ($C_{quart.}$, C_{Ar}), 153.9 ($C_{quart.}$, C_{Ar}), 151.4 ($C_{quart.}$, C_{Ar}), 147.4 (+, C_{Ar}H), 136.8 (+, C_{Ar}H), 125.6 (+, C_{Ar}H), 122.4 ($C_{quart.}$, C_{Ar}), 109.5 (+, C_{Ar}H), 107.0 ($C_{quart.}$, C_{Ar}), 106.5 (+, C_{Ar}H), 51.8 ($C_{quart.}$, C_{CP}), 41.6 (–, CH_2), 37.8 (–, 2 × CH_2), 36.6 (–, CH_2), 27.6 (–, CH_2), 23.4 (–, CH_2), 23.3 (–, 2 × CH_2), 14.1 (+, CH_3) ppm. – **IR** (KBr): ṽ = 2927 (w), 1710 (m), 1615 (m), 1421 (m), 1343 (w), 1288 (w), 1254 (w), 1178 (w), 1052 (m), 1009 (w), 908 (w), 842 (w), 779 (w), 728 (m), 674 (w), 598 (w), 556 (vw), 521 (w), 472 (w), 424 (vw) cm^{-1}. – **MS** (70 eV, EI): *m/z* (%) = 377 (12) $[M]^+$, 320 (53) $[M - C_4H_9]^+$, 284 (44), 256 (39), 69 (100). – **HRMS** ($C_{24}H_{27}O_3N$): calc. 377.1991, found 377.1990.

7-(1-Butylcyclohexyl)-5-hydroxy-3-(pyridin-2-ylmethyl)-2*H*-chromen-2-one (96g-H)

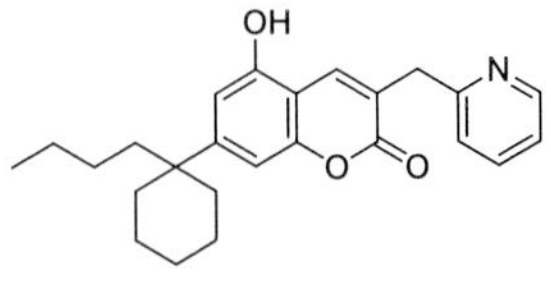

According to **GP8**, 19.0 mg of 5-methoxycoumarin **96g-Me** (47.0 µmol, 4.50 equiv.) was dissolved in 1 mL of dry dichloromethane. The solution was cooled to –78 °C and 0.21 mL of boron tribromide (1 M in dichloromethane, 211 µmol, 5.00 equiv.) was added dropwise. The mixture was stirred for 30 min at this temperature and then allowed to warm to room temperature. The reaction was quenched after 16 h at 0 °C by addition of sodium bicarbonate. The aqueous layer was extracted with 3 × 15 mL of dichloromethane and the combined organic layers were washed with brine, dried over sodium sulfate and the volatiles were removed under reduced pressure. The crude product was then purified via flash column chromatography (CH/EtOAc 1:1) to give the product as 8.9 mg (49%) of a yellow oil.

***R*f** (CH/EtOAc 1:2): 0.35. – ^{1}**H NMR** (400 MHz, $CDCl_3$): δ = 11.82 (bs, 1H, O*H*), 8.62 (s, 1H, H_{Ar}), 8.47 – 8.41 (m, 1H, H_{Ar}), 7.83 – 7.73 (m, 2H, 2 × H_{Ar}), 7.29 (ddd, *J* = 7.0, 5.1, 1.7 Hz, 1H, H_{Ar}), 6.63 – 6.57 (m, 2H, 2 × H_{Ar}), 4.06 (s, 2H, BnCH_2), 1.82 (d, *J* = 11.9 Hz, 2H, CH_2), 1.46 – 1.29 (m, 6H, 3 × CH_2), 1.27 – 1.20 (m, 4H, 2 × CH_2), 1.10 – 0.99 (m, 2H, CH_2), 0.86 – 0.76 (m, 2H, CH_2), 0.69 (t, *J* = 7.3 Hz, 3H, CH_3) ppm. – ^{13}**C NMR** (100 MHz, $CDCl_3$): δ = 162.5 ($C_{quart.}$, *C*OO), 158.0 ($C_{quart.}$, C_{Ar}), 154.9 ($C_{quart.}$, C_{Ar}), 154.7 ($C_{quart.}$, C_{Ar}), 153.5 ($C_{quart.}$, C_{Ar}), 146.6 (+, C_{Ar}H), 139.4 (+, C_{Ar}H), 139.1 (+, C_{Ar}H), 126.6 (+, C_{Ar}H), 123.0 (+, C_{Ar}H), 121.0 ($C_{quart.}$, C_{Ar}), 108.9 (+, C_{Ar}H), 107.4 ($C_{quart.}$, C_{Ar}), 106.1 (+, C_{Ar}H), 42.0 ($C_{quart.}$, C_{CH}), 38.9 (–, CH_2), 36.4 (–, 2 × CH_2), 29.9 (–, CH_2), 26.6 (–, CH_2), 25.8 (–, 2 × CH_2), 23.4 (–, CH_2), 22.5 (– , CH_2), 14.2 (+, CH_3) ppm. – **IR** (KBr): ṽ = 2925 (w), 2855 (w), 1710 (w), 1617 (w), 1570 (w),

1420 (w), 1341 (w), 1290 (w), 1255 (w), 1184 (w), 1079 (w), 1058 (w), 1009 (w), 908 (vw), 840 (w), 768 (w), 729 (w), 673 (vw), 636 (vw), 604 (vw), 528 (vw), 409 (vw) cm^{-1}. – **MS** (70 eV, EI): *m/z* (%) = 391 (61) $[M]^+$, 334 (39) $[M-C_4H_9]^+$, 57 (100) $[C_4H_9]^+$. – **HRMS** ($C_{25}H_{29}O_3N$): calc. 391.2147, found 391.2146.

7-(1-Butylcyclohexyl)-5-hydroxy-3-(pyridin-3-ylmethyl)-2*H*-chromen-2-one (96h-H)

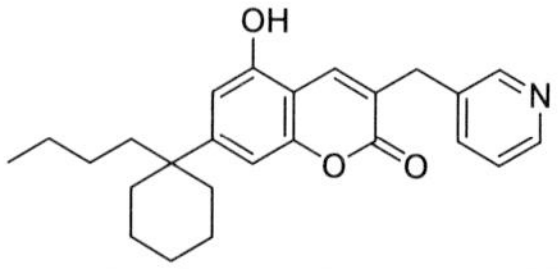

According to **GP8**, 11.0 mg of 5-methoxycoumarin **96h-Me** (27 µmol, 4.50 equiv.) was dissolved in 1 mL of dry dichloromethane. The solution was cooled to –78 °C and 0.13 mL of boron tribromide (1 M in dichloromethane, 126 µmol, 5.00 equiv.) was added dropwise. The mixture was stirred for 30 min at this temperature and then allowed to warm to room temperature. The reaction was quenched after 16 h at 0 °C by addition of sodium bicarbonate. The aqueous layer was extracted with 3 × 15 mL of dichloromethane and the combined organic layers were washed with brine, dried over sodium sulfate and the volatiles were removed under reduced pressure. The crude product was then purified via flash column chromatography (CH/EtOAc 1:1) to give the product as 9.3 mg (88%) of a yellow solid.

***R*f** (CH/EtOAc 1:2): 0.21. – **^{1}H NMR** (400 MHz, $CDCl_3$): δ = 8.52 (d, *J* = 2.1 Hz, 1H, H_{Ar}), 8.47 (dd, *J* = 5.1, 1.5 Hz, 1H, H_{Ar}), 7.92 (s, 1H, H_{Ar}), 7.88 (dt, *J* = 7.9, 1.9 Hz, 1H, H_{Ar}), 7.36 (dd, *J* = 7.9, 4.9 Hz, 1H, H_{Ar}), 6.80 (d, *J* = 1.4 Hz, 1H, H_{Ar}), 6.70 (d, *J* = 1.5 Hz, 1H, H_{Ar}), 3.89 (s, 2H, $BnCH_2$), 2.01 – 1.91 (m, 2H, CH_2), 1.55 – 1.39 (m, 4H, 2 × CH_2), 1.36 – 1.26 (m, 6H, 3 × CH_2), 1.14 – 1.04 (m, 2H, CH_2), 0.92 – 0.81 (m, 2H, CH_2), 0.74 (t, *J* = 7.3 Hz, 3H, CH_3) ppm. – **^{13}C NMR** (100 MHz, $CDCl_3$): δ = 162.2 ($C_{quart.}$, COO), 154.6 ($C_{quart.}$, C_{Ar}), 153.7 ($C_{quart.}$, C_{Ar}), 153.5 ($C_{quart.}$, C_{Ar}), 148.9 (+, $C_{Ar}H$), 146.6 (+, $C_{Ar}H$), 138.7 (+, $C_{Ar}H$), 135.9 (+, $C_{Ar}H$), 135.5 ($C_{quart.}$, C_{Ar}), 124.4 (+, $C_{Ar}H$), 124.1 ($C_{quart.}$, C_{Ar}), 109.4 (+, $C_{Ar}H$), 107.2 ($C_{quart.}$, C_{Ar}), 107.0 (+, $C_{Ar}H$), 42.1 ($C_{quart.}$, C_{CH}), 36.4 (–, 2 × CH_2), 34.4 (–, CH_2), 32.1 (–, CH_2), 29.9 (–, CH_2), 25.8 (–, CH_2), 23.4 (–, CH_2), 22.5 (–, 2 × CH_2), 14.1 (+, CH_3) ppm. – **IR** (KBr): $\tilde{\nu}$ = 2922 (vw), 2852 (vw), 1722 (w), 1616 (w), 1573 (vw), 1451 (vw), 1421 (w), 1339 (w), 1293 (vw), 1252 (vw), 1173 (vw), 1102 (vw), 1081 (vw), 1047v 1031 (vw), 838 (vw), 797 (vw), 777 (vw), 741 (vw), 708 (vw), 642 (vw), 610 (vw), 545 (vw), 451 (vw), 409 (vw) cm^{-1}. – **MS** (70 eV, EI): *m/z* (%) = 391 (45) $[M]^+$, 95 (100). – **HRMS** ($C_{25}H_{30}O_3N = [M+H]^+$): calc. 392.2220, found 392.2221.

7-(1-Butylcyclohexyl)-5-hydroxy-3-(pyridin-4-ylmethyl)-2*H*-chromen-2-one (96i-H)

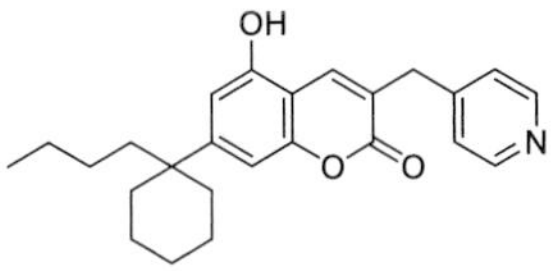

According to **GP8**, 33.6 mg of 5-methoxycoumarin **96i-Me** (83.0 µmol, 4.50 equiv.) was dissolved in 3 mL of dry dichloromethane. The solution was cooled to –78 °C and 0.37 mL of boron tribromide (1 M in dichloromethane, 37.0 µmol, 5.00 equiv.) was added dropwise. The mixture was stirred for 30 min at this temperature and then allowed to warm to room temperature. The reaction was quenched after 16 h at 0 °C by addition of sodium bicarbonate. The aqueous layer was extracted with 3 × 15 mL of dichloromethane and the combined organic layers were washed with brine, dried over sodium sulfate and the volatiles were removed under reduced pressure. The crude product was then purified via flash column chromatography (CH/EtOAc 1:1) to give the product as 12.6 mg (39%) of a yellow solid.

***R*f** (CH/EtOAc 1:2): 0.18. – **^{1}H NMR** (400 MHz, $CDCl_3$): δ = 8.49 (d, *J* = 5.1 Hz, 2H, 2 × H_{Ar}), 7.86 (s, 1H, H_{Ar}), 7.36 (d, *J* = 5.1 Hz, 2H, 2 × H_{Ar}), 6.81 (d, *J* = 1.4 Hz, 1H, H_{Ar}), 6.71 (d, *J* = 1.5 Hz, 1H, H_{Ar}), 3.92 (s, 2H, BnCH_2), 1.99 – 1.87 (m, 2H, CH_2), 1.75 – 1.57 (m, 4H, 2 × CH_2), 1.57 – 1.38 (m, 4H, 2 × CH_2), 1.35 – 1.26 (m, 2H, CH_2), 1.21 – 1.05 (m, 2H, CH_2), 0.90 – 0.81 (m, 2H, CH_2), 0.74 (t, *J* = 7.3 Hz, 3H, CH_3) ppm. – **^{13}C NMR** (100 MHz, $CDCl_3$): δ = 162.3 ($C_{quart.}$, COO), 154.6 ($C_{quart.}$, C_{Ar}), 153.9 ($C_{quart.}$, C_{Ar}), 153.8 ($C_{quart.}$, C_{Ar}), 149.7 ($C_{quart.}$, C_{Ar}), 148.7 (+, 2 × C_{Ar}H), 136.3 (+, C_{Ar}H), 125.3 (+, 2 × C_{Ar}H), 123.1 ($C_{quart.}$, C_{Ar}), 109.3 (+, C_{Ar}H), 107.0 ($C_{quart.}$, C_{Ar}), 106.8 (+, C_{Ar}H), 42.1 ($C_{quart.}$, C_{CH}), 36.3 (–, 2 × CH_2), 30.3 (–, CH_2), 29.9 (–, CH_2), 27.0 (–, CH_2), 25.8 (–, CH_2), 23.4 (–, CH_2), 22.5 (–, 2 × CH_2), 14.1 (+, CH_3) ppm. – **IR** (KBr): ṽ = 2920 (m), 2850 (m), 1715 (w), 1617 (m), 1449 (w), 1420 (m), 1339 (w), 1254 (w), 1181 (w), 1061 (w), 1009 (w), 844 (w), 778 (w), 733 (vw), 675 (vw), 601 (vw), 526 (vw), 476 (vw) cm^{-1}. – **MS** (70 eV, EI): *m/z* (%) = 391 (53) $[M]^+$, 334 (96) $[M - C_4H_9]^+$, 69 (100). – **HRMS** ($C_{25}H_{29}O_3N$): calc. 391.2140, found 391.2142.

3-(Furan-2-ylmethyl)-5-methoxy-7-pentyl-2*H*-chromen-2-one (98)

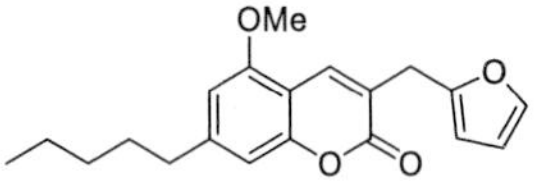

According to GP5, 150 mg of salicylic aldehyde **29a** (0.675 mmol, 1.00 equiv.), 205 mg of potassium carbonate (1.49 mmol, 2.20 equiv.), 289 mg of acrylaldehyde **97** (1.69 mmol, 2.50 equiv.) and 180 mg of 1,3-dimethylimidazolium dimethyl phosphate (**31**) (0.810 mmol, 1.20 equiv.) were suspended in 3.00 mL of toluene. The reaction mixture was heated at 110 °C for 50 min via microwave irradiation. After cooling to room temperature, 5 mL of water was added and the mixture was extracted with 3 × 15 mL of ethyl acetate. The mixture was stirred at room temperature for 30 min in 80 mL of 1 M sodium hydroxide-solution resulting in a dispersion. The remaining solid was dissolved in ethyl acetate and the organic layer was separated and washed again with 40 mL of 1 M sodium hydroxide-solution and then dried over sodium sulfate. Removal of the volatiles and purification via flash column chromatography (CH/EtOAc 20:1) gave 66.9 mg (30%) of the pure product as a brown solid.

R_f (CH/EtOAc 20:1): 0.17. – **^{1}H NMR** (400 MHz, $CDCl_3$): δ = 7.80 (s, 1H, 4-C*H*), 7.36 (dd, *J* = 1.9, 0.9 Hz, 1H, H_{Ar}), 6.73 (s, 1H, H_{Ar}), 6.50 (s, 1H, H_{Ar}), 6.33 (dd, *J* = 3.2, 1.9 Hz, 1H, H_{Ar}), 6.19 (dd, *J* = 3.1, 0.9 Hz, 1H, H_{Ar}), 3.89 (s, 2H, $BnCH_2$), 3.89 (s, 3H, OCH_3), 2.67 – 2.60 (m, 2H, CH_2), 1.68 – 1.58 (m, 2H, CH_2), 1.40 – 1.27 (m, 4H, 2 × CH_2), 0.89 (t, *J* = 6.8 Hz, 3H, CH_3) ppm. – **^{13}C NMR** (100 MHz, $CDCl_3$): δ = 161.9 ($C_{quart.}$, *C*OO), 155.7 ($C_{quart.}$, C_{Ar}), 154.5 ($C_{quart.}$, C_{Ar}), 151.8 ($C_{quart.}$, C_{Ar}), 148.1 ($C_{quart.}$, C_{Ar}), 141.9 (+, $C_{Ar}H$), 135.2 (+, $C_{Ar}H$), 123.0 ($C_{quart.}$, C_{Ar}), 110.6 (+, $C_{Ar}H$), 108.6 (+, $C_{Ar}H$), 108.1 ($C_{quart.}$, C_{Ar}), 107.4 (+, $C_{Ar}H$), 105.7 (+, $C_{Ar}H$), 56.0 (+, OCH_3), 36.7 (–, CH_2), 31.5 (–, CH_2), 30.8 (–, CH_2), 29.3 (–, CH_2), 22.6 (–, CH_2), 14.1 (+, CH_3) ppm. – **IR** (KBr): $\tilde{\nu}$ = 2927 (m), 2857 (w), 1707 (s), 1615 (s), 1504 (m), 1445 (m), 1426 (m), 1301 (w), 1258 (m), 1183 (m), 1142 (m), 1114 (m), 1056 (m), 1005 (m), 941 (w), 870 (w), 833 (w), 809 (w), 789 (m), 726 (m), 665 (w), 638 (w), 598 (m), 574 (m), 472 (w), 405 (w) cm^{-1}. – **MS** (70 eV, EI): *m/z* (%) = 326 (100) $[M]^+$, 298 (19), 297 (41), 270 (16), 241 (11), 222 (10), 214 (13), 191 (10), 166 (46). – **HRMS** ($C_{20}H_{22}O_3N$): calc. 326.1518, found 326.1520.

5.2.6 Synthesis and Characterization 3-Alkylcoumarins (Chapter 3.2.4)

6-Methoxy-5,7,8-trimethyl-2*H*-chromen-2-one (27aa)

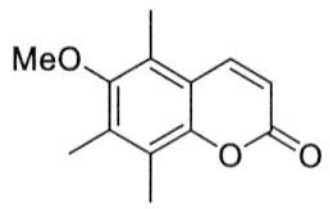

According to **GP7**, 55 mg of salicylic aldehyde **29h** (0.28 mmol, 1.00 equiv.), 94 µL of acetic acid anhydride **100a** (101 mg, 0.99 mmol, 3.50 equiv.) and 1.96 mg of potassium carbonate (10 µmol, 0.05 equiv.) were placed in a microwave vial and heated at 180 °C for 65 min at 300 W microwave irradiation. The resulting mixture was allowed to cool to room temperature, poured onto crushed ice and the pH was adjusted to ~7 with sodium bicarbonate. The mixture was then extracted with 3 × 5 mL of ethyl acetate and the combined organic layers were dried over sodium sulfate. Removal of the volatiles under reduced pressure and purification via flash column chromatography (CH/EtOAc 5:1) resulted in 50.5 mg (82%) of a white solid.

***R*f** (CH/EtOAc 5:1): 0.23. – **^{1}H NMR** (400 MHz, $CDCl_3$): δ = 7.89 (d, *J* = 9.7 Hz, 1H, 4-C*H*), 6.37 (d, *J* = 9.7 Hz, 1H, 3-C*H*), 3.68 (s, 3H, OCH_3), 2.42 (s, 3H, CH_3), 2.34 (s, 3H, CH_3), 2.32 (s, 3H, CH_3) ppm. – **^{13}C NMR** (100 MHz, $CDCl_3$): δ = 161.4 ($C_{quart.}$, *C*OO), 153.1 ($C_{quart.}$, C_{Ar}), 149.4 ($C_{quart.}$, C_{Ar}), 141.0 (+, C_{Ar}H), 135.4 ($C_{quart.}$, C_{Ar}), 124.8 ($C_{quart.}$, C_{Ar}), 123.8 ($C_{quart.}$, C_{Ar}), 116.2 ($C_{quart.}$, C_{Ar}), 114.9 (+, C_{Ar}H), 60.7 (+, OCH_3), 13.5 (+, CH_3), 11.9 (+, CH_3), 11.2 (+, CH_3) ppm. – **IR** (KBr): ṽ = 2923 (w), 1711 (m), 1594 (m), 1556 (w), 1449 (m), 1379 (w), 1329 (w), 1273 (m), 1200 (w), 1140 (w), 1091 (m), 1067 (m), 999 (m), 937 (w), 891 (w), 825 (m), 767 (w), 668 (w), 650 (w), 584 (m), 423 (w) cm^{-1}. – **MS** (70 eV, EI): *m/z* (%) = 218 (86) $[M]^+$, 204 (14), 203 (100) $[M-CH_3]^+$, 175 (21). – **HRMS** ($C_{13}H_{14}O_3$): calc. 218.0937, found 218.0939. – **Elemental analysis**: $C_{13}H_{14}O_3$: calc. C 71.54, H 6.47, found C 71.47, H 6.55.

6-Methoxy-3,5,7,8-tetramethyl-2*H*-chromen-2-one (27ab)

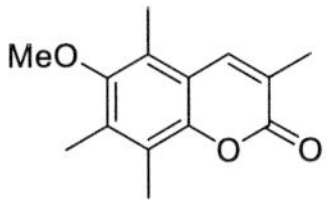

According to GP7, 150 mg of salicylic aldehyde **29h** (0.77 mmol, 1.00 equiv.), 340 µL propionic acid anhydride **100b** (351 mg, 2.70 mmol, 3.50 equiv.) and 5.34 mg of potassium carbonate (40.0 µmol, 0.05 equiv.) were placed in a microwave vial and heated at 180 °C for 65 min at 300 W microwave irradiation. The resulting mixture was allowed to cool to room temperature, poured onto crushed ice and the pH was adjusted to ~7 with sodium bicarbonate. The mixture was then extracted with 3 × 5 mL of ethyl acetate and the combined organic layers were dried over sodium sulfate. Removal of the volatiles under reduced pressure and purification via flash column chromatography (CH/EtOAc 20:1) resulted in 150 mg (89%) of a white solid.

***R*f** (CH/EtOAc 20:1): 0.13. – **MP**: 128.3 °C – **^{1}H NMR** (400 MHz, $CDCl_3$): δ = 7.67 (q, J = 1.4 Hz, 1H, 4-C*H*), 3.68 (s, 3H, OCH_3), 2.40 (s, 3H, CH_3), 2.34 (s, 3H, CH_3), 2.30 (s, 3H, CH_3), 2.22 (d, J = 1.3 Hz 3H, 3-CH_3) ppm. – **^{13}C NMR** (100 MHz, $CDCl_3$): δ = 162.6 ($C_{quart.}$, *C*OO), 152.9 ($C_{quart.}$, C_{Ar}), 148.5 ($C_{quart.}$, C_{Ar}), ($C_{quart.}$, C_{Ar}), 136.8 (+, 4-*C*H), 133.7 ($C_{quart.}$, C_{Ar}), 123.9 ($C_{quart.}$, C_{Ar}), 123.8 ($C_{quart.}$, C_{Ar}), 116.8 ($C_{quart.}$, C_{Ar}), 60.7 (+, CH_3), 17.5 (+, CH_3), 13.3 (+, CH_3), 11.9 (+, CH_3), 11.3 (+, CH_3) ppm. – **IR** (KBr): v˜ = 2920 (w), 1705 (m), 1596 (s), 1442 (s), 1381 (w), 1365 (w), 1331 (w), 1287 (m), 1202 (m), 1140 (w), 1088 (m), 1045 (w), 998 (m), 930 (m), 883 (w), 768 (w), 656 (w), 619 (w), 560 (w), 468 (vw), 388 (w) cm^{-1}. – **MS** (70 eV, EI): *m/z* (%) = 232 (100) $[M]^+$, 218 (15) 217 (98) $[M-CH_3]^+$, 189 (20), 133 (5), 43

(15). – **HRMS** ($C_{14}H_{16}O_3$): calc. 232.1094, found 232.1094. – **Elemental analysis**: $C_{14}H_{16}O_3$: calc. C 72.39, H 6.94, found C 72.19.41, H 6.89.

3,5,7,8-Tetramethyl-2-oxo-2*H*-chromen-6-yl propionate (27ab-Ac)

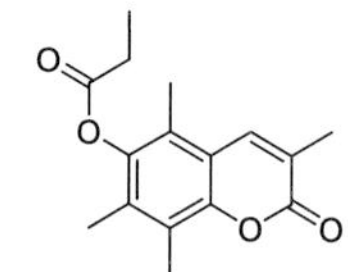

According to **GP7**, 200 mg of salicylic aldehyde **87** (1.11 mmol, 1.00 equiv.), 0.50 mL of propionic anhydride **100b** (505 mg, 3.88 mmol, 3.50 equiv.) and 7.67 mg of potassium carbonate (60 µmol, 0.05 equiv.) were placed in a microwave vial and heated at 180 °C for 65 min at 250 W microwave irradiation. The resulting mixture was allowed to cool to room temperature, poured onto crushed ice and the pH was adjusted to ~7 with sodium bicarbonate. The mixture was then extracted with ethyl acetate, the combined organic layers were dried over sodium sulfate and the volatiles were removed under reduced pressure. Purification via flash column chromatography CH/EtOAc 5:1 resulted in 152.1 mg (50%) of an off-white solid.

***R*f** (CH/EtOAc 5:1): 0.19. – **MP**: 168.1 °C – **^{1}H NMR** (400 MHz, $CDCl_3$): $\delta = 7.68$ (q, $J = 1.3$ Hz, 1H, 4-C*H*), 2.37 (s, 3H, C*H*3), 2.24 (s, 3H, C*H*3), 2.23 (d, $J = 1.3$ Hz, 3H, 3-C*H*3), 2.14 (s, 3H, C*H*3), 1.34 (t, $J = 7.6$ Hz, 3H, CH_2CH_3) ppm. – **^{13}C NMR** (100 MHz, $CDCl_3$): $\delta = 172.6$ ($C_{quart.}$) 162.3 ($C_{quart.}$), 149.8 ($C_{quart.}$), 144.0 ($C_{quart.}$), 136.6 (+, *C*H), 132.5 ($C_{quart.}$), 124.5 ($C_{quart.}$), 123.5 ($C_{quart.}$), 123.4 ($C_{quart.}$), 116.7 ($C_{quart.}$), 27.6(–, CH_2CH_3), 27.6(–, CH_2CH_3), 17.5 (+, CH_3), 13.71 (+, CH_3), 12.0 (+, CH_3), 11.7 (+, CH_3), 9.5 (+, CH_3) ppm. – **IR** (KBr): $\tilde{v}$ = 2919 (w), 1745 (m), 1703 (s), 1604 (m), 1445 (w), 1356 (m), 1284 (m), 1203 (w), 1163 (s), 1097 (m), 1001 (m), 931 (w), 883 (w), 805 (w), 765 (m), 658 (w), 633 (w), 582 (w), 444 (vw), 392 (w) m^{-1}. – **MS** (70 eV, EI): *m/z* (%) = 274 (12) $[M]^+$, 219 (13), 218 (100). – **HRMS** ($C_{16}H_{18}O_4$): calc. 274.1200, found 274.1201. – **Elemental analysis**: $C_{16}H_{18}O_4$: calc. C 70.06, H 6.61 found C 69.76 H 6.76.

5-Isopropyl-8-methyl-2*H*-chromen-2-one (27ac)

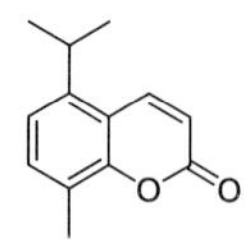

According to **GP7**, 150 mg of salicylic aldehyde **29i** (84 µmol, 1.00 equiv.), 280 µL of acetic acid anhydride **100a** (301 mg, 2.95 mmol, 3.50 equiv.) and 5.82 mg of potassium carbonate (40 µmol, 0.05 equiv.) were placed in a microwave vial and heated at 180 °C for 65 min at 300 W microwave irradiation. The resulting mixture was allowed to cool to room temperature, poured onto crushed ice and the pH was adjusted to ~7 with sodium bicarbonate. The mixture was then extracted with 3 × 5 mL of ethyl acetate and the combined organic layers were dried over sodium sulfate. Removal of the volatiles under reduced pressure and purification via flash column chromatography (CH/EtOAc 5:1) resulted in 169.6 mg (quant.) of a yellow oil.

***R*f** (CH/EtOAc 20:1): 0.12. – **^{1}H NMR** (400 MHz, $CDCl_3$): δ = 8.03 (d, *J* = 9.9 Hz, 1H, 4-C*H*), 7.34 (d, *J* = 7.8 Hz, 1H, H_{Ar}), 7.11 (d, *J* = 7.8 Hz, 1H, H_{Ar}), 6.43 (d, *J* = 9.9 Hz, 1H, 3-C*H*), 3.37 (hept, *J* = 6.9 Hz, 1H, C*H*$(CH_3)_2$), 2.42 (d, *J* = 0.8 Hz, 3H, CH_3), 1.30 (d, *J* = 6.9 Hz, 6H, CH$(CH_3)_2$) ppm. – **^{13}C NMR** (100 MHz, $CDCl_3$): δ = 161.0 ($C_{quart.}$, *C*OO), 153.0 ($C_{quart.}$, C_{Ar}), 144.3 ($C_{quart.}$, C_{Ar}), 140.2 ($C_{quart.}$, C_{Ar}), 133.3 (+, C_{Ar}H), 123.9 (+, C_{Ar}H), 120.4 (+, C_{Ar}H), 116.2 ($C_{quart.}$, C_{Ar}), 115.6 (+, C_{Ar}H), 28.5 (+, *C*H$(CH_3)_2$), 23.8 (+, CH$(CH_3)_2$), 15.5 (+, CH_3) ppm. – **IR** (KBr): ṽ = 2964 (w), 1723 (m), 1595 (m), 1485 (w), 1460 (w), 1386 (vw), 1365 (vw), 1257 (w), 1240 (w), 1188 (w), 1154 (w), 1128 (w), 1051 (w), 1000 (w), 901 (w), 862 (vw), 825 (w), 777 (vw), 677 (w), 636 (vw), 595 (vw), 527 (vw) cm^{-1}. – **MS** (70 eV, EI): *m/z* (%) = 202 (48) $[M]^+$, 188 (14), 187 (100) $[M - CH_3]^+$, 176 (10), 161 (15), 159 (11), 150 (19), 135 (26), 115 (13), 91 (13). – **HRMS** ($C_{13}H_{14}O_2$): calc. 202.0988, found 202.0988.

5-Isopropyl-3,8-dimethyl-2*H*-chromen-2-one (27ad)

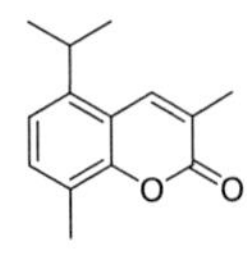

According to GP7, 195 mg of salicylic aldehyde **29i** (1.09 mmol, 1.00 equiv.), 490 µl propionic acid anhydride **100b** (498 mg, 3.83 mmol, 3.50 equiv.) and 7.56 mg of potassium carbonate (50.0 µmol, 0.05 equiv.) were placed in a microwave vial and heated at 180 °C for 65 min at 300 W microwave irradiation. The resulting mixture was allowed to cool to room temperature, poured onto crushed ice and the pH was adjusted to ~7 with sodium bicarbonate. The mixture was then extracted with diethyl ether and the organic phase was dried over sodium sulfate. Removal of the volatiles under reduced pressure resulted in 194 mg (82%) of an off-white solid.

***R*f** (CH/EtOAc 5:1): 0.55. – **MP**: 97.4 °C – **^{1}H NMR** (400 MHz, $CDCl_3$): δ = 7.82 (q, *J* = 1.6 Hz, 1H, 4-C*H*), 7.26 (d, *J* = 7.6 Hz, 1H, H_{Ar}), 7.08 (d, *J* = 7.8 Hz, 1H, H_{Ar}), 3.37 (hept., *J* = 6.9 Hz, 1H, C*H*$(CH_3)_2$),2.41 (s, 3H, ArCH_3), 2.24 (d, *J* = 1.3 Hz, 3H, 3-C(CH_3), 1.30 (d, *J* = 6.8 Hz, 6H, CH$(CH_3)_2$) ppm. – **^{13}C NMR** (100 MHz, $CDCl_3$): δ = 15.6 (+, CH_3), 17.7 (+, CH_3), 23.81 (+, 2 × CH_3, CH$(CH_3)_2$), 28.4 (+, *C*H$(CH_3)_2$), 116.9 ($C_{quart.}$, C-3), 120.2 (+, C_{Ar}H),123.4 ($C_{quart.}$, C_{Ar}), 124.6 ($C_{quart.}$, C_{Ar}), 131.9 (+, C_{Ar}H), 136.1 (+, C_{Ar}H), 143.3 ($C_{quart.}$, C_{Ar}), 152.2 ($C_{quart.}$, C_{Ar}), 152.2 ($C_{quart.}$, C_{Ar}), 162.3 ($C_{quart.}$, *C*-2) ppm. – **IR** (KBr): ṽ = 2959 (w),

1703 (m), 1632 (w), 1600 (w), 1487 (w), 1450 (w), 1380 (w), 1361 (w), 1299 (vw), 1257 (vw), 1195 (m), 1090 (m), 1053 (w), 996 (m), 935 (w), 821 (w), 776 (w), 744 (w), 716 (vw), 687 (vw), 649 (w), 631 (w), 544 (w), 408 (w) cm^{-1}. – **MS** (70 eV, EI): *m/z* (%) = 216 (45) $[M]^+$, 202 (14), 201 (100) $[M - CH_3]^+$, 173 (7) $[M - C_3H_7]^+$ – **HRMS** ($C_{14}H_{16}O_2$): calc. 216.1145, found 216.1144.

3-Ethyl-5-isopropyl-8-methyl-2*H*-chromen-2-one (27ae)

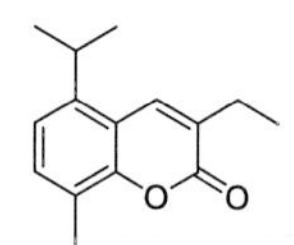

According to **GP7**, 150 mg of salicylic aldehyde **29i** (84 µmol, 1.00 equiv.), 480 µL butyric acid anhydride **100c** (466 mg, 2.95 mmol, 3.50 equiv.) and 5.82 mg of potassium carbonate (40 µmol, 0.05 equiv.) were placed in a microwave vial and heated at 180 °C for 65 min at 300 W microwave irradiation. The resulting mixture was allowed to cool to room temperature, poured onto crushed ice and the pH was adjusted to ~7 with sodium bicarbonate. The mixture was then extracted with 3 × 5 mL of ethyl acetate and the combined organic layers were dried over sodium sulfate. Removal of the volatiles under reduced pressure and purification via flash column chromatography (CH/EtOAc 5:1) resulted in 193.8 mg (quant.) of an off-white solid.

***R*f** (CH/EtOAc 20:1): 0.42. – **^{1}H NMR** (400 MHz, $CDCl_3$): δ = 7.78 (s, 1H, 4-C*H*), 7.27 (d, *J* = 7.7 Hz, 1H, H_{Ar}), 7.08 (d, *J* = 7.8 Hz, 1H, H_{Ar}), 3.39 (hept, *J* = 6.9 Hz, 1H, C*H*$(CH_3)_2$), 2.63 (qd, *J* = 7.5, 1.2 Hz, 2H, CH_2), 2.41 (s, 3H, CH_3), 1.31 (d, *J* = 6.9 Hz, 6H, $CH(CH_3)_2$), 1.27 (t, *J* = 7.5 Hz, 3H, CH_2CH_3) ppm. – **^{13}C NMR** (100 MHz, $CDCl_3$): δ = 161.8 ($C_{quart.}$, *C*OO), 152.0 ($C_{quart.}$, C_{Ar}), 143.4 ($C_{quart.}$, C_{Ar}), 134.4 (+, $C_{Ar}H$), 131.8 (+, $C_{Ar}H$), 130.1 ($C_{quart.}$, C_{Ar}), 123.4 ($C_{quart.}$, C_{Ar}), 120.1 (+, $C_{Ar}H$), 116.9 ($C_{quart.}$, C_{Ar}), 28.4 (+, *C*H$(CH_3)_2$), 24.4 (–, CH_2), 23.8 (+, $CH(CH_3)_2$), 15.6 (+, CH_3), 12.8 (+, CH_3) ppm. – **IR** (KBr): ṽ = 2961 (m), 2926 (m), 2869 (w), 1705 (s), 1600 (m), 1486 (w), 1456 (m), 1384 (w), 1361 (w), 1300 (w), 1248 (w), 1193 (m), 1156 (w), 1099 (m), 1042 (m), 1000 (m), 944 (m), 932 (m), 818 (m), 783 (m), 760 (w), 688 (w), 636 (m), 552 (w), 540 (w), 498 (w), 417 (w) cm^{-1}. – **MS** (70 eV, EI): *m/z* (%) = 230 (60) $[M]^+$, 216 (16), 215 (100) $[M - CH_3]^+$, 181 (17), 179 (13), 178 (97), 177 (15), 162 (14), 160 (54), 159 (14), 145 (33), 119 (10), 91 (14). – **HRMS** ($C_{15}H_{18}O_2$): calc. 230.1301, found 230.1300.

3-Propyl-5-isopropyl-8-methyl-2*H*-chromen-2-one (27af)

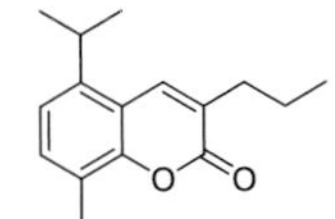

According to **GP7**, 150 mg of salicylic aldehyde **29i** (84 µmol, 1.00 equiv.), 540 µL valeric acid anhydride **100d** (549 mg, 2.95 mmol, 3.50 equiv.) and 5.82 mg of potassium carbonate (40 µmol, 0.05 equiv.) were placed in a microwave vial and heated at 180 °C for 65 min at 300 W microwave irradiation. The resulting mixture was allowed to cool to room temperature, poured onto crushed ice and the pH was adjusted to ~7 with sodium bicarbonate. The mixture was then extracted with 3 × 5 mL of ethyl acetate and the combined organic layers were dried over sodium sulfate. Removal of the volatiles under reduced pressure and purification via flash column chromatography (CH/EtOAc 5:1) resulted in 173.5 mg (84%) of an off-white solid.

***R*f** (CH/EtOAc 20:1): 0.35. – **^{1}H NMR** (400 MHz, $CDCl_3$): δ = 7.77 (s, 1H, 4-C*H*), 7.26 (d, *J* = 7.8 Hz, 1H, H_{Ar}), 7.08 (d, *J* = 7.8 Hz, 1H, H_{Ar}), 3.38 (hept, *J* = 6.9 Hz, 1H, C*H*$(CH_3)_2$), 2.60 – 2.53 (m, 2H, CH_2), 2.41 (d, *J* = 0.8 Hz, 3H, CH_3), 1.76 – 1.62 (m, 2H, CH_2), 1.30 (d, *J* = 6.9 Hz, 6H, CH(CH_3)$_2$), 1.00 (t, *J* = 7.3 Hz, 3H, CH_3) ppm. – **^{13}C NMR** (100 MHz, $CDCl_3$): δ = 161.9 ($C_{quart.}$, *C*OO), 152.1 ($C_{quart.}$, C_{Ar}), 143.4 ($C_{quart.}$, C_{Ar}), 135.4 (+, C_{Ar}H), 131.9 (+, C_{Ar}H), 128.6 ($C_{quart.}$, C_{Ar}), 123.4 ($C_{quart.}$, C_{Ar}), 120.1 (+, C_{Ar}H), 116.8 ($C_{quart.}$, C_{Ar}), 33.4 (–, CH_2), 28.4 (+, *C*H$(CH_3)_2$), 23.8 (+, CH(CH_3)$_2$), 21.6 (–, CH_2), 15.6 (+, CH_3), 13.9 (+, CH_3) ppm. – **IR** (KBr): ṽ = 2961 (w), 2931 (w), 2870 (w), 1705 (m), 1599 (w), 1486 (w), 1457 (w), 1430 (w), 1384 (w), 1362 (w), 1307 (w), 1247 (w), 1192 (w), 1155 (w), 1104 (w), 1064 (m), 1015 (w), 996 (m), 932 (w), 856 (w), 819 (m), 788 (w), 686 (w), 634 (w), 615 (w), 553 (w), 539 (w), 409 (vw) cm^{-1}. – **MS** (70 eV, EI): *m/z* (%) = 244 (75) $[M]^+$, 230 (16), 229 (100) $[M – CH_3]^+$, 216 (36), 215 (56), 201 (43), 181 (21), 178 (59), 172 (10), 171 (12), 160 (27), 145 (19), 131 (26). – **HRMS** ($C_{16}H_{20}O_2$): calc. 244.1458, found 244.1460.

3-Butyl-5-isopropyl-8-methyl-2*H*-chromen-2-one (27ag)

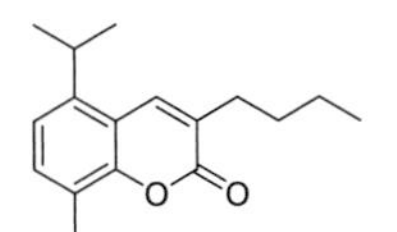

According to **GP7**, 150 mg of salicylic aldehyde **29i** (84 µmol, 1.00 equiv.), 680 µL hexanoic acid anhydride **100e** (631 mg, 2.95 mmol, 3.50 equiv.) and 5.82 mg of potassium carbonate (40 µmol, 0.05 equiv.) were placed in a microwave vial and heated at 180 °C for 65 min at 300 W microwave irradiation. The resulting mixture was allowed to cool to room temperature, poured onto crushed ice and the pH was adjusted to ~7 with sodium bicarbonate. The mixture was then extracted with 3 × 5 mL of ethyl acetate and the combined organic layers were dried over sodium sulfate. Removal of the volatiles under reduced pressure and purification via flash column chromatography (CH/EtOAc 5:1) resulted in 157.6 mg (72%) of a colorless oil.

***R*f** (CH/EtOAc 20:1): 0.39. – **^{1}H NMR** (400 MHz, $CDCl_3$): δ = 7.77 (s, 1H, 4-C*H*), 7.26 (d, *J* = 7.8 Hz, 1H, H_{Ar}), 7.08 (d, *J* = 7.8 Hz, 1H, H_{Ar}), 3.38 (hept, *J* = 6.8 Hz, 1H, $CH(CH_3)_2$), 2.64 – 2.54 (m, 2H, CH_2), 2.41 (s, 3H, CH_3), 1.69 – 1.55 (m, 2H, CH_2), 1.48 – 1.36 (m, 2H, CH_2), 1.30 (d, *J* = 6.9 Hz, 6H, $CH(CH_3)_2$), 0.96 (t, *J* = 7.3 Hz, 3H, CH_3) ppm. – **^{13}C NMR** (100 MHz, $CDCl_3$): δ = 161.9 ($C_{quart.}$, *C*OO), 152.1 ($C_{quart.}$, C_{Ar}), 143.4 ($C_{quart.}$, C_{Ar}), 135.2 (+, $C_{Ar}H$), 131.8 (+, $C_{Ar}H$), 128.8 ($C_{quart.}$, C_{Ar}), 123.4 ($C_{quart.}$, C_{Ar}), 120.1 (+, $C_{Ar}H$), 116.9 ($C_{quart.}$, C_{Ar}), 31.1 (–, CH_2), 30.5 (–, CH_2), 28.4 (+, $CH(CH_3)_2$), 23.8 (+, $CH(CH_3)_2$), 22.5 (–, CH_2), 15.6 (+, CH_3), 14.0 (+, CH_3) ppm. – **IR** (KBr): $\tilde{\nu}$ = 2957 (w), 2926 (w), 2869 (w), 1714 (m), 1599 (w), 1485 (w), 1458 (w), 1382 (w), 1364 (w), 1296 (w), 1247 (w), 1187 (w), 1109 (w), 1062 (w), 997 (w), 820 (w), 784 (w), 733 (vw), 634 (w), 540 (vw) cm^{-1}. – **MS** (70 eV, EI): *m/z* (%) = 258 (46) $[M]^+$, 244 (4), 243 (24) $[M - CH_3]^+$, 229 (28) $[M - C_2H_5]^+$, 217 (11), 216 (78) $[M - C_3H_6]^+$, 215 (100) $[M - C_3H_7]^+$, 201 (30) $[M - C_4H_9]^+$, 181 (31). – **HRMS** ($C_{17}H_{22}O_2$): calc. 258.1614, found 258.1615.

3-Butyl-5-isopropyl-8-methyl-2*H*-chromen-2-one (27ah)

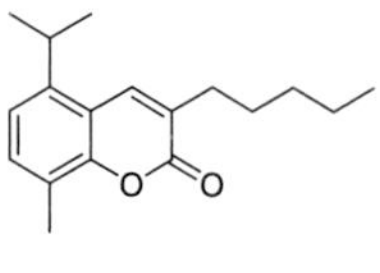

According to **GP7**, 150 mg of salicylic aldehyde **29i** (84 µmol, 1.00 equiv.), 780 µL heptanoic acid anhydride **100f** (714 mg, 2.95 mmol, 3.50 equiv.) and 5.82 mg of potassium carbonate (40 µmol, 0.05 equiv.) were placed in a microwave vial and heated at 180 °C for 65 min at 300 W microwave irradiation. The resulting mixture was allowed to cool to room temperature, poured onto crushed ice and the pH was adjusted to ~7 with sodium bicarbonate. The mixture was then extracted with 3 × 5 mL of ethyl acetate and the combined organic layers were dried over sodium sulfate. Removal of the volatiles under reduced pressure and purification via flash column chromatography (CH/EtOAc 20:1) resulted in 95.3 mg (42%) of a colorless oil.

***R*f** (CH/EtOAc 20:1): 0.40. – **^{1}H NMR** (400 MHz, $CDCl_3$): δ = 7.77 (s, 1H, 4-C*H*), 7.26 (d, *J* = 7.8 Hz, 1H, H_{Ar}), 7.08 (d, *J* = 7.8 Hz, 1H, H_{Ar}), 3.38 (hept, *J* = 6.8 Hz, 1H, $CH(CH_3)_2$), 2.63 – 2.52 (m, 2H, CH_2), 2.41 (d, *J* = 0.7 Hz, 3H, CH_3), 1.71 – 1.57 (m, 2H, CH_2), 1.37 (h, *J* = 3.8 Hz, 4H, 2 × CH_2), 1.30 (d, *J* = 6.9 Hz, 6H, $CH(CH_3)_2$), 0.94 – 0.86 (m, 3H, CH_3)

ppm. – ^{13}C **NMR** (100 MHz, $CDCl_3$): δ = 161.9 ($C_{quart.}$, *C*OO), 152.0 ($C_{quart.}$, C_{Ar}), 143.4 ($C_{quart.}$, C_{Ar}), 135.2 (+, $C_{Ar}H$), 131.8 (+, $C_{Ar}H$), 128.8 ($C_{quart.}$, C_{Ar}), 123.4 ($C_{quart.}$, C_{Ar}), 120.1 (+, $C_{Ar}H$), 116.9 ($C_{quart.}$, C_{Ar}), 31.6 (–, CH_2), 31.3 (–, CH_2), 28.4 (+, $CH(CH_3)_2$), 28.1 (–, CH_2), 23.8 (+, $CH(CH_3)_2$), 22.6 (–, CH_2), 15.6 (+, CH_3), 14.2 (+, CH_3) ppm. – **IR** (KBr): ṽ = 2956 (w), 2925 (w), 2858 (w), 1714 (s), 1599 (m), 1485 (w), 1458 (w), 1383 (w), 1364 (w), 1296 (w), 1252 (w), 1171 (w), 1110 (w), 1066 (m), 997 (m), 915 (w), 820 (m), 784 (w), 757 (w), 728 (w), 657 (w), 634 (w), 541 (w) cm^{-1}. – **MS** (70 eV, EI): *m/z* (%) = 272/273 (31/6) $[M]^+$, 257 (8) $[M-CH_3]^+$, 243 (9) $[M-C_2H_5]^+$, 230 (18), 229 (100) $[M-C_3H_7]^+$, 216 (45), 215 (100) $[M-C_4H_9]^+$, 201 (30) $[M-C_5H_{11}]^+$, 113 (8). – **HRMS** ($C_{18}H_{24}O_2$): calc. 272.1771, found 272.1773.

3-Butyl-5-isopropyl-8-methyl-2*H*-chromen-2-one (27ai)

According to **GP7**, 150 mg of salicylic aldehyde **29i** (84 µmol, 1.00 equiv.), 880 µL ethantic acid anhydride **100g** (797 mg, 2.95 mmol, 3.50 equiv.) and 5.82 mg of potassium carbonate (40 µmol, 0.05 equiv.) were placed in a microwave vial and heated at 180 °C for 65 min at 300 W microwave irradiation. The mixture was allowed to cool to room temperature, poured onto crushed ice and the pH was adjusted to ~7 with sodium bicarbonate. The mixture was then extracted with 3 × 5 mL of ethyl acetate and the combined organic layers were dried over sodium sulfate. Removal of the volatiles under reduced pressure and purification via flash column chromatography (CH/EtOAc 20:1) resulted in 137.5 mg (57%) of a colorless oil.

***R*f** (CH/EtOAc 20:1): 0.49. – ^{1}H **NMR** (400 MHz, $CDCl_3$): δ = 7.77 (s, 1H, 4-C*H*), 7.26 (d, *J* = 7.8 Hz, 1H, H_{Ar}), 7.08 (d, *J* = 7.8 Hz, 1H, H_{Ar}), 3.38 (hept, *J* = 6.9 Hz, 1H, $CH(CH_3)_2$), 2.63 – 2.53 (m, 2H, CH_2), 2.41 (s, 3H, CH_3), 1.71 – 1.56 (m, 2H, CH_2), 1.44 – 1.30 (m, 6H, 3 × CH_2), 1.30 (d, *J* = 6.9 Hz, 6H, $CH(CH_3)_2$), 0.93 – 0.83 (m, 3H, CH_3) ppm. – ^{13}C **NMR** (100 MHz, $CDCl_3$): δ = 161.9 ($C_{quart.}$, *C*OO), 152.1 ($C_{quart.}$, C_{Ar}), 143.4 ($C_{quart.}$, C_{Ar}), 135.2 (+, $C_{Ar}H$), 131.8 (+, $C_{Ar}H$), 128.9 ($C_{quart.}$, C_{Ar}), 123.4 ($C_{quart.}$, C_{Ar}), 120.1 (+, $C_{Ar}H$), 116.9 ($C_{quart.}$, C_{Ar}), 31.8 (–, CH_2), 31.3 (–, CH_2), 29.1 (–, CH_2), 28.4 (+, $CH(CH_3)_2$), 28.4 (–, CH_2), 23.8 (+, $CH(CH_3)_2$), 22.8 (–, CH_2), 15.6 (+, CH_3), 14.2 (+, CH_3) ppm. – **IR** (KBr): ṽ = 2956 (m), 2924 (m), 2855 (w), 1715 (s), 1600 (m), 1485 (w), 1458 (w), 1383 (w), 1364 (w), 1297 (w), 1251 (w), 1194 (w), 1171 (w), 1111 (w), 1068 (m), 997 (m), 918 (w), 820 (w), 784 (w), 725 (w), 634 (w), 540 (w), 402 (vw) cm^{-1}. – **MS** (70 eV, EI): *m/z* (%) = 272 (31) $[M]^+$, 257 (8) $[M - CH_3]^+$, 243 (9) $[M - C_2H_5]^+$, 230 (18), 229 (100) $[M - C_3H_7]^+$, 216 (45), 215 (100) $[M - C_4H_9]^+$, 201 (30) $[M - C_5H_{11}]^+$, 113 (8). – **HRMS** ($C_{19}H_{26}O_2$): calc. 286.1927, found 286.1927.

3,5-Diisopropyl-8-methyl-2*H*-chromen-2-one (27aj)

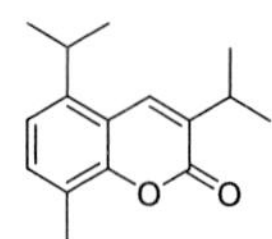

According to **GP7**, 150 mg of salicylic aldehyde **29i** (84 µmol, 1.00 equiv.), 580 µL isovaleric acid anhydride **100h** (549 mg, 2.95 mmol, 3.50 equiv.) and 5.82 mg of potassium carbonate (40 µmol, 0.05 equiv.) were placed in a microwave vial and heated at 180 °C for 180 min at 300 W microwave irradiation. The resulting mixture was allowed to cool to room temperature, poured onto crushed ice and the pH was adjusted to ~7 with sodium bicarbonate. The mixture was then extracted with 3 × 5 mL of ethyl acetate and the combined organic layers were dried over sodium sulfate. Removal of the volatiles under reduced pressure and purification via flash column chromatography (CH/EtOAc 20:1) resulted in 160.2 mg (78%) of a colorless oil.

***R*f** (CH/EtOAc 20:1): 0.20. – **¹H NMR** (400 MHz, $CDCl_3$): δ = 7.77 (s, 1H, 4-C*H*), 7.26 (d, *J* = 7.5 Hz, 1H, H_{Ar}), 7.08 (d, *J* = 7.8 Hz, 1H, H_{Ar}), 3.40 (hept, *J* = 6.8 Hz, 1H, C*H*$(CH_3)_2$), 3.15 (hept, *J* = 7.0 Hz, 1H, C*H*$(CH_3)_2$), 2.41 (d, *J* = 0.8 Hz, 3H, CH_3), 1.31 (d, *J* = 6.9 Hz, 6H, $CH(CH_3)_2$), 1.28 (d, *J* = 6.9 Hz, 6H, $CH(CH_3)_2$) ppm. – **¹³C NMR** (100 MHz, $CDCl_3$): δ = 161.4 ($C_{quart.}$, *C*OO), 151.8 ($C_{quart.}$, C_{Ar}), 143.5 ($C_{quart.}$, C_{Ar}), 134.4 ($C_{quart.}$, C_{Ar}), 132.5 (+, $C_{Ar}H$), 131.9 (+, $C_{Ar}H$), 123.3 ($C_{quart.}$, C_{Ar}), 120.1 (+, $C_{Ar}H$), 116.8 ($C_{quart.}$, C_{Ar}), 29.0 (+, *C*H$(CH_3)_2$), 28.5 (+, *C*H$(CH_3)_2$), 23.8 (+, $CH(CH_3)_2$), 21.7 (+, $CH(CH_3)_2$), 15.6 (+, CH_3) ppm. – **IR** (KBr): ṽ = 2960 (m), 2870 (w), 1714 (s), 1599 (m), 1485 (w), 1461 (m), 1385 (w), 1363 (w), 1294 (w), 1251 (w), 1190 (m), 1154 (m), 1107 (w), 1064 (m), 1017 (m), 993 (m), 917 (w), 868 (vw), 820 (m), 756 (m), 761 (w), 712 (vw), 638 (w), 540 (w), 409 (vw) cm^{-1}. – **MS** (70 eV, EI): *m/z* (%) = 244 (71) $[M]^+$, 230 (17), 229 (100) $[M - CH_3]^+$, 215 (10), 201 (16) $[M - C_3H_7]^+$, 181 (12), 150 (13), 131 (11). – **HRMS** ($C_{16}H_{20}O_2$): calc. 244.1458, found 244.1456.

6-Bromo-5-isopropyl-8-methyl-2*H*-chromen-2-one (27ak) [177]

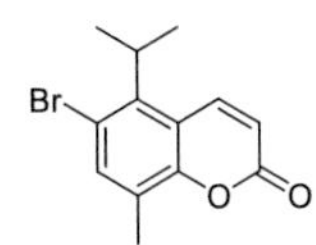

According to **GP7**, 51.5 mg of salicylic aldehyde **29k** (0.200 mmol, 1.00 equiv.), 100 µL acetic acid anhydride **100a** (1.06 mmol, 5.3 equiv.) and 2.30 mg of potassium carbonate (16.6 µmol, 0.05 equiv.) were placed in a microwave vial and heated at 180 °C for 65 min at 300 W microwave irradiation. The resulting mixture was allowed to cool to room temperature, poured onto water and the pH adjusted to ~7 with sodium bicarbonate. The mixture was then extracted with 3 × 5 mL of ethyl acetate and the combined organic layers were dried over sodium sulfate. Removal of the volatiles under reduced pressure and purification via flash column chromatography (CH/EtOAc 10:1) resulted in 58.9 mg (quant.) of an off-white solid.

***R*f** (CH/EtOAc 10:1): 0.21. – **^{1}H NMR** (400 MHz, $CDCl_3$): δ = 8.21 (d, *J* = 9.95 Hz, 1H, 4-C*H*), 7.57 (s, 1H, 7-CH_{Ar}), 6.42 (d, *J* = 10.09 Hz, 1H, 3-C*H*), 3.90 (b, 1H, C*H*(CH_3)$_2$), 2.38 (s, 3H, CH_3), 1.44 (d, *J* = 7.24 Hz, 6H CH(CH_3)$_2$) ppm. – **^{13}C NMR** (100 MHz, $CDCl_3$): δ = 160.0 ($C_{quart.}$, C_{Ar}OO), 153.3 ($C_{quart.}$, C_{Ar}), 142.0 (+, C_{Ar}H), 141.0 ($C_{quart.}$, C_{Ar}CH(CH_3)$_2$), 136.9 (+, C_{Ar}H), 126.3 ($C_{quart.}$, C_{Ar}), 119.4 ($C_{quart.}$, C_{Ar}), 118.0 ($C_{quart.}$, C_{Ar}), 115.6 (+, C_{Ar}H), 29.8 (+, *C*H(CH_3)$_2$), 22.6 (+, (CH_3)$_2$CH), 15.3 (+, CH_3) ppm. – **IR** (KBr): $\tilde{\nu}$ = 2960 (vs), 2923 (m), 2852 (vs), 1737 (vw), 1722 (vs), 1583 (s), 1443 (m), 1426 (vs), 1369 (s), 1258 (m), 1183 (m), 1122 (w), 1050 (s), 990 (s), 898 (m), 876 (s), 839 (w), 799 (m), 775 (s), 702 (vs), 678 (m), 656 (s), 542 (m), 478 (s), 444 (s), 414 (s) cm^{-1}. – **MS** (70 eV, EI): *m/z* (%) = 282/280 (96/97) [M]$^+$, 268/266 (11/13), 267/265 (100/98) [M – CH_3]$^+$, 240 (5), 239/237 (7/7) [M – C_3H_7]$^+$, 231 (6), 230 (5), 215 (10), 201 (6) [M – Br]$^+$, 186 (36) [M – CH_3Br]$^+$, 181 (16), 173 (7), 159 (8), 158 (26), 157 (8), 131 (19), 129 (20), 128 (28), 127 (10), 119 (6), 115 (16), 77 (7), 69 (35), 58 (8), 57 (6). – **HRMS** ($C_{13}H_{13}O_2{}^{79}Br$): calc. 280.0093, found 280.0093.

6-Bromo-5-isopropyl-3,8-dimethyl-2*H*-chromen-2-one (27al)[177]

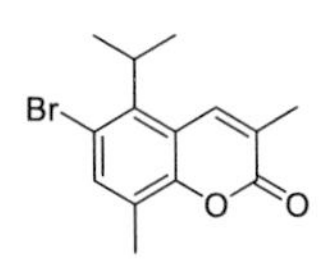

According to **GP7**, 50 mg of salicylic aldehyde **29k** (190 µmol, 1.00 equiv.), 120 µL propionic acid anhydride **100b** (940 µmol, 3.50 equiv.) and 2.40 mg of potassium carbonate (17.4 µmol, 0.05 equiv.) were placed in a microwave vial and heated at 180 °C for 65 min at 300 W microwave irradiation. The resulting mixture was allowed to cool to room temperature, poured onto crushed ice and the pH was adjusted to ~7 with sodium bicarbonate. The mixture was then extracted with 3 × 5 mL of ethyl acetate and the combined organic layers were dried over sodium sulfate. Removal of the volatiles under reduced pressure and purification via flash column chromatography (CH/EtOAc 10:1) resulted in 56.2 mg (quant.) of an off-white solid.

***R*f** (CH/EtOAc 10:1): 0.34. – **^{1}H NMR** (400 MHz, $CDCl_3$): δ = 7.99 (s, 1H, 4-C*H*), 7.50 (s, 1H, 7-CH_{Ar}), 3.90 (b, 1H, C*H*$(CH_3)_2$), 2.38 (s, 3H, CH_3), 2.25 (s, 3H, CH_3), 1.44 (d, *J* = 7.13 Hz, 6H, CH(CH_3)$_2$) ppm. – **^{13}C NMR** (100 MHz, $CDCl_3$): δ = 161.3 ($C_{quart.}$, C_{Ar}OO), 152.4 ($C_{quart.}$, C_{Ar}), 141.0 (+, C_{Ar}H), 136.8 ($C_{quart.}$, C_{Ar}), 135.5 (+, C_{Ar}H), 125.9 ($C_{quart.}$, C_{Ar}), 124.7 ($C_{quart.}$, C_{Ar}), 119.4 ($C_{quart.}$, $C_{Ar}CH_3$), 118.8 ($C_{quart.}$, C_{Ar}), 34.1 (+, *C*H$(CH_3)_2$), 22.5 (+, $(CH_3)_2$CH), 17.8 (+, $C_{Ar}CH_3$), 15.4 (+, $C_{Ar}CH_3$) ppm. – **IR** (KBr): $\tilde{\nu}$ = 2923 (m), 1712 (vw), 1587 (s), 1451 (m), 1429 (s), 1362 (s), 1299 (s), 1265 (m), 1188 (w), 1154 (s), 1089 (s), 1056 (vs), 1011 (s), 924 (s), 870 (s), 837 (m), 773 (m), 744 (s), 727 (vs), 640 (s), 626 (vs), 607 (s), 562 (s), 545 (s), 478 (vs), 426 (s), 393 (s) cm^{-1}. – **MS** (70 eV, EI): *m/z* (%) = = 294/296 (7/6) $[M]^+$, 281/279 (8/7) $[M - CH_3]^+$, 231/229 (7/6), 230/228 (54/55), 216/214 (10/11), 215/213 (100/100), 135 (10), 134 (62), 133 (8), 116 (12), 115 (17), 105 (9), 91 (11), 79 (5), 77 (10), 69 (6). – **HRMS** ($C_{14}H_{15}O_2{}^{79}Br$): calc. 294.0250, found 294.0251.

6-Bromo-3-ethyl-5-isopropyl-8-methyl-2*H*-chromen-2-one (27am)[177]

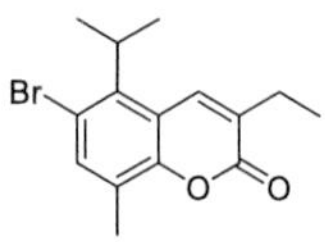

According to **GP7**, 52.5 mg of salicylic aldehyde **29k** (204 µmol, 1.00 equiv.), 120 µL butyric acid anhydride **100c** (736 µmol, 3.61 equiv.) and 2.30 mg of potassium carbonate (21.7 µmol, 0.106 equiv.) were placed in a microwave vial and heated at 180 °C for 65 min at 300 W microwave irradiation. The resulting mixture was allowed to cool to room temperature, poured onto crushed ice and the pH was adjusted to ~7 with sodium bicarbonate. The mixture was then extracted with 3 × 5 mL of ethyl acetate and the combined organic layers were dried over sodium sulfate. Removal of the volatiles under reduced pressure and purification via flash column chromatography (CH/EtOAc 20:1) resulted in 66.2 mg (quant.) of a yellow solid.

***R*f** (CH/EtOAc 20:1): 0.32. – **^{1}H NMR** (400 MHz, $CDCl_3$): δ = 7.97 (s, 1H, 4-C*H*), 7.50 (s, 1H, H_{Ar}), 3.91 (b, 1H, C*H*$(CH_3)_2$), 2.63 (q, *J* = 7.39 Hz, 2H, CH_2CH_3), 2.38 (s, 3H, CH_3), 1.45 (d, *J* = 7.25 Hz, 6 H CH(CH_3)$_2$), 1.28 (t, *J* = 7.25 Hz, 3H, CH_2CH_3) ppm. – **^{13}C NMR** (100 MHz, $CDCl_3$): δ = 161.0 ($C_{quart.}$, C_{Ar}OO), 152.1 ($C_{quart.}$, C_{Ar}), 135.7 (+, C_{Ar}H), 135.1 ($C_{quart.}$, C_{Ar}), 130.2 (+, C_{Ar}H), 127.0 ($C_{quart.}$, C_{Ar}), 125.9 ($C_{quart.}$, C_{Ar}), 118.8 ($C_{quart.}$, C_{Ar}), 117.1 ($C_{quart.}$, $C_{Ar}CH_2CH_3$), 34.1 (+, *C*H$(CH_3)_2$), 24.3 (+, $(CH_3)_2$CH), 22.5 (–, CH_2CH_3), 15.4 (+, $C_{Ar}CH_3$), 12.5 (+, CH_2CH_3) ppm. – **IR** (KBr): $\tilde{\nu}$ = 2963 (s), 2927 (s), 1708 (vw), 1587 (s), 1450 (m),

1381 (s), 1295 (s), 1263 (s), 1188 (m), 1100 (m), 1071 (vs), 1044 (s), 998 (s), 947 (s), 869 (s), 840 (s), 780 (m), 745 (s), 722 (s), 657 (s), 621 (s), 566 (s), 547 (s) cm^{-1}. – **MS** (70 eV, EI): *m/z* (%) = 310/308 (28/33) $[M]^+$, 295/293 (26/36) $[M-CH_3]^+$, 291 (5), 284/282 (9/9) $[M-C_2H_5]^+$, 281 (9), 262 (10), 243 (11), 231/229 (20/6), 230/228 (35/36), 215/213 (45/49), 214 (10), 181 (72), 169 (6), 162 (10), 134 (24), 131 (62), 119 (14), 116 (6), 115 (10), 105 (6), 100 (10), 91 (6), 77 (4), 71 (8), 69 (100). – **HRMS** ($C_{15}H_{17}O_2{}^{79}Br$): calc. 308.0406, found 308.0404.

6-Bromo-3-propyl-5-isopropyl-8-methyl-2*H*-chromen-2-one (27an)[177]

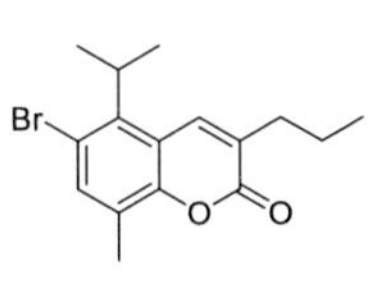

According to **GP7**, 50.0 mg of salicylic aldehyde **29k** (190 µmol, 1.00 equiv.), 130 µL valeric acid anhydride **100d** (712 µmol, 3.75 equiv.) and 2.60 mg of potassium carbonate (18.8 µmol, 0.10 equiv.) were placed in a microwave vial and heated at 180 °C for 65 min at 300 W microwave irradiation. The resulting mixture was allowed to cool to room temperature, poured onto crushed ice and the pH was adjusted to ~7 with sodium bicarbonate. The mixture was then extracted with 3 × 5 mL of ethyl acetate and the combined organic layers were dried over sodium sulfate. Removal of the volatiles under reduced pressure and purification via flash column chromatography (CH/EtOAc 20:1) resulted in 59.3 mg (95%) of a yellow oil.

***R*f** (CH/EtOAc 20:1): 0.28. – **¹H NMR** (400 MHz, $CDCl_3$): δ = 7.95 (s, 1H, 4-C*H*), 7.49 (d, 1H, 7-CH_{Ar}), 3.90 (b, 1H, C*H*$(CH_3)_2$), 2.57 (t, *J* = 7.51 Hz, 2H, $CH_2CH_2CH_3$), 2.37 (s, 3H, CH_3), 1.69 (dt, *J* = 7.22, 7.22 Hz, 2H, $CH_2CH_2CH_3$), 1.44 (d, *J* = 7.72 Hz, 6H $CH(CH_3)_2$), 1.01 (t, *J* = 7.21 Hz, 3H, CH_2CH_3) ppm. – **¹³C NMR** (100 MHz, $CDCl_3$): δ = 161.0 ($C_{quart.}$, $C_{Ar}OO$), 152.2 ($C_{quart.}$, C_{Ar}), 141.0 (+, $C_{Ar}H$), 136.1 ($C_{quart.}$, C_{Ar}), 135.6 (+, $C_{Ar}H$), 128.6 ($C_{quart.}$, C_{Ar}), 125.9 ($C_{quart.}$, C_{Ar}), 119.2 ($C_{quart.}$, C_{Ar}), 118.7 ($C_{quart.}$, C_{Ar}), 34.1 (+, $CH(CH_3)_2$), 21.4 (–, $CH_2CH_2CH_3$), 22.5 (+, $(CH_3)_2CH$), 21.4 (–, $CH_2CH_2CH_3$), 15.3 (+, $C_{Ar}CH_3$), 13.8 (+, $CH_2CH_2CH_3$) ppm. – **IR** (KBr): $\tilde{\nu}$ = 2963 (s), 2927 (s), 1708 (vw), 1587 (s), 1450 (m), 1381 (s), 1295 (s), 1263 (s), 1188 (m), 1100 (m), 1071 (vs), 1044 (s), 998 (s), 947 (s), 869 (s), 840 (s), 780 (m), 745 (s), 722 (s), 657 (s), 621 (s), 566 (s), 547 (s) cm^{-1}. – **MS** (70 eV, EI): *m/z* (%) = 324/322 (97/100) $[M]^+$, 310/308 (10/12), 309/307 (73/65) $[M-CH_3]^+$, 296/294 (28/25), 295/293 (33/35) $[M-C_2H_5]^+$, 282/280 (7/7), 281/279 (43/37) $[M-C_3H_7]^+$, 267/265 (10/12), 262 (5), 251 (5), 243 (13) $[M-Br]^+$, 231 (11), 214 (5), 213 (5), 200 (10), 199 (11),

186 (6), 185 (10), 181 (51), 172 (8), 171 (32), 170 (6), 169 (13), 162 (7), 158 (8), 157 (6), 155 (9), 153 (5), 145 (6), 143 (11), 142 (10), 141 (19), 131 (53), 129 (13), 128 (26), 127 (7), 119 (11), 115 (17), 100 (13), 91 (5), 85 (9), 77 (5), 69 (93), 57 (6), 55 (5). – **HRMS** ($C_{16}H_{19}O_2{}^{79}Br$): calc. 322.0563, found 322.0561.

6-Bromo-3-hexyl-5-isopropyl-8-methyl-2*H*-chromen-2-one (27aq)[177]

According to **GP7**, 50.0 mg of salicylic aldehyde **29k** (190 µmol, 1.00 equiv.), 220 µL octanoic acid anhydride **100g** (740 µmol, 3.89 equiv.) and 2.40 mg of potassium carbonate (17.4 µmol, 0.09 equiv.) were placed in a microwave vial and heated at 180 °C for 65 min at 300 W microwave irradiation. The was allowed to cool to room temperature, poured onto crushed ice and the pH was adjusted to ~7 with sodium bicarbonate. The mixture was then extracted with 3 × 5 mL of ethyl acetate and the combined organic layers were dried over sodium sulfate. Removal of the volatiles under reduced pressure and purification via flash column chromatography (CH/EtOAc 20:1) resulted in 55.3 mg (86%) of a yellow oil.

***R*f** (CH/EtOAc 20:1): 0.50. – **^{1}H NMR** (400 MHz, $CDCl_3$): δ = 7.95 (s, 1H, 4-C*H*), 7.50 (s, 1H, 7-CH_{Ar}), 3.90 (b, 1H, C*H*$(CH_3)_2$), 2.59 (t, *J* = 7.47 Hz, 2H, C$H_2(CH_3)_4CH_3$), 2.38 (s, 3H, CH_3), 1.70 – 1.60 (m, 2H, $CH_2CH_2(CH_3)_3CH_3$), 1.45 (d, *J* = 7.67 Hz, 6H, CH(CH_3)$_2$), 1.37 – 1.25 (m, 6H, 3 × CH_2), 0.91 – 0.87 (m, 3H, $(CH_2)_5CH_3$) ppm. – **^{13}C NMR** (100 MHz, $CDCl_3$): δ = 160.9 ($C_{quart.}$, $C_{Ar}OO$), 152.2 ($C_{quart.}$, C_{Ar}), 141.0 (+, $C_{Ar}H$), 136.5 ($C_{quart.}$, C_{Ar}), 135.9 (+, $C_{Ar}H$), 135.7 ($C_{quart.}$, $C_{Ar}CH_3$), 128.9 ($C_{quart.}$, C_{Ar}), 125.9 ($C_{quart.}$, $C_{Ar}Br$), 118.8 ($C_{quart.}$, C_{Ar}), 34.0 (+, $CH(CH_3)_2$), 31.7 (–, CH_2), 31.2 (–, CH_2), 29.0 (–, CH_2), 28.1 (–, CH_2), 22.7 (+, $(CH_3)_2CH$), 22.3 (–, CH_2CH_3), 15.3 (+, $C_{Ar}CH_3$), 14.1 (+, CH_3) ppm. – **IR** (KBr): $\tilde{\nu}$ = 2955 (vs), 2925 (m), 2854 (s), 1724 (vw), 1588 (s), 1454 (s), 1368 (vs), 1264 (s), 1175 (s), 1112 (vs), 1069 (s), 871 (vs), 785 (s), 398 (7) cm^{-1}. – **MS** (70 eV, EI): *m/z* (%) = 366/364 (19/19) $[M]^+$, 331 (8), 324/322 (18/17), 323/321 (97/100) $[M - C_3H_7]^+$, 319 (13), 309 (9), 307 (9), 296/294 (24/24), 295/293 (14/14) $[M - C_5H_{11}]^+$, 281/279 (24/14) $[M - C_6H_{13}]^+$, 269 (16), 267 (6), 265 (7), 243 (7), 231 (19), 230 (20), 228 (21), 219 (21), 215 (19), 214 (11), 213 (19), 181 (35), 172 (6), 171 (12), 169 (29), 145 (59), 142 (5), 134 (10), 131 (34), 128 (9), 127 (34), 119 (23), 115 (5), 100 (7), 98 (5), 69 (88), 58 (9), 57 (22). – **HRMS** ($C_{19}H_{25}O_2{}^{79}Br$): calc. 364.1032, found 364.1031.

6-Bromo-3,5-diisopropyl-8-methyl-2*H*-chromen-2-one (27ar)[177]

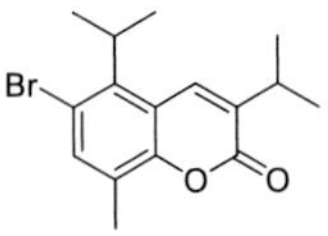

According to **GP7**, 50.0 mg of salicylic aldehyde **29k** (190 µmol, 1.00 equiv.), 150 µL isovaleric acid anhydride **100h** (757 µmol, 3.98 equiv.) and 2.30 mg of potassium carbonate (16.6 µmol, 0.09 equiv.) were placed in a microwave vial and heated at 180 °C for 185 min at 300 W microwave irradiation. The resulting mixture was allowed to cool to room temperature, poured onto crushed ice and the pH was adjusted to ~7 with sodium bicarbonate. The mixture was then extracted with 3 × 5 mL of ethyl acetate and the combined organic layers washed with sodium bicarbonate and brine and dried over sodium sulfate. Removal of the volatiles under reduced pressure and purification via flash column chromatography (CH/EtOAc 60:1 to 20:1) resulted in 45.4 mg (71%) of a white solid.

***R*f** (CH/EtOAc 20:1): 0.44. – **^{1}H NMR** (400 MHz, $CDCl_3$): δ = 7.98 (s, 1H, 4-C*H*), 7.50 (s, 1H, H_{Ar}), 3.91 (b, 1H, C*H*($CH_3)_2$), 3.19 – 3.06 (m, C*H*($CH_3)_2$), 2.38 (s, 3H, CH_3), 1.45 (d, *J* = 7.02 Hz, 6H, CH($CH_3)_2$), 1.28 (d, *J* = 7.32 Hz, 6H, CH($CH_3)_2$) ppm. – **^{13}C NMR** (100 MHz, $CDCl_3$): δ = 160.5 ($C_{quart.}$, C_{Ar}OO), 151.9 ($C_{quart.}$, C_{Ar}), 141.1 ($C_{quart.}$, C_{Ar}), 135.8 (+, C_{Ar}H), 135.4 ($C_{quart.}$, C_{Ar}), 134.4 (+, C_{Ar}H), 133.7 ($C_{quart.}$, C_{Ar}), 125.8 ($C_{quart.}$, C_{Ar}), 118.8 ($C_{quart.}$, C_{Ar}), 34.1 (+, *C*H($CH_3)_2$), 29.0 (+, *C*H($CH_3)_2$), 22.5 (+, CH($CH_3)_2$), 21.7 (+, CH($CH_3)_2$), 15.3 (+, $C_{Ar}CH_3$) ppm. – **IR** (KBr): $\tilde{\nu}$ = 2961 (s), 2927 (vs), 1706 (vw), 1586 (s), 1450 (s), 1388 (s), 1295 (s), 1260 (s), 1192 (s), 1159 (s), 1101 (s), 1066 (s), 1019 (m), 915 (s), 867 (s), 833 (s), 784 (m), 712 (s), 643 (s), 560 (s), 410 (vs), 393 (vs) cm^{-1}. – **MS** (70 eV, EI): *m/z* (%) = 324/322 (100/98) $[M]^+$, 310/308 (17/18), 309/307 (100/100) $[M - CH_3]^+$, 298 (9), 296 (17), 295 (12), 294 (12), 293 (12), 282/280 (5/5), 281/279 (25/21) $[M - C_3H_7]^+$, 269 (5), 267 (6), 265 (8), 243 (6), 231 (7), 219 (7), 213 (6), 185 (13), 181 (18), 172 (5), 169 (13), 158 (8), 142 (6), 142 (8), 131 (18), 129 (9), 128 (10), 119 (9), 115 (7), 85 (9), 69 (40). – **HRMS** ($C_{16}H_{19}O_2{}^{79}Br$): calc. 322.0563, found 322.0564.

6-Bromo-8-isopropyl-5-methyl-2*H*-chromen-2-one (27as)[177]

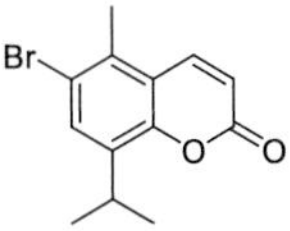

According to **GP7**, 49.1 mg of salicylic aldehyde **29m** (191 µmol, 1.00 equiv.), 100 µL acetic acid anhydride **100a** (1.06 mmol, 5.30 equiv.) and 4.20 mg of potassium carbonate (30.4 µmol, 0.16 equiv.) were placed in a microwave vial and heated at 180 °C for 65 min at 300 W microwave irradiation. The resulting mixture was allowed to cool to room temperature, poured onto crushed ice and the pH was adjusted to ~7 with sodium bicarbonate. The mixture was then extracted with 3 × 5 mL of ethyl acetate and the combined organic layers washed with sodium bicarbonate-solution and brine and dried over sodium sulfate. Removal of the volatiles under reduced pressure and purification via flash column chromatography (CH/EtOAc 10:1) resulted in 45.8 mg (85%) of an off-white solid.

***R*f** (CH/EtOAc 10:1): 0.16. – **^{1}H NMR** (400 MHz, $CDCl_3$): δ = 7.95 (d, *J* = 9.94 Hz, 1H, 4-C*H*), 7.58 (s, 1H, 7-CH_{Ar}), 6.44 (d, *J* = 9.74 Hz, 1H, 3-C*H*), 3.55 (hept, *J* = 6.91 Hz, 1H, C*H*(CH_3)$_2$), 2.56 (s, 3H, CH_3), 1.27 (d, *J* = 6.96 Hz, 6 H CH(CH_3)$_2$) ppm. – **^{13}C NMR** (100 MHz, $CDCl_3$): δ = 160.1 ($C_{quart.}$, C_{Ar}OO), 151.2 ($C_{quart.}$, C_{Ar}), 141.1 (+, C_{Ar}H), 136.3 ($C_{quart.}$, C_{Ar}), 132.8 (+, C_{Ar}H), 132.7 ($C_{quart.}$, C_{Ar}), 120.5 ($C_{quart.}$, C_{Ar}), 118.7 ($C_{quart.}$, C_{Ar}), 116.6 (+, C_{Ar}H), 26.5 (+, *C*H(CH_3)$_2$), 22.6 (+, (*C*H$_3$)$_2$CH), 18.4 (+, CH_3) ppm. – **IR** (KBr): $\tilde{\nu}$ = 2963 (s), 2926 (s), 1768 (vs), 1708 (vw), 1579 (m), 1438 (s), 1384 (s), 1294 (s), 1249 (s), 1205 (vs), 1184 (w), 1166 (s), 1127 (m), 1050 (s), 997 (s), 902 (m), 877 (s), 833 (m), 782 (s), 769 (m), 717 (s), 692 (m), 571 (s), 503 (s), 476 (s), 440 (s), 402 (s) cm^{-1}. – **MS** (70 eV, EI): *m/z* (%) = 282/280 (62/60) $[M]^+$, 268/266 (13/14), 267/265 (100/99) $[M-CH_3]^+$, 258/256 (10/10), 243 (9), 231 (11), 201 (6) $[M-Br]^+$, 186 (16), 181 (34), 173 (6), 169 (7), 158 (31), 157 (14), 143 (5), 142 (21), 141 (5), 131 (35), 130 (6), 129 (19), 128 (20), 127 (7), 119 (10), 115 (18), 100 (7), 77 (6), 69 (60). – **HRMS** ($C_{13}H_{13}O_2Br$): calc. 280.0093, found 280.0092.

6-Bromo-8-isopropyl-3,5-dimethyl-2*H*-chromen-2-one (27at)[177]

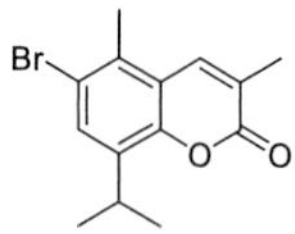

According to **GP7**, 48.6 mg of salicylic aldehyde **29m** (188 µmol, 1.00 equiv.), 120 µL propionic acid anhydride **100b** (941 µmol, 5.01 equiv.) and 3.80 mg of potassium carbonate (27.5 µmol, 0.15 equiv.) were placed in a microwave vial and heated at 180 °C for 65 min at 300 W microwave irradiation. The resulting mixture was allowed to cool to room temperature, poured onto crushed ice and the pH was adjusted to ~7 with sodium bicarbonate. The mixture was then extracted with 3 × 5 mL of ethyl acetate and the combined organic layers were washed with sodium bicarbonate and brine and dried over sodium sulfate. Removal of the volatiles under reduced pressure and purification via flash column chromatography (CH/EtOAc 40:1 to 10:1) resulted in 44.3 mg (80%) of an off-white solid.

***R*f** (CH/EtOAc 20:1): 0.27. – **^{1}H NMR** (400 MHz, $CDCl_3$): δ = 7.74 (s, 1H, 4-C*H*), 7.51 (s, 1H, 7-CH_{Ar}), 3.55 (hept, *J* = 6.89 Hz, 1H, C*H*(CH_3)$_2$), 2.56 (s, 3H, CH_3), 2.24 (s, 3H, CH_3), 1.26 (d,

J = 6.89 Hz, 6 H CH(CH_3)$_2$ ppm. – **^{13}C NMR** (100 MHz, $CDCl_3$): δ = 161.5 ($C_{quart.}$, C_{Ar}OO), 150.2 ($C_{quart.}$, C_{Ar}), 136.9 (+, C_{Ar}H), 135.9 ($C_{quart.}$, C_{Ar}), 131.6 ($C_{quart.}$, C_{Ar}), 131.4 (+, C_{Ar}H), 125.8 ($C_{quart.}$, C_{Ar}), 120.3 ($C_{quart.}$, C_{Ar}), 119.4 (+, C_{Ar}), 26.5 (+, CH(CH_3)$_2$), 22.7 (+, CH(CH_3)$_2$), 18.4 (+, CH_3), 17.6 (+, CH_3) ppm. – **IR** (KBr): ṽ = 3069 (vs), 2959 (m), 2923 (vs), 2922 (vs), 1696 (vw), 1633 (vs), 1586 (m), 1445 (s), 1381 (s), 1344 (s), 1294 (vs), 1261 (m), 1192 (m), 1087 (m), 1049 (s), 999 (s), 928 (s), 873 (s), 840 (s), 768 (s), 743 (m), 706 (s), 654 (s), 626 (s), 613 (s), 570 (s), 527 (s), 408 (s) cm^{-1}. – **MS** (70 eV, EI): *m/z* (%) = 296/294 (45/44) $[M]^+$, 282/280 (11/9), 281/279 (75/64) $[M - CH_3]^+$, 269 (10), 258 (21), 257 (8), 256 (40), 255 (11), 254 (20), 246 (32), 244 (33), 243 (11), 241 (5), 231 (18), 228 (7), 226 (5), 219 (17), 200 (9), 181 (46), 172 (24), 171 (14), 169 (20), 162 (5), 157 (6), 156 (7), 131 (44), 129 (13), 128 (14), 127 (5), 119 (22), 115 (8), 100 (9), 91 (6), 77 (7), 71 (7), 69 (100), 58 (9), 57 (68), 55 (6). – **HRMS** ($C_{14}H_{15}O_2Br$): calc. 294.0250, found 294.0249.

6-Bromo-3-ethyl-8-isopropyl-5-methyl-2*H*-chromen-2-one (27au)[177]

According to **GP7**, 47.6 mg of salicylic aldehyde **29m** (185 µmol, 1.00 equiv.), 130 µL butyric acid anhydride **100c** (797 µmol, 4.31 equiv.) and 4.50 mg of potassium carbonate (32.6 µmol, 0.18 equiv.) were placed in a microwave vial and heated at 180 °C for 65 min at 300 W microwave irradiation. The resulting mixture was allowed to cool to room temperature, poured onto crushed ice and the pH was adjusted to ~7 with sodium bicarbonate. The mixture was then extracted with 3 × 5 mL of ethyl acetate and the combined organic layers were washed with sodium bicarbonate and brine and dried over sodium sulfate. Removal of the volatiles under reduced pressure and purification via flash column chromatography (CH/EtOAc 60:1 to 20:1) resulted in 42.0 mg (73%) of an off-white solid.

***R*f** (CH/EtOAc 10:1): 0.44. – **^{1}H NMR** (400 MHz, $CDCl_3$): δ = 7.70 (s, 1H, 4-CH), 7.51 (s, 1H, H_{Ar}), 3.62 – 3.49 (m, 1H, CH(CH_3)$_2$), 2.67 – 2.59 (m, 2H, CH_2), 2.56 (s, 3H, CH_3), 1.30 – 1.22 (m, 9H, CH_2CH_3, CH(CH_3)$_2$) ppm. – **^{13}C NMR** (100 MHz, $CDCl_3$): δ = 161.1 ($C_{quart.}$, C_{Ar}OO), 150.0 ($C_{quart.}$, C_{Ar}), 135.9 ($C_{quart.}$, C_{Ar}), 135.2 (+, C_{Ar}H), 131.8 ($C_{quart.}$, C_{Ar}), 131.4 (+, C_{Ar}H), 131.3 ($C_{quart.}$, $C_{Ar}CH_3$), 120.3 ($C_{quart.}$, $C_{Ar}CH_2CH_3$), 119.4 ($C_{quart.}$, C_{Ar}), 26.5 (+, CH(CH_3)$_2$), 24.4 (–, CH_2CH_3), 22.7 (+, CH(CH_3)$_2$), 18.5 (+, CH_3), 12.7 (+, CH_3) ppm. – **IR** (KBr): ṽ = 2961 (m), 2931 (vs), 1700 (vw), 1586 (m), 1443 (m), 1381 (s), 1346 (s), 1291 (s), 1256 (m), 1188 (s), 1099 (s), 1037 (m), 1011 (vs), 933 (s), 874 (s), 836 (s), 798 (vs), 777 (s), 739 (s), 705 (vs),

650 (s), 614 (s), 573 (s), 526 (s), 479 (vs), 394 (s) cm^{-1}. – **MS** (70 eV, EI): *m/z* (%) = 310/308 (70/87) $[M]^+$, 306 (14), 296/294 (10/12), 295/293 (72/87) $[M-CH_3]^+$, 291 (4), 281 (11), 269 (16), 267 (6), 265 (5), 257 (5), 256/254 (15/14), 255 (7), 246/244 (13/13), 243 (7), 231 (20), 230 (12), 229 (8) $[M-Br]^+$, 219 (20), 215 (8), 214 (7), 213 (8), 186 (18), 185 (6), 181 (38), 171 (18), 170 (5), 169 (30), 141 (7), 131 (40), 129 (6), 128 (11), 119 (27), 115 (8), 100 (7), 91 (5), 71 (40), 69 (100), 58 (7). – **HRMS** ($C_{15}H_{17}O_2Br$): calc. 308.0406, found 308.0406.

6-Bromo-3-propyl-8-isopropyl-5-methyl-2*H*-chromen-2-one (27av)[177]

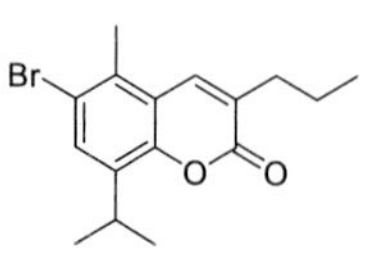

According to **GP7**, 48.9 mg of salicylic aldehyde **29m** (190 µmol, 1.00 equiv.), 150 µL valeric acid anhydride **100d** (812 µmol, 4.32 equiv.) and 3.50 mg of potassium carbonate (25.3 µmol, 0.13 equiv.) were placed in a microwave vial and heated at 180 °C for 65 min at 300 W microwave irradiation. The resulting mixture was allowed to cool to room temperature, poured onto crushed ice and the pH was adjusted to ~7 with sodium bicarbonate. The mixture was then extracted with 3 × 5 mL of ethyl acetate and the combined organic layers washed with sodium bicarbonate and brine and dried over sodium sulfate. Removal of the volatiles under reduced pressure and purification via flash column chromatography (CH/EtOAc 60:1 to 20:1) resulted in 8.0 mg (13%) of a yellow oil.

R_f (CH/EtOAc 20:1): 0.35. – **^{1}H NMR** (400 MHz, $CDCl_3$): δ = 7.69 (s, 1H, 4-C*H*), 7.52 (s, 1H, H_{Ar}), 3.56 (hept, *J* = 7.19 Hz, 1H, C*H*$(CH_3)_2$), 2.55 – 2.59 (m, 5H, CH_3, $CH_2CH_2CH_3$), 1.75 – 1.60 (m, 2H, $CH_2CH_2CH_3$), 1.27 (d, *J* = 6.69 Hz, 6H, $CH(CH_3)_2$), 1.01 (t, 3H, CH_2CH_3) ppm. – **^{13}C NMR** (100 MHz, $CDCl_3$): δ = 161.1 ($C_{quart.}$, $C_{Ar}OO$), 150.1 ($C_{quart.}$, C_{Ar}), 136.2 (+, $C_{Ar}H$), 135.9 ($C_{quart.}$, C_{Ar}), 131.8 ($C_{quart.}$, C_{Ar}), 131.4 (+, $C_{Ar}H$), 129.9 ($C_{quart.}$, C_{Ar}), 120.3 ($C_{quart.}$, C_{Ar}), 119.4 ($C_{quart.}$, C_{Ar}), 33.4 (–, $CH_2CH_2CH_3$), 26.5 (+, $CH(CH_3)_2$), 22.7 (+, $CH(CH_3)_2$), 21.6 (–, CH_2CH_3), 18.5 (+, CH_3), 14.0 (+, CH_3) ppm. – **IR** (KBr): $\tilde{\nu}$ = 2958 (m), 2928 (vs), 2869 (s), 1704 (vw), 1586 (m), 1462 (m), 1382 (s), 1345 (vs), 1294 (s), 1256 (m), 1214 (s), 1180 (m), 1103 (s), 1068 (s), 1044 (s), 1029 (vs), 976 (s), 931 (s), 899 (vs), 880 (s), 837 (s), 776 (s), 740 (s), 666 (s), 612 (s), 568 (s), 525 (m), 401 (s) cm^{-1}. – **MS** (70 eV, EI): *m/z* (%) = 324/322 (46/42) $[M]^+$, 323 (9), 312 (13), 310/308 (37/63), 309/307 (41/42) $[M-CH_3]^+$, 306 (31), 296/294 (13/11), 295/293 (23/37) $[M-C_2H_5]^+$, 291 (9), 281/279 (11/7) $[M-C_3H_7]^+$, 267 (7), 258/256 (14/19), 243 (22) $[M-Br]^+$, 241 (15), 231 (16), 230 (82), 229 (14), 228 (89), 215 (39), 214 (8), 213 (37), 181 (35), 171 (13), 169 (8), 157 (8), 148 (9), 143 (7), 134 (15), 133 (7), 132

(7), 131 (43), 129 (10), 128 (13), 119 (14), 116 (7), 115 (18), 105 (12), 103 (7), 100 (8), 91 (12), 85 (96), 83 (8), 77 (10), 71 (14), 69 (80), 58 (7), 57 (100), 55 (20). – **HRMS** ($C_{16}H_{19}O_2Br$): calc. 322.0563, found 322.0563.

8-Isopropyl-3,5-dimethyl-2*H*-chromen-2-one (27ba)

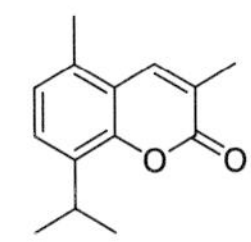

According to GP7, 150 mg of salicylic aldehyde **29l** (840 µmol, 1.00 equiv.), 380 µl propionic acid anhydride **100b** (383 mg, 2.95 mmol, 3.50 equiv.) and 5.82 mg of potassium carbonate (40.0 µmol, 0.05 equiv.) were placed in a microwave vial and heated at 180 °C for 65 min at 300 W microwave irradiation. The resulted mixture was allowed to cool to room temperature, poured onto crushed ice and the pH was adjusted to ~7 with sodium bicarbonate. The mixture was then extracted with diethyl ether and the organic phase dried over sodium sulfate and the volatiles were removed under reduced pressure to result in 182 mg (quant.) of an off-white solid.

***R*f** (CH/EtOAc 20:1): 0.33. – **MP**: 148.5 °C – **^{1}H NMR** (400 MHz, $CDCl_3$): δ = 7.71 (q, *J* = 1.3 Hz, 1H, 4-C*H*), 7.27 (d, *J* = 7.7 Hz, 1H, Ar*H*), 7.03 (d, *J* = 7.8 Hz, 1H, Ar*H*), 3.59 (hept., *J* = 6.9 Hz, 1H, C*H*(CH_3)$_2$), 2.47 (s, 3H, ArC*H*$_3$), 2.24 (d, *J* = 1.3 Hz, 3H, C-3(C*H*$_3$), 1.27 (d, *J* = 11.2 Hz, 6H, CH(C*H*$_3$)$_2$) ppm. – **^{13}C NMR** (100 MHz, $CDCl_3$): δ = 162.4 ($C_{quart.}$, *C*-2), 151.1 ($C_{quart.}$, C_{Ar}O), 136.9 (+, *C*H, C-4), 134.2 ($C_{quart.}$, C_{Ar}), 132.3 ($C_{quart.}$, C_{Ar}), 127.3 (+, C_{Ar}H), 125.5 (+, C_{Ar}H), 124.6 ($C_{quart.}$, C_{Ar}), 118.2 ($C_{quart.}$, C-3), 26.4 (+, *C*H(CH_3)$_2$), 22.9 (+, 2 × *C*H$_3$, CH(*C*H$_3$)$_2$), 18.3 (+, *C*H$_3$), 17.5 (+, *C*H$_3$) ppm. – **IR** (KBr): $\tilde{\nu}$ = 2958 (w), 2925 (vw), 1709 (w), 1594 (w), 1489 (vw), 1448 (w), 1377 (w), 1360 (w), 1293 (vw), 1259 (w), 1186 (w), 1087 (w), 1062 (w), 1046 (w), 1003 (w), 897 (w), 835 (w), 771 (w), 746 (w), 719 (vw), 633 (vw), 607 (w), 480 (w), 410 (vw) cm^{-1}. – **MS** (70 eV, EI): *m/z* (%) = 216 (40) [M]$^+$, 202 (15), 201 (100) [M – CH_3]$^+$, 173 (5), 130 (7), 129 (11), 128 (13), 115 (11), 91 (6). – **HRMS** ($C_{14}H_{16}O_2$): calc. 216.1145, found 216.1145. – **Elemental analysis**: $C_{14}H_{16}O_2$ (216.18): calc. C 77.75, H 7.46, found C 77.34, H 7.47.

5-Isopropyl-8-methyl-2*H*-chromen-2-one (27bb)

According to **GP7**, 66 mg of salicylic aldehyde **29j** (320 µmol, 1.00 equiv.), 150 µL of acetic acid anhydride **100a** (170 mg, 1.67 mmol, 3.50 equiv.) and 2.19 mg of potassium carbonate (20 µmol, 0.05 equiv.) were placed in a microwave vial and heated at 180 °C for 65 min at 300 W microwave irradiation. The resulting mixture was allowed to cool to room temperature, poured onto crushed ice and the pH was adjusted to ~7 with sodium bicarbonate. The mixture was then extracted with 3 × 5 mL of ethyl acetate and the combined organic layers were dried over sodium sulfate. Removal of the volatiles under reduced pressure and purification via flash column chromatography (CH/EtOAc 5:1) resulted in 70.0 mg (95%) of a yellow solid.

***R*f** (CH/EtOAc 5:1): 0.28. – **^{1}H NMR** (400 MHz, $CDCl_3$): δ = 8.11 (d, *J* = 10.0 Hz, 1H, 4-C*H*), 6.97 (s, 1H, 7-CH_{Ar}), 6.39 (d, *J* = 10.0 Hz, 1H, 3-C*H*), 3.84 (s, 3H, OCH_3), 3.67 – 3.45 (m, 1H, C*H*(CH_3)$_2$), 2.42 (s, 3H, CH_3), 1.37 (d, *J* = 7.1 Hz, 6H, CH(CH_3)$_2$) ppm. – **^{13}C NMR** (100 MHz, $CDCl_3$): δ = 161.0 ($C_{quart.}$, *C*OO), 153.8 ($C_{quart.}$, C_{Ar}), 147.4 ($C_{quart.}$, C_{Ar}), 141.0 (+, 4-C_{Ar}H), 130.7 ($C_{quart.}$, C_{Ar}), 124.4 ($C_{quart.}$, C_{Ar}), 118.0 (+, C_{Ar}H), 117.3 ($C_{quart.}$, C_{Ar}), 115.9 (+, C_{Ar}H), 56.4 (+, OCH_3), 26.7 (+, *C*H(CH_3)$_2$), 21.6 (+, CH(CH_3)$_2$), 15.9 (+, CH_3) ppm. – **IR** (KBr): ṽ = 2921 (w), 1709 (m), 1592 (w), 1575 (w), 1454 (w), 1391 (w), 1374 (w), 1295 (w), 1255 (w), 1226 (w), 1198 (w), 1166 (w), 1118 (m), 1048 (w), 1013 (w), 989 (w), 921 (w), 850 (w), 831 (m), 773 (w), 731 (w), 686 (w), 665 (w), 635 (w), 596 (w), 548 (w), 475 (vw), 409 (vw) cm^{-1}. – **MS** (70 eV, EI): *m/z* (%) = 232 (69) [M]$^+$, 218 (15), 217 (100) [M – CH_3]$^+$, 202 (17), 162 (6). – **HRMS** ($C_{14}H_{16}O_2$): calc. 232.1094, found 232.1092. – **Elemental analysis**: $C_{14}H_{16}O_3$: calc. C 72.40, H 6.94, found C 72.08, H 6.93.

5-Isopropyl-6-methoxy-3,8-dimethyl-2*H*-chromen-2-one (27bc)

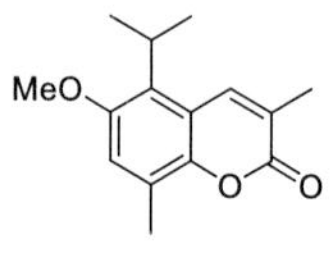

MeO O O

According to **GP7**, 49 mg of salicylic aldehyde **29j** (240 µmol, 1.00 equiv.), 110 µL propionic acid anhydride **100b** (107 mg, 0.82 mmol, 3.50 equiv.) and 1.63 mg of potassium carbonate (10 µmol, 0.05 equiv.) were placed in a microwave vial and heated at 180 °C for 65 min at 300 W microwave irradiation. The resulting mixture was allowed to cool to room temperature, poured onto crushed ice and the pH was adjusted to ~7 with sodium bicarbonate. The mixture was then extracted with 3 × 5 mL of ethyl acetate and the combined organic layers were dried over sodium sulfate. Removal of the volatiles under reduced pressure and purification via flash column chromatography (CH/EtOAc 5:1) resulted in 70.0 mg (74%) of a yellow solid.

***R*f** (CH/EtOAc 10:1): 0.22. – **^{1}H NMR** (400 MHz, $CDCl_3$): δ = 7.90 (s, 1H, 4-C*H*), 6.90 (s, 1H, H_{Ar}), 3.83 (s, 3H, OCH_3), 3.68 – 3.40 (m, 2H, C*H*(CH_3)$_2$), 2.42 (s, 3H, CH_3), 2.23 (s, 3H, CH_3), 1.37 (d, *J* = 7.1 Hz, 6H, CH(CH_3)$_2$) ppm. – **^{13}C NMR** (100 MHz, $CDCl_3$): δ = 162.3 ($C_{quart.}$, *C*OO), 153.9 ($C_{quart.}$, C_{Ar}), 146.7 ($C_{quart.}$, C_{Ar}), 136.9 (+, C_{Ar}H), 129.7 ($C_{quart.}$, C_{Ar}), 124.8 ($C_{quart.}$, C_{Ar}), 124.0 ($C_{quart.}$, C_{Ar}), 117.9 ($C_{quart.}$, C_{Ar}), 116.8 (+, C_{Ar}H), 56.4 (+, OCH_3), 26.8 (+,

*C*H(CH$_3$)$_2$), 21.6 (+, CH(*C*H$_3$)$_2$), 17.8 (+, *C*H$_3$), 15.9 (+, *C*H$_3$) ppm. – **IR** (KBr): $\tilde{\nu}$ = 2959 (w), 1705 (m), 1597 (m), 1462 (m), 1350 (m), 1304 (m), 1252 (w), 1226 (w), 1186 (m), 1170 (m), 1138 (w), 1094 (m), 1051 (m), 1007 (m), 907 (m), 843 (m), 771 (w), 748 (w), 682 (w), 641 (w), 590 (w), 557 (w), 465 (vw), 390 (w) cm^{-1}. – **MS** (70 eV, EI): *m/z* (%) = 246 (24) [M]$^+$, 236 (19), 231 (36) [M – CH$_3$]$^+$, 201 (14), 192 (10), 181 (22), 180 (100). – **HRMS** ($C_{15}H_{18}O_3$): calc. 246.1250, found 232.1253.

3-Ethyl-5-isopropyl-6-methoxy-8-methyl-2*H*-chromen-2-one (27bd)

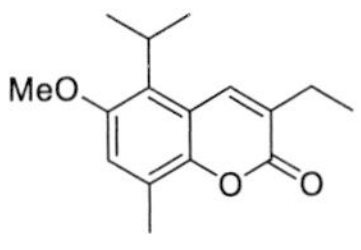

According to **GP7**, 66 mg of salicylic aldehyde **29j** (320 µmol, 1.00 equiv.), 170 µL butyric acid anhydride **100c** (175 mg, 1.11 mmol, 3.50 equiv.) and 2.19 mg of potassium carbonate (20 µmol, 0.05 equiv.) were placed in a microwave vial and heated at 180 °C for 65 min at 300 W microwave irradiation. The resulting mixture was allowed to cool to room temperature, poured onto crushed ice and the pH was adjusted to ~7 with sodium bicarbonate. The mixture was then extracted with 3 × 5 mL of ethyl acetate and the combined organic layers were dried over sodium sulfate. Removal of the volatiles under reduced pressure and purification via flash column chromatography (CH/EtOAc 10:1) resulted in 68.6 mg (83%) of an off-white solid.

***R*$_f$** (CH/EtOAc 5:1): 0.46. – **^{1}H NMR** (400 MHz, CDCl$_3$): δ = 7.86 (s, 1H, 4-C*H*), 6.90 (s, 1H, *H*$_{Ar}$), 3.83 (s, 3H, OC*H*$_3$), 3.67 – 3.50 (m, 1H, C*H*(CH$_3$)$_2$), 2.62 (qd, *J* = 7.5, 1.2 Hz, 2H, C*H*$_2$), 2.42 (s, 3H, C*H*$_3$), 1.38 (d, *J* = 7.1 Hz, 6H, CH(C*H*$_3$)$_2$), 1.26 (t, *J* = 7.4 Hz, 3H, C*H*$_3$) ppm. – **^{13}C NMR** (100 MHz, CDCl$_3$): δ = 161.8 (C$_{quart.}$, *C*OO), 153.8 (C$_{quart.}$, *C*$_{Ar}$), 146.5 (C$_{quart.}$, *C*$_{Ar}$), 135.2 (+, *C*$_{Ar}$H), 130.2 (C$_{quart.}$, *C*$_{Ar}$), 129.9 (C$_{quart.}$, *C*$_{Ar}$), 123.9 (C$_{quart.}$, *C*$_{Ar}$), 118.0 (C$_{quart.}$, *C*$_{Ar}$), 116.8 (+, *C*$_{Ar}$H), 56.4 (+, O*C*H$_3$), 26.7 (+, *C*H(CH$_3$)$_2$), 24.5 (–, *C*H$_2$), 21.6 (+, CH(*C*H$_3$)$_2$), 15.9 (+, *C*H$_3$), 12.8 (+, *C*H$_3$) ppm. – **IR** (KBr): $\tilde{\nu}$ = 2963 (vw), 1708 (w), 1590 (w), 1453 (w), 1370 (vw), 1332 (w), 1291 (w), 1249 (w), 1230 (w), 1184 (w), 1100 (w), 1045 (w), 1008 (w), 967 (w), 904 (w), 858 (w), 843 (w), 768 (w), 653 (vw), 593 (w), 554 (vw), 468 (vw), 389 (vw) cm^{-1}. – **MS** (70 eV, EI): *m/z* (%) = 260 (28) [M]$^+$, 250 (16), 245 (28) [M – CH$_3$]$^+$, 234 (20), 219 (7), 181 (30), 180 (100), 165 (39), 131 (15). – **HRMS** ($C_{16}H_{20}O_3$): calc. 260.1407, found 260.1405.

5-Isopropyl-6-methoxy-8-methyl-3-propyl-2*H*-chromen-2-one (27be)

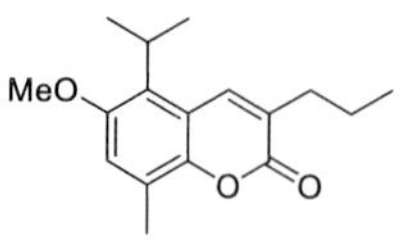

MeO O O

According to **GP7**, 66 mg of salicylic aldehyde **29j** (320 µmol, 1.00 equiv.), 200 µL valeric acid anhydride **100d** (207 mg, 1.11 mmol, 3.50 equiv.) and 2.19 mg of potassium carbonate (20 µmol, 0.05 equiv.) were placed in a microwave vial and heated at 180 °C for 65 min at 300 W microwave irradiation. The resulting mixture was allowed to cool to room temperature, poured onto crushed ice and the pH was adjusted to ~7 with sodium bicarbonate. The mixture was then extracted with 3 × 5 mL of ethyl acetate and the combined organic layers were dried over sodium sulfate. Removal of the volatiles under reduced pressure and purification via flash column chromatography (CH/EtOAc 5:1) resulted in the product with remaining valeric acid that was removed by taking up the compound in ethyl acetate and washing with 3 × 300 mL of sodium bicarbonate. Drying over sodium sulfate and removal of the volatiles under reduced pressure resulted in 45.1 mg (55%) of an yellow oil.

***R*f** (CH/EtOAc 5:1): 0.46. – **^{1}H NMR** (400 MHz, $CDCl_3$): δ = 7.86 (s, 1H, 4-C*H*), 6.90 (s, 1H, H_{Ar}), 3.83 (s, 3H, OCH_3), 3.69 – 3.48 (m, 1H, C*H*$(CH_3)_2$), 2.60 – 2.51 (m, 2H, CH_2), 2.41 (s, 3H, CH_3), 1.75 – 1.62 (m, 2H, CH_2), 1.37 (d, J = 7.1 Hz, 6H, CH(CH_3)$_2$), 1.00 (t, J = 7.3 Hz, 3H, CH_3) ppm. – **^{13}C NMR** (100 MHz, $CDCl_3$): δ = 161.8 ($C_{quart.}$, *C*OO), 153.7 ($C_{quart.}$, C_{Ar}), 146.6 ($C_{quart.}$, C_{Ar}), 136.2 (+, C_{Ar}H), 129.8 ($C_{quart.}$, C_{Ar}), 128.7 ($C_{quart.}$, C_{Ar}), 123.9 ($C_{quart.}$, C_{Ar}), 117.9 ($C_{quart.}$, C_{Ar}), 116.8 (+, C_{Ar}H), 56.4 (+, OCH_3), 33.4 (–, CH_2), 26.6 (+, *C*H$(CH_3)_2$), 21.7 (–, CH_2), 21.6 (+, CH$(CH_3)_2$), 15.9 (+, CH_3), 13.9 (+, CH_3) ppm. – **IR** (KBr): ṽ = 2957 (w), 2871 (vw), 1714 (m), 1596 (w), 1455 (w), 1359 (vw), 1309 (w), 1249 (vw), 1225 (vw), 1188 (w), 1107 (w), 1053 (w), 1012 (w), 920 (w), 850 (vw), 774 (vw), 639 (vw), 595 (vw) cm^{-1}. – **MS** (70 eV, EI): *m/z* (%) = 274 (94) $[M]^+$, 260 (22), 259 (100) $[M - CH_3]^+$, 245 (12) $[M - C_2H_5]^+$, 231 (9), 187 (8). – **HRMS** ($C_{17}H_{22}O_3$): calc. 274.1563, found 274.1462.

3-Butyl-5-isopropyl-6-methoxy-8-methyl-2*H*-chromen-2-one (27bf)

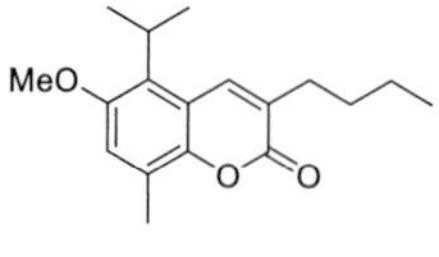

MeO O O

According to **GP7**, 66 mg of salicylic aldehyde **29j** (320 µmol, 1.00 equiv.), 250 µL hexanoic acid anhydride **100e** (238 mg, 1.11 mmol, 3.50 equiv.) and 2.19 mg of potassium carbonate (20 µmol, 0.05 equiv.) were placed in a microwave vial and heated at 180 °C for 65 min at 300 W microwave irradiation. The resulting mixture was allowed to cool to room temperature, poured onto crushed ice and the pH was adjusted to ~7 with sodium bicarbonate. The mixture was then extracted with 3 × 5 mL of ethyl acetate and the combined organic layers were dried over sodium sulfate. Removal of the volatiles under reduced pressure and purification via flash column chromatography (CH/EtOAc 5:1) resulted in 47.0 mg (57%) of an yellow oil.

***R*f** (CH/EtOAc 10:1): 0.28. – **^{1}H NMR** (400 MHz, $CDCl_3$): δ = 7.85 (s, 1H, 4-C*H*), 6.90 (s, 1H, H_{Ar}), 3.83 (s, 3H, OCH_3), 3.67 – 3.49 (m, 1H, C*H*$(CH_3)_2$), 2.63 – 2.52 (m, 2H, CH_2), 2.42 (s, 3H, CH_3), 1.69 – 1.57 (m, 2H, CH_2), 1.47 – 1.39 (m, 2H, CH_2), 1.38 (d, *J* = 7.1 Hz, 6H, $CH(CH_3)_2$), 0.96 (t, *J* = 7.3 Hz, 3H, CH_3) ppm. – **^{13}C NMR** (100 MHz, $CDCl_3$): δ = 161.8 ($C_{quart.}$, *C*OO), 153.8 ($C_{quart.}$, C_{Ar}), 146.6 ($C_{quart.}$, C_{Ar}), 136.0 (+, $C_{Ar}H$), 129.8 ($C_{quart.}$, C_{Ar}), 129.0 ($C_{quart.}$, C_{Ar}), 123.9 ($C_{quart.}$, C_{Ar}), 117.9 ($C_{quart.}$, C_{Ar}), 116.8 (+, $C_{Ar}H$), 56.4 (+, OCH_3), 31.2 (–, CH_2), 30.5 (–, CH_2), 26.6 (+, $CH(CH_3)_2$), 22.6 (–, CH_2), 21.7 (+, $CH(CH_3)_2$), 15.9 (+, CH_3), 14.0 (+, CH_3) ppm. – **IR** (KBr): $\tilde{\nu}$ = 2954 (w), 2870 (w), 1711 (s), 1596 (m), 1455 (m), 1359 (w), 1310 (m), 1248 (w), 1226 (w), 1186 (m), 1137 (w), 1111 (m), 1051 (m), 1013 (m), 910 (w), 849 (w), 785 (w), 640 (w), 596 (w) cm^{-1}. – **MS** (70 eV, EI): *m/z* (%) = 288/289 (100/20) $[M]^+$, 274 (15), 273 (78) $[M-CH_3]^+$, 259 (13) $[M-C_2H_5]^+$, 246 (27), 245 (51) $[M-C_3H_7]^+$, 231 (23), 287 (10), 181 (17), 131 (19), 99 (12). – **HRMS** ($C_{18}H_{24}O_3$): calc. 288.1720, found 288.1720.

5-Isopropyl-6-methoxy-8-methyl-3-pentyl-2*H*-chromen-2-one (27bg)

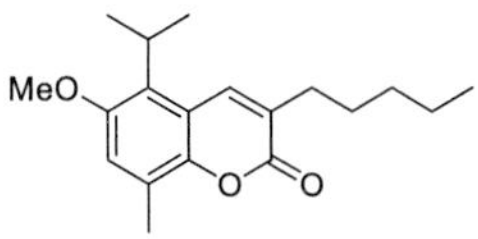

According to **GP7**, 66 mg of salicylic aldehyde **29j** (320 µmol, 1.00 equiv.), 290 µL heptanoic acid anhydride **100f** (269 mg, 1.11 mmol, 3.50 equiv.) and 2.19 mg of potassium carbonate (20 µmol, 0.05 equiv.) were placed in a microwave vial and heated at 180 °C for 65 min at 300 W microwave irradiation. The resulting mixture was allowed to cool to room temperature, poured onto crushed ice and the pH was adjusted to ~7 with sodium bicarbonate. The mixture was then extracted with 3 × 5 mL of ethyl acetate and the combined organic layers were dried over sodium sulfate. Removal of the volatiles under reduced pressure and purification via flash column chromatography (CH/EtOAc 10:1) resulted in 78.8 mg (82%) of a yellow oil.

***R*f** (CH/EtOAc 10:1): 0.39. – **^{1}H NMR** (400 MHz, $CDCl_3$): δ = 7.85 (s, 1H, 4-C*H*), 6.90 (s, 1H, H_{Ar}), 3.83 (s, 3H, OCH_3), 3.68 – 3.48 (m, 1H, C*H*$(CH_3)_2$), 2.61 – 2.52 (m, 2H, CH_2), 2.42 (s, 3H, CH_3), 1.74 – 1.57 (m, 2H, CH_2), 1.44 – 1.29 (m, 10H, 2 × CH_2, $CH(CH_3)_2$), 0.96 – 0.86 (m,

3H, CH_3) ppm. – **^{13}C NMR** (100 MHz, $CDCl_3$): δ = 161.9 ($C_{quart.}$, COO), 153.8 ($C_{quart.}$, C_{Ar}), 146.6 ($C_{quart.}$, C_{Ar}), 136.0 (+, $C_{Ar}H$), 129.8 ($C_{quart.}$, C_{Ar}), 129.0 ($C_{quart.}$, C_{Ar}), 123.9 ($C_{quart.}$, C_{Ar}), 118.0 ($C_{quart.}$, C_{Ar}), 116.8 (+, $C_{Ar}H$), 56.4 (+, OCH_3), 31.6 (–, CH_2), 31.4 (–, CH_2), 28.1 (–, CH_2), 26.6 (+, $CH(CH_3)_2$), 22.6, (–, CH_2) 21.7 (+, $CH(CH_3)_2$), 15.9 (+, CH_3), 14.2 (+, CH_3) ppm. – **IR** (KBr): ṽ = 2926 (w), 2870 (w), 1714 (m), 1597 (w), 1456 (w), 1359 (vw), 1310 (w), 1225 (w), 1182 (w), 1112 (w), 1052 (w), 1013 (w), 914 (w), 848 (vw), 775 (vw), 597 (vw) cm^{-1}. – **MS** (70 eV, EI): *m/z* (%) = 302 (100) $[M]^+$, 288 (13), 287 (62) $[M - CH_3]^+$, 273 (8) $[M - C_2H_5]^+$, 260 (17), 259 (89) $[M - C_3H_7]^+$, 246 (24), 245 (14), 231 (27), 187 (11), 181 (18), 131 (19), 100 (20). – **HRMS** ($C_{19}H_{26}O_3$): calc. 302.1876, found 302.1875.

5-Isopropyl-6-methoxy-8-methyl-3-hexyl-2*H*-chromen-2-one (27bh)

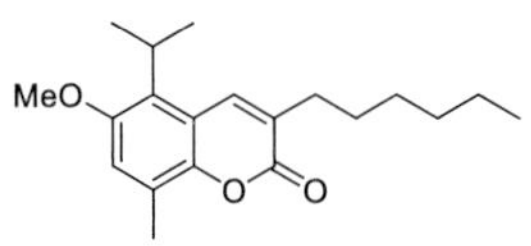

According to **GP7**, 66 mg of salicylic aldehyde **29j** (320 µmol, 1.00 equiv.), 330 µL ethantic acid anhydride **100g** (300 mg, 1.11 mmol, 3.50 equiv.) and 2.19 mg of potassium carbonate (20 µmol, 0.05 equiv.) were placed in a microwave vial and heated at 180 °C for 65 min at 300 W microwave irradiation. The mixture was allowed to cool to room temperature, poured onto crushed ice and the pH was adjusted to ~7 with sodium bicarbonate. The mixture was then extracted with 3 × 5 mL of ethyl acetate and the combined organic layers were dried over sodium sulfate. Removal of the volatiles under reduced pressure and purification via flash column chromatography (CH/EtOAc 10:1) resulted in 61.1 mg (61%) of a yellow oil.

***R*f** (CH/EtOAc 10:1): 0.38. – **^{1}H NMR** (400 MHz, $CDCl_3$): δ = 7.85 (s, 1H, 4-CH), 6.90 (s, 1H, H_{Ar}), 3.83 (s, 3H, OCH_3), 3.66 – 3.49 (m, 1H, $CH(CH_3)_2$), 2.61 – 2.53 (m, 2H, CH_2), 2.42 (s, 3H, CH_3), 1.69 – 1.58 (m, 2H, CH_2), 1.41 – 1.36 (m, 8H, CH_2, $CH(CH_3)_2$), 1.35 – 1.27 (m, 2 × 4H, CH_2), 0.95 – 0.84 (m, 3H, CH_3) ppm. – **^{13}C NMR** (100 MHz, $CDCl_3$): δ = 161.9 ($C_{quart.}$, COO), 153.8 ($C_{quart.}$, C_{Ar}), 146.6 ($C_{quart.}$, C_{Ar}), 136.0 (+, $C_{Ar}H$), 129.8 ($C_{quart.}$, C_{Ar}), 129.0 ($C_{quart.}$, C_{Ar}), 123.9 ($C_{quart.}$, C_{Ar}), 118.0 ($C_{quart.}$, C_{Ar}), 116.8 (+, $C_{Ar}H$), 56.4 (+, OCH_3), 31.8 (–, CH_2), 31.4 (–, CH_2), 29.1 (–, CH_2), 28.3 (–, CH_2), 26.7 (+, $CH(CH_3)_2$), 22.8 (–, CH_2), 21.7 (+, $CH(CH_3)_2$), 15.9 (+, CH_3), 14.2 (+, CH_3) ppm. – **IR** (KBr): ṽ = 2925 (w), 2855 (w), 1711 (s), 1597 (w), 1456 (m), 1359 (w), 1310 (m), 1249 (w), 1225 (w), 1178 (m), 1113 (w), 1072 (m), 1052 (m), 1013 (m), 916 (w), 849 (w), 786 (w), 640 (w), 597 (w) cm^{-1}. – **MS** (70 eV, EI): *m/z* (%) = 316 (100) $[M]^+$, 302 (10), 301 (46) $[M - CH_3]$, 274 (16), 273 (94) $[M - C_3H_7]^+$, 259

(12) $[M - C_4H_9]^+$, 246 (23), 245 (14), 231 (28), 187 (10), 181 (25), 180 (24). – **HRMS** ($C_{20}H_{28}O_3$): calc. 316.2033, found 316.2034.

3,5-Diisopropyl-6-methoxy-8-methyl-2*H*-chromen-2-one (27bi)

According to **GP7**, 66 mg of salicylic aldehyde **29j** (320 µmol, 1.00 equiv.), 220 µL isovaleric acid anhydride **100h** (207 mg, 1.11 mmol, 3.50 equiv.) and 2.19 mg of potassium carbonate (20 µmol, 0.05 equiv.) were placed in a microwave vial and heated at 180 °C for 65 min at 300 W microwave irradiation. The resulting mixture was allowed to cool to room temperature, poured onto crushed ice and the pH was adjusted to ~7 with sodium bicarbonate. The mixture was then extracted with 3 × 5 mL of ethyl acetate and the combined organic layers were dried over sodium sulfate. Removal of the volatiles under reduced pressure and purification via flash column chromatography (CH/EtOAc 10:1) resulted in 87.7 mg (quant.) of a yellow solid.

***R*f** (CH/EtOAc 10:1): 0.36. – **^{1}H NMR** (400 MHz, $CDCl_3$): δ = 7.86 (s, 1H, 4-C*H*), 6.90 (s, 1H, H_{Ar}), 3.83 (s, 3H, OCH_3), 3.69 – 3.53 (m, 1H, C*H*$(CH_3)_2$), 3.13 (hept, *J* = 6.8 Hz, 1H, C*H*$(CH_3)_2$), 2.41 (s, 3H, CH_3), 1.38 (d, *J* = 7.1 Hz, 6H, $CH(CH_3)_2$), 1.27 (d, *J* = 6.9 Hz, 6H, $CH(CH_3)_2$) ppm. – **^{13}C NMR** (100 MHz, $CDCl_3$): δ = 161.4 ($C_{quart.}$, *C*OO), 153.7 ($C_{quart.}$, C_{Ar}), 146.3 ($C_{quart.}$, C_{Ar}), 134.5 ($C_{quart.}$, C_{Ar}), 133.4 (+, $C_{Ar}H$), 130.0 ($C_{quart.}$, C_{Ar}), 123.8 ($C_{quart.}$, C_{Ar}), 117.9 ($C_{quart.}$, C_{Ar}), 116.8 (+, $C_{Ar}H$), 56.5 (+, OCH_3), 29.1 (+, $CH(CH_3)_2$), 26.5 (+, $CH(CH_3)_2$), 21.7 (+, $CH(CH_3)_2$), 21.7 (+, $CH(CH_3)_2$), 15.9 (+, CH_3) ppm. – **IR** (KBr): ṽ = 2954 (m), 1704 (s), 1597 (m), 1465 (m), 1378 (w), 1358 (w), 1326 (m), 1288 (m), 1250 (m), 1232 (m), 1190 (m), 1158 (m),1137 (m), 1070 (m), 1054 (m), 1010 (m), 990 (m), 911 (m), 869 (m), 784 (m), 751 (w), 709 (w), 675 (w), 653 (w), 595 (w), 442 (vw), 404 (w) cm^{-1}. – **MS** (70 eV, EI): *m/z* (%) = 274 (31) $[M]^+$, 260 (7), 259 (40) $[M - CH_3]^+$, 231 (6) $[M - C_3H_7]^+$, 224 (15), 209 (14), 194 (11), 193 (39), 190 (38), 189 (11), 181 (15), 180 (14), 179 (24), 176 (12), 175 (100), 165 (10), 147 (25). – **HRMS** ($C_{17}H_{22}O_3$): calc. 274.1569, found 274.1570.

5-Methoxy-7-pentyl-2*H*-chromen-2-one (27bj)

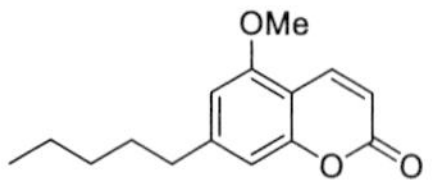

According to **GP7**, 100 mg of salicylic aldehyde **29a** (0.51 mmol, 1.00 equiv.), 170 µL acetic acid anhydride **100a** (183 mg, 0.183 mmol, 3.50 equiv.) and 3.56 mg of potassium carbonate (30 µmol, 0.05 equiv.) were placed in a microwave vial and heated at 180 °C for 65 min at 300 W microwave irradiation. The resulting mixture was allowed to cool to room temperature, poured onto crushed ice and the pH was adjusted to ~7 with sodium bicarbonate. The mixture was then extracted with 3 × 5 mL of ethyl acetate and the combined organic layers were dried over sodium sulfate. Removal of the volatiles under reduced pressure and purification via flash column chromatography (CH/EtOAc 15:1) resulted in 82 mg (67%) of the pure product **27bj** as a white solid, as well as 21 mg (15%) of the side product **27bj-Ac** as a white solid.

THU-C 067: ***R_f*** (CH/EtOAc 5:1): 0.27. – **MP**: 58.3 °C – **^{1}H NMR** (400 MHz, $CDCl_3$): δ = 8.02 (d, J = 9.7 Hz, 1H, 4-C*H*), 6.74 (s, 1H, H_{Ar}), 6.52 (s, 1H, H_{Ar}), 6.25 (d, J = 9.7 Hz, 1H, 3-C*H*), 3.91 (s, 3H, OCH_3), 2.64 (t, J = 7.7 Hz, 2H, CH_2), 1.69 – 1.59 (m, 2H, CH_2), 1.40 – 1.27 (m, 4H, 2 × CH_2), 0.89 (t, J = 6.5 Hz, 3H, CH_3) ppm. – **^{13}C NMR** (100 MHz, $CDCl_3$): δ = 161.4 ($C_{quart.}$, C_{Ar}), 156.0 ($C_{quart.}$, C_{Ar}), 155.3 ($C_{quart.}$, C_{Ar}), 149.0 ($C_{quart.}$, C_{Ar}), 138.7 (+, $C_{Ar}H$), 113.5 (+, $C_{Ar}H$), 109.0 (+, $C_{Ar}H$), 107.7 ($C_{quart.}$, C_{Ar}), 105.8 (+, $C_{Ar}H$), 56.0 (+, OCH_3), 36.8 (–, CH_2), 31.5 (–, CH_2), 30.8 (–, CH_2), 22.6 (–, CH_2), 14.1 (+, CH_3) ppm. – **IR** (KBr): ṽ = 2922 (w), 2861 (w), 1715 (m), 1609 (m), 1490 (w), 1463 (m), 1419 (m), 1340 (w), 1230 (m), 1193 (w), 1117 (m), 1090 (m), 882 (w), 831 (m), 797 (w), 747 (w), 697 (w), 671 (w), 630 (w), 588 (w), 551 (w), 509 (w), 482 (w) 551 (w) cm^{-1}. – **MS** (70 eV, EI): *m/z* (%) = 246 (36) $[M]^+$, 204 (12), 203 (8), 191 (13), 190 (100), 189 (14), 161 (10), 160 (9). – **HRMS** ($C_{15}H_{18}O_3$): calc. 246.1250, found 246.1249. – **Elemental analysis**: $C_{15}H_{18}O_3$: calc. C 73.15, H 7.37, found C 73.25, H 7.55.

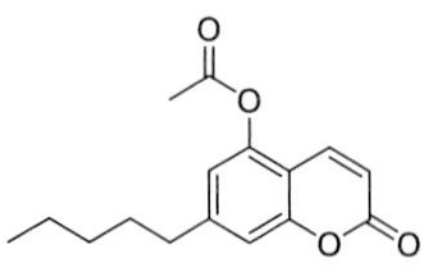

Side Product 27bj-Ac*:* ***R_f*** (CH/EtOAc 5:1): 0.12. – **^{1}H NMR** (400 MHz, $CDCl_3$): δ = 7.68 (d, J = 9.7 Hz, 1H, 4-C*H*), 7.03 (s, 1H, H_{Ar}), 6.90 (s, 1H, H_{Ar}), 6.36 (d, J = 9.7 Hz, 1H, 3-C*H*), 2.69 – 2.65 (m, 2H, CH_2), 2.39 (s, 3H, $COOCH_3$), 1.78 – 1.50 (m, 2H, CH_2), 1.44 – 1.24 (m, 4H, 2 × CH_2), 0.89 (t, J = 6.8 Hz, 3H, CH_3) ppm. – **^{13}C NMR** (100 MHz, $CDCl_3$): δ = 168.8 ($C_{quart.}$), 160.5 ($C_{quart.}$), 154.8 ($C_{quart.}$), 148.6 ($C_{quart.}$), 146.9 (+, $C_{Ar}H$), 137.4 (+, $C_{Ar}H$), 118.1 (+, *C*H), 115.8 (+, $C_{Ar}H$), 114.2 (+, $C_{Ar}H$), 110.6 ($C_{quart.}$), 36.2 (–, CH_2), 31.4 (–, CH_2), 30.5 (–, CH_2), 22.5 (–, CH_2), 21.0 (+, $COOCH_3$), 14.1 (+, CH_3) ppm. – **IR** (KBr): ṽ = 3084 (vw), 2921 (w), 2855 (w), 1557 (m), 1719 (s), 1625 (m), 1554 (w), 1497 (w), 1429 (m), 1375 (m), 1197 (s), 1117 (m), 1053 (m), 1013 (m), 884 (m), 836 (s), 768 (w), 743 (w), 683 (w), 649 (w), 601 (w), 574 (w), 526 (w) 480 (w), 410 (w) cm^{-1}. – **MS** (70 eV, EI): *m/z* (%) = 274 (19) [M]$^+$, 233 (10), 232 (71), 190 (12), 177 (11), 176 (100), 175 (12), 146 (8). – **HRMS** ($C_{16}H_{18}O_4$): calc. 274.1200, found 274.1199. – **Elemental analysis**: $C_{16}H_{18}O_4$: calc. C 70.06, H 6.61, found C 70.27, H 6.72.

5-Methoxy-3-methyl-7-pentyl-2*H*-chromen-2-one (27bk)

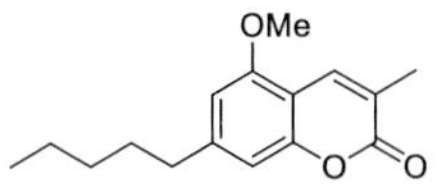

According to GP7, 200 mg of salicylic aldehyde **29a** (0.90 mmol, 1.00 equiv.), 0.40 mL of propionic acid anhydride **100b** (410 mg, 3.15 mmol, 3.50 equiv.) and 6.2 mg of potassium carbonate (40 µmol, 0.05 equiv.) were placed in a microwave vial and heated at 180 °C for 65 min at 300 W microwave irradiation. The resulting mixture was allowed to cool to room temperature, quenched with sodium bicarbonate and the pH was adjusted to ~7 with sodium bicarbonate. The mixture was then extracted with 3 × 15 mL of ethyl acetate, washed with 3 × 15 mL of water and the combined organic layers were dried over sodium sulfate. Removal of the volatiles under reduced pressure and purification via flash column chromatography (CH/EtOAc 20:1) resulted in 186 mg (80%) of an off-white solid.

***R*f** (CH/EtOAc 20:1): 0.26. – **^{1}H NMR** (400 MHz, $CDCl_3$): δ = 7.85 – 7.79 (m, 1H, 4-C*H*), 6.72 (s, 1H, H_{Ar}), 6.50 (s, 1H, H_{Ar}), 3.90 (s, 3H, OCH_3), 2.63 (dd, *J* = 8.6, 6.8 Hz, 2H, CH_2), 2.17 (d, *J* = 1.3 Hz, 3H, CH_3), 1.68 – 1.56 (m, 2H, CH_2), 1.40 – 1.25 (m, 4H, 2 × CH_2), 0.93 – 0.84 (m, 3H, $C_4H_8CH_3$) ppm. – **^{13}C NMR** (100 MHz, $CDCl_3$): δ = 162.8 ($C_{quart.}$, *C*OO), 155.3 ($C_{quart.}$, C_{Ar}), 154.4 ($C_{quart.}$, C_{Ar}), 147.3 ($C_{quart.}$, C_{Ar}), 134.5 (+, 4-*C*H), 122.6 ($C_{quart.}$, C_{Ar}), 108.6 (+, $C_{Ar}H$), 108.2 ($C_{quart.}$, C_{Ar}), 105.6 (+, $C_{Ar}H$), 56.0 (+, OCH_3), 36.7 (–, CH_2), 31.5 (–, CH_2), 30.9 (–, CH_2), 22.6 (–, CH_2), 17.3 (–, CH_2), 14.1(+, CH_3) ppm. – **IR** (KBr): ṽ = 2924 (w), 2856 (w), 1713 (s), 1614 (s), 1574 (m), 1495 (m), 1452 (m), 1425 (m), 1352 (w), 1295 (w), 1254 (m), 1179 (m), 1144 (w), 1115 (m), 1063 (m), 998 (m), 918 (w), 831 (w), 767 (w), 730 (w), 571 (w) cm^{-1}. – **MS** (70 eV, EI): *m/z* (%) = 260 (53) [M]$^+$, 231 (8), 219 (6), 218 (9), 217 (11), 205 (14), 204 (100), 203 (22), 189 (5), 181 (17), 176 (7), 175 (12), 169 (6), 166 (13), 161 (9), 131 (20), 119 (9), 69 (38). – **HRMS** ($C_{16}H_{20}O_3$): calc. 260.1407, found 260.1407. – **Elemental analysis**: $C_{16}H_{20}O_3$: calc. C 73.82, H 7.74, found C 73.32, H 7.79.

3-Ethyl-5-methoxy-7-pentyl-2*H*-chromen-2-one (27bl)

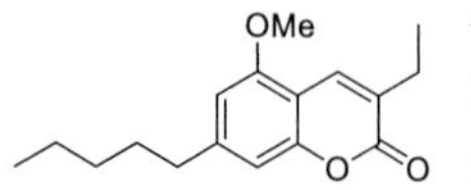

According to **GP7**, 700 mg of salicylic aldehyde **29a** (3.15 mmol, 1.00 equiv.), 1.80 mL of butyric acid anhydride **100c** (1744 mg, 11.0 mmol, 3.50 equiv.) and 21.8 mg of potassium carbonate (160 µmol, 0.05 equiv.) were placed in a microwave vial and heated at 180 °C for 65 min at 300 W microwave irradiation. The resulting mixture was allowed to cool to room temperature, quenched with sodium bicarbonate and the pH was adjusted to ~7 with sodium bicarbonate. The mixture was then extracted with 3 × 15 mL of ethyl acetate, washed with 3 × 15 mL of water and the combined organic layers were dried over sodium sulfate. Removal of the volatiles under reduced pressure and purification via flash column chromatography (CH/EtOAc 20:1) resulted in 542 mg (63%) of an off-white solid.

***R*f** (CH/EtOAc 5:1): 0.58. – **^{1}H NMR** (400 MHz, $CDCl_3$): δ = 7.81 (q, *J* = 1.1 Hz, 1H, 4-C*H*), 6.72 (s, 1H, H_{Ar}), 6.51 (s, 1H, H_{Ar}), 3.91 (s, 3H, OCH_3), 2.67 – 2.60 (m, 2H, CH_2), 2.57 (qd, *J* = 7.4, 1.3 Hz, 2H, CH_2), 1.69 – 1.57 (m, 2H, CH_2), 1.39 – 1.27 (m, 4H, 2 × CH_2), 1.23 (t, *J* = 7.4 Hz, 3H, CH_2CH_3), 0.94 – 0.84 (m, 3H, $C_4H_8CH_3$) ppm. – **^{13}C NMR** (100 MHz, $CDCl_3$): δ = 162.3 ($C_{quart.}$, *C*OO), 155.4 ($C_{quart.}$, C_{Ar}), 154.2 ($C_{quart.}$, C_{Ar}), 147.3 ($C_{quart.}$, C_{Ar}), 132.8 (+, 4-*C*H), 128.2 ($C_{quart.}$, C_{Ar}), 108.6 (+, C_{Ar}H), 108.2 ($C_{quart.}$, C_{Ar}), 105.6 (+, C_{Ar}H), 56.0 (+, OCH_3), 36.7 (–, CH_2), 31.5 (–, CH_2), 30.9 (–, CH_2), 24.1 (–, CH_2), 22.6 (–, CH_2), 14.1 (+, CH_3), 12.6 (+, CH_3) ppm. – **IR** (KBr): ṽ = 2929 (m), 2855 (w), 1705 (s), 1615 (s), 1498 (m), 1449 (m), 1426 (m), 1354 (w), 1291 (w), 1251 (m), 1177 (m), 1144 (m), 1113 (m), 1079 (m), 1045 (m), 963 (m), 924 (m), 871 (w), 834 (m), 755 (w), 719 (m), 646 (w), 556 (w) cm^{-1}. – **MS** (70 eV, EI): *m/z* (%) = 274 (34) $[M]^+$, 262 (7) 259 (9) $[M - CH_3]^+$, 243 (9), 232 (6), 203 (7), 169 (10), 162 (8), 119 (15), 93 (6). – **HRMS** ($C_{17}H_{22}O_3$): calc. 274.1563, found 274.1562. – **Elemental analysis**: $C_{17}H_{22}O_3$: calc. C 74.42, H 8.08, found C 74.40, H 8.02.

3-Propyl-5-methoxy-7-pentyl-2*H*-chromen-2-one (27bm)

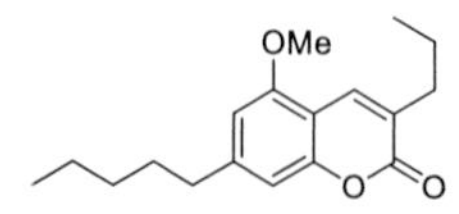

According to **GP7**, 700 mg of salicylic aldehyde **29a** (3.15 mmol, 1.00 equiv.), 2.16 mL of valeric acid anhydride **100d** (2052 mg, 11.0 mmol, 3.50 equiv.) and 21.8 mg of potassium carbonate (160 µmol, 0.05 equiv.) were placed in a microwave vial and heated at 180 °C for 65 min at 300 W microwave irradiation. The resulting mixture was allowed to cool to room temperature, quenched with sodium bicarbonate and the pH was adjusted to ~7 with sodium bicarbonate.

The mixture was then extracted with 3 × 15 mL of ethyl acetate, washed with 3 × 15 mL of water and the combined organic layers were dried over sodium sulfate. Removal of the volatiles under reduced pressure and purification via flash column chromatography (CH/EtOAc 20:1) and subsequent recrystallisation from cyclohexane resulted in 732 mg (81%) of colorless crystals.

***R*f** (CH/EtOAc 5:1): 0.63. – **^{1}H NMR** (400 MHz, $CDCl_3$): δ = 7.80 (q, *J* = 1.0 Hz, 1H, 4-C*H*), 6.72 (dd, *J* = 1.4, 0.7 Hz, 1H, H_{Ar}), 6.51 (d, *J* = 1.3 Hz, 1H, H_{Ar}), 3.91 (s, 3H, OCH_3), 2.68 – 2.59 (m, 2H, CH_2), 2.51 (ddd, *J* = 8.7, 6.5, 1.1 Hz, 2H, CH_2), 1.72 – 1.57 (m, 4H, 2 × CH_2), 1.40 – 1.25 (m, 4H, 2 × CH_2), 0.98 (t, *J* = 7.4 Hz, 3H, CH_3), 0.92 – 0.84 (m, 3H, CH_3) ppm. – **^{13}C NMR** (100 MHz, $CDCl_3$): δ = 162.4 ($C_{quart.}$, *C*OO), 155.4 ($C_{quart.}$, C_{Ar}), 154.3 ($C_{quart.}$, C_{Ar}), 147.3 ($C_{quart.}$, C_{Ar}), 133.8 (+, 4-*C*H), 126.7 ($C_{quart.}$, C_{Ar}), 108.6 (+, C_{Ar}H), 108.2 (+, C_{Ar}H), 105.6 ($C_{quart.}$, C_{Ar}), 56.0 (+, OCH_3), 36.7 (–, CH_2), 33.1 (–, CH_2), 31.5 (–, CH_2), 30.9 (–, CH_2), 22.6 (–, CH_2), 21.6 (–, CH_2), 14.1 (+, CH_3), 13.9 (+, CH_3) ppm. – **IR** (KBr): $\tilde{\nu}$ = 2955 (m), 2928 (m), 2857 (w),1707 (s), 1615 (s), 1497 (w), 1459 (m), 1428 (m), 1355 (w), 1288 (w), 1256 (m), 1173 (m), 1142 (m), 1115 (m), 1051 (m), 1028 (m), 1005 (m), 930 (m), 869 (m), 853 (m), 838 (m), 800 (w), 775 (w), 713 (m), 643 (w), 588 (w), 555 (w), 539 (w). cm^{-1}. – **MS** (70 eV, EI): *m/z* (%) = 288 (87) $[M]^+$, 273 (20) $[M\ CH_3]^+$, 260 (24), 259 (100) $[M - C_2H_5]^+$, 245 (45), 233 (7), 232 (46), 231 (12), 204 (6), 203 (8), 202 (6), 174 (9), 173 (10), 143 (11), 142 (10), 131 (9), 130 (8), 129 (10), 117 (9), 116 (12), 69 (14). – **HRMS** ($C_{18}H_{24}O_3$): calc. 288.1720, found 288.1719. – **Elemental analysis**: $C_{18}H_{24}O_3$: calc. C 74.97, H 8.39, found C 74.78, H 8.33.

3-Butyl-5-methoxy-7-pentyl-2*H*-chromen-2-one (27bn)

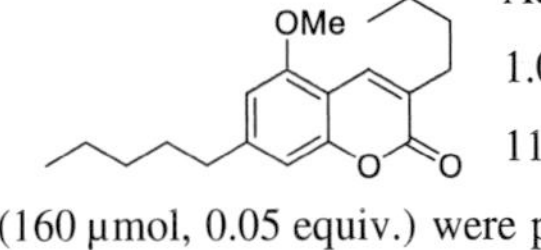

According to GP7, 700 mg of salicylic aldehyde **29a** (3.15 mmol, 1.00 equiv.), 2.16 mL of hexanoic acid anhydride **100e** (2362 mg, 11.0 mmol, 3.50 equiv.) and 21.8 mg of potassium carbonate (160 µmol, 0.05 equiv.) were placed in a microwave vial and heated at 180 °C for 65 min at 300 W microwave irradiation. The resulting mixture was allowed to cool to room temperature, quenched with sodium bicarbonate and the pH was adjusted to ~7 with sodium bicarbonate. The mixture was then extracted with 3 × 15 mL of ethyl acetate, washed with 3 × 15 mL of water and the combined organic layers were dried over sodium sulfate. Removal of the volatiles under reduced pressure and purification via flash column chromatography (CH/EtOAc 20:1) in 40 mg of colorless crystals that were washed with cyclohexane as well as 680 mg of an yellow oil that contained 90% of the product and was used in subsequent reactions without further purification. In total 652 mg (68%) of the product was obtained.

***R*f** (CH/EtOAc 5:1): 0.75. – **^{1}H NMR** (400 MHz, $CDCl_3$): δ = 7.80 (s, 1H, 4-C*H*), 6.73 (d, *J* = 1.2 Hz, 1H, H_{Ar}), 6.51 (d, *J* = 1.2 Hz, 1H, H_{Ar}), 3.92 (s, 3H, OCH_3), 2.67 – 2.60 (m, 2H, CH_2), 2.58 – 2.50 (m, 2H, CH_2), 1.62 (dt, *J* = 9.6, 7.3 Hz, 4H, 2 × CH_2), 1.46 – 1.28 (m, 6H, 3 × CH_2), 0.95 (t, *J* = 7.3 Hz, 3H, CH_3), 0.92 – 0.87 (m, 3H, CH_3) ppm. – **^{13}C NMR** (100 MHz, $CDCl_3$): δ = 162.4 ($C_{quart.}$, *C*OO), 155.4 ($C_{quart.}$, C_{Ar}), 154.3 ($C_{quart.}$, C_{Ar}), 147.3 ($C_{quart.}$, C_{Ar}), 133.7 (+, 4-*C*H), 126.9 ($C_{quart.}$, C_{Ar}), 108.6(+, C_{Ar}H), 108.2 ($C_{quart.}$, C_{Ar}), 105.6(+, C_{Ar}H), 56.0 (+, O*C*H_3), 36.7 (–, CH_2), 31.5 (–, CH_2), 30.9 (–, CH_2), 30.8 (–, CH_2), 30.5 (–, CH_2), 22.7 (–, CH_2), 22.6 (–, CH_2), 14.2 (+, CH_3), 14.1 (+, CH_3) ppm. – **IR** (KBr): ṽ = 2914 (w), 283 (w), 1708 (m), 1614 (m), 1573 (w), 1498 (w), 1457 (m), 1429 (m), 1376 (w), 1354 (w), 1292 (w), 1262 (w), 1183 (w), 1146 (w), 1112 (m), 1086 (m), 1050 (m), 999 (w), 934 (w), 883 (w), 847 (m), 780 (m), 744 (w), 725 (w), 672 (w), 639 (w), 581 (w), 552 (w), 481 (w) cm^{-1}. – **MS** (70 eV, EI): *m/z* (%) = 302 (28) $[M]^+$, 281 (8), 273 (12), 262 (10), 260 (27), 259 (30), 246 (10), 243 (11), 204 (12), 162 (9), 119 (13), 100 (14). – **HRMS** ($C_{19}H_{26}O_3$): calc. 302.1876, found 302.1877. – **Elemental analysis**: $C_{19}H_{26}O_3$: calc. C 75.46, H 8.67, found C 75.01, H 8.64.

7-(1-Butylcyclopentyl)-5-methoxy-2*H*-chromen-2-one (27bo)

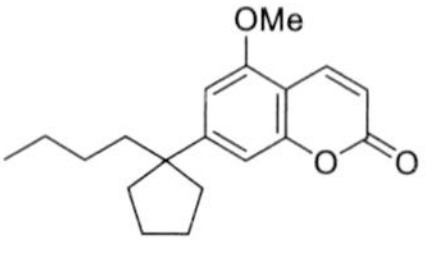

According to **GP7**, 1500 mg of salicylic aldehyde **29b** (5.43 mmol, 1.00 equiv.), 1.90 mL of acetic acid anhydride **100a** (1939 mg, 19.0 mmol, 3.50 equiv.) and 37.5 mg of potassium carbonate (270 µmol, 0.05 equiv.) were placed in a microwave vial and heated at 180 °C for 65 min at 300 W microwave irradiation. The resulting mixture was allowed to cool to room temperature, poured onto crushed ice and the pH was adjusted to ~7 with sodium bicarbonate. The mixture was then extracted with 3 × 50 mL of ethyl acetate and the combined organic layers were dried over sodium sulfate. Removal of the volatiles under reduced pressure and purification via flash column chromatography (CH/EtOAc 15:1) resulted in 1.31 g (93%) of the pure product as a white solid.

***R*f** (CH/EtOAc 5:1): 0.43. – **^{1}H NMR** (400 MHz, $CDCl_3$): δ = 8.03 (d, *J* = 9.6 Hz, 1H, 4-C*H*), 6.83 (d, *J* = 1.6 Hz, 1H, H_{Ar}), 6.62 (d, *J* = 1.5 Hz, 1H, H_{Ar}), 6.27 (d, *J* = 9.7 Hz, 1H, 3-C*H*), 3.92 (s, 3H, O-CH_3), 1.95– 1.54 (m, 10H, 5 × CH_2), 1.24 – 1.10 (m, 2H, CH_2), 0.86 – 1.03 (m, 2H, CH_2), 0.78 (t, *J* = 7.3 Hz, 3H, CH_3) ppm. – **^{13}C NMR** (100 MHz, $CDCl_3$): δ = 161. 42 ($C_{quart.}$,

*C*OO), 155.58 ($C_{quart.}$, $C_{Ar}OCH_3$) 155.44 ($C_{quart.}$, C_{Ar}), 154.84 ($C_{quart.}$, C_{Ar}), 138.55 (+, $C_{Ar}H$), 113.47 (+, $C_{Ar}H$), 107.85 (+, $C_{Ar}H$), 107. 36 ($C_{quart.}$, C_{Ar}- C_{Ar}), 104.10 (+, $C_{Ar}H$), 55.86 (+, O-CH_3), 52.02 ($C_{quart.}$, R- *C*-C_{Ar}), 41.51 (–, CH_2), 37.68 (–, 2x CH_2), 27. 45 (–, CH_2), 23.23 (–, CH_2), 23.19 (–, 2 × CH_2), 13.98 (+, CH_3) ppm. – **IR** (KBr): ṽ = 2928 (w), 2854 (w), 1725 (w),1612 (m), 1555 (w), 1488 (w), 1461 (w), 1415 (w), 1235 (w), 1118 (m), 927 (w), 896 (w), 833 (m), 763 (w), 695 (w), 618 (w), 576 (vw), 517 (vw) cm^{-1}. – **MS** (70 eV, EI): *m/z* (%) = 300 (40) $[M]^+$, 245 (10), 244 (63), 243 (100) $[M - C_4H_9]^+$, 218 (12), 204 (22), 190 (9), 189 (56), 67 (12). – **HRMS** ($C_{19}H_{24}O_3$): calc. 300.1719, found 300.1720.

7-(1-Butylcyclopentyl)-5-methoxy-3-methyl-2*H*-chromen-2-one (27bp)

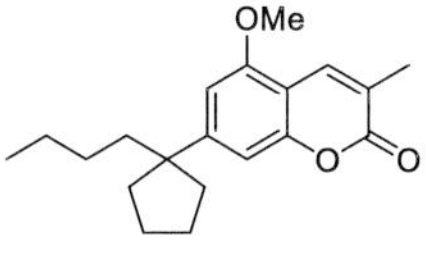

According to GP7, 200 mg of salicylic aldehyde **29b** (0.72 mmol, 1.00 equiv.), 0.32 mL of propionic acid anhydride **100b** (330 mg, 2.53 mmol, 3.50 equiv.) and 5.0 mg of potassium carbonate (40 µmol, 0.05 equiv.) were placed in a microwave vial and heated at 180 °C for 65 min at 300 W microwave irradiation. The resulting mixture was allowed to cool to room temperature, quenched with sodium bicarbonate and the pH was adjusted to ~7 with sodium bicarbonate. The mixture was then extracted with 3 × 15 mL of ethyl acetate, washed with 3 × 15 mL of water, 3 × 15 mL of sodium bicarbonate and the combined organic layers were dried over sodium sulfate. Removal of the volatiles under reduced pressure and purification via flash column chromatography (CH/EtOAc 20:1) resulted in 196 mg (86%) of an off-white solid.

R_f (CH/EtOAc 50:1): 0.16. – **MP**: 63.5 °C – **^{1}H NMR** (400 MHz, $CDCl_3$): δ = 7.84 (d, *J* = 1.5 Hz, 1H, 4-C*H*), 6.82 (d, *J* = 1.4 Hz, 1H, H_{Ar}), 6.61 (d, *J* = 1.4 Hz, 1H, H_{Ar}), 3.91 (s, 3H, OCH_3), 2.18 (d, *J* = 1.3 Hz, 3H, CH_3), 1.94 – 1.77 (m, 4H, 2 × CH_2), 1.77 – 1.54 (m, 6H, 3 × CH_2), 1.15 (p, *J* = 7.1 Hz, 2H, CH_2), 0.94 (dtd, *J* = 12.1, 9.3, 8.8, 5.7 Hz, 2H, CH_2), 0.78 (t, *J* = 7.3 Hz, 3H, $C_3H_6CH_3$) ppm. – **^{13}C NMR** (100 MHz, $CDCl_3$): δ = 162.8 ($C_{quart.}$, *C*OO), 155.0 ($C_{quart.}$, C_{Ar}), 154.1 ($C_{quart.}$, C_{Ar}), 153.8 ($C_{quart.}$, C_{Ar}), 134.5 (+, 4-*C*H), 122.7 ($C_{quart.}$, C_{Ar}), 108.0 ($C_{quart.}$, C_{Ar}), 107.6 (+, $C_{Ar}H$), 104.1 (+, $C_{Ar}H$), 55.9 (+, OCH_3), 52.0 ($C_{quart.}$, C_{CP}), 41.7 (–, CH_2), 37.8 (–, CH_2), 27.6 (–, 2 × CH_2), 23.4 (–, CH_2), 23.3 (–, 2 × CH_2), 17.3 (+, CH_3), 14.1 (+, CH_3) ppm. – **IR** (KBr): ṽ = 2951 (m), 2869 (w), 1725 (m), 1710 (m), 1612 (m), 1571 (m), 1494 (w), 1444 (m), 1415 (m), 1377 (w), 1348 (w), 1288 (w), 1257 (m), 1178 (m), 1108 (m), 1062 (m), 999 (m), 913 (m), 841 (m), 762 (m), 728 (m), 687 (w), 621 (w), 558 (w) cm^{-1}. – **MS**

(70 eV, EI): *m/z* (%) = 314 (35) $[M]^+$, 262 (8), 259 (5), 258 (32), 257 (76), 232 (5), 219 (6), 218 (7), 204 (5), 203 (30), 191 (10), 169 (8), 162 (9), 119 (14), 100 (13), 93 (6), 69 (100). – **HRMS** ($C_{20}H_{26}O_3$): calc. 314.1875, found 314.1876. – **Elemental analysis**: $C_{20}H_{26}O_3$: calc. C 76.40, H 8.34, found C 76.41, H 8.32.

7-(1-Butylcyclopentyl)-3-ethyl-5-methoxy-2*H*-chromen-2-one (27bq)

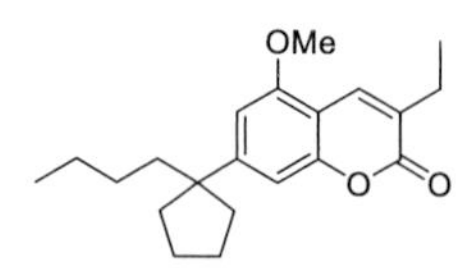

According to **GP7**, 200 mg of salicylic aldehyde **29b** (0.72 mmol, 1.00 equiv.), 0.41 mL of butyric acid anhydride **100c** (400 mg, 2.53 mmol, 3.50 equiv.) and 5.0 mg of potassium carbonate (40 µmol, 0.05 equiv.) were placed in a microwave vial and heated at 180 °C for 65 min at 300 W microwave irradiation. The resulting mixture was allowed to cool to room temperature, quenched with sodium bicarbonate and the pH was adjusted to ~7 with sodium bicarbonate. The mixture was then extracted with 3 × 15 mL of ethyl acetate, washed with 3 × 15 mL of water, 3 × 15 mL of sodium bicarbonate and the combined organic layers were dried over sodium sulfate. Removal of the volatiles under reduced pressure and purification via flash column chromatography (CH/EtOAc 50:1) resulted in 207 mg (87%) of an off-white solid.

$\boldsymbol{R_f}$ (CH/EtOAc 50:1): 0.23. – **MP**: 58.0 °C – ^{1}**H NMR** (400 MHz, $CDCl_3$): δ = 7.81 (d, J = 1.4 Hz, 1H, 4-CH), 6.82 (d, J = 1.4 Hz, 1H, H_{Ar}), 6.61 (d, J = 1.5 Hz, 1H, H_{Ar}), 3.92 (s, 3H, OCH_3), 2.57 (qd, J = 7.5, 1.3 Hz, 2H, CH_2), 1.95 – 1.76 (m, 4H, 2 × CH_2), 1.76 – 1.53 (m, 6H, 3 × CH_2), 1.23 (t, J = 7.4 Hz, 3H, CH_2CH_3), 1.15 (p, J = 7.1 Hz, 2H, CH_2), 0.94 (dtd, J = 12.1, 9.4, 8.8, 5.7 Hz, 2H, CH_2), 0.78 (t, J = 7.3 Hz, 3H, $C_3H_6CH_3$) ppm. – ^{13}**C NMR** (100 MHz, $CDCl_3$): δ = 162.4 ($C_{quart.}$, COO), 155.1 ($C_{quart.}$, C_{Ar}), 153.9 ($C_{quart.}$, C_{Ar}), 153.8 (+, 4-CH), 132.7 ($C_{quart.}$, C_{Ar}), 128.3 ($C_{quart.}$, C_{Ar}), 108.0 ($C_{quart.}$, C_{Ar}), 107.6 (+, C_{Ar}H), 104.0 (+, C_{Ar}H), 55.9 (+, OCH_3), 52.0 ($C_{quart.}$, C_{CP}), 41.7 (–, CH_2), 37.8 (–, 2 × CH_2), 27.6 (–, CH_2), 24.1 (–, CH_2), 23.4 (–, CH_2), 23.3 (–, 2 × CH_2), 14.1 (+, CH_3), 12.7 (+, CH_3) ppm. – **IR** (KBr): $\tilde{\nu}$ = 2955 (m), 2872 (m), 1713 s), 1613 (s), 1569 (m), 1494 (w), 1440 (m), 1413 (m), 1346 (w), 1290 (w), 1247 (m), 1167 (m), 1107 (s), 1079 (m), 1048 (m), 958 (m), 918 (m), 841 (m), 753 (w), 716 (m), 678 (w), 558 (w) cm^{-1}. – **MS** (70 eV, EI): *m/z* (%) = 328 (45) $[M]^+$, 319 (8), 281 (6), 273 (7), 272 (42), 271 (100), 269 (11), 246 (6), 243 (5), 232 (10), 231 (13), 229 (5), 219 (14), 218 (6), 217 (37), 215 (6), 205 (14), 181 (14), 169 (18), 131 (18), 119 (14), 69 (43). – **HRMS** ($C_{21}H_{28}O_3$): calc.

328.2031, found 328.2033. – **Elemental analysis**: $C_{21}H_{28}O_3$: calc. C 76.79, H 8.59, found C 76.74, H 8.56.

7-(1-Butylcyclopentyl)-5-methoxy-3-propyl-2*H*-chromen-2-one (27br)

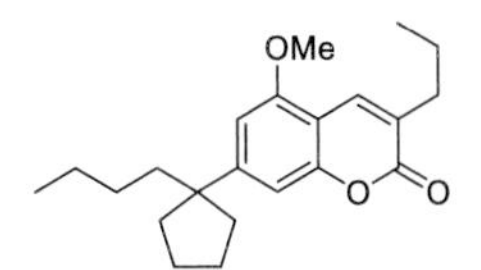

According to **GP7**, 200 mg of salicylic aldehyde **29b** (0.72 mmol, 1.00 equiv.), 0.50 mL of valeric acid anhydride **100d** (472 mg, 2.53 mmol, 3.50 equiv.) and 5.0 mg of potassium carbonate (40 µmol, 0.05 equiv.) were placed in a microwave vial and heated at 180 °C for 65 min at 300 W microwave irradiation. The resulting mixture was allowed to cool to room temperature, quenched with sodium bicarbonate and the pH was adjusted to ~7 with sodium bicarbonate. The mixture was then extracted with 3 × 15 mL of ethyl acetate, washed with 3 × 15 mL of water, 3 × 15 mL of sodium bicarbonate and the combined organic layers were dried over sodium sulfate. Removal of the volatiles under reduced pressure and purification via flash column chromatography (CH/EtOAc 50:1) resulted in 227 mg (92%) of an off-white solid.

***R*f** (CH/EtOAc 50:1): 0.29. – **MP**: 70.7 °C – **^{1}H NMR** (400 MHz, $CDCl_3$): δ = 7.80 (s, 1H, 4-C*H*), 6.82 (d, *J* = 1.3 Hz, 1H, H_{Ar}), 6.61 (d, *J* = 1.4 Hz, 1H, H_{Ar}), 3.92 (s, 3H, OCH_3), 2.52 (td, *J* = 7.6, 1.1 Hz, 2H), 1.96 – 1.76 (m, 4H, 2 × CH_2), 1.76 – 1.53 (m, 8H, 4 × CH_2), 1.15 (p, *J* = 7.3 Hz, 2H, CH_2), 0.98 (t, *J* = 7.3 Hz, 3H, CH_3), 0.97 – 0.89 (m, 2H, CH_2), 0.78 (t, *J* = 7.3 Hz, 3H, CH_3) ppm. – **^{13}C NMR** (100 MHz, $CDCl_3$): δ = 162.4 ($C_{quart.}$, *C*OO), 155.1 ($C_{quart.}$, C_{Ar}), 154.0 ($C_{quart.}$, C_{Ar}), 153.8 ($C_{quart.}$, C_{Ar}), 133.7 (+, 4-*C*H), 126.8 ($C_{quart.}$, C_{Ar}), 107.9 ($C_{quart.}$, C_{Ar}), 107.6 (+, C_{Ar}H), 104.0 (+, C_{Ar}H), 55.9 (+, OCH_3), 52.0 ($C_{quart.}$, C_{CP}), 41.7 (–, CH_2), 37.8 (–, 2 × CH_2), 33.1 (–, CH_2), 27.6 (–, CH_2), 23.4 (–, CH_2), 23.3 (–, 2 × CH_2), 21.6 (–, CH_2), 14.1 (+, CH_3), 13.9 (+, CH_3) ppm. – **IR** (KBr): ṽ = 2954 (m), 2925 (m), 2869 (m), 1712 (m), 1612 (s), 1571 (m), 1494 (w), 1454 (m), 1414 (m), 1351 (w), 1288 (w), 1246 (m), 1167 (m), 1104 (m), 1051 (m), 1026 (m), 923 (m), 902 (w), 841 (m), 772 (w), 714 (w), 557 (w) cm^{-1}. – **MS** (70 eV, EI): *m/z* (%) = 342 (53) $[M]^+$, 313 (5), 287 (7), 286 (41), 285 (100), 257 (6), 256 (5), 246 (5), 231 (27), 229 (5), 219 (9), 215 (11). – **HRMS** ($C_{22}H_{30}O_3$): calc. 342.2192, found 342.2189. – **Elemental analysis**: $C_{22}H_{30}O_3$: calc. C 77.16, H 8.83, found C 76.84, H 8.79.

7-(1-Butylcyclopentyl)-5-methoxy-3-propyl-2*H*-chromen-2-one (27bs)

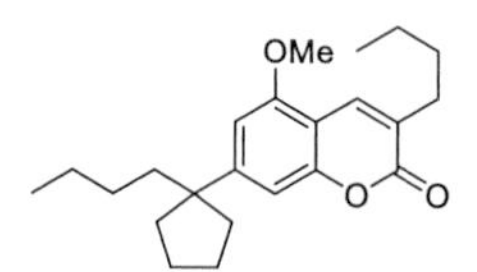

According to GP7, 200 mg of salicylic aldehyde **29b** (0.72 mmol, 1.00 equiv.), 0.58 mL of hexanoic acid anhydride **100e** (543 mg, 2.53 mmol, 3.50 equiv.) and 5.0 mg of potassium carbonate (40 µmol, 0.05 equiv.) were placed in a microwave vial and heated at 180 °C for 65 min at 300 W microwave irradiation. The resulting mixture was allowed to cool to room temperature, quenched with sodium bicarbonate and the pH was adjusted to ~7 with sodium bicarbonate. The mixture was then extracted with 3 × 15 mL of ethyl acetate, washed with 3 × 15 mL of water, 3 × 15 mL of sodium bicarbonate and the combined organic layers were dried over sodium sulfate. Removal of the volatiles under reduced pressure and purification via flash column chromatography (CH/EtOAc 75:1) resulted in 81 mg (31%) of an off-white solid.

***R*f** (CH/EtOAc 50:1): 0.32. – **MP**: 83.4 °C – **^{1}H NMR** (400 MHz, $CDCl_3$): δ = 7.80 (q, *J* = 0.9 Hz, 1H, 4-C*H*), 6.82 (d, *J* = 1.4 Hz, 1H, H_{Ar}), 6.61 (d, *J* = 1.4 Hz, 1H, H_{Ar}), 3.92 (s, 3H, OCH_3), 2.54 (ddd, *J* = 8.9, 6.6, 1.1 Hz, 2H, CH_2), 1.95 – 1.76 (m, 4H, 2 × CH_2), 1.76 – 1.54 (m, 8H, 4 × CH_2), 1.40 (h, *J* = 7.4 Hz, 2H, CH_2), 1.15 (p, *J* = 7.1 Hz, 2H, CH_2), 1.00 – 0.87 (m, 5H, CH_3, CH_2), 0.78 (t, *J* = 7.3 Hz, 3H, CH_3) ppm. – **^{13}C NMR** (100 MHz, $CDCl_3$): δ = 162.5 ($C_{quart.}$, *C*OO), 155.1($C_{quart.}$, C_{Ar}), 153.9($C_{quart.}$, C_{Ar}), 153.8($C_{quart.}$, C_{Ar}), 133.6 (+, 4-*C*H), 127.1($C_{quart.}$, C_{Ar}), 108.0 (+, C_{Ar}H), 107.6 (+, C_{Ar}H), 104.0 (+, OCH_3), 55.9 ($C_{quart.}$, C_{CP}), 52.0 (–, CH_2), 41.7 (–, CH_2), 37.9 (–, 2 × CH_2), 30.8 (–, CH_2), 30.5 (–, CH_2), 27.6 (–, CH_2), 23.4 (–, CH_2), 23.3 (–, 2 × CH_2), 22.6 (–, CH_2), 14.1 (+, CH_3), 14.0 (+, CH_3) ppm. – **IR** (KBr): $\tilde{\nu}$ = 2951 (m), 2868 (m), 1712 (m), 1613 (m), 1571 (w), 1491 (w), 1441 (m), 1354 (w), 1290 (w), 1239 (m), 1168 (w), 1103 (s), 1073 (m), 1045 (m), 993 (m), 907 (w), 838 (m), 800 (w), 714 (m), 641 (w), 558 (w), 427 (w) cm^{-1}. – **MS** (70 eV, EI): *m/z* (%) = 356 (59) $[M]^+$, 314 (8), 313 (10), 301 (7), 300 (42), 299 (100), 257 (6), 256 (6), 245 (24), 233 (10), 229 (5), 216 (5), 215 (13). – **HRMS** ($C_{23}H_{32}O_3$): calc. 356.2346, found 356.2347. – **Elemental analysis**: $C_{23}H_{32}O_3$: calc. C 77.49, H 9.05, found C 77.25, H 9.08.

7-(1-Butylcyclohexyl)-5-methoxy-2*H*-chromen-2-one (27bt)

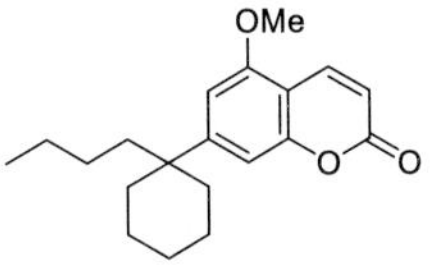

According to **GP7**, 200 mg of salicylic aldehyde **29c** (0.69 mmol, 1.00 equiv.), 0.24 mL of acetic acid anhydride **100a** (246 mg, 2.41 mmol, 3.50 equiv.) and 4.8 mg of potassium carbonate (270 µmol, 0.05 equiv.) were placed in a microwave vial and heated at 180 °C for 65 min at 300 W microwave irradiation. The resulting mixture was allowed to cool to room temperature, poured onto crushed ice and the pH was adjusted to ~7 with sodium bicarbonate. The mixture was then extracted with 3 × 50 mL of ethyl acetate and the combined organic layers were dried over sodium sulfate. Removal of the volatiles under reduced pressure and purification via flash column chromatography (CH/EtOAc 15:1) resulted in 121 mg (56%) of a colorless oil.

***R*f** (CH/EtOAc 50:1): 0.14. – **¹H NMR** (300 MHz, $CDCl_3$): δ = 8.04 (d, *J* = 10.1 Hz, 1H, 4-C*H*), 6.89 (s, 1H, H_{Ar}), 6.67 (s, 1H, H_{Ar}), 6.28 (d, *J* = 9.8 Hz, 1H, 3-C*H*), 3.92 (s, 3H, O-CH_3), 2.10 – 1.94 (m, 2H, CH_2), 1.71 – 1.27 (m, 12H, 6 × CH_2), 1.16 (dt, *J* = 14.6, 7.4 Hz, 2H, CH_2), 0.91 (d, *J* = 10.0 Hz, 2H, CH_2), 0.78 (t, *J* = 7.2 Hz, 3H, CH_3) ppm. – **¹³C NMR** (100 MHz, $CDCl_3$): δ = 161.5 ($C_{quart.}$, *C*OO), 155.9 ($C_{quart.}$, C_{Ar}), 155.2 ($C_{quart.}$, C_{Ar}), 154.2 ($C_{quart.}$, C_{Ar}), 138.6(+, 4-C_{Ar}H), 113.7 (+, C_{Ar}H), 108.2 (+, C_{Ar}H), 107.4 ($C_{quart.}$, C_{Ar}), 104.0 (+, 3-C_{Ar}H), 56.0 (+, O*C*H₃), 43.5 ($C_{quart.}$, C_{CH}), 42.4 (–, *C*H₂), 36.4 (–, 2 × *C*H₂), 26.5 (–, *C*H₂), 25.8 (–, *C*H₂), 23.4 (–, *C*H₂), 22.6 (–, 2 × *C*H₂), 14.1 (+, *C*H₃) ppm. – **IR** (KBr): ṽ = 2926 (m), 2856 (m), 1726 (s), 1611 (s), 1556 (m), 1494 (m), 1453 (m), 1411 (s), 1337 (w), 1235 (m), 1192 (w), 1112 (s), 1093 (s), 920 (m), 893 (m), 825 (s), 733 (w), 695 (w), 667 (w), 629 (w), 611 (w), 464 (w) cm^{-1}. – **MS** (70 eV, EI): *m/z* (%) = 314 (9) $[M]^+$, 262 (8), 258 (9), 257 (14), 219 (6), 189 (14), 119 (15), 100 (12), 93 (5), 69 (100). – **HRMS** ($C_{20}H_{26}O_3$): calc. 314.1876, found 314.1878. – **Elemental analysis**: $C_{20}H_{26}O_3$: calc. C 76.40, H 8.34, found C 76.58, H 8.41.

7-(1-Butylcyclohexyl)-5-methoxy-3-methyl-2*H*-chromen-2-one (27bu)

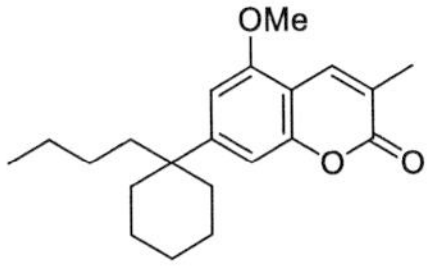

According to **GP7**, 200 mg of salicylic aldehyde **29c** (0.69 mmol, 1.00 equiv.), 0.31 mL of propionic acid anhydride **100c** (314 mg, 2.41 mmol, 3.50 equiv.) and 4.8 mg of potassium carbonate (270 µmol, 0.05 equiv.) were placed in a microwave vial and heated at 180 °C for 65 min at 300 W microwave irradiation. The resulting mixture was allowed to cool to room temperature, poured onto crushed ice and the pH was adjusted to ~7 with sodium bicarbonate. The mixture was then extracted with 3 × 50 mL of ethyl acetate and the combined organic layers were dried over sodium sulfate. Removal of the volatiles under reduced pressure and purification via flash column chromatography (CH/EtOAc 20:1) resulted in 194 mg (77%) of an off-white solid.

***R*f** (CH/EtOAc 50:1): 0.17. – **MP**: 84.8 °C – **^{1}H NMR** (400 MHz, $CDCl_3$): δ = 7.84 (d, *J* = 1.6 Hz, 1H, 4-C*H*), 6.88 (d, *J* = 1.5 Hz, 1H, H_{Ar}), 6.65 (d, *J* = 1.5 Hz, 1H, H_{Ar}), 3.91 (s, 3H, O-CH_3), 2.19 (d, *J* = 1.4 Hz, 3H, CH_3), 2.07 – 1.96 (m, 2H, CH_2), 1.66 – 1.31 (m, 10H, 5 × CH_2), 1.19 – 1.06 (m, 2H, CH_2), 0.89 (dtd, *J* = 12.1, 9.4, 8.8, 5.7 Hz, 2H, CH_2), 0.76 (t, *J* = 7.3 Hz, 3H, $C_4H_8CH_3$) ppm. – **^{13}C NMR** (100 MHz, $CDCl_3$): δ = 162.8 ($C_{quart.}$, *C*OO), 155.2 ($C_{quart.}$, C_{Ar}), 154.3 ($C_{quart.}$, C_{Ar}), 152.4 ($C_{quart.}$, C_{Ar}), 134.4 (+, C_{Ar}H), 122.9 ($C_{quart.}$, C_{Ar}), 107.9 (+, C_{Ar}H), 107.9 ($C_{quart.}$, C_{Ar}), 103.9 (+, C_{Ar}H), 55.9 (+, OCH_3), 43.6 ($C_{quart.}$, C_{CH}), 42.3 (–, CH_2), 36.4 (–, 2 × CH_2), 26.6 (–, CH_2), 25.8 (–, CH_2), 23.4 (–, CH_2), 22.5 (–, 2 × CH_2), 17.3 (+, CH_3), 14.1 (+, CH_3) ppm. – **IR** (KBr): ṽ = 2925 (w), 2855 (w), 1727 (w), 1714 (w), 1612 (w), 1571 (w), 1495 (w), 1446 (w), 1415 (w), 1377 (vw), 1256 (w), 1183 (w), 1113 (w), 1066 (w), 996 (w), 913 (w), 842 (w), 762 (w), 728 (w), 683 (vw), 634 (vw), 610 (vw), 559 (vw) cm^{-1}. – **MS** (70 eV, EI): *m/z* (%) = 328 (66) $[M]^+$, 273 (7), 272 (44), 271 (87), 232 (6), 230 (5), 229 (6), 218 (6), 217 (9), 204 (16), 203 (100), 191 (22), 175 (10), 81 (10), 57 (17). – **HRMS** ($C_{21}H_{28}O_3$): calc. 328.2033, found 328.2032. – **Elemental analysis**: $C_{21}H_{28}O_3$: calc. C 76.79, H 8.59, found C 76.54, H 8.64.

7-(1-Butylcyclohexyl)-5-methoxy-3-ethyl-2*H*-chromen-2-one (27bv)

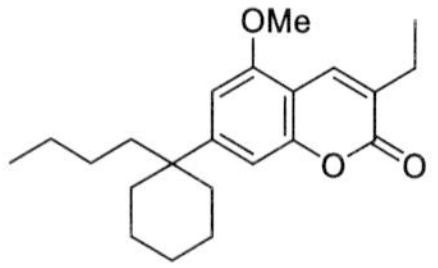

According to **GP7**, 200 mg of salicylic aldehyde **29c** (0.69 mmol, 1.00 equiv.), 0.39 mL of butyric acid anhydride **100c** (381 mg, 2.41 mmol, 3.50 equiv.) and 4.8 mg of potassium carbonate (270 µmol, 0.05 equiv.) were placed in a microwave vial and heated at 180 °C for 65 min at 300 W microwave irradiation. The resulting mixture was allowed to cool to room temperature, poured onto crushed ice and the pH was adjusted to ~7 with sodium bicarbonate. The mixture was then extracted with 3 × 50 mL of ethyl acetate and the combined organic layers were dried over sodium sulfate. Removal of the volatiles under reduced pressure and purification via flash column chromatography (CH/EtOAc 50:1) resulted in 219 mg (93%) of a white solid.

***R*f** (CH/EtOAc 50:1): 0.26. – **MP**: 78.5 °C – **^{1}H NMR** (400 MHz, $CDCl_3$): δ = 7.82 (s, 1H, 4-C*H*), 6.89 (d, *J* = 1.5 Hz, 1H, H_{Ar}), 6.66 (d, *J* = 1.5 Hz, 1H, H_{Ar}), 3.93 (s, 3H, O-CH_3), 2.59 (qd, *J* = 7.4, 1.3 Hz, 2H, CH_2), 2.01 (t, *J* = 8.2 Hz, 2H, CH_2), 1.70 – 1.30 (m, 10H, 5 × CH_2), 1.24 (t, *J* = 7.4 Hz, 3H, CH_3), 1.18 – 1.07 (m, 2H, CH_2), 0.90 (dtd, *J* = 11.7, 8.6, 5.5 Hz, 2H, CH_2), 0.77 (t, *J* = 7.3 Hz, 3H, $C_3H_6CH_3$) ppm. – **^{13}C NMR** (100 MHz, $CDCl_3$): δ = 162.7 ($C_{quart.}$, *C*OO), 155.7 ($C_{quart.}$, C_{Ar}), 154.5 ($C_{quart.}$, C_{Ar}), 152.7 ($C_{quart.}$, C_{Ar}), 133.0 (+, 4-C_{Ar}H), 128.8 ($C_{quart.}$, C_{Ar}), 108.2 ($C_{quart.}$, C_{Ar}), 108.2 (+, C_{Ar}H), 104.2 (+, C_{Ar}H), 56.3 (+, OCH_3), 43.8 ($C_{quart.}$, C_{CH}), 42.6 (–, CH_2), 36.8 (–, 2 × CH_2), 26.9 (–, CH_2), 26.1 (–, CH_2), 24.4 (–, CH_2), 23.7 (–, CH_2), 22.9 (–, 2 × CH_2), 14.4 (+, CH_3), 13.0 (+, CH_3) ppm. – **IR** (KBr): ṽ = 2926 (m), 2858 (w), 1711 (m), 1612 (m), 1568 (w), 1495 (w), 1442 (m), 1413 (m), 1343 (w), 1288 (w),1248 (m),1165 (w),1105 (m), 1077 (m), 1046 (m), 999 (w), 956 (m), 918 (w), 840 (w), 753 (w), 717 (w), 683 (w), 654 (w), 558 (w) cm^{-1}. – **MS** (70 eV, EI): *m/z* (%) = 342 (75) $[M]^+$, 287 (8), 286 (51), 285 (100) $[M – C_4H_9]^+$, 246 (7), 243 (6), 232 (6), 231 (8), 229 (6), 218 (16), 217 (98), 205 (25), 189 (9), 81 (15), 57 (40). – **HRMS** ($C_{22}H_{30}O_3$): calc. 342.2189, found 342.2178. – **Elemental analysis**: $C_{22}H_{30}O_3$: calc. C 77.16, H 8.83, found C 76.99, H 8.91.

7-(1-Butylcyclohexyl)-5-methoxy-3-propyl-2*H*-chromen-2-one (27bw)

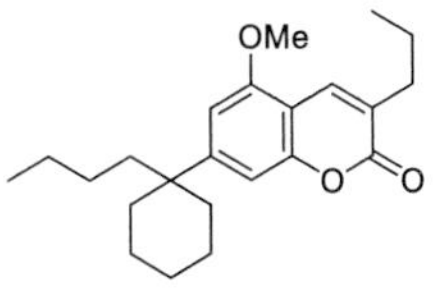

According to **GP7**, 200 mg of salicylic aldehyde **29c** (0.69 mmol, 1.00 equiv.), 0.47 mL of valeric acid anhydride **100d** (449 mg, 2.41 mmol, 3.50 equiv.) and 4.8 mg of potassium carbonate (270 µmol, 0.05 equiv.) were placed in a microwave vial and heated at 180 °C for 65 min at 300 W microwave irradiation. The resulting mixture was allowed to cool to room temperature, poured onto crushed ice and the pH was adjusted to ~7 with sodium bicarbonate. The mixture was then extracted with 3 × 50 mL of ethyl acetate and the combined organic layers were dried over sodium sulfate. Removal of the volatiles under reduced pressure and purification via flash column chromatography (CH/EtOAc 50:1) resulted in 167 mg (68%) of a white solid.

***R*f** (CH/EtOAc 50:1): 0.29. – **MP**: 67.9 °C – **^{1}H NMR** (400 MHz, $CDCl_3$): δ = 7.81 (s, 1H, 4-C*H*), 6.88 (s, 1H, H_{Ar}), 6.66 (s, 1H, H_{Ar}), 3.91 (d, *J* = 4.0 Hz, 3H, OCH_3), 2.53 (t, *J* = 7.6 Hz, 2H, CH_2), 2.01 (s, 2H, CH_2), 1.76 – 1.30 (m, 6 × 12H, CH_2), 1.14 (td, *J* = 14.5, 7.2 Hz, 2H, CH_2), 0.99 (t, *J* = 7.2 Hz, 3H, CH_3), 0.97 – 0.81 (m, 2H, CH_2), 0.77 (t, *J* = 7.2 Hz, 3H, CH_3) ppm. – **^{13}C NMR** (100 MHz, $CDCl_3$): δ = 162.4 ($C_{quart.}$, *C*OO), 155.4 ($C_{quart.}$, C_{Ar}), 154.3

($C_{quart.}$, C_{Ar}), 152.4 ($C_{quart.}$, C_{Ar}), 133.7 (+, 4-C_{Ar}H), 127.0 ($C_{quart.}$, C_{Ar}), 107.9 ($C_{quart.}$, C_{Ar}), 107.9 (+, C_{Ar}H), 103.9 (+, C_{Ar}H), 55.9 (+, OCH_3), 43.5 ($C_{quart.}$, C_{CH}), 42.3 (–, CH_2), 36.5 (–, 2 × CH_2), 33.1 (–, CH_2), 26.6 (–, CH_2), 25.8 (–, CH_2), 23.4 (–, CH_2), 22.5 (–, 2 × CH_2), 21.6 (–, CH_2), 14.1 (+, CH_3), 13.9 (+, CH_3) ppm.– **IR** (KBr): ṽ = 2925 (m), 2855 (m), 1709 (m), 1612 (e), 1571 (w), 1495 (w), 1453 (m), 1414 (m), 1347 (w), 1322 (w), 1287 (w), 1247 (m), 1166 (m), 1104 (m), 1051 (m), 1026 (m), 1000 (w), 916 (w), 848 (m), 774 (w), 715 (w), 684 (w), 653 (w), 558 (w), 478 (w) cm^{-1}. – **MS** (70 eV, EI): *m/z* (%) = 356 (84) $[M]^+$, 327 (6), 301 (10), 300 (50), 299 (100) $[M - C_4H_9]^+$, 260 (7), 246 (5), 245 (7), 232 (11), 231 (65), 229 (6), 219 (21), 202 (12), 81 (6). – **HRMS** ($C_{23}H_{32}O_3$): calc. 356.2346, found 356.2347. – **Elemental analysis**: $C_{23}H_{32}O_3$: calc. C 77.49, H 9.05, found C 77.39, H 9.06.

7-(1-Butylcyclohexyl)-5-methoxy-3-butyl-2*H*-chromen-2-one (27bx)

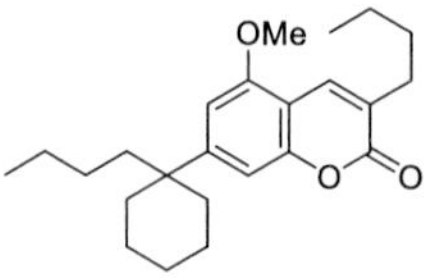

According to **GP7**, 200 mg of salicylic aldehyde **29c** (0.69 mmol, 1.00 equiv.), 0.56 mL of hexanoic acid anhydride **100e** (517 mg, 2.41 mmol, 3.50 equiv.) and 4.8 mg of potassium carbonate (270 µmol, 0.05 equiv.) were placed in a microwave vial and heated at 180 °C for 65 min at 300 W microwave irradiation. The resulting mixture was allowed to cool to room temperature, poured onto crushed ice and the pH was adjusted to ~7 with sodium bicarbonate. The mixture was then extracted with 3 × 50 mL of ethyl acetate and the combined organic layers were dried over sodium sulfate. Removal of the volatiles under reduced pressure and purification via flash column chromatography (CH/EtOAc 100:1) resulted in 210 mg (82%) of an off-white solid.

***R*f** (CH/EtOAc 50:1): 0.31. – **MP**: 143.8 °C – **^{1}H NMR** (400 MHz, $CDCl_3$): δ = 7.81 (s, 1H, 4-C*H*), 6.90 – 6.86 (m, 1H, H_{Ar}), 6.66 (d, *J* = 1.5 Hz, 1H, H_{Ar}), 3.92 (s, 3H, OCH_3), 2.55 (t, *J* = 7.7 Hz, 2H, CH_2), 2.11 – 1.94 (m, 2H, CH_2), 1.70 – 1.30 (m, 14H, 7 × CH_2), 1.13 (p, *J* = 7.3 Hz, 2H, CH_2), 1.00 – 0.85 (m, 5H, CH_2, CH_3), 0.77 (t, *J* = 7.3 Hz, 3H, CH_3) ppm. – **^{13}C NMR** (100 MHz, $CDCl_3$): δ = 162.4 ($C_{quart.}$, COO), 155.4 ($C_{quart.}$, C_{Ar}), 154.2 ($C_{quart.}$, C_{Ar}), 152.4 ($C_{quart.}$, C_{Ar}), 133.5 (+, 4-C_{Ar}H), 127.2($C_{quart.}$, C_{Ar}), 108.0 ($C_{quart.}$, C_{Ar}), 107.9 (+, C_{Ar}H), 103.9 (+, C_{Ar}H), 55.9 (+, OCH_3), 43.6($C_{quart.}$, C_{CH}), 42.3 (–, CH_2), 36.5 (–, 2 × CH_2), 30.8 (–, CH_2), 30.5 (–, CH_2), 26.6 (–, CH_2), 25.8 (–, CH_2), 23.4 (–, CH_2), 22.6 (–, 2 × CH_2), 14.1 (+, CH_3), 14.0 (+, CH_3) ppm. – **IR** (KBr): ṽ = 2926 (m), 2854 (w), 1714 (m), 1613 (m), 1570 (w), 1495 (w), 1453 (m), 1413 (m), 1376 (w), 1344 (w), 1291 (w), 1245 (m), 1163 (w), 1104 (m),

1073 (w), 1044 (m), 991 (m), 943 (w), 906 (w), 834 (w), 798 (w), 760 (w), 712 (w), 684 (w), 653 (w), 558 (w), 494 (vw), 429 (vw) cm^{-1}. – **MS** (70 eV, EI): *m/z* (%) = 370 (96) $[M]^+$, 328 (7), 327 (8), 315 (10), 314 (45), 313 (100) $[M - C_4H_9]^+$, 274 (6), 271 (5), 259 (5), 246 (10), 245 (54), 233 (19), 203 (7), 202 (11), 189 (5), 81 (7). – **HRMS** ($C_{24}H_{34}O_3$): calc. 370.2502, found 370.2502. – **Elemental analysis**: $C_{24}H_{34}O_3$: calc. C 77.80, H 9.25, found C 77.78, H 9.43.

6-Hydroxy-5-isopropyl-8-methyl-2*H*-chromen-2-one (27bb-H)

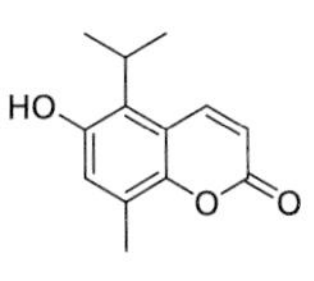

According to **GP8**, 71.2 mg of 5-methoxycoumarin **27bb** (307 µmol, 1.00 equiv.) were dissolved in 1 mL of dry dichloromethane. The solution was cooled to –78 °C and 1.38 mL of boron tribromide (1 M in dichloromethane, 1.38 mmol, 4.50 equiv.) were added dropwise. The mixture was stirred for 30 min at this temperature and then allowed to warm to room temperature. The reaction was quenched after 16 h at 0 °C by addition of sodium bicarbonate. The aqueous layer was extracted with 3 × 15 mL of dichloromethane and the combined organic layers were washed with brine, dried over sodium sulfate and the volatiles were removed under reduced pressure. The crude product was then purified via flash column chromatography (CH/EtOAc 5:1) to give the product as 59.8 mg (89%) of an off-white solid.

***R*$_f$** (CH/EtOAc 2:1): 0.40. – **^{1}H NMR** (400 MHz, $CDCl_3$): δ = 8.12 (d, *J* = 10.0 Hz, 1H, 4-C*H*), 6.87 (s, 1H, 7-C*H*$_{Ar}$), 6.42 (d, *J* = 10.0 Hz, 1H, 3-C*H*), 5.56 (s, 1H, O*H*), 3.54 (hept, *J* = 7.0 Hz, 1H, C*H*$(CH_3)_2$), 2.36 (s, 3H, C*H*$_3$), 1.42 (d, *J* = 7.1 Hz, 6H, CH(C*H*$_3$)$_2$) ppm. – **^{13}C NMR** (100 MHz, $CDCl_3$): δ = 161.5 ($C_{quart.}$, *C*OO), 150.1 ($C_{quart.}$, *C*$_{Ar}$), 147.6 ($C_{quart.}$, *C*$_{Ar}$), 141.3 (+, *C*$_{Ar}$H), 127.9 ($C_{quart.}$, *C*$_{Ar}$), 124.8 ($C_{quart.}$, *C*$_{Ar}$), 122.3 (+, *C*$_{Ar}$H), 117.5 ($C_{quart.}$, *C*$_{Ar}$), 115.6 (+, *C*$_{Ar}$H), 27.1 (+, *C*H$(CH_3)_2$), 21.4 (+, CH(*C*H$_3$)$_2$), 15.4 (+, *C*H$_3$) ppm. – **IR** (KBr): $\tilde{\nu}$ = 3338 (w), 2960 (w), 1694 (m), 1579 (m), 1455 (w), 1397 (w), 1378 (w), 1291 (m), 1192 (w), 1170 (w), 1114 (w), 1059 (w), 988 (w), 949 (w), 871 (w), 832 (m), 773 (w), 725 (vw), 687 (vw), 634 (m), 603 (w), 548 (w), 434 (w) cm^{-1}. – **MS** (70 eV, EI): *m/z* (%) = 218 (68) $[M]^+$, 204 (8), 203 (100) $[M - CH_3]^+$, 188 (8), 175 (9). – **HRMS** ($C_{13}H_{14}O_3$): calc. 218.0937, found 218.0939.

6-Hydroxy-5-isopropyl-3,8-dimethyl-2*H*-chromen-2-one (27bc-H)

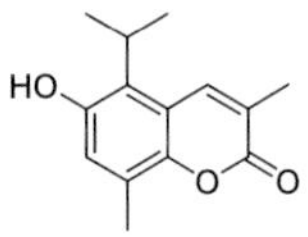

According to **GP8**, 32.9 mg of 5-methoxycoumarin **27bc** (134 µmol, 1.00 equiv.) were dissolved in 1 mL of dry dichloromethane. The solution was cooled to –78 °C and 0.60 mL of boron tribromide (1 M in dichloromethane, 0.60 mmol, 4.50 equiv.) were added dropwise. The mixture was stirred for 30 min at this temperature and then allowed to warm to room temperature. The reaction was quenched after 16 h at 0 °C by addition of sodium bicarbonate. The aqueous layer was extracted with 3 × 15 mL of dichloromethane and the combined organic layers were washed with brine, dried over sodium sulfate and the volatiles were removed under reduced pressure. The crude product was then purified via flash column chromatography (CH/EtOAc 5:1) to give the product as 30.4 mg (98%) of an off-white solid.

***R*f** (CH/EtOAc 5:1): 0.16. – **^{1}H NMR** (400 MHz, $CDCl_3$): δ = 7.89 (d, *J* = 1.5 Hz, 1H, 4-C*H*), 6.76 (s, 1H, *7-CH*$_{Ar}$), 5.01 (s, 1H, O*H*), 3.53 (hept, *J* = 6.6 Hz, 1H, C*H*(CH_3)$_2$), 2.36 (s, 3H, C*H*$_3$), 2.24 (s, 3H, C*H*$_3$), 1.42 (d, *J* = 7.1 Hz, 6H, CH(C*H*$_3$)$_2$) ppm. – **^{13}C NMR** (100 MHz, $CDCl_3$): δ = 162.4 ($C_{quart.}$, *C*OO), 149.8 ($C_{quart.}$, *C*$_{Ar}$), 146.9 ($C_{quart.}$, *C*$_{Ar}$), 136.9 (+, 4-*C*$_{Ar}$H), 126.8 ($C_{quart.}$, *C*$_{Ar}$), 124.7 ($C_{quart.}$, *C*$_{Ar}$), 124.4 ($C_{quart.}$, *C*$_{Ar}$), 121.0 (+, 7-*C*$_{Ar}$H), 118.1 ($C_{quart.}$, *C*$_{Ar}$), 26.8 (+, *C*H(CH_3)$_2$), 21.4 (+, CH(*C*H$_3$)$_2$), 17.8 (+, *C*H$_3$), 15.5 (+, *C*H$_3$) ppm. – **IR** (KBr): ṽ = 3285 (w), 2921 (w), 2870 (w), 1682 (m), 1627 (w), 1589 (m), 1453 (w), 1364 (w), 1297 (m), 1249 (w), 1194 (m), 1135 (w), 1110 (w), 1035 (w), 1007 (w), 944 (w), 907 (w), 874 (m), 768 (m), 744 (w), 711 (vw), 687 (w), 635 (w), 601 (vw), 551 (w), 419 (vw) cm^{-1}. – **MS** (70 eV, EI): *m/z* (%) = 232 (57) $[M]^+$, 231 (12), 218 (15), 217 (100) $[M – CH_3]^+$, 189 (11) $[M – C_3H_7]^+$. – **HRMS** ($C_{14}H_{16}O_3$): calc. 232.1099, found 232.1099.

3-Ethyl-6-hydroxy-5-isopropyl-8-methyl-2*H*-chromen-2-one (27bd-H)

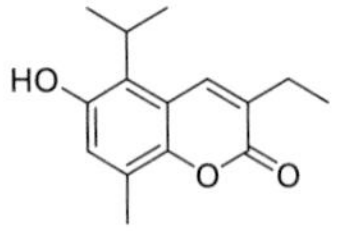

According to **GP8**, 46.6 mg of 5-methoxycoumarin **27bd** (179 µmol, 1.00 equiv.) were dissolved in 1 mL of dry dichloromethane. The solution was cooled to –78 °C and 0.81 mL of boron tribromide (1 M in dichloromethane, 0.81 mmol, 4.50 equiv.) were added dropwise. The mixture was stirred for 30 min at this temperature and then allowed to warm to room temperature. The reaction was quenched after 16 h at 0 °C by addition of sodium bicarbonate. The aqueous layer was extracted with 3 × 15 mL of dichloromethane and the combined organic layers were washed with brine, dried over sodium sulfate and the volatiles were removed under reduced pressure. The crude product was then purified via flash column chromatography (CH/EtOAc 5:1) to give the product as 39.7 mg (90%) of an off-white solid.

***R*f** (CH/EtOAc 5:1): 0.16. – **^{1}H NMR** (400 MHz, $CDCl_3$): δ = 7.87 (s, 1H, 4-C*H*), 6.77 (s, 1H, 7-CH_{Ar}), 5.11 (s, 1H, O*H*), 3.56 (hept, *J* = 8.4, 7.8 Hz, 1H, C*H*(CH_3)$_2$), 2.63 (qd, *J* = 7.4, 1.2 Hz, 2H, CH_2), 2.36 (s, 3H, CH_3), 1.43 (d, *J* = 7.1 Hz, 6H, CH(CH_3)$_2$), 1.27 (t, *J* = 7.4 Hz, 3H, CH_3) ppm. – **^{13}C NMR** (100 MHz, $CDCl_3$): δ = 162.0 ($C_{quart.}$, *C*OO), 149.8 ($C_{quart.}$, C_{Ar}), 146.7 ($C_{quart.}$, C_{Ar}), 135.3 (+, 4-C_{Ar}H), 130.2 ($C_{quart.}$, C_{Ar}), 127.0 ($C_{quart.}$, C_{Ar}), 124.3 ($C_{quart.}$, C_{Ar}), 121.0 (+, 7-C_{Ar}H), 118.1 ($C_{quart.}$, C_{Ar}), 26.8 (+, *C*H(CH_3)$_2$), 24.5 (–, CH_2), 21.5 (+, CH(CH_3)$_2$), 15.5 (+, CH_3), 12.8 (+, CH_3) ppm. – **IR** (KBr): ṽ = 3293 (w), 2919 (w), 2867 (w), 1680 (m), 1627 (w), 1588 (m), 1456 (w), 1394 (w), 1372 (w), 1329 (w), 1301 (w), 1284 (m), 1248 (w), 1192 (m), 1133 (w), 1107 (w), 1052 (w), 994 (w), 972 (w), 948 (w), 911 (w), 874 (w), 797 (w), 766 (w), 723 (w), 709 (w), 696 (w), 638 (w), 605 (w) cm^{-1}. – **MS** (70 eV, EI): *m/z* (%) = 246 (62) $[M]^+$, 232 (16), 231 (100) $[M-CH_3]^+$, 203 (7) $[M-C_3H_7]^+$. – **HRMS** ($C_{15}H_{18}O_3$): calc. 246.1256, found 246.1256.

3-Propyl-6-hydroxy-5-isopropyl-8-methyl-2*H*-chromen-2-one (27be-H)

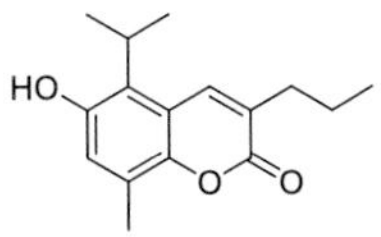

According to **GP8**, 17.6 mg of 5-methoxycoumarin **27be** (64.0 µmol, 1.00 equiv.) were dissolved in 1 mL of dry dichloromethane. The solution was cooled to –78 °C and 0.29 mL of boron tribromide (1 M in dichloromethane, 290 µmol, 4.50 equiv.) were added dropwise. The mixture was stirred for 30 min at this temperature and then allowed to warm to room temperature. The reaction was quenched after 16 h at 0 °C by addition of sodium bicarbonate. The aqueous layer was extracted with 3 × 15 mL of dichloromethane and the combined organic layers were washed with brine, dried over sodium sulfate and the volatiles were removed under reduced pressure. The crude product was then purified via flash column chromatography (CH/EtOAc 5:1) to give the product as 11.6 mg (70%) of an off-white solid.

***R*f** (CH/EtOAc 2:1): 0.43. – **^{1}H NMR** (400 MHz, $CDCl_3$): δ = 7.88 (s, 1H, 4-C*H*), 6.83 (s, 1H, 7-CH_{Ar}), 5.73 (s, 1H, O*H*), 3.56 (hept, *J* = 6.7 Hz, 1H, C*H*(CH_3)$_2$), 2.57 (ddd, *J* = 8.7, 6.4, 1.0 Hz, 2H, CH_2), 2.33 (s, 3H, CH_3), 1.75 – 1.63 (m, 2H, CH_2), 1.43 (d, *J* = 7.1 Hz, 6H, CH(CH_3)$_2$), 1.01 (t, *J* = 7.3 Hz, 3H, CH_3) ppm. – **^{13}C NMR** (100 MHz, $CDCl_3$): δ = 162.4 ($C_{quart.}$, *C*OO), 150.1 ($C_{quart.}$, C_{Ar}), 146.7 ($C_{quart.}$, C_{Ar}), 136.6 (+, 4-C_{Ar}H), 128.4 ($C_{quart.}$, C_{Ar}), 127.0 ($C_{quart.}$, C_{Ar}), 124.2 ($C_{quart.}$, C_{Ar}), 121.1 (+, 7-C_{Ar}H), 118.1 ($C_{quart.}$, C_{Ar}), 33.4 (–, CH_2),

26.7 (+, *C*H(CH$_3$)$_2$), 21.7 (–, *C*H$_2$), 21.5 (+, CH(*C*H$_3$)$_2$), 15.4 (+, *C*H$_3$), 13.9 (+, *C*H$_3$) ppm. – **IR** (KBr): ṽ = 3227 (w), 2960 (w), 1686 (m), 1665 (m), 1627 (w), 1586 (m), 1457 (w), 1392 (w), 1354 (w), 1303 (m), 1276 (m), 1188 (m), 1120 (w), 1072 (w), 991 (w), 953 (w), 906 (m), 869 (m), 778 (w), 732 (w), 612 (m), 557 (w), 403 (vw) cm^{-1}. – **MS** (70 eV, EI): *m/z* (%) = 260 (86) [M]$^+$, 246 (18), 245 (100) [M – CH$_3$]$^+$, 232 (10), 231 (21) [M – C$_2$H$_5$]$^+$, 217 (17) [M – C$_3$H$_7$]$^+$. – **HRMS** ($C_{16}H_{20}O_3$): calc. 260.1407, found 260.1406.

3-Butyl-6-hydroxy-5-isopropyl-8-methyl-2*H*-chromen-2-one (27bf-H)

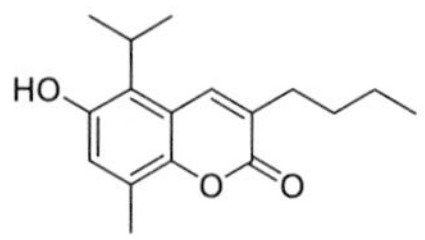

According to **GP8**, 31.4 mg of 5-methoxycoumarin **27bf** (109 µmol, 1.00 equiv.) were dissolved in 1 mL of dry dichloromethane. The solution was cooled to –78 °C and 0.49 mL of boron tribromide (1 M in dichloromethane, 0.49 mmol, 4.50 equiv.) were added dropwise. The mixture was stirred for 30 min at this temperature and then allowed to warm to room temperature. The reaction was quenched after 16 h at 0 °C by addition of sodium bicarbonate. The aqueous layer was extracted with 3 × 15 mL of dichloromethane and the combined organic layers were washed with brine, dried over sodium sulfate and the volatiles were removed under reduced pressure. The crude product was then purified via flash column chromatography (CH/EtOAc 5:1) to give the product as 29.9 mg (quant.) of a yellow solid.

***R*$_f$** (CH/EtOAc 2:1): 0.45. – **MP**: 166.6 °C – **^{1}H NMR** (400 MHz, CDCl$_3$): δ = 7.89 (s, 1H, 4-C*H*), 6.84 (s, 1H, 7-C*H*$_{Ar}$), 5.85 (s, 1H, O*H*), 3.57 (hept, *J* = 6.4 Hz, 1H, C*H*(CH$_3$)$_2$), 2.64 – 2.54 (m, 2H, C*H*$_2$), 2.33 (s, 3H, C*H*$_3$), 1.69 – 1.59 (m, 2H, C*H*$_2$), 1.48 – 1.36 (m, 8H, CH(C*H*$_3$)$_2$, C*H*$_2$), 0.96 (t, *J* = 7.3 Hz, 3H, C*H*$_3$) ppm. – **^{13}C NMR** (100 MHz, CDCl$_3$): δ = 162.5 ($C_{quart.}$, *C*OO), 150.2 ($C_{quart.}$, *C*$_{Ar}$), 146.6 ($C_{quart.}$, *C*$_{Ar}$), 136.5 (+, 4-*C*$_{Ar}$H), 128.6 ($C_{quart.}$, *C*$_{Ar}$), 127.0 ($C_{quart.}$, *C*$_{Ar}$), 124.2 ($C_{quart.}$, *C*$_{Ar}$), 121.1 (+, 7-*C*$_{Ar}$H), 118.1 ($C_{quart.}$, *C*$_{Ar}$), 31.1 (–, *C*H$_2$), 30.5 (–, *C*H$_2$), 26.7 (+, *C*H(CH$_3$)$_2$), 22.6 (–, *C*H$_2$), 21.5 (+, CH(*C*H$_3$)$_2$), 15.4 (+, *C*H$_3$), 14.0 (+, *C*H$_3$) ppm. – **IR** (KBr): ṽ = 3243 (w), 2923 (w), 2869 (w), 1673 (w), 1624 (vw), 1585 (w), 1456 (w), 1389 (w), 1304 (w), 1193 (w), 1168 (w), 1124 (w), 1072 (w), 1036 (w), 994 (vw), 954 (vw), 931 (w), 870 (w), 823 (vw), 782 (w), 733 (vw), 686 (w), 665 (w), 632 (w), 603 (w), 554 (vw) cm^{-1}. – **MS** (70 eV, EI): *m/z* (%) = 274 (94) [M]$^+$, 260 (12), 259 (67) [M – CH$_3$]$^+$, 245 (27) [M – C$_2$H$_5$]$^+$, 232 (61), 231 (100) [M – C$_3$H$_7$]$^+$, 217 (34) [M – C$_4$H$_9$]$^+$, 188 (13). – **HRMS**

($C_{17}H_{22}O_3$): calc. 274.1563, found 274.1565. – **Elemental analysis**: $C_{17}H_{22}O_3$: calc. C 74.42, H 8.08, found C 74.33, H 7.99.

3-Pentyl-6-hydroxy-5-isopropyl-8-methyl-2*H*-chromen-2-one (27bg)

According to **GP8**, 31.4 mg of 6-methoxycoumarin **27bg** (104 µmol, 1.00 equiv.) were dissolved in 1 mL of dry dichloromethane. The solution was cooled to –78 °C and 0.47 mL of boron tribromide (1 M in dichloromethane, 0.47 mmol, 4.50 equiv.) were added dropwise. The mixture was stirred for 30 min at this temperature and then allowed to warm to room temperature. The reaction was quenched after 16 h at 0 °C by addition of sodium bicarbonate. The aqueous layer was extracted with 3 × 15 mL of dichloromethane and the combined organic layers were washed with brine, dried over sodium sulfate and the volatiles were removed under reduced pressure. The crude product was then purified via flash column chromatography (CH/EtOAc 5:1) to give the product as 26.2 mg (88%) of a yellow solid.

***R*f** (CH/EtOAc 5:1): 0.23. – **MP**: 152.8 °C – **^{1}H NMR** (400 MHz, $CDCl_3$): δ = 7.87 (s, 1H, 4-C*H*), 6.81 (s, 1H, 7-CH_{Ar}), 5.48 (s, 1H, O*H*), 3.56 (hept, *J* = 6.2 Hz, 1H, C*H*(CH_3)$_2$), 2.63 – 2.55 (m, 2H, CH_2), 2.34 (s, 3H, CH_3), 1.69 – 1.61 (m, 2H, CH_2), 1.43 (d, *J* = 7.1 Hz, 6H, CH(CH_3)$_2$), 1.40 – 1.32 (m, 4H, 2 × CH_2), 0.96 – 0.84 (m, 3H, CH_3) ppm. – **^{13}C NMR** (100 MHz, $CDCl_3$): δ = 162.3 ($C_{quart.}$, *C*OO), 150.0 ($C_{quart.}$, C_{Ar}), 146.7 ($C_{quart.}$, C_{Ar}), 136.3 (+, 4-C_{Ar}H), 128.8 ($C_{quart.}$, C_{Ar}), 126.9 ($C_{quart.}$, C_{Ar}), 124.3 ($C_{quart.}$, C_{Ar}), 121.0 (+, 7-C_{Ar}H), 118.1 ($C_{quart.}$, C_{Ar}), 31.6 (–, CH_2), 31.3 (–, CH_2), 28.1 (–, CH_2), 26.7 (+, *C*H(CH_3)$_2$), 22.6 (–, CH_2), 21.5 (+, CH(*C*H_3)$_2$), 15.4 (+, CH_3), 14.2 (+, CH_3) ppm. – **IR** (KBr): ṽ = 3377 (m), 2953 (w), 2916 (m), 2867 (w), 1677 (m), 1587 (m), 1445 (w), 1390 (w), 1372 (w), 1308 (m), 1292 (m), 1187 (m), 1170 (w), 1120 (m), 1071 (m), 991 (w), 945 (w), 913 (w), 865 (m), 840 (w), 793 (w), 770 (w), 724 (w), 684 (w), 770 (w), 724 (w), 684 (w), 607 (m), 433 (vw), 402 (vw) cm^{-1}. – **MS** (70 eV, EI): *m/z* (%) = 288 (62) $[M]^+$, 274 (6), 273 (29) $[M – CH_3]^+$, 259 (11) $[M – C_2H_5]^+$, 246 (19), 245 (100) $[M – C_3H_7]^+$, 232 (33), 231 (20) $[M – C_4H_9]^+$, 217 (26) $[M – C_5H_{11}]^+$, 188 (11). – **HRMS** ($C_{18}H_{24}O_3$): calc. 288.1720, found 288.1719.

3-Hexyl-6-hydroxy-5-isopropyl-8-methyl-2*H*-chromen-2-one (27bh-H)

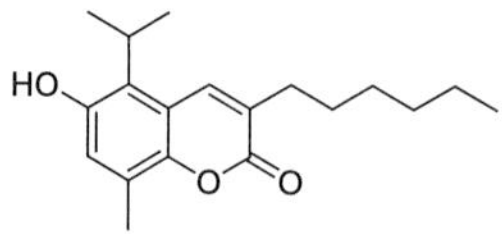

According to **GP8**, 33.4 mg of 6-methoxycoumarin **27bh** (106 µmol, 1.00 equiv.) were dissolved in 1 mL of dry dichloromethane. The solution was cooled to –78 °C and 0.47 mL of boron tribromide (1 M in dichloromethane, 0.47 mmol, 4.50 equiv.) were added dropwise. The mixture was stirred for 30 min at this temperature and then allowed to warm to room temperature. The reaction was quenched after 16 h at 0 °C by addition of sodium bicarbonate. The aqueous layer was extracted with 3 × 15 mL of dichloromethane and the combined organic layers were washed with brine, dried over sodium sulfate and the volatiles were removed under reduced pressure. The crude product was then purified via flash column chromatography (CH/EtOAc 5:1) to give the product as 24.0 mg (75%) of a yellow solid.

***R*f** (CH/EtOAc 5:1): 0.20. – **¹H NMR** (400 MHz, $CDCl_3$): δ = 7.90 (s, 1H, 4-C*H*), 6.86 (d, J = 0.9 Hz, 1H, 7-CH_{Ar}), 6.05 (s, 1H, O*H*), 3.64 – 3.49 (m, 1H, C*H*$(CH_3)_2$), 2.65 – 2.55 (m, 2H, CH_2), 2.33 (s, 3H, CH_3), 1.70 – 1.59 (m, 2H, CH_2), 1.43 (d, *J* = 7.1 Hz, 6H, CH$(CH_3)_2$), 1.40 – 1.23 (m, 6H, 3 × CH_2), 0.93 – 0.83 (m, 3H, CH_3) ppm. – **¹³C NMR** (100 MHz, $CDCl_3$): δ = 162.6 ($C_{quart.}$, *C*OO), 150.3 ($C_{quart.}$, C_{Ar}), 146.5 ($C_{quart.}$, C_{Ar}), 136.7 (+, 4-C_{Ar}H), 128.5 ($C_{quart.}$, C_{Ar}), 127.0 ($C_{quart.}$, C_{Ar}), 124.2 ($C_{quart.}$, C_{Ar}), 121.1 (+, 7-C_{Ar}H), 118.1 ($C_{quart.}$, C_{Ar}), 31.8 (–, CH_2), 31.3 (–, CH_2), 29.1 (–, CH_2), 28.4 (–, CH_2), 26.7 (+, *C*H$(CH_3)_2$), 22.8 (–, CH_2), 21.5 (+, CH$(CH_3)_2$), 15.4 (+, CH_3), 14.2 (+, CH_3) ppm. – **IR** (KBr): ṽ = 3260 (vw), 2922 (w), 2853 (vw), 1675 (w), 1624 (vw), 1585 (w), 1458 (vw), 1390 (vw), 1298 (w), 1197 (w), 1125 (vw), 1081 (vw), 1037 (vw), 997 (vw), 951 (vw), 919 (vw), 873 (vw), 783 (vw), 685 (vw), 664 (vw), 632 (vw), 603 (vw), 557 (vw), 397 (vw) cm^{-1}. – **MS** (70 eV, EI): *m/z* (%) = 302 (52) $[M]^+$, 287 (19) $[M - CH_3]^+$, 260 (18), 259 (100) $[M - C_3H_7]^+$, 245 (13) $[M - C_4H_9]^+$, 232 (33), 231 (16) $[M - C_5H_{11}]^+$, 217 (25) $[M - C_6H_{13}]^+$, 188 (9). – **HRMS** ($C_{19}H_{26}O_3$): calc. 302.1876, found 302.1876.

5-Hydroxy-7-pentyl-2*H*-chromen-2-one (27bj-H)

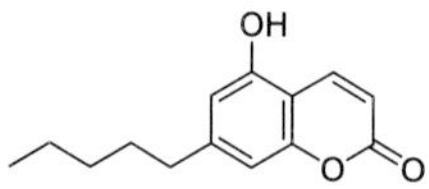

According to **GP8**, 76 mg of of 5-methoxycoumarin **27bj** (309 µmol, 1.00 equiv.) were dissolved in 5 mL of dry dichloromethane. The solution was cooled to –78 °C and 1.54 mL of boron tribromide (1 M in dichloromethane, 1.54 mmol, 5.00 equiv.) were added dropwise. The mixture was stirred for 30 min at this temperature and then allowed to warm to room temperature. The reaction was quenched after 16 h at 0 °C by addition of sodium bicarbonate. The aqueous layer was extracted with 3 × 15 mL of dichloromethane and the combined organic layers were washed with brine, dried over sodium sulfate and the volatiles were removed under reduced pressure. The crude product was then purified via flash column chromatography (CH/EtOAc 5:1) to give the product as 49 mg (69%) of a white solid.

***R*f** (CH/EtOAc 5:1): 0.17. – **MP**: 143.4 °C – **^{1}H NMR** (400 MHz, $CDCl_3$): δ = 8.16 (d, *J* = 9.6 Hz, 1H, 4-C*H*), 7.42 (s, 1H, O*H*), 6.70 (d, *J* = 1.3 Hz, 1H, H_{Ar}), 6.64 (d, *J* = 1.4 Hz, 1H, H_{Ar}), 6.31 (d, *J* = 9.6 Hz, 1H, 3-C*H*), 2.70 – 2.48 (m, 2H, CH_2), 1.72 – 1.52 (m, 2H, CH_2), 1.36 – 1.20 (m, 4H, 2 × CH_2), 0.95 – 0.80 (m, 3H, CH_3) ppm. – **^{13}C NMR** (100 MHz, $CDCl_3$): δ = 162.9 ($C_{quart.}$, *C*OO), 155.1 ($C_{quart.}$, C_{Ar}), 153.4 ($C_{quart.}$, C_{Ar}), 149.4 ($C_{quart.}$, C_{Ar}), 139.9 (+, 4-C_{Ar}H), 112.7 (+, C_{Ar}H), 110.8 (+, C_{Ar}H),108.5 (+, C_{Ar}H), 107.1 ($C_{quart.}$, C_{Ar}), 36.4 (–, CH_2), 31.5 (–, CH_2), 30.7 (–, CH_2), 22.6 (–, CH_2), 14.1 (+, CH_3) ppm. – **IR** (KBr): ṽ = 3202 (w), 2958 (w), 2927 (w), 2857 (w), 1683 (m), 1612 (m), 1513 (w), 1468 (w), 1434 (m), 1408 (m), 1350 (m), 1281 (m), 1243 (m), 1196 (w), 1121 (m), 1061 (m), 999 (w), 905 (w), 862 (w), 836 (m), 824 (m), 698 (m), 636 (m), 557 (w), 494 (w), 427 (w) cm^{-1}. – **MS** (70 eV, EI): *m/z* (%) = 232 (28) $[M]^+$, 190 (11), 189 (9), 177 (16), 176 (100), 175 (20), 148 (18), 147 (18), 91 (13), 65 (5). – **HRMS** ($C_{14}H_{16}O_3$): calc. 232.1094, found 232.1095. – **Elemental analysis**: $C_{14}H_{16}O_3$: calc. C 72.39, H 6.94, found C 72.44, H 6.83.

5-Hydroxy-3-methyl-7-pentyl-2*H*-chromen-2-one (27bk-H)

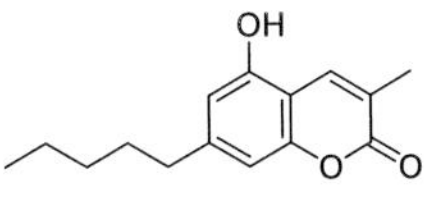

OH

O O

According to **GP8**, 75 mg of 5-methoxycoumarin **27bk** (288 µmol, 1.00 equiv.) were dissolved in 5 mL of dry dichloromethane. The solution was cooled to –78 °C and 1.44 mL of boron tribromide (1 M in dichloromethane, 1.44 mmol, 5.00 equiv.) were added dropwise. The mixture was stirred for 30 min at this temperature and then allowed to warm to room temperature. The reaction was quenched after 16 h at 0 °C by addition of sodium bicarbonate. The aqueous layer was extracted with 3 × 15 mL of dichloromethane and the combined organic layers were washed with brine, dried over sodium sulfate and the volatiles were removed under reduced pressure. The crude product was then purified via flash column chromatography (CH/EtOAc 5:1) to give the product as 61 mg (86%) of a white solid.

***R*f** (CH/EtOAc 5:1): 0.22. – **MP**: 130.1 °C – **^{1}H NMR** (400 MHz, $CDCl_3$): δ = 7.94 (s, 1H, 4-C*H*), 6.98 (s, 1H, O*H*), 6.69 (d, *J* = 1.3 Hz, 1H, H_{Ar}), 6.61 (d, *J* = 1.4 Hz, 1H, H_{Ar}), 2.62 – 2.50

(m, 2H, CH_2), 2.21 (s, 3H, CH_3), 1.66 – 1.54 (m, 2H, CH_2), 1.38 – 1.19 (m, 4H, 2 × CH_2), 0.93 – 0.82 (m, 3H, CH_3) ppm. – **^{13}C NMR** (100 MHz, $CDCl_3$): δ = 163.8 ($C_{quart.}$, COO), 154.4 ($C_{quart.}$, C_{Ar}), 152.4 ($C_{quart.}$, C_{Ar}), 147.6 ($C_{quart.}$, C_{Ar}), 135.5 (+, 4-C_{Ar}H), 122.1 ($C_{quart.}$, C_{Ar}), 110.6 (+, C_{Ar}H), 108.3 (+, C_{Ar}H), 107.5 ($C_{quart.}$, C_{Ar}), 36.3 (–, CH_2), 31.5 (–, CH_2), 30.7 (–, CH_2), 22.6 (–, CH_2), 17.2 (+, CH_3), 14.1 (+, CH_3) ppm. – **IR** (KBr): ṽ = 3299 (w), 2951 (w), 2920 (w), 2854 (w), 1679 (w), 1623 (m), 1582 (w), 1518 (w), 1437 (w), 1372 (w), 1359 (w), 1284 (w), 1249 (w), 1189 (w), 1133 (w), 1104 (w), 1067 (w), 998 (w), 980 (w), 914 (w), 875 (w), 847 (w), 818 (w), 765 (w), 730 (w), 673 (w), 650 (w), 590 (w), 567 (w) cm^{-1}. – **MS** (70 eV, EI): *m/z* (%) = 246 (48) $[M]^+$, 204 (8), 203 (11), 191 (13), 190 (100), 189 (26), 162 (13), 161 (27), 147 (5), 91 (5). – **HRMS** ($C_{15}H_{18}O_3$): calc. 246.1250, found 246.1249. – **Elemental analysis**: $C_{15}H_{18}O_3$: calc. C 73.15, H 7.37, found C 72.68, H 7.37.

3-Ethyl-5-hydroxy-7-pentyl-2*H*-chromen-2-one (27bl-H)

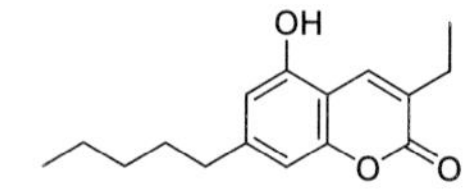

According to **GP8**, 94 mg of 5-methoxycoumarin **27bl** (343 µmol, 1.00 equiv.) were dissolved in 5 mL of dry dichloromethane. The solution was cooled to –78 °C and 1.44 mL of boron tribromide (1 M in dichloromethane, 1.44 mmol, 5.00 equiv.) were added dropwise. The mixture was stirred for 30 min at this temperature and then allowed to warm to room temperature. The reaction was quenched after 16 h at 0 °C by addition of sodium bicarbonate. The aqueous layer was extracted with 3 × 15 mL of dichloromethane and the combined organic layers were washed with brine, dried over sodium sulfate and the volatiles were removed under reduced pressure. The crude product was then purified via flash column chromatography (CH/EtOAc 5:1) to give the product as 80 mg (89%) of a yellow solid.

R_f (CH/EtOAc 5:1): 0.28. – **MP**: 123.5 °C – **^{1}H NMR** (400 MHz, $CDCl_3$): δ = 7.94 (s, 1H, O*H*), 7.39 (s, 1H, 4-C*H*), 6.68 (d, *J* = 1.3 Hz, 1H, H_{Ar}), 6.65 (d, *J* = 1.3 Hz, 1H, H_{Ar}), 2.60 (qd, *J* = 8.3, 7.9, 6.3 Hz, 4H, 2 × CH_2), 1.65 – 1.53 (m, 2H, CH_2), 1.35 – 1.21 (m, 7H, 2 × CH_2, CH_3), 0.90 – 0.81 (m, 3H, CH_3) ppm. – **^{13}C NMR** (100 MHz, $CDCl_3$): δ = 163.6 ($C_{quart.}$, COO), 154.1 ($C_{quart.}$, C_{Ar}), 152.8 ($C_{quart.}$, C_{Ar}), 147.6 ($C_{quart.}$, C_{Ar}), 134.0 (+, 4-C_{Ar}H), 127.5 ($C_{quart.}$, C_{Ar}), 110.6 (+, C_{Ar}H), 108.1 (+, C_{Ar}H), 107.6 ($C_{quart.}$, C_{Ar}), 36.3 (–, CH_2), 31.5 (–, CH_2), 30.7 (–, CH_2), 24.0 (–, CH_2), 22.6 (–, CH_2), 14.1 (+, CH_3), 12.6 (+, CH_3) ppm. – **MS** (70 eV, EI): *m/z* (%) = 260 (61) $[M]^+$, 247 (5), 246 (5), 245 (15), 233 (8), 232 (8), 219 (10), 218 (10), 217 (13), 207 (7), 206 (17), 205 (16), 204 (100), 203 (8), 189 (10), 181 (7), 178 (5), 175 (10), 164

(6), 161 (8), 160 (6), 147 (5), 131 (7), 69 (11). – **HRMS** ($C_{16}H_{20}O_3$): calc. 260.1407, found 260.1409.

5-Hydroxy-7-pentyl-3-propyl-2*H*-chromen-2-one (27bm-H)

According to **GP8**, 97 mg of 5-methoxycoumarin **27bm** (336 µmol, 1.00 equiv.) were dissolved in 5 mL of dry dichloromethane. The solution was cooled to –78 °C and 1.68 mL of boron tribromide (1 M in dichloromethane, 1.68 mmol, 5.00 equiv.) were added dropwise. The mixture was stirred for 30 min at this temperature and then allowed to warm to room temperature. The reaction was quenched after 16 h at 0 °C by addition of sodium bicarbonate. The aqueous layer was extracted with 3 × 15 mL of dichloromethane and the combined organic layers were washed with brine, dried over sodium sulfate and the volatiles were removed under reduced pressure. The crude product was then purified via flash column chromatography (CH/EtOAc 5:1) to give the product as 82 mg (89%) of yellow crystals.

***R*f** (CH/EtOAc 5:1): 0.33. – **MP**: 122.5 °C – **^{1}H NMR** (400 MHz, $CDCl_3$): δ = 7.90 (s, 1H, 4-C*H*), 6.94 (s, 1H, O*H*), 6.69 (d, *J* = 1.5 Hz, 1H, H_{Ar}), 6.61 (d, *J* = 1.3 Hz, 1H, H_{Ar}), 2.64 – 2.50 (m, 4H, 2 × CH_2), 1.74 – 1.64 (m, 2H, CH_2), 1.64 – 1.54 (m, 2H, CH_2), 1.37 – 1.22 (m, 4H, 2 × CH_2), 0.99 (t, *J* = 7.3 Hz, 3H, CH_3), 0.91 – 0.82 (m, 3H, CH_3) ppm. – **^{13}C NMR** (100 MHz, $CDCl_3$): δ = 163.4 ($C_{quart.}$, *C*OO), 154.3 ($C_{quart.}$, C_{Ar}), 152.5 ($C_{quart.}$, C_{Ar}), 147.6 ($C_{quart.}$, C_{Ar}), 134.7 (+, 4-C_{Ar}H), 126.2 ($C_{quart.}$, C_{Ar}), 110.5 (+, C_{Ar}H), 108.3 (+, C_{Ar}H), 107.5 ($C_{quart.}$, C_{Ar}), 36.3 (–, *C*H_2), 33.0 (–, *C*H_2), 31.5 (–, *C*H_2), 30.7 (–, *C*H_2), 22.6 (–, *C*H_2), 21.6 (–, *C*H_2), 14.1 (+, *C*H_3), 13.9 (+, *C*H_3) ppm. – **IR** (KBr): $\tilde{\nu}$ = 3189 (w), 2954 (w), 2928 (m), 2854 (w), 1670 (s), 1617 (s), 1516 (w), 1439 (m), 1353 (m), 1292 (m), 1251 (m), 1174 (w), 1138 (m), 1113 (m), 1078 (m), 1058 (m), 974 (w), 943 (w), 867 (m), 852 (w), 817 (m), 778 (w), 750 (w), 716 (m), 578 (w), 527 (w), 487 (w), 419 (w) cm^{-1}. – **MS** (70 eV, EI): *m/z* (%) = 274 (100) $[M]^+$, 260 (5), 259 (24), 246 (26), 245 (85), 232 (5), 231 (5), 219 (9), 218 (63), 217 (9), 190 (9), 189 (8). – **HRMS** ($C_{17}H_{22}O_3$): calc. 274.1563, found 274.1562. – **Elemental analysis**: $C_{17}H_{22}O_3$: calc. C 74.42, H 8.08, found C 74.30, H 8.18.

5-Hydroxy-7-pentyl-3-butyl-2*H*-chromen-2-one (27bn-H)

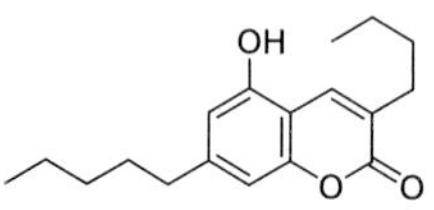

According to **GP8**, 143 mg of 5-methoxycoumarin **27bn** (473 µmol, 1.00 equiv.) were dissolved in 5 mL of dry dichloromethane. The solution was cooled to –78 °C and 2.36 mL of boron tribromide (1 M in dichloromethane, 2.36 mmol, 5.00 equiv.) were added dropwise. The mixture was stirred for 30 min at this temperature and then allowed to warm to room temperature. The reaction was quenched after 16 h at 0 °C by addition of sodium bicarbonate. The aqueous layer was extracted with 3 × 15 mL of dichloromethane and the combined organic layers were washed with brine, dried over sodium sulfate and the volatiles were removed under reduced pressure. The crude product was then purified via flash column chromatography (CH/EtOAc 5:1) to give the product as 106 mg (78%) of yellow crystalline solid.

***R*f** (CH/EtOAc 5:1): 0.39. – **MP**: 126.8 °C – **^{1}H NMR** (400 MHz, $CDCl_3$): δ = 7.92 (s, 1H, 4-C*H*), 7.20 – 7.02 (m, 1H, O*H*), 6.68 (d, *J* = 1.2 Hz, 1H, H_{Ar}), 6.65 – 6.60 (m, 1H, H_{Ar}), 2.65 – 2.50 (m, 4H, 2 × CH_2), 1.68 – 1.54 (m, 4H, 2 × CH_2), 1.48 – 1.35 (m, 2H, CH_2), 1.35 – 1.23 (m, 4H, 2 × CH_2), 0.94 (t, *J* = 7.3 Hz, 3H, CH_3), 0.90 – 0.82 (m, 3H, CH_3) ppm. – **^{13}C NMR** (100 MHz, $CDCl_3$): δ = 163.5 ($C_{quart.}$, *C*OO), 154.2 ($C_{quart.}$, C_{Ar}), 152.6 ($C_{quart.}$, C_{Ar}), 147.6 ($C_{quart.}$, C_{Ar}), 134.7 (+, 4-C_{Ar}H), 126.4 ($C_{quart.}$, C_{Ar}), 110.6 (+, C_{Ar}H), 108.2 (+, C_{Ar}H), 108.2 ($C_{quart.}$, C_{Ar}), 107.6 ($C_{quart.}$, C_{Ar}), 36.3 (–, CH_2), 31.5 (–, CH_2), 30.7 (–, CH_2), 30.7 (–, CH_2), 30.5 (–, CH_2), 22.6 (–, CH_2), 22.5 (–, CH_2), 14.1 (+, CH_3), 14.0 (+, CH_3) ppm. – **IR** (KBr): $\tilde{\nu}$ = 3180 (w), 2954 (m), 2926 (m), 2856 (w), 1669 (s), 1621 (s), 1580 (m), 1440 (m),1349 (m), 1275 (m), 1248 (m), 1183 (m), 1141 (m), 1119 (w), 1057 (m), 860 (w), 826 (w), 779 (w), 717 (w), 564 (w), 524 (w), 430 (w) cm^{-1}. – **MS** (70 eV, EI): *m/z* (%) = 288/289 (65/13) $[M]^+$, 271 (8), 260 (9), 259 (38), 247 (16), 246 (100), 245 (71), 233 (6), 232 (30), 231 (10), 217 (5), 204 (7), 203 (8), 190 (47), 189 (13), 161 (15), 160 (12), 147 (5), 131 (8), 69 (12). – **HRMS** ($C_{18}H_{24}O_3$): calc. 288.1720, found 288.1719. – **Elemental analysis**: $C_{18}H_{24}O_3$: calc. C 74.97, H 8.39, found C 74.98, H 8.53.

7-(1-Butylcyclopentyl)-5-hydroxy-2*H*-chromen-2-one (27bo-H)

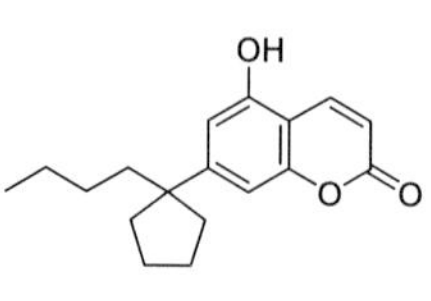

According to **GP8**, 122 mg of 5-methoxycoumarin **27bo** (404 µmol, 1.00 equiv.) was dissolved in 5 mL of dry dichloromethane. The solution was cooled to –78 °C and 2.02 mL of boron tribromide (1 M in dichloromethane, 2.02 mmol,

5.00 equiv.) was added dropwise. The mixture was stirred for 30 min at this temperature and then allowed to warm to room temperature. The reaction was quenched after 16 h at 0 °C by addition of sodium bicarbonate. The aqueous layer was extracted with 3 × 15 mL of dichloromethane and the combined organic layers were washed with brine, dried over sodium sulfate and the volatiles were removed under reduced pressure. The crude product was then purified via flash column chromatography (CH/EtOAc 20:1) to give the product as 104 mg (90%) of yellow solid.

***R*f** (CH/EtOAc 5:1): 0.13. – **MP**: 157.0 °C – **^{1}H NMR** (400 MHz, $CDCl_3$): δ = 8.16 (d, *J* = 9.6 Hz, 1H, 4-C*H*), 7.33 (s, 1H, O*H*), 6.79 (d, *J* = 1.4 Hz, 1H, H_{Ar}), 6.75 (d, *J* = 1.5 Hz, 1H, H_{Ar}), 6.32 (d, *J* = 9.6 Hz, 1H, 3-C*H*), 1.98 – 1.49 (m, 10H, 5 × CH_2), 1.20 – 1.07 (m, 2H, CH_2), 1.00 – 0.87 (m, 2H, CH_2), 0.76 (t, *J* = 7.3 Hz, 3H, CH_3) ppm. – **^{13}C NMR** (100 MHz, $CDCl_3$): δ = 162.9 ($C_{quart.}$, *C*OO), 156.0 ($C_{quart.}$, C_{Ar}), 154.9 ($C_{quart.}$, C_{Ar}), 153.1 ($C_{quart.}$, C_{Ar}), 139.8 (+, 4-C_{Ar}H), 112.9 (+, C_{Ar}H), 109.5 (+, C_{Ar}H), 107.4 (+, 3-C_{Ar}H), 106.8 ($C_{quart.}$, C_{Ar}), 51.9 ($C_{quart.}$, C_{CP}), 41.6 (–, CH_2), 37.7 (–, 2 × CH_2), 27.6 (–, CH_2), 23.4 (–, CH_2), 23.3 (–, 2 × CH_2), 14.1 (+, CH_3) ppm. – **IR** (KBr): ṽ = 3135 (w), 2953 (m), 2923 (m), 2869 (w), 1675 (w), 1610 (s), 1454 (w), 1407 (m), 1348 (m), 1295 (m), 1241 (m), 1191 (w), 1122 (m), 1058 (m), 936 (w), 909 (w), 865 (w), 841 (w), 819 (m), 729 (m), 662 (w), 614 (w), 525 (w), 484 (m), 423 (w) cm^{-1}. – **MS** (70 eV, EI): *m/z* (%) = 286 (32) $[M]^+$, 231 (9), 230 (58), 229 (100) $[M – C_4H_9]^+$, 204 (9), 202 (5), 201 (5), 190 (15), 189 (7), 175 (44), 163 (10), 131 (5), 69 (6), 67 (12). – **HRMS** ($C_{18}H_{22}O_3$): calc. 286.1563, found 286.1563. – **Elemental analysis**: $C_{18}H_{22}O_3$: calc. C 75.50, H 7.74, found C 75.61, H 7.85.

7-(1-Butylcyclopentyl)-5-hydroxy-3-methyl-2*H*-chromen-2-one (27bp-H)

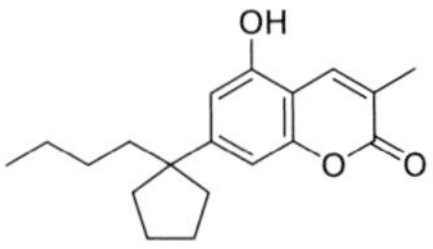

According to **GP8**, 84 mg of 5-methoxycoumarin **27bp** (267 µmol, 1.00 equiv.) were dissolved in 5 mL of dry dichloromethane. The solution was cooled to –78 °C and 1.34 mL of boron tribromide (1 M in dichloromethane, 1.34 mmol, 5.00 equiv.) were added dropwise. The mixture was stirred for 30 min at this temperature and then allowed to warm to room temperature. The reaction was quenched after 16 h at 0 °C by addition of sodium bicarbonate. The aqueous layer was extracted with 3 × 15 mL of dichloromethane and the combined organic layers were washed with brine, dried over sodium sulfate and the volatiles were removed under reduced pressure. The crude product was then purified via flash column chromatography (CH/EtOAc 20:1) to give the product as 70 mg (87%) of a yellow solid.

***R*f** (CH/EtOAc 5:1): 0.18. – **MP**: 178.6 °C – **^{1}H NMR** (400 MHz, $CDCl_3$): δ = 7.93 (d, *J* = 1.6 Hz, 1H, 4-C*H*), 6.79 (d, *J* = 1.5 Hz, 1H, H_{Ar}), 6.69 (d, *J* = 1.5 Hz, 1H, H_{Ar}), 6.64 (s, 1H, O*H*), 2.21 (d, *J* = 1.3 Hz, 3H, CH_3), 1.92 – 1.50 (m, 10H, 5 × CH_2), 1.19 – 1.08 (m, 2H, CH_2), 0.98 – 0.86 (m, 2H, CH_2), 0.76 (t, *J* = 7.3 Hz, 3H, CH_3) ppm. – **^{13}C NMR** (100 MHz, $CDCl_3$): δ = 163.7 ($C_{quart.}$, *C*OO), 154.1 ($C_{quart.}$, C_{Ar}), 154.1 ($C_{quart.}$, C_{Ar}), 151.9 ($C_{quart.}$, C_{Ar}), 135.2 (+, 4-C_{Ar}H), 122.4 ($C_{quart.}$, C_{Ar}), 109.2 (+, C_{Ar}H), 107.3 (+, C_{Ar}H), 107.2 ($C_{quart.}$, C_{Ar}), 51.7 ($C_{quart.}$, C_{CP}), 41.7 (–, CH_2), 37.8 (–, 2 × CH_2), 27.6 (–, CH_2), 23.4 (–, CH_2), 23.3 (–, 2 × CH_2), 17.2 (+, CH_3), 14.1 (+, CH_3) ppm. – **IR** (KBr): ṽ = 3319 (w), 2925 (w), 2869 (w), 1684 (m), 1618 (m), 1577 (w), 1453 (w), 1422 (m), 1372 (w), 1286 (w), 1183 (w), 1100 (w), 1075 (w), 1005 (w), 918 (w), 870 (vw), 839 (w), 765 (w), 729 (w), 623 (w), 550 (w), 529 (w), 433 (w) cm^{-1}. – **MS** (70 eV, EI): *m/z* (%) = 300/301 (35/7) $[M]^+$, 245 (6), 244 (39), 243 (100) $[M - C_4H_9]^+$, 218 (5), 190 (5), 189 (34), 177 (10), 67 (8). – **HRMS** ($C_{19}H_{24}O_3$): calc. 300.1719, found 300.1720. – **Elemental analysis**: $C_{19}H_{24}O_3$: calc. C 75.97, H 8.05, found C 75.90, H 8.21.

7-(1-Butylcyclopentyl)-3-ethyl-5-hydroxy-2*H*-chromen-2-one (27bq-H)

According to **GP8**, 84 mg of 5-methoxycoumarin **27bq** (256 µmol, 1.00 equiv.) were dissolved in 5 mL of dry dichloromethane. The solution was cooled to –78 °C and 1.28 mL of boron tribromide (1 M in dichloromethane, 1.28 mmol, 5.00 equiv.) were added dropwise. The mixture was stirred for 30 min at this temperature and then allowed to warm to room temperature. The reaction was quenched after 16 h at 0 °C by addition of sodium bicarbonate. The aqueous layer was extracted with 3 × 15 mL of dichloromethane and the combined organic layers were washed with brine, dried over sodium sulfate and the volatiles were removed under reduced pressure. The crude product was then purified via flash column chromatography (CH/EtOAc 20:1) to give the product as 62 mg (77%) of a white solid.

***R*f** (CH/EtOAc 5:1): 0.23. – **MP**: 168.9 °C – **^{1}H NMR** (400 MHz, $CDCl_3$): δ = 7.89 (d, *J* = 1.2 Hz, 1H, 4-C*H*), 6.79 (d, *J* = 1.4 Hz, 1H, H_{Ar}), 6.68 (d, *J* = 1.5 Hz, 1H, H_{Ar}), 6.42 (s, 1H,

O*H*), 2.60 (qd, J = 7.5, 1.3 Hz, 2H, C*H*$_2$), 1.91 – 1.51 (m, 10H, 5 × C*H*$_2$), 1.25 (t, J = 7.4 Hz, 3H, C*H*$_3$), 1.19 – 1.08 (m, 2H, C*H*$_2$), 0.98 – 0.86 (m, 2H, C*H*$_2$), 0.76 (t, J = 7.3 Hz, 3H, C*H*$_3$) ppm. – **^{13}C NMR** (100 MHz, CDCl$_3$): δ = 163.1 (C$_{quart.}$, *C*OO), 154.0 (C$_{quart.}$, *C*$_{Ar}$), 153.9 (C$_{quart.}$, *C*$_{Ar}$), 151.9 (C$_{quart.}$, *C*$_{Ar}$), 133.3 (+, 4-*C*$_{Ar}$H), 128.1 (C$_{quart.}$, *C*$_{Ar}$), 109.1 (+, *C*$_{Ar}$H), 107.3 (+, *C*$_{Ar}$H), 107.2 (C$_{quart.}$, *C*$_{Ar}$), 51.7 (C$_{quart.}$, *C*$_{CP}$), 41.7 (–, *C*H$_2$), 37.8 (–, 2 × *C*H$_2$), 27.6 (–, *C*H$_2$), 24.1 (–, *C*H$_2$), 23.4 (–, 2 × *C*H$_2$), 23.3 (–, *C*H$_2$), 14.1 (+, *C*H$_3$), 12.7 (+, *C*H$_3$) ppm. – **IR** (KBr): ṽ = 3345 (w), 2956 (w), 2870 (w), 1684 (m), 1618 (m), 1576 (w), 1511 (w), 1457 (w), 1424 (w), 1372 (w), 1346 (w), 1282 (w), 1102 (w), 1054 (w), 969 (w), 925 (w), 875 (w), 842 (w), 720 (w), 633 (w), 549 (w), 529 (w), 454 (w), 411 (vw) cm^{-1}. – **MS** (70 eV, EI): *m/z* (%) = 314 (34) [M]$^+$, 259 (6), 258 (39), 257 (100) [M – C$_4$H$_9$]$^+$, 243 (5), 218 (7), 203 (28), 191 (9), 181 (5), 131 (5), 69 (7), 67 (7). – **HRMS** (C$_{20}$H$_{26}$O$_3$): calc. 314.1876, found 314.1876. – **Elemental analysis**: C$_{20}$H$_{26}$O$_3$: calc. C 76.40, H 8.34, found C 76.22, H 8.52.

7-(1-Butylcyclopentyl)-3-propyl-5-hydroxy-2*H*-chromen-2-one (27br-H)

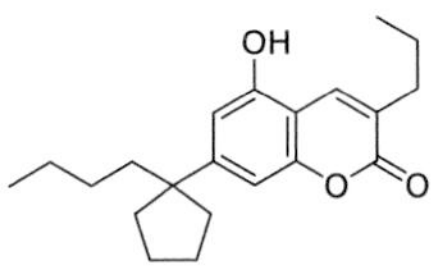

According to **GP8**, 139 mg of 5-methoxycoumarin **27br** (406 µmol, 1.00 equiv.) were dissolved in 5 mL of dry dichloromethane. The solution was cooled to –78 °C and 2.03 mL of boron tribromide (1 M in dichloromethane, 2.03 mmol, 5.00 equiv.) were added dropwise. The mixture was stirred for 30 min at this temperature and then allowed to warm to room temperature. The reaction was quenched after 16 h at 0 °C by addition of sodium bicarbonate. The aqueous layer was extracted with 3 × 15 mL of dichloromethane and the combined organic layers were washed with brine, dried over sodium sulfate and the volatiles were removed under reduced pressure. The crude product was then purified via flash column chromatography (CH/EtOAc 20:1) to give the product as 130 mg (97%) of a brown solid.

***R*$_f$** (CH/EtOAc 5:1): 0.30. – **MP**: 139.5 °C – **^{1}H NMR** (400 MHz, CDCl$_3$): δ = 7.93 (s, 1H, 4-C*H*), 7.13 (s, 1H, O*H*), 6.78 (d, J = 1.4 Hz, 1H, *H*$_{Ar}$), 6.74 (d, J = 1.5 Hz, 1H, *H*$_{Ar}$), 2.60 – 2.51 (m, 2H, C*H*$_2$), 1.91 – 1.50 (m, 12H, 6 × C*H*$_2$), 1.18 – 1.07 (m, 2H, C*H*$_2$), 0.98 (t, J = 7.4 Hz, 3H, C*H*$_3$), 0.96 – 0.86 (m, 2H, C*H*$_2$), 0.75 (t, J = 7.3 Hz, 3H, C*H*$_3$) ppm. – **^{13}C NMR** (100 MHz, CDCl$_3$): δ = 163.6 (C$_{quart.}$, *C*OO), 154.2 (C$_{quart.}$, *C*$_{Ar}$), 154.0 (C$_{quart.}$, *C*$_{Ar}$), 152.3 (C$_{quart.}$, *C*$_{Ar}$), 134.8 (+, 4-*C*$_{Ar}$H), 126.2 (C$_{quart.}$, *C*$_{Ar}$), 109.3 (+, *C*$_{Ar}$H), 107.3 (+, *C*$_{Ar}$H), 107.1 (C$_{quart.}$, *C*$_{Ar}$), 51.7 (C$_{quart.}$, *C*$_{CP}$), 41.7 (–, *C*H$_2$), 37.8 (–, 2 × *C*H$_2$), 33.0 (–, *C*H$_2$), 27.6 (–, *C*H$_2$), 23.4 (–, *C*H$_2$),

23.3 (–, 2 × CH_2), 21.6 (–, CH_2), 14.1 (+, CH_3), 13.9 (+, CH_3) ppm. – **IR** (KBr): ṽ = 3301 (w), 2953 (w), 2869 (w), 1680 (m), 1616 (s), 1573 (w), 1514 (w), 1424 (m), 1390 (w), 1346 (m), 1298 (w), 1252 (w), 1178 (w), 1115 (w), 1056 (m), 938 (w), 904 (w), 869 (w), 837 (m), 775 (w), 715 (m), 634 (w), 526 (w), 435 (w) cm^{-1}. – **MS** (70 eV, EI): *m/z* (%) = 328 (44) $[M]^+$, 314 (7), 273 (7), 272 (40), 271 (100) $[M - C_4H_9]^+$, 258 (13), 257 (19), 256 (12), 255 (13), 246 (5), 244 (5), 243 (13), 232 (9), 231 (6), 218 (7), 217 (25), 216 (5), 215 (6), 206 (5), 205 (10), 201 (9), 128 (8). – **HRMS** ($C_{21}H_{28}O_3$): calc. 328.2033, found 328.2035. – **Elemental analysis:** $C_{21}H_{28}O_3$: calc. C 76.79, H 8.59, found C 76.91, H 8.58.

7-(1-Butylcyclopentyl)-3-butyl-5-hydroxy-2*H*-chromen-2-one (27bs-H)

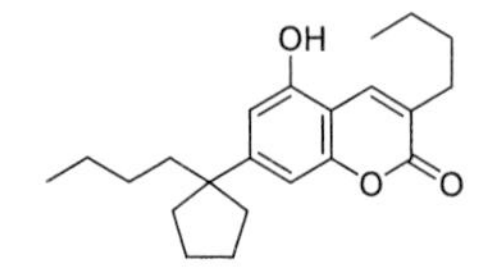

According to **GP8**, 84 mg of 5-methoxycoumarin **27bs** (236 µmol, 1.00 equiv.) were dissolved in 5 mL of dry dichloromethane. The solution was cooled to –78 °C and 1.18 mL of boron tribromide (1 M in dichloromethane, 1.18 mmol, 5.00 equiv.) were added dropwise. The mixture was stirred for 30 min at this temperature and then allowed to warm to room temperature. The reaction was quenched after 16 h at 0 °C by addition of sodium bicarbonate. The aqueous layer was extracted with 3 × 15 mL of dichloromethane and the combined organic layers were washed with brine, dried over sodium sulfate and the volatiles were removed under reduced pressure. The crude product was then purified via flash column chromatography (CH/EtOAc 20:1) to give the product as 68 mg (85%) of a brown solid.

***R*f** (CH/EtOAc 5:1): 0.33. – **MP**: 162.5 °C – **^{1}H NMR** (400 MHz, $CDCl_3$): δ = 7.90 (s, 1H, 4-C*H*), 6.78 (d, *J* = 1.4 Hz, 1H, H_{Ar}), 6.75 (s, 1H, O*H*), 6.70 (d, *J* = 1.5 Hz, 1H, H_{Ar}), 2.64 – 2.48 (m, 2H, CH_2), 1.94 – 1.51 (m, 12H, 6 × CH_2), 1.49 – 1.34 (m, 2H, CH_2), 1.19 – 1.06 (m, 2H, CH_2), 0.99 – 0.85 (m, 5H, CH_2, CH_3), 0.76 (t, *J* = 7.3 Hz, 3H, CH_3) ppm. – **^{13}C NMR** (100 MHz, $CDCl_3$): δ = 163.3 ($C_{quart.}$, *C*OO), 154.1 ($C_{quart.}$, C_{Ar}), 154.0 ($C_{quart.}$, C_{Ar}), 152.1 ($C_{quart.}$, C_{Ar}), 134.4 (+, 4-C_{Ar}H), 126.7 ($C_{quart.}$, C_{Ar}), 109.2 (+, C_{Ar}H), 107.2 (+, C_{Ar}H), 107.2 ($C_{quart.}$, C_{Ar}), 51.7 ($C_{quart.}$, C_{CP}), 41.7 (–, CH_2), 37.8 (–, 2 × CH_2), 30.7 (–, CH_2), 30.5 (–, CH_2), 27.6 (–, CH_2), 23.4 (–, CH_2), 23.3 (–, 2 × CH_2), 22.6 (–, CH_2), 14.1 (+, CH_3), 14.0 (+, CH_3) ppm. – **IR** (KBr): ṽ = 3181 (w), 2956 (w), 2927 (w), 2855 (w), 1670 (m), 1616 (m), 1575 (w), 1454 (w), 1423 (m), 1346 (w), 1295 (w), 1250 (w), 1183 (w), 1123 (w), 1102 (w), 1068 (w), 938 (w), 895 (vw), 866 (w), 840 (w), 784 (w), 744 (w), 728 (w), 675 (w), 627 (vw), 520 (w),

443 (vw), 427 (vw) cm^{-1}. – **MS** (70 eV, EI): *m/z* (%) = 342 (61) [M]$^+$, 300 (10), 299 (5), 287 (7), 286 (42), 285 (100) [M – C_4H_9]$^+$, 260 (5), 246 (5), 243 (7), 231 (20), 219 (8), 181 (5), 69 (9), 67 (6). – **HRMS** ($C_{22}H_{30}O_3$): calc. 342.2189, found 342.2188.

7-(1-Butylcyclohexyl)-5-hydroxy-2*H*-chromen-2-one (27bt-H)

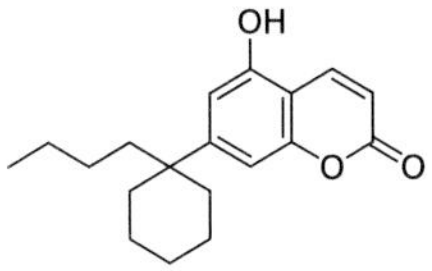

According to **GP8**, 70 mg of 5-methoxycoumarin **27bt** (223 µmol, 1.00 equiv.) were dissolved in 5 mL of dry dichloromethane. The solution was cooled to –78 °C and 1.11 mL of boron tribromide (1 M in dichloromethane, 1.11 mmol, 5.00 equiv.) were added dropwise. The mixture was stirred for 30 min at this temperature and then allowed to warm to room temperature. The reaction was quenched after 16 h at 0 °C by addition of sodium bicarbonate. The aqueous layer was extracted with 3 × 15 mL of dichloromethane and the combined organic layers were washed with brine, dried over sodium sulfate and the volatiles were removed under reduced pressure. The crude product was then purified via flash column chromatography (CH/EtOAc 5:1) to give the product as 59 mg (87%) of a white solid.

***R*$_f$** (CH/EtOAc 5:1): 0.18. – **MP**: 159.6 °C – **^{1}H NMR** (400 MHz, $CDCl_3$): δ = 8.16 (d, *J* = 9.6 Hz, 1H, 4-C*H*), 7.37 (s, 1H, O*H*), 6.84 (d, *J* = 1.5 Hz, 1H, H_{Ar}), 6.79 (d, *J* = 1.5 Hz, 1H, H_{Ar}), 6.32 (d, *J* = 9.6 Hz, 1H, 3-C*H*), 2.06 – 1.91 (m, 2H, CH_2), 1.67 – 1.22 (m, 10H, 5 × CH_2), 1.18 – 1.04 (m, 2H, CH_2), 0.96 – 0.81 (m, 2H, CH_2), 0.75 (t, *J* = 7.3 Hz, 3H, CH_3) ppm. – **^{13}C NMR** (100 MHz, $CDCl_3$): δ = 162.9 ($C_{quart.}$, *C*OO), 155.1 ($C_{quart.}$, C_{Ar}), 154.7 ($C_{quart.}$, C_{Ar}), 153.3 ($C_{quart.}$, C_{Ar}), 139.8 (+, 4-C_{Ar}H), 113.0 (+, 3-C_{Ar}H), 109.5 (+, C_{Ar}H), 107.6 (+, C_{Ar}H), 106.7 ($C_{quart.}$, C_{Ar}), 43.7 (–, CH_2), 42.2 ($C_{quart.}$, C_{CH}), 36.3 (–, 2 × CH_2), 26.5 (–, CH_2), 25.8 (–, CH_2), 23.4 (–, CH_2), 22.5 (–, 2 × CH_2), 14.1 (+, CH_3) ppm. – **IR** (KBr): ṽ = 3128 (w), 2925 (w), 2853 (w), 1678 (m), 1611 (m), 1562 (w), 1406 (m), 1333 (w), 1292 (w), 1242 (m), 1202 (w), 1123 (m), 1098 (w), 973 (w), 935 (w), 908 (w), 863 (w), 843 (w), 818 (m), 729 (w), 671 (w), 647 (w), 604 (w), 528 (vw), 484 (w), 421 (vw) cm^{-1}. – **MS** (70 eV, EI): *m/z* (%) = 300 (43) [M]$^+$, 245 (9), 244 (56), 243 (100) [M – C_4H_9]$^+$, 204 (15), 190 (12), 189 (10), 181 (7), 176 (13), 175 (82), 163 (11), 131 (6), 81 (11), 69 (9). – **HRMS** ($C_{19}H_{24}O_3$): calc. 300.1720, found 300.1723. – **Elemental analysis**: $C_{19}H_{24}O_3$: calc. C 75.97, H 8.05, found C 75.83, H 8.09.

7-(1-Butylcyclohexyl)-5-hydroxy-3-methyl-2*H*-chromen-2-one (27bu-H)

According to **GP8**, 70 mg of 5-methoxycoumarin **27bu** (213 µmol, 1.00 equiv.) were dissolved in 5 mL of dry dichloromethane. The solution was cooled to –78 °C and 1.07 mL of boron tribromide (1 M in dichloromethane, 1.07 mmol, 5.00 equiv.) were added dropwise. The mixture was stirred for 30 min at this temperature and then allowed to warm to room temperature. The reaction was quenched after 16 h at 0 °C by addition of sodium bicarbonate. The aqueous layer was extracted with 3 × 15 mL of dichloromethane and the combined organic layers were washed with brine, dried over sodium sulfate and the volatiles were removed under reduced pressure. The crude product was then purified via flash column chromatography (CH/EtOAc 5:1) to give the product as 62 mg (92%) of a white solid.

***R*f** (CH/EtOAc 5:1): 0.29. – **MP**: 190.7 °C – **^{1}H NMR** (400 MHz, $CDCl_3$): δ = 7.94 (s, 1H, 4-C*H*), 6.83 (d, *J* = 1.4 Hz, 1H, H_{Ar}), 6.78 (s, 1H, O*H*), 6.75 (d, *J* = 1.5 Hz, 1H, H_{Ar}), 2.22 (d, *J* = 1.3 Hz, 3H, CH_3), 2.03 – 1.93 (m, 2H, CH_2), 1.60 – 1.24 (m, 10H, 5 × CH_2), 1.18 – 1.01 (m, 2H, CH_2), 0.94 – 0.82 (m, 2H, CH_2), 0.75 (t, *J* = 7.3 Hz, 3H, CH_3) ppm. – **^{13}C NMR** (100 MHz, $CDCl_3$): δ = 163.8 ($C_{quart.}$, *C*OO), 154.4 ($C_{quart.}$, C_{Ar}), 152.7 ($C_{quart.}$, C_{Ar}), 152.2 ($C_{quart.}$, C_{Ar}), 135.3 (+, 4-C_{Ar}H), 122.4 ($C_{quart.}$, C_{Ar}), 109.2 (+, C_{Ar}H), 107.5 (+, C_{Ar}H), 107.1 ($C_{quart.}$, C_{Ar}), 43.8 (–, CH_2), 42.0 ($C_{quart.}$, C_{CH}), 36.3 (–, 2 × CH_2), 26.6 (–, CH_2), 25.8 (–, CH_2), 23.4 (–, CH_2), 22.5 (–, 2 × CH_2), 17.2 (+, CH_3), 14.1 (+, CH_3) ppm. – **IR** (KBr): ṽ = 3345 (w), 2924 (w), 2853 (w), 1689 (m), 1619 (m), 1576 (w), 1450 (w), 1422 (m), 1369 (w), 1339 (w), 1286 (w), 1186 (w), 1094 (m), 1071 (m), 1007 (w), 976 (w), 920 (w), 871 (w), 840 (w), 765 (w), 730 (w), 686 (w), 620 (w), 599 (w), 551 (w), 528 (w), 429 (w) cm^{-1}. – **MS** (70 eV, EI): *m/z* (%) = 314 (50) $[M]^+$, 259 (6), 258 (38), 257 (100) $[M - C_4H_9]^+$, 218 (7), 215 (6), 189 (10), 204 (7), 203 (7), 190 (12), 189 (78), 177 (14), 161 (5), 81 (9). – **HRMS** ($C_{20}H_{26}O_3$): calc. 314.1876, found 314.1875. – **Elemental analysis**: $C_{20}H_{26}O_3$: calc. C 76.40, H 8.34, found C 76.22, H 8.51.

7-(1-Butylcyclohexyl)-5-hydroxy-3-ethyl-2*H*-chromen-2-one (27bv-H)

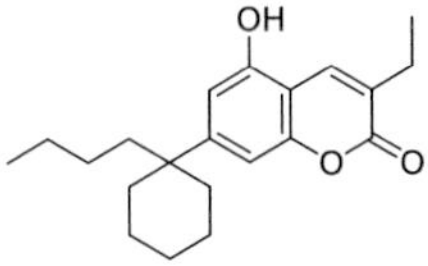

According to **GP8**, 134 mg of 5-methoxycoumarin **27bv** (391 µmol, 1.00 equiv.) were dissolved in 5 mL of dry dichloromethane. The solution was cooled to –78 °C and 1.96 mL of boron tribromide (1 M in dichloromethane, 1.96 mmol, 5.00 equiv.) were added dropwise. The mixture was stirred for 30 min at this temperature and then allowed to warm to room temperature. The reaction was quenched after 16 h at 0 °C by addition of sodium bicarbonate. The aqueous layer was extracted with 3 × 15 mL of dichloromethane and the combined organic layers were washed with brine, dried over sodium sulfate and the volatiles were removed under reduced pressure. The crude product was then purified via flash column chromatography (CH/EtOAc 5:1) to give the product as 116 mg (91%) of a white solid.

***R*f** (CH/EtOAc 5:1): 0.32. – **MP**: 134.3 °C – **^{1}H NMR** (400 MHz, $CDCl_3$): δ = 7.89 (s, 1H, 4-C*H*), 6.84 (d, *J* = 1.4 Hz, 1H, H_{Ar}), 6.72 (d, *J* = 1.6 Hz, 1H, H_{Ar}), 6.41 (s, 1H, O*H*), 2.61 (qd, *J* = 7.5, 1.3 Hz, 2H, CH_2), 1.58 – 1.28 (m, 12H, 6 × CH_2), 1.26 (t, *J* = 7.4 Hz, 3H, CH_3), 1.18 – 1.05 (m, 2H, CH_2), 0.94 – 0.82 (m, 2H, CH_2), 0.75 (t, *J* = 7.3 Hz, 3H, CH_3) ppm. – **^{13}C NMR** (100 MHz, $CDCl_3$): δ = 163.1 ($C_{quart.}$, COO), 154.2 ($C_{quart.}$, C_{Ar}), 152.6 ($C_{quart.}$, C_{Ar}), 152.1 ($C_{quart.}$, C_{Ar}), 133.3 (+, 4-C_{Ar}H), 128.2 ($C_{quart.}$, C_{Ar}), 109.0 (+, C_{Ar}H), 107.6 (+, C_{Ar}H), 107.1 ($C_{quart.}$, C_{Ar}), 43.8 (–, CH_2), 42.0 ($C_{quart.}$, C_{CH}), 36.4 (–, 2 × CH_2), 26.6 (–, CH_2), 25.8 (–, CH_2), 24.1 (–, CH_2), 23.4 (–, CH_2), 22.5 (–, 2 × CH_2), 14.1 (+, CH_3), 12.7 (+, CH_3) ppm. – **IR** (KBr): ṽ = 3360 (w), 2919 (m), 2849 (w), 1681 (m), 1619 (m), 1575 (w), 1510 (w), 1448 (w), 1423 (m), 1373 (w), 1274 (w), 1185 (w), 1170 (w), 1099 (m), 1069 (m), 963 (w), 927 (w), 873 (w), 846 (m), 804 (vw), 757 (vw), 719 (w), 685 (w), 597 (w), 550 (w), 528 (w), 461 (w), 426 (w) cm^{-1}. – **MS** (70 eV, EI): *m/z* (%) = 328 (51) $[M]^+$, 273 (7), 272 (41), 271 (100) $[M - C_4H_9]^+$, 232 (7), 218 (5), 217 (6), 204 (11), 203 (67), 191 (14), 181 (5), 69 (7). – **HRMS** ($C_{21}H_{28}O_3$): calc. 328.2033, found 328.2032. – **Elemental analysis**: $C_{21}H_{28}O_3$: calc. C 76.79, H 8.59, found C 76.59, H 8.86.

7-(1-Butylcyclohexyl)-5-hydroxy-3-propyl-2*H*-chromen-2-one (27bw-H)

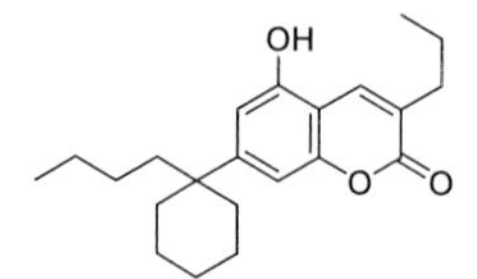

According to **GP8**, 131 mg of 5-methoxycoumarin **27bw** (367 µmol, 1.00 equiv.) were dissolved in 5 mL of dry dichloromethane. The solution was cooled to –78 °C and 1.84 mL of boron tribromide (1 M in dichloromethane, 1.84 mmol, 5.00 equiv.) were added dropwise. The mixture was stirred for 30 min at this temperature and then allowed to warm to room temperature. The reaction was quenched after 16 h at 0 °C by addition of sodium bicarbonate. The aqueous layer was extracted with 3 × 15 mL of dichloromethane and the combined organic layers were washed with brine, dried over sodium sulfate and the volatiles were removed under reduced pressure. The crude product was then purified via flash column chromatography (CH/EtOAc 5:1) to give the product as 113 mg (90%) of a white solid.

***R*f** (CH/EtOAc 5:1): 0.36. – **MP**: 144.0 °C – **^{1}H NMR** (400 MHz, $CDCl_3$): δ = 7.89 (s, 1H, 4-C*H*), 6.84 (d, *J* = 1.4 Hz, 1H, H_{Ar}), 6.72 (d, *J* = 1.5 Hz, 1H, H_{Ar}), 6.43 (s, 1H, O*H*), 2.55 (td, *J* = 7.6, 1.1 Hz, 2H, CH_2), 2.04 – 1.93 (m, 2H, CH_2), 1.75 – 1.63 (m, 2H, CH_2), 1.59 – 1.27 (m, 10H, 5 × CH_2), 1.17 – 1.05 (m, 2H, CH_2), 0.99 (t, *J* = 7.3 Hz, 3H, CH_3), 0.94 – 0.82 (m, 2H, CH_2), 0.75 (t, *J* = 7.3 Hz, 3H, CH_3) ppm. – **^{13}C NMR** (100 MHz, $CDCl_3$): δ = 163.5 ($C_{quart.}$, *C*OO), 154.6 ($C_{quart.}$, C_{Ar}), 153.0 ($C_{quart.}$, C_{Ar}), 152.4 ($C_{quart.}$, C_{Ar}), 134.6 (+, 4-C_{Ar}H), 127.0 ($C_{quart.}$, C_{Ar}), 109.3 (+, C_{Ar}H), 107.9 (+, C_{Ar}H), 107.4 ($C_{quart.}$, C_{Ar}), 44.1 (–, CH_2), 42.3 ($C_{quart.}$, C_{CH}), 36.7 (–, 2 × CH_2), 33.4 (–, CH_2), 26.9 (–, 2 × CH_2), 26.1 (–, CH_2), 23.7 (–, CH_2), 22.8 (–, CH_2), 21.9 (–, CH_2), 14.4 (+, CH_3), 14.3 (+, CH_3) ppm. – **IR** (KBr): $\tilde{\nu}$ = 3214 (vw), 2926 (w), 2854 (w), 1677 (m), 1615 (m), 1575 (w), 1452 (w), 1422 (w), 1343 (w), 1294 (w), 1254 (w), 1185 (w), 1171 (w), 1102 (w), 1064 (w), 939 (w), 866 (w), 836 (w), 718 (w), 687 (w), 529 (w), 471 (vw) cm^{-1}. – **MS** (70 eV, EI): *m/z* (%) = 342 (44) $[M]^+$, 287 (8), 286 (43), 285 (100) $[M - C_4H_9]^+$, 232 (7), 231 (8), 218 (7), 218 (46), 201 (12), 181 (6), 69 (9). – **HRMS** ($C_{22}H_{30}O_3$): calc. 342.2189, found 342.2188. – **Elemental analysis**: $C_{22}H_{30}O_3$: calc. C 77.16, H 8.83, found C 76.84, H 8.85.

7-(1-Butylcyclohexyl)-5-hydroxy-3-butyl-2*H*-chromen-2-one (27bx-H)

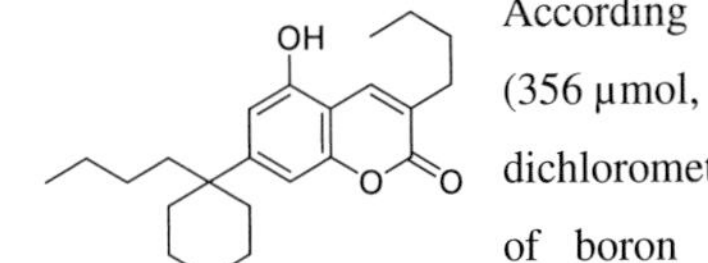

According to **GP8**, 116 mg of 5-methoxycoumarin **27bx** (356 µmol, 1.00 equiv.) were dissolved in 5 mL of dry dichloromethane. The solution was cooled to –78 °C and 1.78 mL of boron tribromide (1 M in dichloromethane, 1.78 mmol, 5.00 equiv.) were added dropwise. The mixture was stirred for 30 min at this temperature and then allowed to warm to room temperature. The reaction was quenched after 16 h at 0 °C by addition of sodium bicarbonate. The aqueous layer was extracted with 3 × 15 mL of dichloromethane and the combined organic layers were washed with brine, dried over sodium sulfate and the volatiles were removed under reduced pressure. The crude product was then purified via flash column chromatography (CH/EtOAc 5:1) to give the product as 116 mg (92%) of a white solid.

***R*f** (CH/EtOAc 5:1): 0.43. – **MP**: 154.0 °C – **^{1}H NMR** (400 MHz, $CDCl_3$): δ = 7.89 (s, 1H, 4-C*H*), 6.83 (d, *J* = 1.4 Hz, 1H, H_{Ar}), 6.73 (d, *J* = 1.5 Hz, 1H, H_{Ar}), 6.54 (s, 1H, O*H*), 2.70 – 2.48 (m, 2H, CH_2), 2.04 – 1.93 (m, 2H, CH_2), 1.68 – 1.58 (m, 2H, CH_2), 1.57 – 1.27 (m, 12H, 6 × CH_2), 1.17 – 1.05 (m, 2H, CH_2), 0.94 (t, *J* = 7.3 Hz, 3H, CH_3), 0.92 – 0.83 (m, 2H, CH_2), 0.75 (t, *J* = 7.3 Hz, 3H, CH_3) ppm. – **^{13}C NMR** (100 MHz, $CDCl_3$): δ = 163.3 ($C_{quart.}$, *C*OO), 154.3 ($C_{quart.}$, C_{Ar}), 152.7 ($C_{quart.}$, C_{Ar}), 152.2 ($C_{quart.}$, C_{Ar}), 134.3 (+, 4-C_{Ar}H), 126.8 ($C_{quart.}$, C_{Ar}), 109.1 (+, C_{Ar}H), 107.5 (+, C_{Ar}H), 107.1 ($C_{quart.}$, C_{Ar}), 43.8 (–, CH_2), 42.0 ($C_{quart.}$, C_{CH}), 36.4 (–, 2 × CH_2), 30.7 (–, CH_2), 30.5 (–, CH_2), 26.6 (–, CH_2), 25.8 (–, CH_2), 23.4 (–, CH_2), 22.6 (–, CH_2), 22.5 (–, 2 × CH_2), 14.1 (+, CH_3), 14.0 (+, CH_3) ppm. – **IR** (KBr): ṽ = 3171 (w), 2925 (w), 2853 (w), 1670 (m), 1613 (m),1573 (w), 1451 (w), 1421 (m), 1345 (w), 1288 (w), 1253 (w), 1185 (w), 1124 (w), 1101 (w), 1066 (w), 939 (w), 862 (w), 842 (w), 782 (w), 745 (w), 728 (w), 608 (vw), 529 (w), 414 (vw) cm^{-1}. – **MS** (70 eV, EI): *m/z* (%) = 356 (71) $[M]^+$, 331 (8), 314 (12), 301 (8), 300 (46), 299 (100) $[M – C_4H_9]^+$, 281 (8), 262 (7), 260 (9). – **HRMS** ($C_{23}H_{32}O_3$): calc. 356.2346, found 356.2347. – **Elemental analysis**: $C_{23}H_{32}O_3$: calc. C 77.49, H 9.05, found C 77.27, H 9.09.

5.2.7 Synthesis and Characterization 3-Arylcoumarins (Chapter 0)

3-Bromo-7-(1-butylcyclopentyl)-5-methoxy-2*H*-chromen-2-on (105)

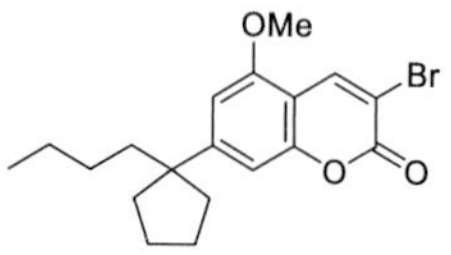

To a solution of 730 mg of Coumarin **(27bo)** (2.43 mmol, 1.00 equiv.), 900 mg Oxone (2.92 mmol, 1.20 equiv.) in 70 mL of dry dichloromethane, 2.92 ml 2 N HBr (5.83 mmol, 2.40 eq) were added dropwise. The mixture was stirred for 2 h at room temperature and 1.01 mL of triethylamine (7.29 mmol, 3.00 equiv.) were added and the mixture was stirred for another 15 min. Then 50 mL of water were added, the aqueous layer was extracted with 3 × 15 mL of ethyl acetate and the combined organic layers were dried over sodium sulfate and the volatiles were removed under reduced pressure. The crude product was then purified via flash column chromatography (CH/EtOAc 10:1) to give 602 mg (83%) of the product as a pale-yellow solid.

***R*f** (CH/EtOAc 10:1): 0.58–– **¹H NMR** (400 MHz, $CDCl_3$): δ = 8.36 (d, J = 0.6 Hz, 1H, 4-C*H*), 6.79 (d, J = 0.7 Hz, 1H, H_{Ar}), 6.63 (d, J = 1.5 Hz, 1H, H_{Ar}), 3.91 (s, 3H, OCH_3), 1.95 – 1.49 (m, 10H, 5 × CH_2), 1.20 – 1.08 (m, 2H, CH_2), 0.98 – 0.83 (m, 2H, CH_2), 0.75 (t, J = 7.3 Hz, 3H, CH_3) ppm. – **¹³C NMR** (100 MHz, $CDCl_3$): δ = 157.6 ($C_{quart.}$, *C*OO), 156.1 ($C_{quart.}$, C_{Ar}), 154.9 ($C_{quart.}$, C_{Ar}), 154.0 ($C_{quart.}$, C_{Ar}), 140.0 (+, 4-C_{Ar}H), 108.4 ($C_{quart.}$, C_{Ar}), 108.0 ($C_{quart.}$, C_{Ar}), 107.5 (+, C_{Ar}H), 104.6 (+, C_{Ar}H), 56.1 (+, OCH_3), 52.9 ($C_{quart.}$, C_{CP}), 41.5 (–, CH_2), 37.7 (–, 2 × CH_2), 27.5 (–, CH_2), 23.2 (–, CH_2), 23.2 (–, 2 × CH_2), 14.0 (+, CH_3) ppm. – **IR** (KBr): $\tilde{\nu}$ = 2949 (m), 2925 (m), 2866 (w), 1730 (m), 1604 (m), 1490 (w), 1455 (m), 1410 (m), 1345 (w), 1289 (w), 1233 (m), 1103 (m), 955 (m), 921 (w), 839 (m), 791 (w), 750 (w), 729 (m), 638 (w), 546 (w). – **MS** (70 eV, EI): *m/z* (%) = 380/378 (43/42) $[M]^+$, 323/321 (98/100) $[M - C_4H_9]^+$, 269/268 (35/37), 255 (8), 69 (7), 67 (12). – **HRMS** ($C_{19}H_{23}O_3{}^{79}Br_1$): calc. 378.0827, found 378.0825.

7-(1-Butylcyclopentyl)-5-methoxy-3-phenyl-2*H*-chromen-2-one (40c-Me)

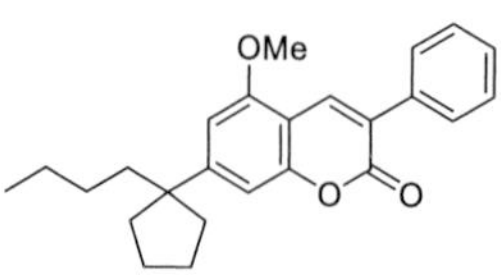

According to **GP9**, 100 mg of 3-bromocoumarin (**105**) (260 µmol, 1.00 equiv.), 64.3 mg of phenyl boronic acid (520 µmol, 2.00 equiv.), 172 mg of cesium carbonate (520 µmol, 2.00 equiv.) and 15.2 mg of tetrakis triphenylphosphine palladium (0) (10.0 µmol, 0.05 equiv.) were put in a sealed vial and 3 mL of dioxane was added. The mixture was degassed with three freeze pump thaw cycles, put under argon atmosphere and was then heated at 90 °C for 16 h. After cooling to room temperature, the reaction was quenched with 20 mL of water and extracted with 3 × 15 mL of ethyl acetate. The combined organic layers were dried over sodium sulfate and the volatiles were removed under reduced pressure. The crude product was then purified via flash column chromatography (CH/EtOAc 40:1) to give 82 mg (82%) of the product as colorless oil.

R_f (CH/EtOAc 10:1): 0.52. – **^{1}H NMR** (400 MHz, $CDCl_3$): δ = 8.17 (s, 1H, 4-C*H*), 7.76 – 7.68 (m, 2H, 2 × H_{Ar}), 7.48 – 7.40 (m, 2H, 2 × H_{Ar}), 7.40 – 7.33 (m, 1H, H_{Ar}), 6.89 (d, *J* = 1.4 Hz, 1H, H_{Ar}), 6.65 (d, *J* = 1.5 Hz, 1H, H_{Ar}), 3.95 (s, 3H, OCH_3), 1.98 – 1.57 (m, 10H, 5 × CH_2), 1.23 – 1.13 (m, 2H, CH_2), 1.03 – 0.90 (m, 2H, CH_2), 0.80 (t, *J* = 7.3 Hz, 3H, CH_3) ppm. – **^{13}C NMR** (100 MHz, $CDCl_3$): δ = 161.5 ($C_{quart.}$, *C*OO), 156.2 ($C_{quart.}$, C_{Ar}), 155.5 ($C_{quart.}$, C_{Ar}), 154.7 ($C_{quart.}$, C_{Ar}), 135.9 ($C_{quart.}$, C_{Ar}), 135.6 (+, 4-C_{Ar}H), 129.0 (+, 2 × C_{Ar}H), 128.8 (+, 3 × C_{Ar}H), 125.7 ($C_{quart.}$, C_{Ar}), 108.6 ($C_{quart.}$, C_{Ar}), 107.9 (+, C_{Ar}H), 104.5 (+, C_{Ar}H), 56.3 (+, OCH_3), 52.5 (–, CH_2), 42.0 ($C_{quart.}$, C_{CP}), 38.2 (–, 2 × CH_2), 28.0 (–, CH_2), 23.7 (–, CH_2), 23.7 (–, 2 × CH_2), 14.5 (+, CH_3) ppm. – **IR** (KBr): ṽ = 2927 (w), 2868 (w), 1760 (w), 1721 (m), 1611 (m), 1563 (w), 1487 (w), 1459 (w), 1415 (w), 1350 (w), 1280 (w),1232 (w), 1212 (w), 1101 (m), 952 (w), 841 (w), 785 (w), 755 (w), 734 (vw), 693 (w), 641 (vw), 591 (w), 557 (vw), 515 (w) cm^{-1}. – **MS** (70 eV, EI): *m/z* (%) = 376 (87) $[M]^+$, 356 (9), 354 (10), 321 (7), 320 (40), 319 (95) $[M - C_4H_9]^+$, 299 (7), 297 (9), 296 (8), 278 (8), 276 (19), 265 (33), 253 (15), 221 (12), 220 (72), 219 (72), 194 (14), 181 (9), 180 (31), 179 (16), 165 (13), 154 (37), 153 (15), 152 (10), 137 (12), 91 (14), 88 (12), 86 (63), 84 (100). – **HRMS** ($C_{25}H_{28}O_3$): calc. 376.2033, found 376.2032.

7-(1-Butylcyclopentyl)-5-methoxy-3-(*o*-tolyl)-2*H*-chromen-2-one (40d-Me)

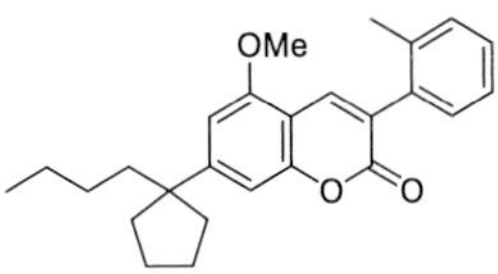

According to **GP9**, 100 mg of 3-bromocoumarin (**105**) (260 µmol, 1.00 equiv.), 71.7 mg of *o*-methyl phenyl boronic acid (520 µmol, 2.00 equiv.), 172 mg of cesium carbonate (520 µmol, 2.00 equiv.) and 15.2 mg of tetrakis triphenylphosphine palladium (0) (10 µmol, 0.05 equiv.) were put in a sealed vial and 3 mL of dioxane was added. The mixture was degassed with three freeze pump thaw cycles, put under argon atmosphere and was then heated at 90 °C for 16 h. After cooling to room temperature, the reaction was quenched with 20 mL of water and extracted with 3 × 15 mL of ethyl acetate.

The combined organic layers were dried over sodium sulfate and the volatiles were removed under reduced pressure. The crude product was then purified via flash column chromatography (CH/EtOAc 40:1) to give 88 mg (90%) of the product as a yellow solid.

***R*f** (CH/EtOAc 5:1): 0.70. – **MP**: 156.3 °C – **^{1}H NMR** (400 MHz, $CDCl_3$): δ = 8.00 (s, 1H, 4-C*H*), 7.34 – 7.21 (m, 4H, 4 × H_{Ar}), 6.90 (dd, J = 1.4, 0.7 Hz, 1H, H_{Ar}), 6.66 (d, J = 1.5 Hz, 1H, H_{Ar}), 3.92 (s, 3H, OCH_3), 2.30 (s, 3H, CH_3), 1.99 – 1.53 (m, 10H, 5 × CH_2), 1.19 (h, J = 7.4 Hz, 2H, CH_2), 1.06 – 0.92 (m, 2H, CH_2), 0.81 (t, J = 7.3 Hz, 3H, CH_3) ppm. – **^{13}C NMR** (100 MHz, $CDCl_3$): δ = 160.8 ($C_{quart.}$, *C*OO), 155.8 ($C_{quart.}$, C_{Ar}), 155.2 ($C_{quart.}$, C_{Ar}), 154.7 ($C_{quart.}$, C_{Ar}), 137.1 ($C_{quart.}$, C_{Ar}), 137.1 (+, 4-C_{Ar}H), 135.5 ($C_{quart.}$, C_{Ar}), 130.4 (+, C_{Ar}H), 130.1 (+, C_{Ar}H), 128.7 (+, C_{Ar}H), 126.8 ($C_{quart.}$, C_{Ar}), 125.9 (+, C_{Ar}H), 107.9 ($C_{quart.}$, C_{Ar}), 107.7 (+, C_{Ar}H), 104.2 (+, C_{Ar}H), 56.0 (+, OCH_3), 52.2 ($C_{quart.}$, C_{CP}), 41.7 (–, CH_2), 37.9 (–, 2 × CH_2), 27.6 (–, CH_2), 23.4 (–, CH_2), 23.4 (–, 2 × CH_2), 20.2 (+, CH_3), 14.2 (+, CH_3) ppm. – **IR** (KBr): $\tilde{\nu}$ = 2922 (w), 2856 (w), 1714 (m), 1612 (m), 1571 (w), 1494 (w), 1457 (w), 1416 (w),1379 (w), 1348 (w), 1287 (w), 1269 (w), 1214 (w), 1111 (m), 953 (w), 939 (w), 850 (w), 788 (w), 757 (w), 724 (w), 680 (w), 603 (w), 564 (w), 492 (vw), 449 (w), 388 (vw) cm^{-1}. – **MS** (70 eV, EI): *m/z* (%) = 390 (100) $[M]^+$, 335 (7), 334 (35), 333 (82) $[M - C_4H_9]^+$, 279 (16), 267 (10), 181 (15), 131 (13), 69 (17). – **HRMS** ($C_{26}H_{30}O_3$): calc. 390.2189, found 390.2190. – **Elemental analysis**: $C_{26}H_{30}O_3$: calc. C 79.97, H 7.74, found C 79.81, H 7.98.

7-(1-Butylcyclopentyl)-5-methoxy-3-(*m*-tolyl)-2*H*-chromen-2-one (40e-Me)

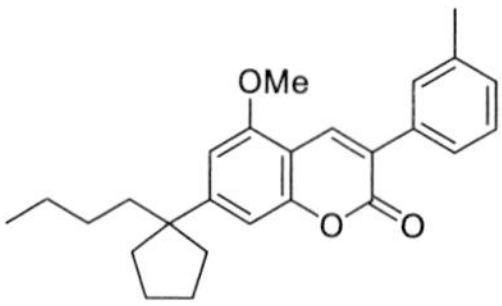

According to **GP9**, 100 mg of 3-bromocoumarin (**105**) (260 µmol, 1.00 equiv.), 71.7 mg of *m*-methyl phenyl boronic acid (520 µmol, 2.00 equiv.), 172 mg of cesium carbonate (520 µmol, 2.00 equiv.) and 15.2 mg of tetrakis triphenylphosphine palladium (0) (10.0 µmol, 0.05 equiv.) were put in a sealed vial and 3 mL of dioxane was added. The mixture was degassed with three freeze pump thaw cycles, put under argon atmosphere and was then heated at 90 °C for 16 h. After cooling to room temperature, the reaction was quenched with 20 mL of water and extracted with 3 × 15 mL of ethyl acetate. The combined organic layers were dried over sodium sulfate and the volatiles were removed under reduced pressure. The crude product was then purified via flash column chromatography (CH/EtOAc 40:1) to give 66 mg (67%) of the product as a yellow oil.

***R*f** (CH/EtOAc 5:1): 0.07. – **^{1}H NMR** (400 MHz, $CDCl_3$): δ = 8.15 (s, 1H, 4-C*H*), 7.56 – 7.47 (m, 2H, 2 × H_{Ar}), 7.32 (t, *J* = 7.6 Hz, 1H, H_{Ar}), 7.19 (d, *J* = 7.6 Hz, 1H, H_{Ar}), 6.88 (d, *J* = 1.4 Hz, 1H, H_{Ar}), 6.65 (d, *J* = 1.4 Hz, 1H, H_{Ar}), 3.95 (s, 3H, OCH_3), 2.42 (s, 3H, CH_3), 1.97 – 1.55 (m, 10H, 5 × CH_2), 1.23 – 1.12 (m, 2H, CH_2), 1.03 – 0.91 (m, 2H, CH_2), 0.80 (t, *J* = 7.3 Hz, 3H, CH_3) ppm. – **^{13}C NMR** (100 MHz, $CDCl_3$): δ = 161.2 ($C_{quart.}$, *C*OO), 155.8 ($C_{quart.}$, C_{Ar}), 155.0 ($C_{quart.}$, C_{Ar}), 154.3 ($C_{quart.}$, C_{Ar}), 138.1 (+, 4-C_{Ar}H), 135.5 ($C_{quart.}$, C_{Ar}), 135.2 (+, C_{Ar}H), 129.3 ($C_{quart.}$, C_{Ar}), 129.3 (+, C_{Ar}H), 128.4 (+, C_{Ar}H), 125.8 (+, C_{Ar}H), 125.5 ($C_{quart.}$, C_{Ar}), 108.3 ($C_{quart.}$, C_{Ar}), 107.6 (+, C_{Ar}H), 104.2 (+, C_{Ar}H), 56.0 (+, OCH_3), 52.2 ($C_{quart.}$, C_{CP}), 41.7 (–, CH_2), 37.8 (–, 2 × CH_2), 27.6 (–, CH_2), 23.4 (–, CH_2), 23.3 (–, 2 × CH_2), 21.7 (+, CH_3), 14.1 (+, CH_3) ppm. – **IR** (KBr): ṽ = 2926 (m), 2868 (w), 1721 (w), 1608 (s), 1563 (s), 1492 (w), 1459 (m), 1414 (m), 1347 (w), 1285 (w), 1230 (m), 1187 (w), 1104 (s), 999 (w), 966 (m), 907 (w), 881 (w), 841 (m), 793 (m), 730 (m), 698 (m), 644 (w), 597 (w), 569 (w), 544 (vw), 448 (w) cm^{-1}. – **MS** (70 eV, EI): *m/z* (%) = 390 (90) $[M]^+$, 365 (8), 364 (16), 351 (10), 334 (25), 333 (82) $[M – C_4H_9]^+$, 310 (17), 308 (8), 307 (14), 288 (19), 279 (27), 277 (10), 276 (52), 261 (21), 221 (26), 220 (58), 219 (13), 181 (28), 180 (15), 164 (65) 86 (64), 84 (100). – **HRMS** ($C_{26}H_{30}O_3$): calc. 390.2189, found 390.2191. – **Elemental analysis**: $C_{26}H_{30}O_3$: calc. C 79.97, H 7.74, found C 79.63, H 7.68.

7-(1-Butylcyclopentyl)-5-methoxy-3-(*p*-tolyl)-2*H*-chromen-2-one (40f-Me)

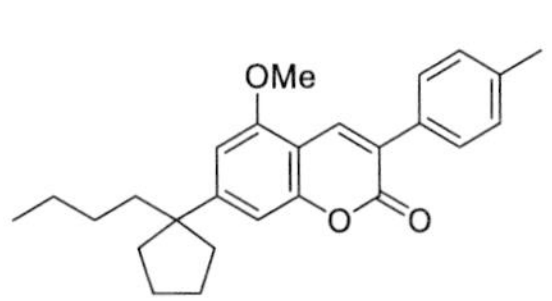

According to **GP9**, 100 mg of 3-bromocoumarin (**105**) (260 µmol, 1.00 equiv.), 71.7 mg of *p*-methyl phenyl boronic acid (520 µmol, 2.00 equiv.), 172 mg of cesium carbonate (520 µmol, 2.00 equiv.) and 15.2 mg of tetrakis triphenylphosphine palladium (0) (10.0 µmol, 0.05 equiv.) were put in a sealed vial and 3 mL of dioxane was added. The mixture was degassed with three freeze pump thaw cycles, put under argon atmosphere and was then heated at 90 °C for 16 h. After cooling to room temperature, the reaction was quenched with 20 mL of water and extracted with 3 × 15 mL of ethyl acetate. The combined organic layers were dried over sodium sulfate and the volatiles were removed under reduced pressure. The crude product was then purified via flash column chromatography (CH/EtOAc 40:1) to give 72 mg (73%) of the product as a yellow solid.

***R*f** (CH/EtOAc 5:1): 0.68. – **^{1}H NMR** (400 MHz, $CDCl_3$): δ = 8.14 (s, 1H, 4-C*H*), 7.62 (d, *J* = 8.2 Hz, 2H, 2 × H_{Ar}), 7.24 (d, *J* = 8.0 Hz, 2H, 2 × H_{Ar}), 6.88 (d, *J* = 1.4 Hz, 1H, H_{Ar}), 6.64

(d, $J = 1.5$ Hz, 1H, H_{Ar}), 3.94 (s, 3H, OCH_3), 2.39 (s, 3H, CH_3), 1.98 – 1.50 (m, 10H, 5 × CH_2), 1.22 – 1.11 (m, 2H, CH_2), 1.03 – 0.89 (m, 2H, CH_2), 0.80 (t, $J = 7.3$ Hz, 3H, CH_3) ppm. – **^{13}C NMR** (100 MHz, $CDCl_3$): δ = 161.3 ($C_{quart.}$, COO), 155.8 ($C_{quart.}$, C_{Ar}), 154.9 ($C_{quart.}$, C_{Ar}), 154.3 ($C_{quart.}$, C_{Ar}), 138.5 ($C_{quart.}$, C_{Ar}), 134.6 (+, 4-C_{Ar}H), 132.6 ($C_{quart.}$, C_{Ar}), 129.2 (+, 2 × C_{Ar}H), 128.5 (+, 2 × C_{Ar}H), 125.3 ($C_{quart.}$, C_{Ar}), 108.4 ($C_{quart.}$, C_{Ar}), 107.6 (+, C_{Ar}H), 104.1 (+, C_{Ar}H), 56.0 (+, OCH_3), 52.2 ($C_{quart.}$, C_{CP}), 41.7 (–, CH_2), 37.9 (–, 2 × CH_2), 27.6 (–, CH_2), 23.4 (–, CH_2), 23.3 (–, 2 × CH_2), 21.4 (+, CH_3), 14.1 (+, CH_3) ppm. – **IR** (KBr): $\tilde{\nu}$ = 2928 (w), 2867 (w), 1759 (vw), 1723 (m), 1608 (w), 1512 (w), 1490 (w), 1455 (w), 1417 (w), 1352 (w), 1282 (w), 1103 (m), 953 (w), 840 (w), 817 (w), 782 (w), 710 (w), 657 (w), 560 (w), 521 (w) cm^{-1}. – **MS** (70 eV, EI): *m/z* (%) = 390 (100) $[M]^+$, 334 (28), 333 (79) $[M - C_4H_9]^+$, 279 (29), 267 (12), 220 (24), 181 (17), 149 (11), 131 (14), 105 (11), 86 (26), 84 (40), 69 (26). – **HRMS** ($C_{26}H_{30}O_3$): calc. 390.2189, found 390.2188.

7-(1-Butylcyclopentyl)-5-methoxy-3-(2-methoxyphenyl)-2*H*-chromen-2-one (40g-Me)

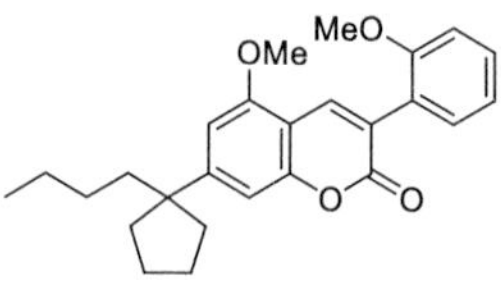

According to **GP9**, 100 mg of 3-bromocoumarin (**105**) (260 µmol, 1.00 equiv.), 80.1 mg of *o*-methoxy phenyl boronic acid (520 µmol, 2.00 equiv.), 172 mg of cesium carbonate (520 µmol, 2.00 equiv.) and 15.2 mg of tetrakis triphenylphosphine palladium (0) (10 µmol, 0.05 equiv.) were put in a sealed vial and 3 mL of dioxane was added. The mixture was degassed with three freeze pump thaw cycles, put under argon atmosphere and was then heated at 90 °C for 16 h. After cooling to room temperature, the reaction was quenched with 20 mL of water and extracted with 3 × 15 mL of ethyl acetate. The combined organic layers were dried over sodium sulfate and the volatiles were removed under reduced pressure. The crude product was then purified via flash column chromatography (CH/EtOAc 40:1) to give 64 mg (63%) of the product as an off-white solid.

***R*f** (CH/EtOAc 5:1): 0.50. – **MP**: 148.9 °C – **^{1}H NMR** (400 MHz, $CDCl_3$): δ = 8.06 (s, 1H, 4-C*H*), 7.39 – 7.32 (m, 2H, 2 × H_{Ar}), 7.05 – 6.95 (m, 2H, 2 × H_{Ar}), 6.89 (d, $J = 1.5$ Hz, 1H, H_{Ar}), 6.64 (d, $J = 1.4$ Hz, 1H, H_{Ar}), 3.92 (s, 3H, OCH_3), 3.83 (s, 3H, OCH_3), 1.99 – 1.53 (m, 10H, 5 × CH_2), 1.24 – 1.13 (m, 2H, CH_2), 1.04 – 0.89 (m, 2H, CH_2), 0.81 (t, $J = 7.3$ Hz, 3H, CH_3) ppm. – **^{13}C NMR** (100 MHz, $CDCl_3$): δ = 160.8 ($C_{quart.}$, COO), 157.5 ($C_{quart.}$, C_{Ar}), 155.8 ($C_{quart.}$, C_{Ar}), 154.8 ($C_{quart.}$, C_{Ar}), 154.6 ($C_{quart.}$, C_{Ar}), 137.0 (+, 4-C_{Ar}H), 131.0 (+, C_{Ar}H), 130.0 (+, C_{Ar}H), 125.1 ($C_{quart.}$, C_{Ar}), 123.8 ($C_{quart.}$, C_{Ar}), 120.7 (+, C_{Ar}H), 111.4 (+, C_{Ar}H), 108.1

($C_{quart.}$, C_{Ar}), 107.7 (+, $C_{Ar}H$), 104.1 (+, $C_{Ar}H$), 56.0 (+, OCH_3), 55.9 (+, OCH_3), 52.2 ($C_{quart.}$, C_{CP}), 41.7 (–, CH_2), 37.9 (–, 2 × CH_2), 27.7 (–, CH_2), 23.4 (–, CH_2), 23.3 (–,2 × CH_2), 14.1 (+, CH_3) ppm. – **IR** (KBr): $\tilde{\nu}$ = 2921 (w), 2855 (w), 1721 (m),1613 (m), 1566 (m), 1495 (m), 1460 (m), 1418 (m), 1348 (w), 1291 (w), 1267 (m), 1246 (m), 1207 (m), 1177 (w), 1107 (m), 1045 (w), 1022 (m), 950 (m), 936 (m), 848 (m), 784 (w), 758 (m), 674 (w), 637 (w), 599 (w), 566 (w), 545 (w), 525 (w), 503 (w) cm^{-1}. – **MS** (70 eV, EI): *m/z* (%) = 406 (100) $[M]^+$, 350 (25), 349 (67) $[M - C_4H_9]^+$, 295 (12), 69 (11). – **HRMS** ($C_{26}H_{30}O_4$): calc. 406.2139, found 406.2137.

7-(1-Butylcyclopentyl)-5-methoxy-3-(3-methoxyphenyl)-2*H*-chromen-2-one (40h-Me)

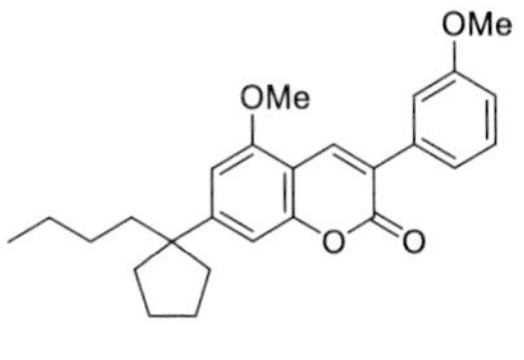

According to **GP9**, 100 mg of 3-bromocoumarin (**105**) (260 µmol, 1.00 equiv.), 80.1 mg of *m*-methoxy phenyl boronic acid (520 µmol, 2.00 equiv.), 172 mg of cesium carbonate (520 µmol, 2.00 equiv.) and 15.2 mg of tetrakis triphenylphosphine palladium (0) (10.0 µmol, 0.05 equiv.) were put in a sealed vial and 3 mL of dioxane was added. The mixture was degassed with three freeze pump thaw cycles, put under argon atmosphere and was then heated at 90 °C for 16 h. After cooling to room temperature, the reaction was quenched with 20 mL of water and extracted with 3 × 15 mL of ethyl acetate. The combined organic layers were dried over sodium sulfate and the volatiles were removed under reduced pressure. The crude product was then purified via flash column chromatography (CH/EtOAc 40:1) to give 89.0 mg (83%) of the product as a yellow oil.

***R*f** (CH/EtOAc 20:1): 0.45. – **^{1}H NMR** (400 MHz, $CDCl_3$): δ = 8.18 (s, 1H, 4-C*H*), 7.38 – 7.27 (m, 3H, 3 × H_{Ar}), 6.92 (ddd, *J* = 8.0, 2.6, 1.4 Hz, 1H, H_{Ar}), 6.88 (d, *J* = 1.3 Hz, 1H, H_{Ar}), 6.65 (d, *J* = 1.4 Hz, 1H, H_{Ar}), 3.94 (s, 3H, OCH_3), 3.85 (s, 3H, OCH_3), 1.98 – 1.80 (m, 4H, 2 × CH_2), 1.80 – 1.55 (m, 6H, 3 × CH_2), 1.23 – 1.12 (m, 2H, CH_2), 1.03 – 0.89 (m, 2H, CH_2), 0.80 (t, *J* = 7.3 Hz, 3H, CH_3) ppm. – **^{13}C NMR** (100 MHz, $CDCl_3$): δ = 161.0 ($C_{quart.}$, *C*OO), 159.6 ($C_{quart.}$, C_{Ar}), 155.8 ($C_{quart.}$, C_{Ar}), 155.2 ($C_{quart.}$, C_{Ar}), 154.3 ($C_{quart.}$, C_{Ar}), 136.8 ($C_{quart.}$, C_{Ar}), 135.4 (+, $C_{Ar}H$), 129.4 (+, $C_{Ar}H$), 125.0 ($C_{quart.}$, C_{Ar}), 121.1 (+, $C_{Ar}H$), 114.3 (+, $C_{Ar}H$), 114.1 (+, $C_{Ar}H$), 108.2 ($C_{quart.}$, C_{Ar}), 107.5 (+, $C_{Ar}H$), 104.2 (+, $C_{Ar}H$), 56.0 (+, OCH_3), 55.5 (+, OCH_3), 52.2 ($C_{quart.}$, C_{CP}), 41.7 (–, CH_2), 37.8 (–, 2 × CH_2), 27.6 (–, CH_2), 23.4 (–, CH_2), 23.3 (–, 2 × CH_2), 14.1 (+, CH_3) ppm. – **IR** (KBr): $\tilde{\nu}$ = 2928 (m), 2869 (w), 1720 (m), 1607 (s), 1490 (m), 1460 (m), 1415 (m), 1347 (w), 1286 (m), 1233 (m), 1201 (m), 1173 (m), 1044 (m), 972 (m), 908 (w), 842 (m), 787 (m), 730 (m), 694 (m), 647 (w), 593 (w), 466 (vw) cm^{-1}. – **MS** (70 eV, EI): *m/z* (%) = 406 (14) $[M]^+$, 349 (11) $[M - C_4H_9]^+$, 300 (29), 244 (52), 243 (75), 204 (16), 189 (43), 181 (9), 177 (13), 149 (18), 88 (10), 86 (65), 84 (100), 69 (19), 67 (15). – **HRMS** ($C_{26}H_{30}O_4$): calc. 406.2139, found 406.2137.

7-(1-Butylcyclopentyl)-5-methoxy-3-(4-methoxyphenyl)-2*H*-chromen-2-one (40b-Me)

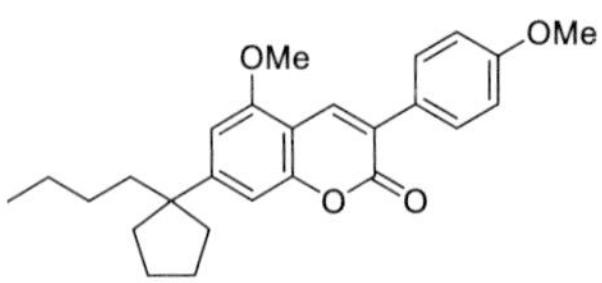

According to **GP9**, 100 mg of 3-bromocoumarin (**105**) (260 µmol, 1.00 equiv.), 80.1 mg of *p*-methoxy phenyl boronic acid (520 µmol, 2.00 equiv.), 172 mg of cesium carbonate (520 µmol, 2.00 equiv.) and 15.2 mg of tetrakis triphenylphosphine palladium (0) (10.0 µmol, 0.05 equiv.) were put in a sealed vial and 3 mL of dioxane was added. The mixture was degassed with three freeze pump thaw cycles, put under argon atmosphere and was then heated at 90 °C for 16 h. After cooling to room temperature, the reaction was quenched with 20 mL of water and extracted with 3 × 15 mL of ethyl acetate. The combined organic layers were dried over sodium sulfate and the volatiles were removed under reduced pressure. The crude product was then purified via flash column chromatography (CH/EtOAc 40:1) to give 79.7 mg (79%) of the product as a yellow oil.

R_f (CH/EtOAc 20:1): 0.45. – **MP**: 105.3 °C – **^{1}H NMR** (400 MHz, $CDCl_3$): δ = 8.11 (s, 1H, 4-C*H*), 7.77 – 7.63 (m, 2H, 2 × H_{Ar}), 7.01 – 6.91 (m, 2H, 2 × H_{Ar}), 6.89 – 6.86 (m, 1H, H_{Ar}), 6.64 (d, *J* = 1.6 Hz, 1H, H_{Ar}), 3.94 (s, 3H, OCH_3), 3.84 (s, 3H, OCH_3), 1.99 – 1.80 (m, 4H, 2 × CH_2), 1.79 – 1.55 (m, 6H, 3 × CH_2), 1.22 – 1.08 (m, 2H, CH_2), 1.04 – 0.90 (m, 2H, CH_2), 0.79 (t, *J* = 7.3 Hz, 3H, CH_3) ppm. – **^{13}C NMR** (100 MHz, $CDCl_3$): δ = 161.4 ($C_{quart.}$, *C*OO), 159.9 ($C_{quart.}$, C_{Ar}), 155.7 ($C_{quart.}$, C_{Ar}), 154.7 ($C_{quart.}$, C_{Ar}), 154.1 ($C_{quart.}$, C_{Ar}), 133.9 (+, 4-C_{Ar}H), 129.9 (+, 2 × C_{Ar}H), 127.9 ($C_{quart.}$, C_{Ar}), 124.9 ($C_{quart.}$, C_{Ar}), 113.9 (+, 2 × C_{Ar}H), 108.4 ($C_{quart.}$, C_{Ar}), 107.5 (+, C_{Ar}H), 104.1 (+, C_{Ar}H), 56.0 (+, OCH_3), 55.5 (+, OCH_3), 52.1 ($C_{quart.}$, C_{CP}), 41.7 (–, CH_2), 37.8 (–, 2 × CH_2), 27.6 (–, CH_2), 23.4 (–, CH_2), 23.3 (–, 2 × CH_2), 14.1 (+, CH_3) ppm. – **IR** (KBr): $\tilde{\nu}$ = 2922 (m), 2856 (w), 1713 (s), 1613 (s), 1512 (m), 1489 (m), 1459 (m), 1416 (m), 1353 (w), 1284 (m), 1248 (m), 1213 (m), 1178 (m), 1107 (s), 1030 (m), 950 (m), 831 (s), 800 (m), 778 (w), 731 (w), 664 (w), 629 (w), 568 (m), 530 (m), 460 (w) cm^{-1}. – **MS** (70 eV, EI): *m/z* (%) = 406 (28) $[M]^+$, 349 (17) $[M - C_4H_9]^+$, 300 (38), 245 (69), 243 (100), 218 (12), 204 (20), 203 (10), 190 (10), 189 (64), 181 (15), 177 (18), 167 (10), 149 (23), 131 (16), 97 (11), 83 (12). – **HRMS** ($C_{26}H_{30}O_4$): calc. 406.2139, found 406.2137.

7-(1-Butylcyclopentyl)-3-(3-(dimethylamino)phenyl)-5-methoxy-2*H*-chromen-2-one (40i-Me)

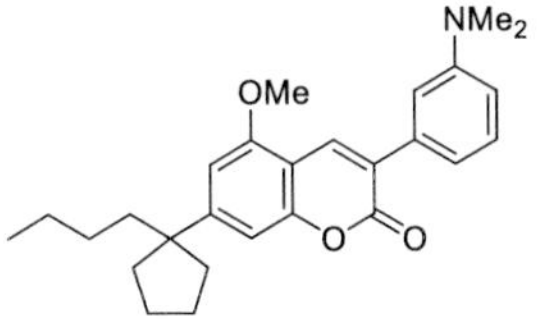

According to **GP9**, 100 mg of 3-bromocoumarin (**105**) (260 μmol, 1.00 equiv.), 87.0 mg of *m*-dimethylamino phenyl boronic acid (520 μmol, 2.00 equiv.), 172 mg of cesium carbonate (520 μmol, 2.00 equiv.) and 15.2 mg of tetrakis triphenylphosphine palladium (0) (10 μmol, 0.05 equiv.) were put in a sealed vial and 3 mL of dioxane was added. The mixture was degassed with three freeze pump thaw cycles, put under argon atmosphere and was then heated at 90 °C for 16 h. After cooling to room temperature, the reaction was quenched with 20 mL of water and extracted with 3 × 15 mL of ethyl acetate. The combined organic layers were dried over sodium sulfate and the volatiles were removed under reduced pressure. The crude product was then purified via flash column chromatography (CH/EtOAc 20:1) to give 69.4 mg (63%) of the product as a yellow solid.

***R*$_f$** (CH/EtOAc 20:1): 0.10. – **^{1}H NMR** (400 MHz, $CDCl_3$): δ = 8.19 – 8.13 (m, 1H, 4-C*H*), 7.29 (t, *J* = 8.0 Hz, 1H, H_{Ar}), 7.09 (dd, *J* = 2.6, 1.6 Hz, 1H, H_{Ar}), 7.02 (dt, *J* = 7.7, 1.1 Hz, 1H, H_{Ar}), 6.88 (d, *J* = 1.8 Hz, 1H, H_{Ar}), 6.81 – 6.74 (m, 1H, H_{Ar}), 6.64 (d, *J* = 1.5 Hz, 1H, H_{Ar}), 3.94 (s, 3H, OCH_3), 2.99 (s, 6H, $N(CH_3)_2$), 1.98 – 1.81 (m, 4H, 2 × CH_2), 1.80 – 1.57 (m, 6H, 3 × CH_2), 1.24 – 1.11 (m, 2H, CH_2), 1.03 – 0.90 (m, 2H, CH_2), 0.80 (t, *J* = 7.3 Hz, 3H, CH_3) ppm. – **^{13}C NMR** (100 MHz, $CDCl_3$): δ = 161.2 ($C_{quart.}$, *C*OO), 155.8 ($C_{quart.}$, C_{Ar}), 154.8 ($C_{quart.}$, C_{Ar}), 154.3 ($C_{quart.}$, C_{Ar}), 150.7 ($C_{quart.}$, C_{Ar}), 136.3 ($C_{quart.}$, C_{Ar}), 135.0 (+, 4-C_{Ar}H), 129.1 (+, C_{Ar}H), 126.2 ($C_{quart.}$, C_{Ar}), 117.1 (+, C_{Ar}H), 113.1 (+, C_{Ar}H), 112.9 (+, C_{Ar}H), 108.3 ($C_{quart.}$, C_{Ar}), 107.5 (+, C_{Ar}H), 104.1 (+, C_{Ar}H), 56.0 (+, OCH_3), 52.2 ($C_{quart.}$, C_{CP}), 41.7 (–, CH_2), 40.9 (+, $N(CH_3)_2$), 37.8 (–, 2 × CH_2), 27.6 (–, CH_2), 23.4 (–, CH_2), 23.3 (–, 2 × CH_2), 14.1 (+, CH_3) ppm. – **IR** (KBr): ṽ = 2927 (m), 2869 (w), 1721 (s), 1597 (m), 1493 (m), 1459 (m), 1415 (m), 1350 (m), 1274 (w), 1213 (m), 1180 (w), 1162 (w), 1111 (m), 1003 (w), 983 (w), 942 (m), 908 (w), 843 (m), 785 (m), 730 (m),694 (m), 646 (w), 594 (w), 553 (w), 462 (vw) cm^{-1}. – **MS** (70 eV, EI): *m/z* (%) = 419 (100) $[M]^+$, 362 (8) $[M – C_4H_9]^+$, 314 (13), 271 (17), 258 (14), 257 (20), 241 (9), 240 (59), 239 (21), 189 (14), 181 (14), 131 (10), 86 (18), 84 (27). – **HRMS** ($C_{27}H_{33}O_3N$): calc. 419.2455, found 419.2456.

7-(1-Butylcyclopentyl)-3-(4-(dimethylamino)phenyl)-5-methoxy-2*H*-chromen-2-one (40j--Me)

OMe NMe$_2$ O O

According to **GP9**, 100 mg of 3-bromocoumarin (**105**) (260 µmol, 1.00 equiv.), 87.0 mg of *p*-dimethylamino phenyl boronic acid (520 µmol, 2.00 equiv.), 172 mg of cesium carbonate (520 µmol, 2.00 equiv.) and 15.2 mg of tetrakis triphenylphosphine palladium (0) (10.0 µmol, 0.05 equiv.) were put in a sealed vial and 3 mL of dioxane was added. The mixture was degassed with three freeze pump thaw cycles, put under argon atmosphere and was then heated at 90 °C for 16 h. After cooling to room temperature, the reaction was quenched with 20 mL of water and extracted with 3 × 15 mL of ethyl acetate. The combined organic layers were dried over sodium sulfate and the volatiles were removed under reduced pressure. The crude product was then purified via flash column chromatography (CH/EtOAc 40:1 to 10:1) to give 100 mg (90%) of the product as a yellow solid.

***R*$_f$** (CH/EtOAc 10:1): 0.16. – **^{1}H NMR** (400 MHz, CDCl$_3$): δ = 8.08 (s, 1H, 4-C*H*), 7.73 – 7.64 (m, 2H, 2 × *H*$_{Ar}$), 6.86 (d, *J* = 1.4 Hz, 1H, *H*$_{Ar}$), 6.81 – 6.73 (m, 2H, 2 × *H*$_{Ar}$), 6.63 (d, *J* = 1.5 Hz, 1H, *H*$_{Ar}$), 3.94 (s, 3H, OC*H*$_3$), 3.00 (s, 6H, N(C*H*$_3$)$_2$), 1.97 – 1.80 (m, 2H, C*H*$_2$), 1.80 – 1.56 (m, 6H, 3 × C*H*$_2$), 1.23 – 1.12 (m, 2H, 2 × C*H*$_2$), 1.04 – 0.90 (m, 2H, C*H*$_2$), 0.80 (t, *J* = 7.3 Hz, 3H, C*H*$_3$) ppm. – **^{13}C NMR** (100 MHz, CDCl$_3$): δ = 161.6 (C$_{quart.}$, *C*OO), 155.5 (C$_{quart.}$, *C*$_{Ar}$), 153.9 (C$_{quart.}$, *C*$_{Ar}$), 153.9 (C$_{quart.}$, *C*$_{Ar}$), 150.6 (C$_{quart.}$, *C*$_{Ar}$), 132.2 (+, *C*$_{Ar}$H), 129.4 (+, 2 × *C*$_{Ar}$H), 125.3 (C$_{quart.}$, *C*$_{Ar}$), 123.3 (C$_{quart.}$, *C*$_{Ar}$), 112.2 (+, 2 × *C*$_{Ar}$H), 108.7 (C$_{quart.}$, *C*$_{Ar}$), 107.5 (+, *C*$_{Ar}$H), 104.0 (+, *C*$_{Ar}$H), 56.0 (+, O*C*H$_3$), 52.1 (C$_{quart.}$, *C*$_{CP}$), 41.7 (–, *C*H$_2$), 40.6 (+, N(*C*H$_3$)$_2$), 37.9 (–, 2 × *C*H$_2$), 27.6 (–, *C*H$_2$), 23.4 (–, *C*H$_2$), 23.3 (–, 2 × *C*H$_2$), 14.1 (+, *C*H$_3$) ppm. – **IR** (KBr): ṽ = 2945 (w), 2857 (w), 1713 (m), 1609 (m), 1522 (w), 1485 (w), 1447 (w), 1416 (w), 1357 (w), 1283 (w), 1229 (w), 1198 (w), 1101 (m), 1064 (w), 1006 (w), 944 (w), 920 (w), 840 (w), 816 (m), 777 (w), 742 (w), 682 (w), 628 (vw), 561 (w), 527 (w), 419 (vw) cm^{-1}. – **MS** (70 eV, EI): *m/z* (%) = 419 (100) [M]$^+$, 362 (7) [M – C$_4$H$_9$]$^+$, 321 (7), 241 (9), 240 (50), 225 (11), 224 (6), 181 (8), 239 (21), 131 (6), 86 (10), 84 (17). – **HRMS** (C$_{27}$H$_{33}$O$_3$N): calc. 419.2455, found 419.2453.

7-(1-Butylcyclopentyl)-5-methoxy-3-(pyridin-4-yl)-2*H*-chromen-2-one (40k-Me)

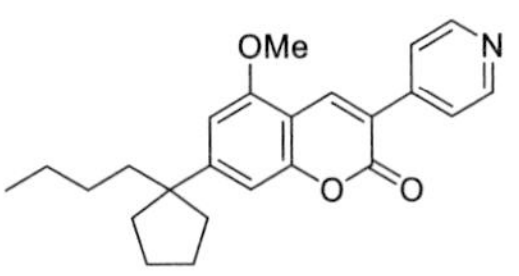

According to **GP9**, 100 mg of 3-bromocoumarin (**105**) (260 µmol, 1.00 equiv.), 64.8 mg of pyridin-4-ylboronic acid (520 µmol, 2.00 equiv.), 172 mg of cesium carbonate (520 µmol, 2.00 equiv.) and 15.2 mg of tetrakis triphenylphosphine palladium (0) (10.0 µmol, 0.05 equiv.) were put in a sealed vial and 3 mL of dioxane was added. The mixture was degassed with three freeze pump thaw cycles, put under argon atmosphere and was then heated at 90 °C for 16 h. After cooling to room temperature, the reaction was quenched with 20 mL of water and extracted with 3 × 15 mL of ethyl acetate. The combined organic layers were dried over sodium sulfate and the volatiles were removed under reduced pressure. The crude product was then purified via flash column chromatography (CH/EtOAc 10:1 to 2:1) to give 22.8 mg (23%) of the product as a as yellow crystalline solid.

***R*$_f$** (CH/EtOAc 2:1): 0.18. – **^{1}H NMR** (400 MHz, $CDCl_3$): δ = 8.67 (d, *J* = 5.2 Hz, 2H, 2 × H_{Ar}), 8.35 (s, 1H, 4-C*H*), 7.77 – 7.70 (m, 2H, 2 × H_{Ar}), 6.89 (d, *J* = 1.4 Hz, 1H, H_{Ar}), 6.67 (d, *J* = 1.4 Hz, 1H, H_{Ar}), 3.97 (s, 3H, OCH_3), 1.98 – 1.81 (m, 4H, 2 × CH_2)1.81 – 1.57 (m, 6H, 3 × CH_2), 1.23 – 1.11 (m, 2H, CH_2), 1.02 – 0.90 (m, 2H, CH_2), 0.80 (t, *J* = 7.3 Hz, 3H, CH_3) ppm. – **^{13}C NMR** (100 MHz, $CDCl_3$): δ = 160.2 ($C_{quart.}$, *C*OO), 157.0 ($C_{quart.}$, C_{Ar}), 156.2 ($C_{quart.}$, C_{Ar}), 154.8 ($C_{quart.}$, C_{Ar}), 149.5 (+, 2 × C_{Ar}H), 143.5 ($C_{quart.}$, C_{Ar}), 137.2 (+, 4-C_{Ar}H), 122.9 (+, 2 × C_{Ar}H), 122.0 ($C_{quart.}$, C_{Ar}), 107.9 ($C_{quart.}$, C_{Ar}), 107.7 (+, C_{Ar}H), 104.5 (+, C_{Ar}H), 56.1 (+, OCH_3), 52.4 ($C_{quart.}$, C_{CP}), 41.7 (–, CH_2), 37.8 (–, 2 × CH_2), 27.6 (–, CH_2), 23.4 (–, CH_2), 23.3 (–, 2 × CH_2), 14.1 (+, CH_3) ppm. – **IR** (KBr): ṽ = 3022 (w), 2953 (w), 2924 (m), 2869 (w), 1718 (m), 1609 (m), 1597 (m), 1564 (w), 1487 (w), 1465 (m), 1413 (m), 1354 (w), 1287 (w), 1237 (w), 1216 (m), 1110 (m), 994 (w), 978 (w), 961 (w), 951 (w), 845 (w), 833 (m), 810 (w), 775 (w), 684 (w), 641 (w), 591 (m), 561 (w), 532 (w) cm^{-1}. – **MS** (70 eV, EI): *m/z* (%) = 377 (86) $[M]^+$, 322 (9), 321 (47), 320 (100) $[M - C_4H_9]^+$, 281 (11), 266 (36), 254 (12), 181 (12), 69 (17). – **HRMS** ($C_{24}H_{27}O_3N$): calc. 377.1985, found 377.1986.

7-(1-Butylcyclopentyl)-3-(3-fluorophenyl)-5-methoxy-2*H*-chromen-2-one (40m-Me)

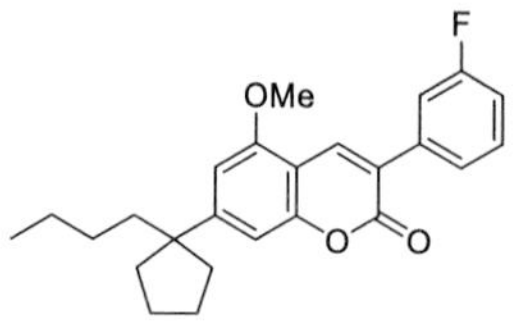

According to **GP9**, 100 mg of 3-bromocoumarin (**105**) (260 µmol, 1.00 equiv.), 73.8 mg of *m*-fluoro phenyl boronic acid (520 µmol, 2.00 equiv.), 172 mg of cesium carbonate (520 µmol, 2.00 equiv.) and 15.2 mg of tetrakis triphenylphosphine palladium (0) (10 µmol, 0.05 equiv.) were put in a sealed vial and 3 mL of dioxane was added. The mixture was degassed with three freeze pump thaw cycles, put under argon atmosphere and was then heated at 90 °C for 16 h. After cooling to room temperature, the reaction was quenched with 20 mL of water and extracted with 3 × 15 mL of ethyl acetate. The combined organic layers were dried over sodium sulfate and the volatiles were removed under reduced pressure. The crude product was then purified via flash column chromatography (CH/EtOAc 20:1) to give 80.1 mg (77%) of the product as a white solid.

***R*f** (CH/EtOAc 20:1): 0.43. – **^{1}H NMR** (400 MHz, $CDCl_3$): δ = 8.20 (d, *J* = 0.6 Hz, 1H, 4-C*H*), 7.56 – 7.44 (m, 2H, 2 × H_{Ar}), 7.39 (td, *J* = 8.0, 6.0 Hz, 1H, H_{Ar}), 7.07 (tdd, *J* = 8.4, 2.6, 1.0 Hz, 1H, H_{Ar}), 6.92 – 6.82 (m, 1H, H_{Ar}), 6.65 (d, *J* = 1.4 Hz, 1H, H_{Ar}), 3.95 (s, 3H, OCH_3), 2.01 – 1.80 (m, 4H, 2 × CH_2), 1.79 – 1.53 (m, 6H, 3 × CH_2), 1.27 – 1.10 (m, 2H, CH_2), 1.03 – 0.89 (m, 2H, CH_2), 0.80 (t, *J* = 7.3 Hz, 3H, CH_3) ppm. – **^{13}C NMR** (100 MHz, $CDCl_3$): δ = 164.0 ($C_{quart.}$, *C*OO), 161.6 ($C_{quart.}$, C_{Ar}), 160.8 ($C_{quart.}$, C_{Ar}), 155.9 (d, $C_{quart.}$, C_{Ar}), 154.4 ($C_{quart.}$, C_{Ar}), 137.6 (d, $C_{quart.}$, C_{Ar}), 135.8 (+, 4-C_{Ar}H), 130.0 (d, +, C_{Ar}H), 124.3 (d, +, C_{Ar}H), 123.9 (d, $C_{quart.}$, C_{Ar}), 115.7 (d, +, C_{Ar}H), 115.5 (d, +, C_{Ar}H), 108.1 ($C_{quart.}$, C_{Ar}), 107.6 (+, C_{Ar}H), 104.3 (+, C_{Ar}H), 56.0 (+, OCH_3), 52.3 ($C_{quart.}$, C_{CP}), 41.7 (–, CH_2), 37.8 (–, 2 × CH_2), 27.6 (–, CH_2), 23.4 (–, CH_2), 23.3 (–, 2 × CH_2), 14.1 (+, CH_3) ppm. – **IR** (KBr): ṽ = 2953 (w), 2859 (w), 1724 (m), 1608 (m), 1581 (w), 1487 (w), 1459 (w), 1415 (w), 1347 (w), 1286 (w), 1262 (w), 1234 (w), 1176 (w), 1095 (w), 982 (w), 936 (vw), 886 (w), 840 (w), 787 (w), 692 (w), 598 (vw), 550 (vw), 520 (w), 469 (vw) cm^{-1}. – **MS** (70 eV, EI): *m/z* (%) = 394 (90) $[M]^+$, 338 (38), 337 (90) $[M – C_4H_9]^+$, 331 (13), 283 (33), 281 (15), 262 (16), 243 (14), 231 (23), 181 (92), 69 (100). – **HRMS** ($C_{25}H_{27}O_3F$): calc. 394.1939, found 394.1937. – **Elemental analysis**: $C_{25}H_{27}O_3F$: calc. C 76.12, H 6.90, found C 76.09, H 7.00.

7-(1-Butylcyclopentyl)-3-(4-fluorophenyl)-5-methoxy-2*H*-chromen-2-one (40n-Me)

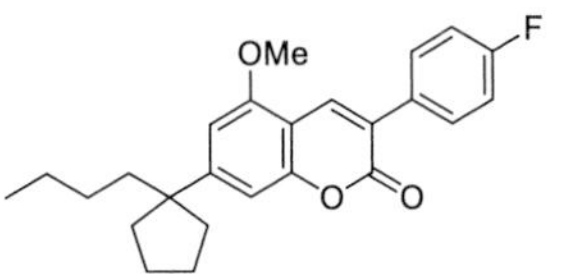

According to **GP9**, 50 mg of 3-bromocoumarin (**105**) (130 µmol, 1.00 equiv.), 36.7 mg of *p*-fluoro phenyl boronic acid (260 µmol, 2.00 equiv.), 85.9 mg of cesium carbonate (260 µmol, 2.00 equiv.) and 15.2 mg of tetrakis triphenylphosphine palladium (0) (10 µmol, 0.10 equiv.) were put in a sealed vial and 3 mL of dioxane was added. The mixture was degassed with three freeze pump thaw cycles, put under argon atmosphere and was then heated at 90 °C for 16 h. After cooling to room temperature, the reaction was quenched with 20 mL of water and extracted with 3 × 15 mL of ethyl acetate. The combined organic layers were dried over sodium sulfate and the volatiles were removed under reduced pressure. The crude product was then purified via flash column chromatography (CH/EtOAc 20:1) to give 24.3 mg (47%) of the product as a white solid.

***R*f** (CH/EtOAc 20:1): 0.43. – **^{1}H NMR** (400 MHz, $CDCl_3$): δ = 8.14 (s, 1H, 4-C*H*), 7.75 – 7.66 (m, 2H, 2 × H_{Ar}), 7.11 (t, *J* = 8.7 Hz, 2H, 2 × H_{Ar}), 6.88 (d, *J* = 1.3 Hz, 1H, H_{Ar}), 6.65 (d, *J* = 1.4 Hz, 1H, H_{Ar}), 3.95 (s, 3H, OCH_3), 1.96 – 1.80 (m, 4H, 2 × CH_2), 1.80 – 1.54 (m, 6H, 3 × CH_2), 1.23 – 1.10 (m, 2H, CH_2), 1.06 – 0.89 (m, 2H, CH_2), 0.80 (t, *J* = 7.3 Hz, 3H, CH_3) ppm. – **^{13}C NMR** (100 MHz, $CDCl_3$): δ = 164.2 ($C_{quart.}$, *C*OO), 161.7 ($C_{quart.}$, C_{Ar}), 161.1 ($C_{quart.}$, C_{Ar}), 155.6 (d, $C_{quart.}$, C_{Ar}), 154.3 ($C_{quart.}$, C_{Ar}), 135.1 (+, 4-C_{Ar}H), 131.5 (d, $C_{quart.}$, C_{Ar}), 130.5 (d, +, 2 × C_{Ar}H), 124.3 ($C_{quart.}$, C_{Ar}), 115.5 (d, +, 2 × C_{Ar}H), 108.2 ($C_{quart.}$, C_{Ar}), 107.6 (+, C_{Ar}H), 104.3 (+, C_{Ar}H), 56.0 (+, OCH_3), 52.2 ($C_{quart.}$, C_{CP}), 41.7 (–, CH_2), 37.8 (–, 2 × CH_2), 27.6 (–, CH_2), 23.4 (–, CH_2), 23.3 (–, 2 × CH_2), 14.1 (+, CH_3) ppm. – **IR** (KBr): ṽ = 2928 (w), 2857 (w), 1760 (w), 1727 (m), 1612 (m), 1566 (w), 1509 (m), 1491 (w), 1455 (w), 1416 (w), 1355 (w), 1283 (w), 1214 (m), 1159 (w), 1112 (m), 1015 (w), 981 (vw), 956 (w), 938 (w), 835 (m), 790 (w), 775 (w), 731 (w), 658 (w), 627 (w), 562 (w), 546 (w), 521 (w), 479 (w) cm^{-1}. – **MS** (70 eV, EI): *m/z* (%) = 394 (56) $[M]^+$, 338 (24), 337 (62) $[M - C_4H_9]^+$, 283 (12), 279 (13), 281 (15), 167 (15), 149 (18), 131 (12), 69 (100). – **HRMS** ($C_{25}H_{27}O_3F$): calc. 394.1939, found 394.1937.

7-(1-Butylcyclopentyl)-5-methoxy-3-(4-(trifluoromethyl)phenyl)-2*H*-chromen-2-one (40o-Me)

According to **GP9**, 85 mg of 3-bromocoumarin (**105**) (220 µmol, 1.00 equiv.), 85.1 mg of *p*-fluoro phenyl boronic acid (450 µmol, 2.00 equiv.), 146 mg of cesium carbonate (450 µmol, 2.00 equiv.) and 13.0 mg of tetrakis triphenylphosphine palladium (0) (10 µmol, 0.10 equiv.) were put in a sealed vial and 3 mL of dioxane was added. The mixture was degassed with three freeze pump thaw cycles, put under argon atmosphere and was then heated at 90 °C for 16 h. After cooling to room temperature, the reaction was quenched with 20 mL of water and extracted with 3 × 15 mL of ethyl acetate. The combined organic layers were dried over sodium sulfate and the volatiles were removed under reduced pressure. The crude product was then purified via flash column chromatography (CH/EtOAc 20:1) to give 54.8 mg (56%) of the product as white solid.

$\boldsymbol{R_f}$ (CH/EtOAc 20:1): 0.50. – **^{1}H NMR** (400 MHz, $CDCl_3$): δ = 8.23 (s, 1H, 4-C*H*), 7.85 (d, *J* = 7.8 Hz, 2H, 2 × H_{Ar}), 7.68 (d, *J* = 8.2 Hz, 2H, 2 × H_{Ar}), 6.89 (d, *J* = 1.3 Hz, 1H, H_{Ar}), 6.66 (d, *J* = 1.4 Hz, 1H, H_{Ar}), 3.96 (s, 3H, OCH_3), 1.99 – 1.80 (m, 4H, 2 × CH_2), 1.80 – 1.57 (m, 6H, 3 × CH_2), 1.26 – 1.11 (m, 2H, CH_2), 1.06 – 0.90 (m, 2H, CH_2), 0.80 (t, *J* = 7.3 Hz, 3H, CH_3) ppm. – **^{13}C NMR** (100 MHz, $CDCl_3$): δ = 160.8 ($C_{quart.}$, *C*OO), 156.1 ($C_{quart.}$, C_{Ar}), 156.0 ($C_{quart.}$, C_{Ar}), 154.6 ($C_{quart.}$, C_{Ar}), 139.0 ($C_{quart.}$, C_{Ar}), 136.4 (+, 4-C_{Ar}H), 130.5 (q, $C_{quart.}$, C_{Ar}), 129.0 (+, 2 × C_{Ar}H), 125.5 (q, +, 2 × C_{Ar}H), 123.8 ($C_{quart.}$, C_{Ar}), 108.1 ($C_{quart.}$, C_{Ar}), 107.7 (+, C_{Ar}H), 104.4 (+, C_{Ar}H), 56.1 (+, OCH_3), 52.3 ($C_{quart.}$, C_{CP}), 41.7 (–, CH_2), 37.8 (–, 2 × CH_2), 27.6 (–, CH_2), 23.4 (–, CH_2), 23.3 (–, 2 × CH_2), 14.1 (+, CH_3) ppm. – **IR** (KBr): $\tilde{\nu}$ = 2930 (w), 2860 (w), 1730 (m), 1609 (m), 1489 (w), 1457 (w), 1411 (w), 1321 (m), 1234 (w), 1163 (m), 1016 (m), 947 (m), 843 (m), 772 (w), 694 (w), 632 (w), 598 (w), 554 (w), 516 (w), 415 (vw) cm^{-1}. – **MS** (70 eV, EI): *m/z (%)* = 444 (100) $[M]^+$, 388 (18), 387 (39) $[M - C_4H_9]^+$, 283 (12), 279 (13), 281 (15), 167 (15), 149 (18), 131 (12), 69 (100). – **HRMS** ($C_{26}H_{27}O_3F_3$): calc. 444.1907, found 444.1908. – **Elemental analysis**: $C_{26}H_{27}O_3F_3$: calc. C 70.26, H 6.12, found C 70.32, H 6.01.

7-(1-Butylcyclopentyl)-5-hydroxy-3-phenyl-2*H*-chromen-2-one (40c-H)

According to **GP8**, 66.3 mg of of 5-methoxycoumarin **40c-Me** (176 µmol, 1.00 equiv.) was dissolved in 5 mL of dry dichloromethane. The solution was cooled to –78 °C and 0.88 mL of boron tribromide (1 M in dichloromethane, 0.880 mmol, 5.00 equiv.) was added dropwise. The mixture was stirred for 30 min at this temperature and then allowed to warm to room temperature. The reaction was quenched after 16 h at 0 °C by addition of sodium bicarbonate. The aqueous layer was extracted with 3 × 15 mL of dichloromethane and the combined organic layers were washed with brine, dried over sodium sulfate and the volatiles were removed under reduced pressure. The crude product was then purified via flash column chromatography (CH/EtOAc 5:1) to give the product as 53.3 mg (85%) of a white solid.

***R*f** (CH/EtOAc 5:1): 0.38. – **MP**: 186.7 °C – **^{1}H NMR** (400 MHz, $CDCl_3$): δ = 8.20 (d, *J* = 0.7 Hz, 1H, 4-C*H*), 7.75 – 7.69 (m, 2H, 2 × H_{Ar}), 7.49 – 7.34 (m, 3H, 3 × H_{Ar}), 6.86 – 6.82 (m, 1H, H_{Ar}), 6.66 (d, *J* = 1.5 Hz, 1H, H_{Ar}), 6.23 (s, 1H, O*H*), 1.93 – 1.52 (m, 10H, 5 × CH_2), 1.22 – 1.08 (m, 2H, CH_2), 1.00 – 0.86 (m, 2H, CH_2), 0.78 (t, *J* = 7.3 Hz, 3H, CH_3) ppm. – **^{13}C NMR** (100 MHz, $CDCl_3$): δ = 161.7 ($C_{quart.}$, *C*OO), 155.3 ($C_{quart.}$, C_{Ar}), 154.4 ($C_{quart.}$, C_{Ar}), 152.5 ($C_{quart.}$, C_{Ar}), 135.6 (+, 4-C_{Ar}H), 135.3 ($C_{quart.}$, C_{Ar}), 128.7 (+, 2 × C_{Ar}H), 128.6 (+, C_{Ar}H), 128.5 (+, 2 × C_{Ar}H), 125.3 ($C_{quart.}$, C_{Ar}), 109.2 (+, C_{Ar}H), 107.5 ($C_{quart.}$, C_{Ar}), 107.5 (+, C_{Ar}H), 51.9 ($C_{quart.}$, C_{CP}), 41.7 (–, CH_2), 37.8 (–, 2 × CH_2), 27.6 (–, CH_2), 23.4 (–, CH_2), 23.3 (–, 2 × CH_2), 14.1 (+, CH_3) ppm. – **IR** (KBr): ṽ = 3173 (w), 2953 (w), 2868 (w), 1729 (w), 1670 (w), 1609 (m), 1491 (w), 1422 (w), 1344 (w), 1290 (w), 1226 (w), 1132 (w), 1073 (w), 966 (w), 908 (vw), 867 (vw), 842 (w), 785 (w), 756 (w), 727 (w), 692 (m), 670 (w), 535 (w), 436 (vw) cm^{-1}. – **MS** (70 eV, EI): *m/z* (%) = 362 (71) [M]$^+$, 306 (32), 305 (100) [M – C_4H_9]$^+$, 251 (20), 239 (7). – **HRMS** ($C_{24}H_{26}O_3$): calc. 362.1876, found 362.1875.

7-(1-Butylcyclopentyl)-5-hydroxy-3-(*o*-tolyl)-2*H*-chromen-2-one (40d-H)

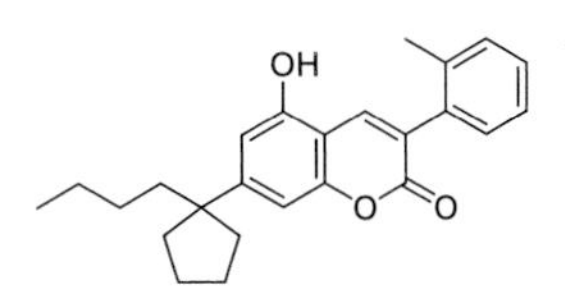

According to **GP8**, 59.4 mg of 5-methoxycoumarin **40d-Me** (152 µmol, 1.00 equiv.) was dissolved in 5 mL of dry dichloromethane. The solution was cooled to –78 °C and 0.76 mL of boron tribromide (1 M in dichloromethane, 0.76 mmol, 5.00 equiv.) was added dropwise. The mixture was stirred for 30 min at this temperature and then allowed to warm to room temperature. The reaction was quenched after 16 h at 0 °C by addition of sodium bicarbonate. The aqueous layer was extracted with 3 × 15 mL of dichloromethane and the combined organic layers were washed with brine, dried over sodium sulfate and the volatiles were removed under reduced pressure. The crude product was then purified via flash column chromatography (CH/EtOAc 5:1) to give the product as 42.1 mg (73%) of a brown solid.

***R*f** (CH/EtOAc 5:1): 0.38. – **^{1}H NMR** (400 MHz, $CDCl_3$): δ = 7.99 (s, 1H, 4-C*H*), 7.36 – 7.19 (m, 4H, 4 × H_{Ar}), 6.90 – 6.83 (m, 1H, H_{Ar}), 6.63 (d, *J* = 1.5 Hz, 1H, H_{Ar}), 5.88 (s, 1H, O*H*), 2.30 (s, 3H, CH_3), 1.93 – 1.52 (m, 10H, 5 × CH_2), 1.23 – 1.08 (m, 2H, CH_2), 1.04 – 0.91 (m, 2H, CH_2), 0.80 (t, *J* = 7.3 Hz, 3H, CH_3) ppm. – **^{13}C NMR** (100 MHz, $CDCl_3$): δ = 161.2 ($C_{quart.}$, *C*OO), 155.3 ($C_{quart.}$, C_{Ar}), 154.8 ($C_{quart.}$, C_{Ar}), 152.3 ($C_{quart.}$, C_{Ar}), 137.2 ($C_{quart.}$, C_{Ar}), 137.1 (+, 4-C_{Ar}H), 135.3 ($C_{quart.}$, C_{Ar}), 130.4 (+, C_{Ar}H), 130.1 (+, C_{Ar}H), 128.7 (+, C_{Ar}H), 126.7 (+, C_{Ar}H), 126.0 ($C_{quart.}$, C_{Ar}), 109.1 (+, C_{Ar}H), 107.7 (+, C_{Ar}H), 107.0 ($C_{quart.}$, C_{Ar}), 51.8 ($C_{quart.}$, C_{CP}), 41.7 (–, CH_2), 37.8 (–, 2 × CH_2), 27.7 (–, CH_2), 23.4 (–, CH_2), 23.3 (–, 2 × CH_2), 20.2 (+, CH_3), 14.2 (+, CH_3) ppm. – **IR** (KBr): $\tilde{\nu}$ = 3279 (vw), 2956 (w), 1683 (w), 1613 (w), 1455 (vw), 1425 (w), 1342 (w), 1295 (w), 1222 (w), 1133 (w), 1104 (w), 1080 (w), 963 (w), 947 (w), 878 (vw), 845 (w), 792 (vw), 755 (w), 725 (w), 691 (w), 547 (vw), 530 (vw), 502 (vw), 430 (vw) cm^{-1}. – **MS** (70 eV, EI): *m/z* (%) = 376 (80) $[M]^+$, 331 (9), 320 (38), 319 (100) $[M - C_4H_9]^+$, 265 (12), 181 (17). – **HRMS** ($C_{25}H_{28}O_3$): calc. 376.2033, found 376.2032.

7-(1-Butylcyclopentyl)-5-hydroxy-3-(*m*-tolyl)-2*H*-chromen-2-one (40e-H)

According to **GP8**, 46.4 mg of 5-methoxycoumarin **40e-Me**(119 µmol, 1.00 equiv.) was dissolved in 5 mL of dry dichloromethane. The solution was cooled to –78 °C and 0.59 mL of boron tribromide (1 M in dichloromethane, 0.59 mmol, 5.00 equiv.) was added dropwise. The mixture was stirred for 30 min at this temperature and then allowed to warm to room temperature. The reaction was quenched after 16 h at 0 °C by addition of sodium bicarbonate. The aqueous layer was extracted with 3 × 15 mL of dichloromethane and the combined organic layers were washed with brine, dried over sodium sulfate and the volatiles were removed under reduced pressure. The crude product was then purified via flash column chromatography (CH/EtOAc 5:1) to give the product as 10.9 mg (24%) yellow crystals.

***R*f** (CH/EtOAc 5:1): 0.38. – **^{1}H NMR** (400 MHz, $CDCl_3$): δ = 8.14 (s, 1H, 4-C*H*), 7.56 – 7.47 (m, 2H, 2 × H_{Ar}), 7.32 (t, *J* = 7.6 Hz, 1H, H_{Ar}), 7.19 (d, *J* = 7.6 Hz, 1H, H_{Ar}), 6.86 (d, *J* = 1.4 Hz, 1H, H_{Ar}), 6.61 (d, *J* = 1.5 Hz, 1H, H_{Ar}), 5.70 (bs, 1H, O*H*), 2.41 (s, 3H, CH_3), 1.96 – 1.51 (m, 10H, 5 × CH_2), 1.22 – 1.11 (m, 2H, CH_2), 1.05 – 0.82 (m, 2H, CH_2), 0.79 (t, *J* = 7.3 Hz, 3H, CH_3) ppm. – **^{13}C NMR** (100 MHz, $CDCl_3$): δ = 161.3 ($C_{quart.}$, *C*OO), 155.1 ($C_{quart.}$, C_{Ar}), 154.5 ($C_{quart.}$, C_{Ar}), 152.2 ($C_{quart.}$, C_{Ar}), 138.1 ($C_{quart.}$, C_{Ar}), 135.3 ($C_{quart.}$, C_{Ar}), 135.0 (+, 4-C_{Ar}H), 129.4 (+, C_{Ar}H), 129.3 (+, C_{Ar}H), 128.4 (+, C_{Ar}H), 125.8 (+, C_{Ar}H), 125.7 ($C_{quart.}$, C_{Ar}), 109.0 (+, C_{Ar}H), 107.7 (+, C_{Ar}H), 107.4 ($C_{quart.}$, C_{Ar}), 51.8 ($C_{quart.}$, C_{CP}), 41.7 (–, CH_2), 37.8 (–, 2 × CH_2), 27.7 (–, CH_2), 23.4 (–, CH_2), 23.3 (–, 2 × CH_2), 21.7 (+, CH_3), 14.2 (+, CH_3) ppm. – **IR** (KBr): $\tilde{\nu}$ = 3267 (vw), 2921 (w), 2853 (w),1681 (w), 1606 (w), 1457 (w), 1426 (w), 1375 (w), 1291 (w), 1266 (w), 1236 (w), 1188 (w), 1123 (w), 1076 (w), 975 (vw), 938 (vw), 908 (vw), 873 (vw), 842 (w), 797 (w), 773 (vw), 733 (vw), 700 (w), 671 (w), 531 (vw), 457 (vw) cm^{-1}. – **MS** (70 eV, EI): *m/z* (%) = 376 (28) $[M]^+$, 320 (13), 319 (32) $[M-C_4H_9]^+$, 307 (26), 69 (100). – **HRMS** ($C_{25}H_{28}O_3$): calc. 376.2033, found 376.2034.

7-(1-Butylcyclopentyl)-5-hydroxy-3-(*p*-tolyl)-2*H*-chromen-2-one (40f-H)

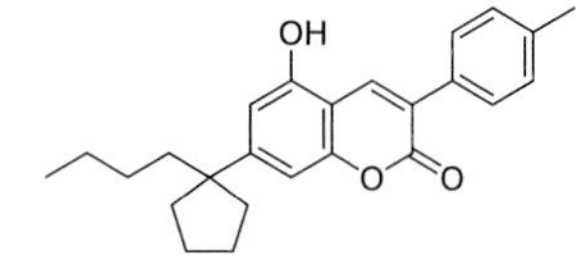

According to **GP8**, 57.0 mg of 5-methoxycoumarin **40f-Me** 146 µmol, 1.00 equiv.) was dissolved in 5 mL of dry dichloromethane. The solution was cooled to –78 °C and 0.73 mL of boron tribromide (1 M in dichloromethane, 0.73 mmol, 5.00 equiv.) was added dropwise. The mixture was stirred for 30 min at this temperature and then allowed to warm to room temperature. The reaction was quenched after 16 h at 0 °C by addition of sodium bicarbonate. The aqueous layer was extracted with 3 × 15 mL of dichloromethane and the combined organic layers were washed with brine, dried over sodium sulfate and the volatiles were removed under reduced pressure. The crude product was then purified via flash column chromatography (CH/EtOAc 5:1) to give the product as 51.0 mg (93%) as a yellow solid.

***R*f** (CH/EtOAc 5:1): 0.37. – **^{1}H NMR** (400 MHz, $CDCl_3$): δ = 8.16 (s, 1H, 4-C*H*), 7.64 – 7.58 (m, 2H, 2 × H_{Ar}), 7.23 (d, *J* = 7.9 Hz, 2H, 2 × H_{Ar}), 6.84 (d, *J* = 1.4 Hz, 1H, H_{Ar}), 6.64 (d, *J* = 1.5 Hz, 1H, H_{Ar}), 6.12 (s, 1H, O*H*), 2.38 (s, 3H, CH_3), 1.95 – 1.49 (m, 10H, 5 × CH_2), 1.23 – 1.08 (m, 2H, CH_2), 1.02 – 0.87 (m, 2H, CH_2), 0.78 (t, *J* = 7.3 Hz, 3H, CH_3) ppm. – **^{13}C NMR** (100 MHz, $CDCl_3$): δ = 161.7 ($C_{quart.}$, *C*OO), 155.0 ($C_{quart.}$, C_{Ar}), 154.3 ($C_{quart.}$, C_{Ar}),

152.4 ($C_{quart.}$, C_{Ar}), 138.6 ($C_{quart.}$, C_{Ar}), 134.8 (+, 4-C_{Ar}H), 132.4 ($C_{quart.}$, C_{Ar}), 129.2 (+, 2 × C_{Ar}H), 128.5 (+, 2 × C_{Ar}H), 125.3 ($C_{quart.}$, C_{Ar}), 109.1 (+, C_{Ar}H), 107.5 ($C_{quart.}$, C_{Ar}), 107.5 (+, C_{Ar}H), 51.8 ($C_{quart.}$, C_{CP}), 41.7 (–, CH_2), 37.8 (–, 2 × CH_2), 27.6 (–, CH_2), 23.4 (–, CH_2), 23.3 (–, 2 × CH_2), 21.4 (+, CH_3), 14.2 (+, CH_3) ppm. – **IR** (KBr): $\tilde{\nu}$ = 3166 (vw), 2949 (vw), 2867 (vw), 1680 (w), 1604 (w), 1511 (vw), 1454 (vw), 1423 (w), 1375 (vw), 1345 (w), 1291 (w), 1229 (vw), 1125 (w), 1077 (w), 965 (vw), 935 (vw), 865 (vw), 837 (vw), 817 (w), 775 (vw), 745 (vw), 714 (w), 554 (vw), 536 (vw), 519 (w), 425 (vw), 389 (vw) cm^{-1}. – **MS** (70 eV, EI): *m/z* (%) = 376 (100) $[M]^+$, 320 (23), 319 (73) $[M - C_4H_9]^+$, 265 (7). – **HRMS** ($C_{25}H_{28}O_3$): calc. 376.2033, found 376.2031.

7-(1-Butylcyclopentyl)-5-hydroxy-3-(2-hydroxyphenyl)-2*H*-chromen-2-one (40g-H)

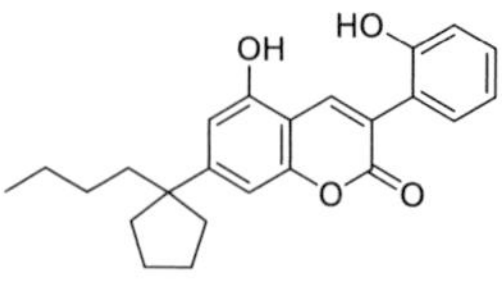

According to **GP8**, 31.4 mg of 5-methoxycoumarin **40g-Me** (77 µmol, 1.00 equiv.) was dissolved in 5 mL of dry dichloromethane. The solution was cooled to –78 °C and 0.39 mL of boron tribromide (1 M in dichloromethane, 0.39 mmol, 5.00 equiv.) was added dropwise. The mixture was stirred for 30 min at this temperature and then allowed to warm to room temperature. The reaction was quenched after 16 h at 0 °C by addition of sodium bicarbonate. The aqueous layer was extracted with 3 × 15 mL of dichloromethane and the combined organic layers were washed with brine, dried over sodium sulfate and the volatiles were removed under reduced pressure. The crude product was then purified via flash column chromatography (CH/EtOAc 5:1) to give the product as 28.4 mg (97%) as an off-white solid.

***R*f** (CH/EtOAc 5:1): 0.23. – **MP**: 191.6 °C – **^{1}H NMR** (400 MHz, $CDCl_3$): δ = 8.27 (s, 1H, 4-C*H*), 7.93 (s, 1H, O*H*), 7.39 – 7.31 (m, 2H, 2 × H_{Ar}), 7.07 (dd, *J* = 8.0, 1.2 Hz, 1H, H_{Ar}), 7.02 (td, *J* = 7.5, 1.3 Hz, 1H, H_{Ar}), 6.90 (d, *J* = 1.3 Hz, 1H, H_{Ar}), 6.68 (d, *J* = 1.5 Hz, 1H, H_{Ar}), 6.16 (bs, 1H, O*H*), 1.95 – 1.52 (m, 10H, 5 × CH_2), 1.22 – 1.10 (m, 2H, CH_2), 1.02 – 0.90 (m, 2H, CH_2), 0.79 (t, *J* = 7.3 Hz, 3H, CH_3) ppm. – **^{13}C NMR** (100 MHz, $CDCl_3$): δ = 164.1 ($C_{quart.}$, COO), 156.5 ($C_{quart.}$, C_{Ar}), 154.8 ($C_{quart.}$, C_{Ar}), 154.0 ($C_{quart.}$, C_{Ar}), 152.6 ($C_{quart.}$, C_{Ar}), 139.5 (+, 4-C_{Ar}H), 130.9 (+, C_{Ar}H), 130.8 (+, C_{Ar}H), 124.1 ($C_{quart.}$, C_{Ar}), 124.0 ($C_{quart.}$, C_{Ar}), 121.6 (+, C_{Ar}H), 119.6 (+, C_{Ar}H), 109.7 (+, C_{Ar}H), 107.6 ($C_{quart.}$, C_{Ar}), 107.5 (+, C_{Ar}H), 52.0 ($C_{quart.}$, C_{CP}), 41.6 (–, CH_2), 37.7 (–, 2 × CH_2), 27.6 (–, CH_2), 23.4 (–, CH_2), 23.3 (–, 2 × CH_2), 14.1 (+, CH_3) ppm. – **IR** (KBr): $\tilde{\nu}$ = 3361 (w), 2953 (m), 2868 (w), 1681 (s), 1620 (s), 1604 (s), 1488 (w),

1444 (m), 1426 (m), 1376 (w), 1344 (w), 1288 (m), 1236 (m), 1214 (m), 1174 (m), 1133 (m), 1099 (m), 1075, (m), 1041 (w), 962 (m), 938 (w), 908 (w), 856 (w), 837 (m), 802 (m), 782 (m), 750 (m), 664 (w), 636 (m) cm^{-1}. – **MS** (70 eV, EI): *m/z* (%) = 378 (100) $[M]^+$, 322 (16), 321 (53) $[M - C_4H_9]^+$, 319 (8). – **HRMS** ($C_{25}H_{28}O_3$): calc. 378.1826, found 378.1827.

7-(1-Butylcyclopentyl)-5-hydroxy-3-(3-hydroxyphenyl)-2*H*-chromen-2-one (40h-H)

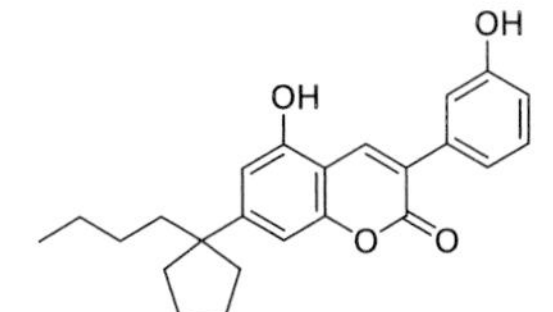

According to **GP8**, 23.8 mg of 5-methoxycoumarin **40h-Me** (59.0 µmol, 1.00 equiv.) was dissolved in 5 mL of dry dichloromethane. The solution was cooled to –78 °C and 0.29 mL of boron tribromide (1 M in dichloromethane, 0.290 mmol, 5.00 equiv.) was added dropwise. The mixture was stirred for 30 min at this temperature and then allowed to warm to room temperature. The reaction was quenched after 16 h at 0 °C by addition of sodium bicarbonate. The aqueous layer was extracted with 3 × 15 mL of dichloromethane and the combined organic layers were washed with brine, dried over sodium sulfate and the volatiles were removed under reduced pressure. The crude product was then purified via flash column chromatography (CH/EtOAc 5:1) to give the product as 13.5 mg (61%) as an off-white solid.

$\boldsymbol{R}_f$ (CH/EtOAc 5:1): 0.13. – **^{1}H NMR** (400 MHz, acetone-D6): δ = 9.40 (s, 1H, O*H*), 8.42 (s, 1H, O*H*), 8.18 (d, *J* = 0.7 Hz, 1H, 4-C*H*), 7.30 – 7.25 (m, 2H, 2 × H_{Ar}), 7.22 (dt, *J* = 7.7, 1.3 Hz, 1H, H_{Ar}), 6.88 (ddd, *J* = 7.9, 2.5, 1.1 Hz, 1H, H_{Ar}), 6.83 (d, *J* = 1.6 Hz, 1H, H_{Ar}), 6.79 (dd, *J* = 1.6, 0.7 Hz, 1H, H_{Ar}), 1.99 – 1.89 (m, 2H, CH_2), 1.89 – 1.81 (m, 2H, CH_2), 1.80 – 1.71 (m, 2H, CH_2), 1.65 (ddt, *J* = 11.4, 4.5, 2.8 Hz, 4H, 2 × CH_2), 1.24 – 1.15 (m, 2H, CH_2), 1.08 – 0.97 (m, 2H, CH_2), 0.79 (t, *J* = 7.3 Hz, 3H, CH_3) ppm. – **^{13}C NMR** (100 MHz, acetone-D6): δ = 160.8 ($C_{quart.}$, *C*OO), 158.2 ($C_{quart.}$, C_{Ar}), 155.8 ($C_{quart.}$, C_{Ar}), 155.5 ($C_{quart.}$, C_{Ar}), 155.1 ($C_{quart.}$, C_{Ar}), 138.0 ($C_{quart.}$, C_{Ar}), 135.6 (+, C_{Ar}H), 130.2 (+, C_{Ar}H), 125.6 ($C_{quart.}$, C_{Ar}), 120.6 (+, C_{Ar}H), 116.5 (+, C_{Ar}H), 116.2 (+, C_{Ar}H), 109.7 (+, C_{Ar}H), 108.1 ($C_{quart.}$, C_{Ar}), 106.7 (+, C_{Ar}H), 52.6 ($C_{quart.}$, C_{CP}), 42.3 (–, CH_2), 38.4 (–, 2 × CH_2), 28.4 (–, CH_2), 24.0 (–, CH_2), 23.9 (–, 2 × CH_2), 14.4 (+, CH_3) ppm. – **IR** (KBr): ṽ = 3140 (w), 2924 (m), 2855 (w), 1681 (m), 1625 (m), 1603 (m), 1554 (w), 1487 (m), 1442 (m), 1378 (w), 1305 (w), 1248 (m), 1207 (w), 1178 (m), 1136 (m), 1079 (w), 1020 (w), 920 (w), 889 (w), 872 (m), 845 (m), 784 (m), 730 (w), 695 (m), 668 (w), 611 (w), 594 (w), 522 (w), 455 (w) cm^{-1}. – **MS** (70 eV, EI): *m/z* (%) = 378 (60) $[M]^+$, 322 (15), 321 (45) $[M - C_4H_9]^+$, 149 (100). – **HRMS** ($C_{25}H_{28}O_3$): calc. 378.1826, found 378.1826.

7-(1-Butylcyclopentyl)-5-hydroxy-3-(4-hydroxyphenyl)-2*H*-chromen-2-one (40b-H)

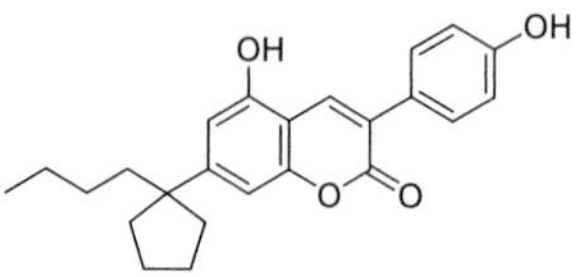

According to **GP8**, 16.7 mg of 5-methoxycoumarin **40b-Me** (41.0 µmol, 1.00 equiv.) was dissolved in 5 mL of dry dichloromethane. The solution was cooled to –78 °C and 0.21 mL of boron tribromide (1 M in dichloromethane, 0.210 mmol, 5.00 equiv.) was added dropwise. The mixture was stirred for 30 min at this temperature and then allowed to warm to room temperature. The reaction was quenched after 16 h at 0 °C by addition of sodium bicarbonate. The aqueous layer was extracted with 3 × 15 mL of dichloromethane and the combined organic layers were washed with brine, dried over sodium sulfate and the volatiles were removed under reduced pressure. The crude product was then purified via flash column chromatography (CH/EtOAc 5:1) to give the product as 14.0 mg (90%) as an off-white solid.

***R*f** (CH/EtOAc 2:1): 0.33. – **^{1}H NMR** (400 MHz, acetone-D6): δ = 9.32 (s, 1H, O*H*), 8.57 (s, 1H, O*H*), 8.11 (s, 1H, 4-C*H*), 7.67 – 7.60 (m, 2H, 2 × H_{Ar}), 6.91 (dd, *J* = 9.1, 2.4 Hz, 2H, 2 × H_{Ar}), 6.81 (d, *J* = 1.6 Hz, 1H, H_{Ar}), 6.77 (d, *J* = 1.5 Hz, 1H, H_{Ar}), 1.97 – 1.88 (m, 2H, CH_2), 1.88 – 1.80 (m, 2H, CH_2), 1.80 – 1.69 (m, 2H, CH_2), 1.68 – 1.58 (m, 4H, 2 × CH_2), 1.21 – 1.12 (m, 2H, CH_2), 1.09 – 0.98 (m, 2H, CH_2), 0.79 (t, *J* = 7.3 Hz, 3H, CH_3) ppm. – **^{13}C NMR** (100 MHz, acetone-D6): δ = 160.2 ($C_{quart.}$, *C*OO), 157.7 ($C_{quart.}$, C_{Ar}), 154.3 ($C_{quart.}$, C_{Ar}), 154.1 ($C_{quart.}$, C_{Ar}), 153.9 ($C_{quart.}$, C_{Ar}), 132.9 (+, 4-C_{Ar}H), 129.8 (+, 2 × C_{Ar}H), 126.9 ($C_{quart.}$, C_{Ar}), 124.6 ($C_{quart.}$, C_{Ar}), 115.0 (+, 2 × C_{Ar}H), 108.6 (+, C_{Ar}H), 107.3 ($C_{quart.}$, C_{Ar}), 105.7 (+, C_{Ar}H), 51.5 ($C_{quart.}$, C_{CP}), 41.3 (–, *C*H$_2$), 37.4 (–, 2 × *C*H$_2$), 27.4 (–, *C*H$_2$), 23.0 (–, *C*H$_2$), 22.9 (–, 2 × *C*H$_2$), 13.4 (+, *C*H$_3$) ppm. – **IR** (KBr): ṽ = 3176 (w), 2923 (w), 2854 (w), 1672 (w), 1608 (w), 1511 (w), 1422 (w), 1343 (w), 1229 (w), 1177 (w), 1132 (w), 1075 (w), 966 (w), 834 (w), 633 (vw), 541 (w), 526 (w) cm^{-1}. – **MS** (70 eV, EI): *m/z* (%) = 378 (60) [M]$^+$, 22 (15), 321 (60) [M – C_4H_9]$^+$, 149 (100). – **HRMS** ($C_{25}H_{28}O_3$): calc. 378.1826, found 378.1826.

7-(1-Butylcyclopentyl)-3-(3-(dimethylamino)phenyl)-5-hydroxy-2*H*-chromen-2-one (40i-H)

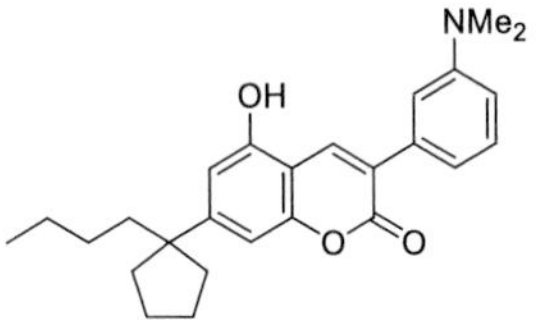

According to **GP8**, 33.9 mg of 5-methoxycoumarin **40i-Me** (81.0 µmol, 1.00 equiv.) was dissolved in 5 mL of dry dichloromethane. The solution was cooled to –78 °C and 0.40 mL of boron tribromide (1 M in dichloromethane, 0.400 mmol, 5.00 equiv.) was added dropwise. The mixture was stirred for 30 min at this temperature and then allowed to warm to room temperature. The reaction was quenched after 16 h at 0 °C by addition of sodium bicarbonate. The aqueous layer was extracted with 3 × 15 mL of dichloromethane and the combined organic layers were washed with brine, dried over sodium sulfate and the volatiles were removed under reduced pressure. The crude product was then purified via flash column chromatography (CH/EtOAc 5:1) to give the product as 9.4 mg (29%) as a yellow solid.

***R*f** (CH/EtOAc 2:1): 0.63. – **^{1}H NMR** (300 MHz, $CDCl_3$): δ = 8.16 (s, 1H, 4-C*H*), 7.30 (t, *J* = 8.0 Hz, 1H, H_{Ar}), 7.13 – 7.03 (m, 2H, 2 × H_{Ar}), 6.86 – 6.78 (m, 2H, 2 × H_{Ar}), 6.60 (d, *J* = 1.5 Hz, 1H, H_{Ar}), 5.91 (bs, 1H, O*H*), 2.99 (s, 6H, $N(CH_3)_2$), 1.93 – 1.51 (m, 10H, 5 × CH_2), 1.21 – 1.10 (m, 2H, CH_2), 1.01 – 0.84 (m, 2H, CH_2), 0.78 (t, *J* = 7.3 Hz, 3H, CH_3) ppm. – **^{13}C NMR** (100 MHz, $CDCl_3$): δ = 161.4 ($C_{quart.}$, *C*OO), 158.7 ($C_{quart.}$, C_{Ar}), 155.0 ($C_{quart.}$, C_{Ar}), 154.4 ($C_{quart.}$, C_{Ar}), 152.3 ($C_{quart.}$, C_{Ar}), 136.3 ($C_{quart.}$, C_{Ar}), 135.1 (+, 4-C_{Ar}H), 129.2 (+, C_{Ar}H), 126.1 ($C_{quart.}$, C_{Ar}), 113.5 (+, C_{Ar}H), 113.4 (+, C_{Ar}H), 109.0 (+, C_{Ar}H), 107.5 (+, C_{Ar}H), 107.4 ($C_{quart.}$, C_{Ar}), 100.1 (+, C_{Ar}H), 51.8 ($C_{quart.}$, C_{CP}), 41.7 (+, $N(CH_3)_2$), 41.2 (–, CH_2), 37.8 (–, 2 × CH_2), 29.9 (–, CH_2), 23.4 (–, CH_2), 23.3 (–, 2 × CH_2), 14.2 (+, CH_3) ppm. – **IR** (KBr): ṽ = 3254 (vw), 2923 (w), 2854 (w), 1685 (w), 1608 (m), 1497 (w), 1454 (w), 1421 (w), 1342 (w), 1285 (w), 1224 (w), 1104 (w), 1076 (w), 984 (w), 944 (w), 844 (w), 786 (w), 743 (vw), 694 (w), 528 (vw), 465 (vw), 438 (vw) cm^{-1}. – **MS** (70 eV, EI): *m/z* (%) = 405 (46) $[M]^+$, 57 (100). – **HRMS** ($C_{26}H_{31}O_3N$): calc. 405.2298, found 405.2296.

7-(1-Butylcyclopentyl)-3-(4-(dimethylamino)phenyl)-5-hydroxy-2*H*-chromen-2-one (40j-H)

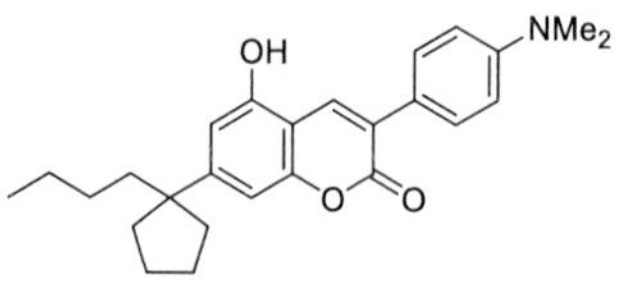

According to **GP8**, 60.9 mg of 5-methoxycoumarin **40j-Me** (145 µmol, 1.00 equiv.) was dissolved in 5 mL of dry dichloromethane. The solution was cooled to –78 °C and 0.73 mL of boron tribromide (1 M in dichloromethane, 0.730 mmol, 5.00 equiv.) was added dropwise. The mixture was stirred for 30 min at this temperature and then allowed to warm to room temperature. The reaction was quenched after 16 h at 0 °C by addition of sodium bicarbonate. The aqueous layer was extracted with 3 × 15 mL of dichloromethane and the combined organic layers were washed with brine, dried over sodium sulfate and the volatiles were removed under reduced pressure. The crude product was then purified via flash column chromatography (CH/EtOAc 5:1) to give the product as 27.5 mg (%) as an off-white solid.

***R*f** (CH/EtOAc 2:1): 0.47. – **¹H NMR** (400 MHz, $CDCl_3$): δ = 8.04 (s, 1H, 4-C*H*), 7.64 – 7.55 (m, 2H, 2 × H_{Ar}), 6.74 (s, 1H, H_{Ar}), 6.71 (d, *J* = 8.4 Hz, 2H, 2 × H_{Ar}), 6.58 (s, 1H, H_{Ar}), 6.52 (bs, 1H, O*H*), 2.91 (s, 6H, $N(CH_3)_2$), 1.90 – 1.42 (m, 10H, 5 × CH_2), 1.14 – 1.02 (m, 2H, CH_2), 0.97 – 0.80 (m, 2H, CH_2), 0.69 (t, *J* = 7.3 Hz, 3H, CH_3) ppm. – **¹³C NMR** (100 MHz, $CDCl_3$): δ = 162.2 ($C_{quart.}$, *C*OO), 154.2 ($C_{quart.}$, C_{Ar}), 153.9 ($C_{quart.}$, C_{Ar}), 152.4 ($C_{quart.}$, C_{Ar}), 150.4 ($C_{quart.}$, C_{Ar}), 132.9 (+, 4-C_{Ar}H), 129.5 (+, 2 × C_{Ar}H), 125.0 ($C_{quart.}$, C_{Ar}), 123.6 ($C_{quart.}$, C_{Ar}), 112.5 (+, 2 × C_{Ar}H), 109.2 (+, C_{Ar}H), 107.9 ($C_{quart.}$, C_{Ar}), 107.2 (+, C_{Ar}H), 51.8 ($C_{quart.}$, C_{CP}), 41.7 (–, CH_2), 40.8 (+, $N(CH_3)_2$), 37.8 (–, 2 × CH_2), 27.6 (–, CH_2), 23.4 (–, CH_2), 23.3 (–, 2 × CH_2), 14.2 (+, CH_3) ppm. – **IR** (KBr): ṽ = 3166 (w), 2924 (w), 2857 (w), 1678 (w), 1608 (m), 1520 (w), 1421 (m), 1350 (w), 1289 (w), 1227 (w), 1202 (w), 1168 (w), 1125 (w), 1077 (w), 1012 (w), 946 (w), 946 (w), 840 (w), 814 (w), 778 (w), 724 (w), 667 (w), 628 (vw), 529 (vw), 483 (w), 428 (vw) cm^{-1}. – **MS** (70 eV, EI): *m/z* (%) = 405 (100) $[M]^+$, 348 (19) $[M - C_4H_9]^+$. – **HRMS** ($C_{26}H_{31}O_3N$): calc. 405.2298, found 405.2300.

7-(1-Butylcyclopentyl)-3-(3-fluorophenyl)-5-hydroxy-2*H*-chromen-2-one (40m-H)

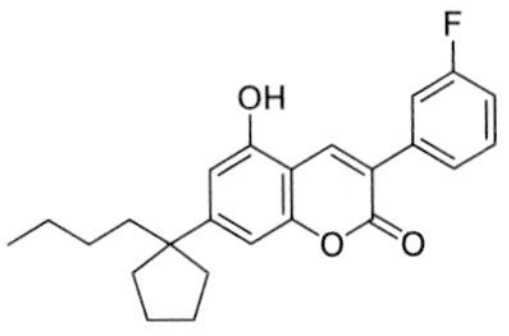

According to **GP8**, 29.0 mg of 5-methoxycoumarin **40m-Me** (74 µmol, 1.00 equiv.) was dissolved in 5 mL of dry dichloromethane. The solution was cooled to –78 °C and 0.37 mL of boron tribromide (1 M in dichloromethane, 0.37 mmol, 5.00 equiv.) was added dropwise. The mixture was stirred for 30 min at this temperature and then allowed to warm to room temperature. The reaction was quenched after 16 h at 0 °C by addition of sodium bicarbonate. The aqueous layer was extracted with 3 × 15 mL of dichloromethane and the combined organic layers were washed with brine, dried over sodium sulfate and the volatiles were removed under reduced pressure. The crude product was then purified via flash column chromatography (CH/EtOAc 20:1 to 5:1) to give the product as 18.0 mg (64%) as an off-white solid.

***R*f** (CH/EtOAc 5:1): 0.57. – **^{1}H NMR** (400 MHz, $CDCL_3$): δ = 8.21 (s, 1H, 4-C*H*), 7.56 – 7.44 (m, 2H, 2 × H_{Ar}), 7.39 (td, *J* = 8.0, 5.9 Hz, 1H, H_{Ar}), 7.07 (tdd, *J* = 8.4, 2.6, 1.0 Hz, 1H, H_{Ar}), 6.86 (d, *J* = 1.4 Hz, 1H, H_{Ar}), 6.65 (d, *J* = 1.5 Hz, 1H, H_{Ar}), 6.04 (bs, 1H, O*H*), 1.93 – 1.51 (m, 10H, 5 × CH_2), 1.21 – 1.11 (m, 2H, CH_2), 1.00 – 0.90 (m, 2H, CH_2), 0.78 (t, *J* = 7.3 Hz, 3H, CH_3) ppm. – **^{13}C NMR** (100 MHz, $CDCL_3$): δ = 164.0 ($C_{quart.}$, *C*OO), 161.6 ($C_{quart.}$, C_{Ar}), 161.1 ($C_{quart.}$, C_{Ar}), 155.2 (d, $C_{quart.}$, C_{Ar}), 152.5 ($C_{quart.}$, C_{Ar}), 137.4 (d, $C_{quart.}$, C_{Ar}), 135.9 (+, 4-C_{Ar}H), 130.0 (d, +, C_{Ar}H), 124.3 (d, +, C_{Ar}H), 124.0 (d, $C_{quart.}$, C_{Ar}), 115.7 (d, +, C_{Ar}H), 115.5 (d, +, C_{Ar}H), 109.2 (+, C_{Ar}H), 107.6 (+, C_{Ar}H), 107.2 ($C_{quart.}$, C_{Ar}), 51.9 ($C_{quart.}$, C_{CP}), 41.7 (–, CH_2), 37.7 (–, 2 × CH_2), 27.7 (–, CH_2), 23.4 (–,CH_2), 23.3 (–, 2 × CH_2), 14.2 (+, CH_3). ppm. – **^{19}F NMR** (375 MHz, $CDCL_3$): δ = 117.3 ppm. – **IR** (KBr): ṽ = 3171 (w), 2924 (m), 1674 (m), 1607 (s), 1488 (w), 1422 (m), 1345 (m), 1261 (w), 1238 (m), 1182 (m), 1077 (m), 946 (w), 867 (w), 843 (w), 788 (m), 687 (m), 533 (w), 517 (w), 472 (w), 431 (vw) cm^{-1}. – **MS** (70 eV, EI): *m/z* (%) = 380 (60) $[M]^+$, 325 (6), 324 (45), 323 (100) $[M – C_4H_9]^+$, 269 (36), 257 (11), 241 (7). – **HRMS** ($C_{24}H_{25}O_3F$): calc. 380.1788, found 380.1789.

7-(1-Butylcyclopentyl)-5-hydroxy-3-(4-(trifluoromethyl)phenyl)-2*H*-chromen-2-one (40o-H)

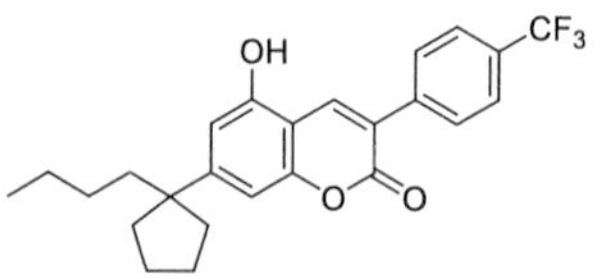

According to **GP8**, 17.7 mg of 5-methoxycoumarin **40m-Me** (40.0 µmol, 1.00 equiv.) was dissolved in 5 mL of dry dichloromethane. The solution was cooled to –78 °C and 0.22 mL of boron tribromide (1 M in dichloromethane, 0.220 mmol, 5.00 equiv.) was added dropwise. The mixture was stirred for 30 min at this temperature and then allowed to warm to room temperature. The reaction was quenched after 16 h at 0 °C by addition of sodium bicarbonate. The aqueous layer was extracted with 3 × 15 mL of dichloromethane and the combined organic layers were washed with brine, dried over sodium sulfate and the volatiles were removed under reduced pressure. The crude product was then purified via flash column chromatography (CH/EtOAc 20:1 to 5:1) to give the product as 18.0 mg (79%) as an off-white solid.

$\boldsymbol{R_f}$ (CH/EtOAc 5:1): 0.19. – **^{1}H NMR** (400 MHz, $CDCL_3$): δ = 8.27 (s, 1H, 4-C*H*), 7.85 (d, *J* = 8.1 Hz, 2H, 2 × H_{Ar}), 7.68 (d, *J* = 8.2 Hz, 2H, 2 × H_{Ar}), 6.86 (d, *J* = 1.4 Hz, 1H, H_{Ar}), 6.69 (d, *J* = 1.5 Hz, 1H, H_{Ar}), 6.45 (bs, 1H, O*H*), 1.94 – 1.52 (m, 10H, 5 × CH_2), 1.22 – 1.11 (m, 2H, CH_2), 1.02 – 0.91 (m, 2H, CH_2), 0.78 (t, *J* = 7.3 Hz, 3H, CH_3) ppm. – **^{13}C NMR** (100 MHz, $CDCL_3$): δ = 161.3 ($C_{quart.}$, *C*OO), 156.2 ($C_{quart.}$, C_{Ar}), 154.6 ($C_{quart.}$, C_{Ar}), 152.8 ($C_{quart.}$, C_{Ar}), 138.9 ($C_{quart.}$, C_{Ar}), 136.7 (+, C_{Ar}H), 130.5 (q, $C_{quart.}$, C_{Ar}) 129.0 (+, 2 × C_{Ar}H), 125.5 (q, +, 2 × C_{Ar}H), 125.4, 123.8 ($C_{quart.}$, C_{Ar}), 109.3 (+, C_{Ar}H), 107.5 (+, C_{Ar}H), 107.3 ($C_{quart.}$, C_{Ar}), 52.0 ($C_{quart.}$, C_{CP}), 41.7 (–, CH_2), 37.7 (–, 2 × CH_2), 27.7 (–, CH_2), 23.4 (–, CH_2), 23.3 (–, 2 × CH_2), 14.1 (+, CH_3) ppm. – **^{19}F NMR** (375 MHz, $CDCL_3$): δ = –67.0 ppm. – **IR** (KBr): ṽ = 3183 (vw), 2928 (w), 1679 (w), 1608 (m), 1426 (w), 1413 (w), 1325 (m), 1296 (w), 1231 (w), 1164 (w), 1122 (m), 1069 (w), 1018 (w), 968 (w), 844 (w), 779 (vw), 697 (vw), 598 (vw), 532 (vw), 460 (w) cm^{-1}. – **MS** (70 eV, EI): *m/z* (%) = 430 (55) $[M]^+$, 374 (37), 373 (100) $[M - C_4H_9]^+$, 320 (8), 319 (37). – **HRMS** ($C_{25}H_{25}O_3F_3$): calc. 430.1756, found 430.1754.

5.2.8 Synthesis and Characterization 3-Styrylcoumarins (Chapter 3.2.6)

(*E*)-7-(1-Butylcyclopentyl)-5-methoxy-3-styryl-2*H*-chromen-2-one (41a)

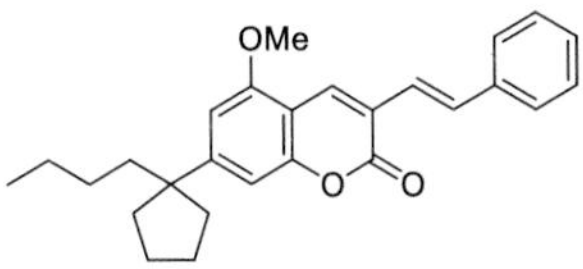

Under argon atmosphere, 50 mg of 3-bromocoumarin (**105**) (130 µmol, 1.00 equiv.), 58.5 mg of (*E*)-styrylboronic acid (400 µmol, 2.00 equiv.), 0.5 mL of 2 M sodium carbonate aqueous solution and 15.0 mg of tetrakis triphenylphosphine palladium (0) (10 µmol, 0.10 equiv.) were put in a sealed vial and 2.5 mL of dioxane and 0.5 mL of water was added. The mixture was degassed with three freeze pump thaw cycles, put under argon atmosphere and was then heated at 90 °C for 16 h. After cooling to room temperature, the reaction was quenched with 20 mL of water and extracted with 3 × 15 mL of ethyl acetate. The combined organic layers were dried over sodium sulfate and the volatiles were removed under reduced pressure. The crude product was then purified via flash column chromatography (CH/EtOAc 20:1) to give 39.1 mg (74%) of the product as a white solid.

$\boldsymbol{R_f}$ (CH/EtOAc 20:1): 0.43. – **^{1}H NMR** (400 MHz, $CDCl_3$): δ = 8.15 (s, 1H, 4-C*H*), 7.63 – 7.51 (m, 3H, H_{DB}, 2 × H_{Ar}), 7.36 (dd, *J* = 8.3, 6.7 Hz, 2H, 2 × H_{Ar}), 7.30 – 7.24 (m, 2H, 2 × H_{Ar}), 7.15 (dd, *J* = 16.3, 0.8 Hz, 1H, H_{DB}), 6.89 – 6.83 (m, 1H, H_{Ar}), 6.64 (d, *J* = 1.4 Hz, 1H, H_{Ar}), 3.97 (s, 3H, OCH_3), 1.96 – 1.80 (m, 4H, 2 × CH_2), 1.79 – 1.57 (m, 6H, 3 × CH_2), 1.23 – 1.12 (m, 2H, CH_2), 1.02 – 0.91 (m, 2H, CH_2), 0.80 (t, *J* = 7.3 Hz, 3H, CH_3) ppm. – **^{13}C NMR** (100 MHz, $CDCl_3$): δ = 161.0 ($C_{quart.}$, *C*OO), 155.6 ($C_{quart.}$, C_{Ar}), 155.0 ($C_{quart.}$, C_{Ar}), 153.7 ($C_{quart.}$, C_{Ar}), 137.4 ($C_{quart.}$, C_{Ar}), 132.4 (+, C_{Ar}H), 132.4 (+, C_{Ar}H), 128.8 (+, 2 × C_{Ar}H), 128.2 (+, C_{Ar}H), 127.0 (+, 2 × C_{Ar}H), 122.9 (+, C_{Ar}H), 122.1 ($C_{quart.}$, C_{Ar}), 108.4 ($C_{quart.}$, C_{Ar}), 107.6 (+, C_{Ar}H), 104.2 (+, C_{Ar}H), 56.0 (+, OCH_3), 52.2 ($C_{quart.}$, C_{CP}), 41.7 (–, CH_2), 37.8 (–, 2 × CH_2), 27.6 (–, CH_2), 23.4 (–, CH_2), 23.4 (–, 2 × CH_2), 14.1 (+, CH_3) ppm. – **IR** (KBr): $\tilde{\nu}$ = 2952 (w), 2927 (w), 2868 (w), 1720 (m), 1613 (m), 1492 (w), 1450 (w), 1415 (w), 1350 (w), 1291 (w), 1238 (w), 1174 (w), 1104 (m), 1053 (w), 966 (w), 909 (w), 842 (w), 748 (w), 733 (w), 691 (w), 558 (vw), 509 (vw) cm^{-1}. – **MS** (70 eV, EI): *m/z* (%) = 402 (100) $[M]^+$, 346 (16), 345 (50) $[M - C_4H_9]^+$, 300 (16), 291 (10), 272 (7), 271 (10). – **HRMS** ($C_{27}H_{30}O_3$): calc. 402.2189, found 402.2188.

5.2.9 Synthesis and Characterization 2,2-Dimethyl-2*H*-chromenes (Chapter 3.2.7)

3-Bromo-2,2-dimethyl-2*H*-chromene (112)

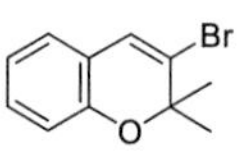

Under argon atmosphere, 100 mg of 3-bromo-2*H*-chromen-2-on **(111)** (444 µmol, 1.00 equiv.), 385 µL NMP (4.00 mmol, 9.00 equiv.) were dissolved in 5 mL of tetrahydrofuran and 320 µL methyl magnesium bromide (3.0 M in diethyl ether, 488 µmol, 2.20 equiv.) were added dropwise at –20 °C. After stirring for an additional 15 min, the reaction was quenched with sodium carbonate-solution and the aqueous layer was extracted with 3 × 5 mL of ethyl acetate. The combined organic Layers were dried over sodium sulfate and purified via flash column chromatography (CH/EtOAc 5:1) to give the product as 75.0 mg (71%) of an off-white solid. Analytical data are consistent with literature.[227]

R_f (CH/EtOAc 5:1): 0.21. – **^{1}H NMR** (300 MHz, $CDCl_3$): δ =7.15 – 7.24 (m, 1H, H_{Ar}) 7.03 – 7.10 (m, 1H, H_{Ar}), 7.84 – 7.94 (m, 3H, H_{Ar}), 1.49 (s, 6H, $(CH_3)_2$) ppm. – **^{13}C-NMR** (100 MHz, $CDCl_3$): δ = 152.2 ($C_{quart.}$, C_{Ar}), 139.2 ($C_{quart.}$, $C_{Ar}O$), 129.6 (+, $C_{Ar}H$), 129.2 (+, $C_{Ar}H$), 126.5 (+, $C_{Ar}H$), 124.8 ($C_{quart.}$, C_{Ar}), 120.8 (+, $C_{Ar}H$), 116.6 (+, $C_{Ar}H$), 76.4 ($C_{quart.}$, $C(CH_3)_2$), 30.0 (+, 2 × CH_3) ppm. – **MS** (EI, 70 eV): m/z (%) = 238/240 (12/3) [M^+], 226 (10), 223/225 (100/98) $[M – CH_3]^+$, 224 (10), 159 (61) $[M – Br]^+$, 144 (13), 115 (12), 91 (21), 89 (12), 59 (13) – **HRMS** ($C_{11}H_{11}O_1Br$): calc. 237.9986, found 237.9988.

5-Methoxy-2,2-dimethyl-3-(2-methylbenzyl)-7-pentyl-2*H*-chromene (117a)

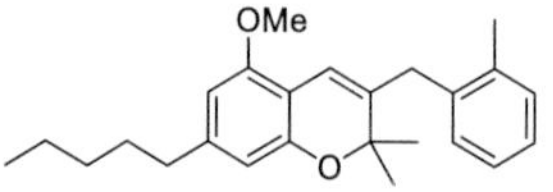

According to **GP10**, 50 mg of coumarin **26e** (143 µmol, 1.00 equiv.) was dissolved in 5 mL of dry tetrahydrofuran and 127 mg of NMP (1.28 mmol, 9.00 equiv.) were added. The mixture was cooled to –15 °C and 0.11 mL of a methyl magnesium bromide solution (313 µmol, 3 M in diethyl ether, 2.2 equiv.) were added dropwise. After 15 min, the mixture was allowed to warm to room temperature and stirred for further 2 h. The reaction mixture was cooled to –10 °C and quenched via the addition of 5 mL 1 N hydrochloric acid. The aqueous layer was extracted with 3 × 5 mL of ethyl acetate and the combined organic layers were washed with 5 mL of sodium carbonate solution, dried over sodium sulfate and the volatiles were removed under reduced pressure. The crude product was then purified via flash column chromatography (CH/EtOAc 15:1) resulted in 40 mg (77%) of the pure product as white solid.

***R*f** (CH/EtOAc 5:1): 0.28. – **^{1}H NMR** (400 MHz, $CDCl_3$): δ = 7.27 – 7.10 (m, 4H, 4 × *H*$_{Ar}$), 6.23 (dd, *J* = 1.5, 0.7 Hz, 1H, *H*$_{Ar}$), 6.11 (d, *J* = 1.4 Hz, 1H, 4-C*H*), 5.89 – 5.79 (m, 1H, *H*$_{Ar}$), 3.69 (s, 3H, OC*H*$_3$), 3.32 (s, 2H, BnC*H*$_2$), 2.46 – 2.37 (m, 2H, C*H*$_2$), 1.10 (s, 3H, C*H*$_3$), 1.57 – 1.44 (m, 2H, C*H*$_2$), 1.38 (s, 6H, 2 × C*H*$_3$), 1.29 – 1.16 (m, 4H, 2 × C*H*$_2$), 0.82 – 0.78 (m, 3H, C*H*$_3$) ppm. – **^{13}C NMR** (100 MHz, $CDCl_3$): δ = 154.7 (C$_{quart.}$), 152.6 (C$_{quart.}$), 144.0 (C$_{quart.}$), 137.1 (C$_{quart.}$), 137.0 (C$_{quart.}$), 136.8 (C$_{quart.}$), 130.6 (+, *C*H), 130.4 (+, *C*H), 126.7 (+, *C*H), 126.1 (+, *C*H), 114.1 (+, *C*H), 109.6 (C$_{quart.}$), 109.1 (+, *C*H), 103.6 (+, *C*H), 78.6 (C$_{quart.}$), 55.6(+, O*C*H$_3$), 36.5(–, *C*H$_2$), 36.0(–, *C*H$_2$), 31.7(–, *C*H$_2$), 31.0(–, *C*H$_2$), 25.8 (+, 2 × *C*H$_3$), 22.7(–, *C*H$_2$), 19.5(+, *C*H$_3$), 14.2(+, *C*H$_3$) ppm. – **MS** (70 eV, EI): *m/z* (%) = 364 (18) [M]$^+$, 350 (24), 349 (100) [M – CH_3]$^+$, 259 (5), 181 (17), 131 (17), 69 (18). – **HRMS** ($C_{25}H_{32}O_2$): calc. 364.2397, found 364.2396.

7-(1-Butylcyclopentyl)-5-methoxy-3-(2-methoxybenzyl)-2,2-dimethyl-2*H*-chromene (117b)

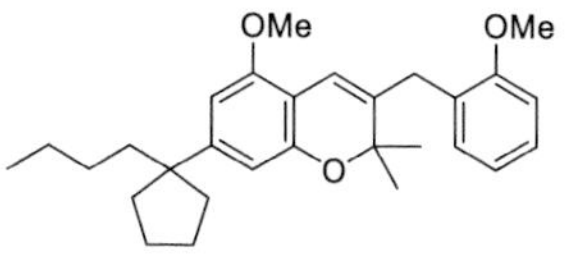

According to **GP10**, 73 mg of coumarin **26h** (174 µmol, 1.00 equiv.) was dissolved in 5 mL of dry tetrahydrofuran and 233 mg NMP (2.36 mmol, 9.00 equiv.) were added. The mixture was cooled to –15 °C and 0.19 mL of a methyl magnesium bromide solution (574 µmol, 3 M in diethyl ether, 2.2 equiv.) were added dropwise. After 15 min, the mixture was allowed to warm to room temperature and stirred for further 2 h. The reaction mixture was cooled to –10 °C and quenched via the addition of 5 mL 1 N hydrochloric acid. The aqueous layer was extracted with 3 × 5 mL of ethyl acetate and the combined organic layers were washed with 5 mL of sodium carbonate solution, dried over sodium sulfate and the volatiles were removed under reduced pressure. The crude product was then purified via flash column chromatography (CH/EtOAc 15:1) resulted in 38 mg (44%) of the pure product as a white solid.

***R*f** (CH/EtOAc 5:1): 0.85. – **MP**: 86.4 °C – **^{1}H NMR** (400 MHz, $CDCl_3$): δ = 7.25 – 7.20 (m, 2H, 2 × *H*$_{Ar}$), 6.95 – 6.86 (m, 2H, 2 × *H*$_{Ar}$), 6.39 (s, 1H, *H*$_{Ar}$), 6.30 (d, J = 1.5 Hz, 1H, 4-C*H*), 6.19 (s, 1H, *H*$_{Ar}$), 3.81 (s, 3H, OC*H*$_3$), 3.74 (s, 3H, OC*H*$_3$), 3.45 (s, 2H, BnC*H*$_2$), 1.91 – 1.81 (m, 2H, C*H*$_2$), 1.79 – 1.59 (m, 6H, 3 × C*H*$_2$), 1.55 – 1.47 (m, 2H, C*H*$_2$), 1.41 (s, 6H, 2 × C*H*$_3$), 1.20 – 1.09 (m, 2H, C*H*$_2$), 1.03 – 0.92 (m, 2H, C*H*$_2$), 0.78 (t, J = 7.3 Hz, 3H, C*H*$_3$) ppm. – **^{13}C NMR** (100 MHz, $CDCl_3$): δ = 157.4 (C$_{quart.}$, *C*$_{Ar}$), 154.3 (C$_{quart.}$, *C*$_{Ar}$), 152.2 (C$_{quart.}$, *C*$_{Ar}$), 150.3(C$_{quart.}$, *C*$_{Ar}$), 138.0 (C$_{quart.}$, *C*$_{Ar}$), 131.0 (+, *C*H), 127.6 (C$_{quart.}$, *C*$_{Ar}$), 127.6 (+, *C*H), 120.6

(+, *C*H), 114.5 (+, *C*H), 110.7 (+, *C*H), 109.5 ($C_{quart.}$, C_{Ar}), 108.1 (+, *C*H), 102.3 (+, *C*H), 78.9 ($C_{quart.}$, *C*(CH_3)$_2$), 55.7 (+, CH_3), 55.6 (+, CH_3), 51.5 ($C_{quart.}$, C_{CP}), 41.8 (–, CH_2), 37.8 (–, 2 × CH_2), 32.1 (–, CH_2), 27.6 (–, CH_2), 25.9 (+, 2 × CH_3), 23.5 (–, CH_2), 23.4 (–, 2 × CH_2), 14.2 (+, CH_3) ppm. – **IR** (KBr): $\tilde{\nu}$ = 2954 (m), 2924 (m), 1643 (w), 1611 (m), 1568 (m), 1492 (w), 1457 (m), 1414 (m), 1381 (w), 1359 (w), 1277 (w), 1242 (m), 1184 (w), 1148 (m), 1118 (s), 1061 (m), 1047 (w), 1032 (m), 920 (w), 880 (w), 846 (w), 831 (w), 755 (m), 674 (w), 573 (w), 528 (vw), 477 (vw) cm^{-1}. – **MS** (70 eV, EI): *m/z* (%) = 434 (20) $[M]^+$, 420 (30), 419 (100), 313 (5). – **HRMS** ($C_{29}H_{38}O_3$): calc. 434.2815, found 434.2815. – **Elemental analysis**: $C_{29}H_{38}O_3$: calc. C 80.14, H 8.81, found C 78.89, H 8.95.

5-Methoxy-3-(2-methoxybenzyl)-2,2-dimethyl-7-pentyl-2*H*-chromene (117c)

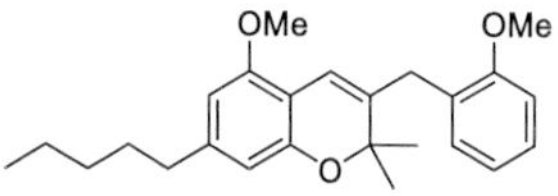

According to **GP10**, 52.5 mg coumarin **26e** (143 µmol, 1.00 equiv.) was dissolved in 5 mL of dry tetrahydrofuran and 128 mg of NMP (1.29 mmol, 9.00 equiv.) were added. The mixture was cooled to –15 °C and 0.11 mL of a methyl magnesium bromide solution (315 µmol, 3 M in diethyl ether, 2.2 equiv.) were added dropwise. After 15 min, the mixture was allowed to warm to room temperature and stirred for further 1 h, then another 0.05 mL of a methyl magnesium bromide solution (165 µmol, 3 N in diethyl ether, 1.0 equiv.) were added dropwise and the stirring continued for 1 h. The reaction mixture was cooled to –10 °C and quenched via the addition of 5 mL of 5 mL 1 N hydrochloric acid. The aqueous layer was extracted with 3 × 5 mL of ethyl acetate and the combined organic layers were washed with 5 mL of sodium carbonate solution, dried over sodium sulfate and the volatiles were removed under reduced pressure. The crude product was then purified via flash column chromatography (CH/EtOAc 15:1) resulted in 25.0 mg (49%) of the pure product as white solid. The product decomposed when exposed to $CDCl_3$ for more than 16 h (e.g. ^{13}C NMR measurement).

***R*$_f$** (CH/EtOAc 5:1): 0.23. – **^{1}H NMR** (300 MHz, $CDCl_3$): δ = 7.33 – 7.19 (m, 2H), 7.00 – 6.86 (m, 2H), 6.37 (d, *J* = 1.4 Hz, 1H), 6.22 (d, *J* = 1.4 Hz, 1H), 5.52 (s, 1H), 3.88 (s, 3H, OCH_3), 3.76 (s, 3H, OCH_3), 3.57 (s, 2H, CH_2), 2.50 (t, *J* = 7.7 Hz, 2H), 1.67 – 1.49 (m, 2H, CH_2), 1.36 (s, 6H), 1.34 – 1.25 (m, 4H, 2 × CH_2), 0.93 – 0.83 (m, 3H, CH_3) ppm. – **IR** (KBr): $\tilde{\nu}$ = 2928 (w), 2855(vw), 1612 (w), 1574 (w), 1493 (w), 1462 (w), 1423 (w), 1360 (vw), 1288 (vw), 1243 (w), 1193 (w), 1140 (w), 1124 (w), 1049 (w),1029 (w), 820 (vw), 752 (w), 558 (vw) cm^{-1}. – **MS**

(70 eV, EI): *m/z* (%) = 380 (18) $[M]^+$, 366 (25), 365 (100) $[M - CH_3]^+$, 259 (6). – **HRMS** ($C_{25}H_{32}O_3$): calc. 380.2346, found 280.2345.

1-Isopropyl-4,5a-dimethyl-5aH,11H-chromeno[2,3-b]chromene (118)

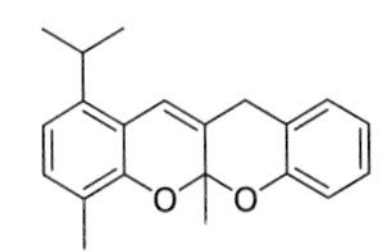

According to **GP10**, 30 mg of coumarin **26c** (97 µmol, 1.00 equiv.) was dissolved in 1.5 mL of dry tetrahydrofuran and 86.8 mg of NMP (875 µmol, 9.00 equiv.) were added. The mixture was cooled to –15 °C and 0.07 mL of a methyl magnesium bromide solution (214 µmol, 3 N in diethyl ether, 2.2 equiv.) was added dropwise. After 15 min, the mixture was allowed to warm to room temperature and stirred for further 2 h. The reaction mixture was cooled to –10 °C and quenched via the addition of 5 mL 1 N hydrochloric acid. The aqueous layer was extracted with 3 × 5 mL of ethyl acetate and the combined organic layers were washed with 5 mL of sodium carbonate solution, dried over sodium sulfate and the volatiles were removed under reduced pressure. The crude product was then purified via flash column chromatography (CH/EtOAc 15:1) resulted in 14 mg (44%) of the pure product as a white solid.

***R*f** (CH/EtOAc 5:1): 0.33. – **^{1}H NMR** (400 MHz, $CDCl_3$): δ = 7.22 – 7.05 (m, 2H, H_{Ar}), 7.05 – 6.86 (m, 3H, H_{Ar}), 6.77 (d, *J* = 7.9 Hz, 1H, H_{Ar}), 6.70 (d, *J* = 2.2 Hz, 1H, H_{Ar}), 3.83 (d, *J* = 17.5 Hz, 1H, CH_2), 3.50 (d, *J* = 17.2 Hz, 1H, CH_2), 3.18 (hept, *J* = 6.8 Hz, 1H, $CH(CH_3)_2$), 2.28 (s, 3H, CH_3), 1.76 (s, 3H, CH_3), 1.22 (dd, *J* = 6.5 Hz, 6H, 2 × CH_3) ppm. – **^{13}C NMR** (100 MHz, $CDCl_3$): δ = 153.2 ($C_{quart.}$), 148.8 ($C_{quart.}$), 141.9 ($C_{quart.}$), 130.3 (+, CH),, 128.8 ($C_{quart.}$), 127.9 (+, CH), 127.8 (+, CH), 123.2 ($C_{quart.}$), 122.5 ($C_{quart.}$), 121.4 (+, CH), 117.6 (+, CH), 117.4 (+, CH), 116.5 ($C_{quart.}$), 116.3 (+, CH), 101.9 ($C_{quart.}$), 32.3 (–, CH_2), 28.2 (+, CH), 24.6, 23.7 (+, CH_3), 23.4 (+, CH_3), 15.6 (+, CH_3) ppm. – **IR** (KBr): ṽ = 2959 (w), 2925 (w), 2868 (w), 1580 (w), 1485 (m), 1455 (m), 1419 (w), 1369 (w), 1296 (w), 1247 (m), 1229 (m), 1207 (m), 1142 (m), 1075 (m), 977 (m), 943 (w), 902 (m), 810 (m), 750 (s), 707 (w), 632 (w), 603 (vw), 535 (vw), 501 (w), 443 (vw) cm^{-1}. – **MS** (70 eV, EI): *m/z* (%) = 306 (44) $[M]^+$, 305 (100) $[M - H]^+$, 304 (11), 292 (7), 291 (29) $[M - CH_3]^+$, 290 (12), 289 (32), 280 (5), 276 (5), 261 (6), 173 (7), 145 (5). – **HRMS** ($C_{21}H_{22}O_2$): calc. 306.1620, found 306.1620.

5.3 Materials and Methods Controlled Release Systems (Chapter 3.3)

Materials

Poly(D,L-lactic-co-glycolic acid) [PLGA] of medical grade quality (Resomer® **PLGA-Et**, LotNr: 1002249, M_w 24 kDa, M_n 11.2 kDa, PDI 2.1; Resomer® **PLGA-COOH** LotNr: 1046458, M_w 22 kDa, M_n 1152 kDa, PDI 1.9) was purchased from Boehringer Ingelheim Chemicals, Inc. (Petersburg, VA, USA). Detergents used in this study were polyvinyl alcohol (PVA, Mowiol 4-88 LotNr.: DE 13 021 607, M_w 10 kDa, 88% hydrolyzed,) and Polysorbate 80 (Tween® 80, SigmaUltra) both from Sigma-Aldrich (St. Louis, MO, USA). All solvents and other chemicals were HPLC or USP grade or higher. Water was purified by a Millipore system.

Prepared solutions

PVA solutions (0.5% (w/w) PVA solution: 2.50 g of PVA and 497.5 g of water; 2% (w/w) PVA solution: 5.00 g of PVA and 245 g of water; 5% (w/w) PVA solution: 5.00 g of PVA and 95 g of water) were prepared as follows. The respective amount of water was placed in a Schott glass bottle and cooled to 0 °C. PVA was added and after 5 min of stirring the temperature raised to 100 °C until all solids were dissolved (ca. 3 h). The mixture was allowed to cool to room temperature under continued stirring for 16 h. Phosphate buffered saline (PBS) solution was prepared by dissolving 8.00 g of sodium chloride, 0.20 g of potassium chloride, 1.78 g of disodium hydrogen phosphate dihydrate and 0.27 g of potassium dihydrogen phosphate in 1.00 L water and adjusting the pH to 7.4. PBST solutions (Polysorbate 80 (Tween® 80) in PBS Buffer) were prepared as (v/v) mixtures.

Solubility

Dissolution of the coumarins in aqueous media with low drug solubility was tested by suspension of the drugs in 2 mL of the medium and incubation at room temperature on a rocking platform shaker for 3 days. The remaining solids were removed by high speed centrifugation and the drug solution analyzed by HPLC.

HPLC Analysis for Drug Quantification

Samples were analyzed on a DIONEX HPLC system (UltiMate 3000 pump, UltiMate 3000 Autosampler, UltiMate 3000 variable wavelength Detector) with a LiChroCART® 125-4 RP-18 column.

Drug quantification was realized by HPLC analysis and comparison of the absorption with the respective calibration curves (Chapter 3.3.1, Figure 20, B). The HPLC detector channel at 310 nm was used for quantification.

A gradient method with 0.1% TFA in water as solvent A and 0.1% TFA in acetonitrile (ACN) as solvent B was employed, starting with 70% solvent B and increasing the proportion to 95% over 12 min runtime at an elution speed of 1.00 mL/min. The injection volume was set to 50 µL sample. Retention time for **T1** was 4.1 min, while **T2** eluted after 6.9 min and **T3** after 7.4 min (Chapter 3.3.1, Figure 20, A). The lowest observed limit of quantification corresponded to a sample concentration of 0.1 µg/mL.

Encapsulation Efficiency and Recovery Assay

In order to determine the encapsulation efficiency (EE) or the remaining drug in microparticle pellets from the release studies, the lyophilized samples were dissolved in 1 mL of dichloromethane, and PLGA was precipitated by the addition of 9 mL of ethanol. After centrifugation (5 min, 14000 rpm, Hettich EBA 12), aliquots of the samples were analyzed via HPLC.

Microparticle/Nanoparticle Preparation and Encapsulation

Preparation of microparticles in the range of 40-80 µm

In a glass tube 15 wt% PLGA (234.7 mg) and 11.4 mg of the respective coumarin was dissolved in 1 mL of dichloromethane. The solution was overlaid with 2 mL of 2% (w/w) PVA-solution and shaken with a Vortex stirrer (IKA MS3 digital) at 3000 rpm for 45 s. The resulting suspension was then poured into a moderately stirred beaker with 40 mL of 0.5% (w/w) PVA solution (tube was rinsed with 1 mL of the bath solution). The suspension was allowed to stir for three hours at room temperature, in which the dichloromethane evaporated and particles were. Filtering through a MILLIPORE filter (0.45 µm polyamide) and lyophilization of the pellet (CHRIST Alpha 2-4) resulted in the dry particles.

Preparation of microparticles in the range of 2-5 µm

In a Falcon tube, 10 wt% PLGA (147.8 mg) and 7.39 mg of the respective coumarin was dissolved in 1 mL of dichloromethane. The solution was overlaid with 2 mL of 2% (w/w) PVA-solution and a suspension prepared by treating of the two-phased mixture with a high-performance dispersing instrument (IKA® T 25 digital Ultra Turrax) at 24000 rpm for 6 s. The resulting suspension was then poured into a moderately stirred beaker with 40 mL of 0.5% (w/w) PVA solution (Falcon tube was rinsed with 1 mL of the bath solution). The suspension was allowed to stir for three hours at room temperature in which the dichloromethane evaporated and particles were formed. The particles were harvested by centrifugation at 1300 rpm (CHRIST AVC 2-25 CD-plus) and washed 3 times by replacing the centrifugate with purified water and resuspension of the pellet. Lyophilization of the pellets (CHRIST Alpha 2-4) resulted in the dry particles.

Preparation of Nanoparticles in the range of 250 nm

In a Falcon tube 5 wt% PLGA (70.0 mg) and 3.5 mg of the respective coumarin was dissolved in 1 mL of dichloromethane. After full dissolution, 4 mL of 2% (w/w) PVA solution were added and a suspension prepared by treating of the two-phased mixture under cooling in an ice bath with an ultrasonic homogenizer (BANDELIN Sonopuls UW 3200, ultrasonic probe) at 52% intensity for up to 60 seconds. The resulting suspension was then poured into a moderately stirred beaker with 35 mL of 0.5% (w/w) PVA solution. The suspension was allowed to stir for three hours, in which the dichloromethane evaporated and particles were formed. The particles were harvested by centrifugation at 7000 rpm and 15 °C (HERAEUS Biofuge Primo R) and washed 3 times by replacing the centrifugate with purified water and resuspension of the pellet. The pellet was resuspended in 10 mL of purified water and 500 µL fractions lyophilized (CHRIST Alpha 2-4) to determine the particle mass density.

Particle characterization

Size - Static light scattering (SLS) measurements were conducted on a MALVERN Mastersizer 2000 in water. Dynamic light scattering (DLS) measurements were recorded on a BECKMAN COULTER Delsa™ Nano C (Doppler Electrophoretic Light Scattering Analyzer) in water.

Imaging – Light microscopic and fluorescence microscopy images were acquired with a LEICA DMI6000. Scanning Electron Microscopy (SEM) was performed using a CARL ZEISS SMT AG Gemini Supra 40 VP microscope with a EVERHART-THOMLEY-DETEKTOR (SE2) or an INLENS-detector, at an acceleration voltage of 3 kV. The samples were spattered with Iridium (4 nm layer) with a LOT QUORUM TECHNOLOGIES DARMSTADT Q150 ES, sputter coater (SE2) or measured as native samples (INLENS).

DSC -Differential scanning calorimetry (DSC; NETZSCH DSC 204 F1, Selb, Germany) in N_2 atmosphere was used to monitor phase transitions and evaluated from the second heating cycle between −100 °C and 150 °C with a heating and cooling rate of 10 K min^{-1}.

Release Studies

Release studies were realized in triplicates by putting a weighted sample (or aliquot of dispersion in case of the nanoparticles) of particles into 2 mL EPPENDORF tubes. The particles were overlayed with 1.5 mL of PBST (PBS buffer with 1 v/v % Polysorbate 80) and dispersed via vortexing (IKA MS3 digital, 30 s, 3000 rpm). The tubes were then incubated at 37.5 °C on a rocking platform shaker. At predefined time points, the tubes were removed and centrifuged in a HETTICH EBA 12 benchtop centrifuge at 14,000 rpm. 1 mL of the supernatant was withdrawn for HPLC analysis and replaced by fresh PBST. The samples were then dispersed again via vortexing and replaced into the incubation set-up. When the release studies were terminated, the remaining particles and solution were lyophilized (CHRIST Alpha 2-4) and a recovery assay was performed to determine the remaining drug.

5.3.1 Additional data Chapter 3.3

UV/Vis measurements

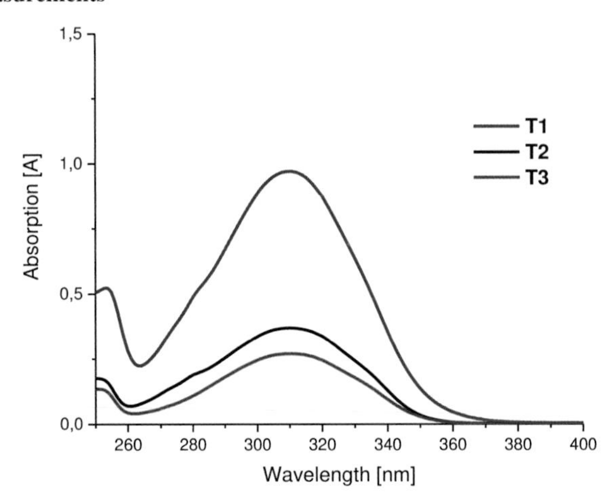

Figure 38: UV-Vis spectra for the three coumarins **T1-T3** measured in acetonitrile, maximum absorption for all compounds is at 310 nm.

Solubility of Coumarins T1-T3

Table 20: Maximum solubility of coumarins **T1-T3** in different solvents and buffer solutions at room temperature.

Solution	max. solubility T1 [µg/mL]	max. solubility T2 [µg/mL]	max. solubility T3 [µg/mL]
water	0.006	0.004	0.043
PBS	0.003	0.006	0.004
0.5% PVA	58.5	0.06	0.12
2% PVA	179	0.08	0.21
0.1% PBST[a]	30.9	1.69	2.79
0.5% PBST[a]	36.8	7.23	3.13
1% PBST[a]	95.5	16.5	18.3
5% PBST[a]	300	111	160

[a] solution of polysorbate 80 (Tween 80®) in PBS Buffer (v/v) ratio.

Fluorescence spectra for compound T1-T3

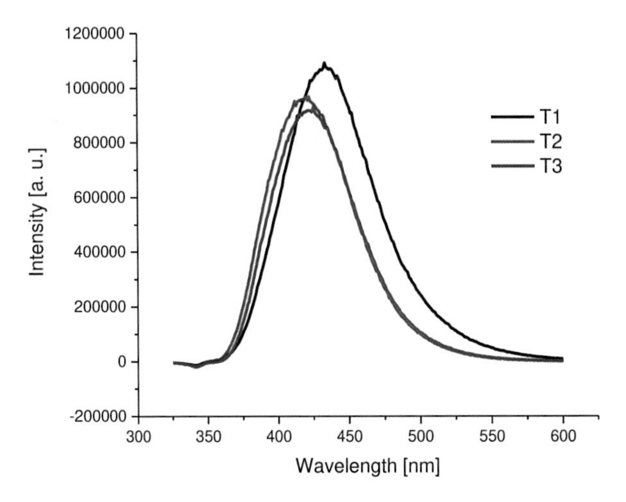

Figure 39: Fluorescence spectra of coumarins **T1-T3** solutions in acetonitrile after excitation at 310 nm. **T1** (62.5 µg/mL) maximum 433 nm, **T2** (0.5 mg/mL) maximum 422 nm, **T3** (0.5 mg/mL) maximum 425 nm.

Fluorescence spectra for compound T1

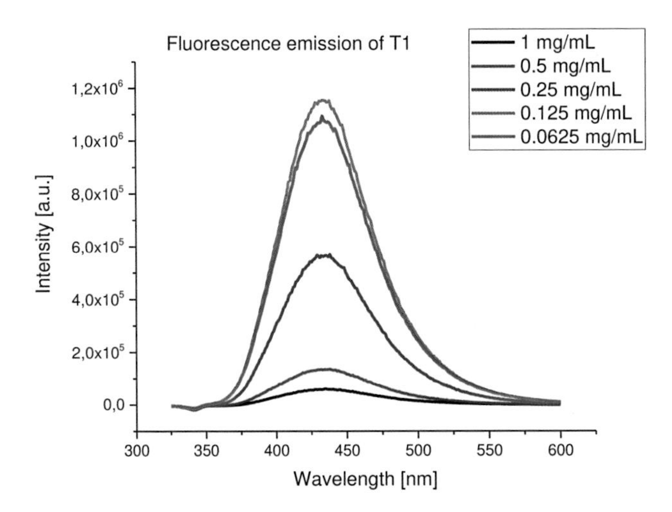

Figure 40: Fluorescence emission spectra of compound **T1** in acetonitrile, that shows self-quenching at high concentrations.

5.4 Crystallographic Data

The crystal structures were measured and elucidated by Dr. Martin NIEGER at the institute for inorganic chemistry, university of Helsinki (Finland).

3-(2-hydroxybenzyl)-5-isopropyl-8-methyl-2H-chromen-2-one – 26c

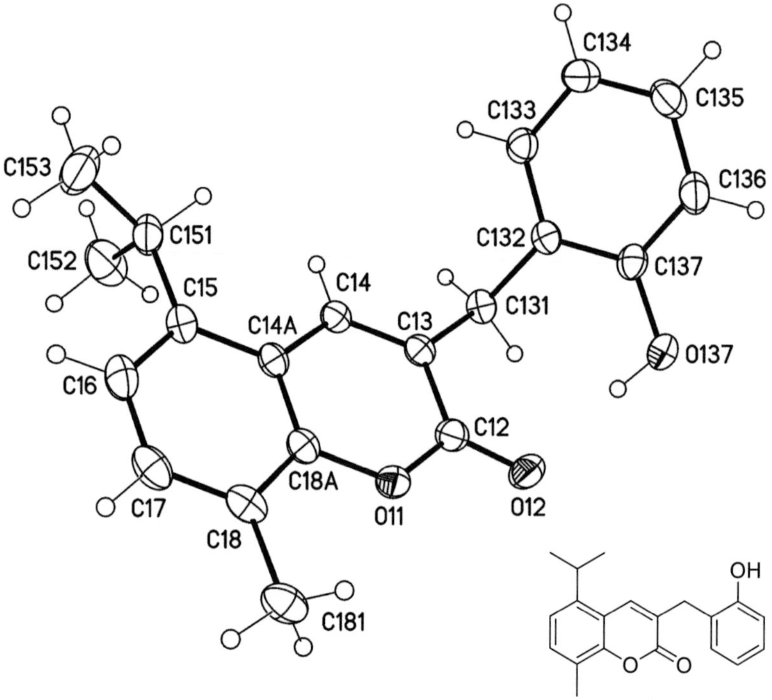

Crystal data

$C_{20}H_{20}O_3$	$F(000) = 1312$
$M_r = 308.36$	$D_x = 1.263$ Mg m^{-3}
Monoclinic, $P2_1/n$ *(no.14)*	Cu *K*a radiation, l = 1.54178 Å
$a = 13.0044$ (4) Å	Cell parameters from 9767 reflections
$b = 18.1553$ (5) Å	q = 4.0–72.0°
$c = 14.2893$ (4) Å	m = 0.67 mm^{-1}
b = 106.009 (1)	$T = 123$ K
$V = 3242.85$ (16) Å^3	Blocks, colourless
$Z = 8$	0.24 × 0.18 × 0.14 mm

Data collection

Bruker D8 VENTURE diffractometer with Photon100 detector	6387 independent reflections
Radiation source: INCOATEC microfocus sealed tube	5752 reflections with I > 2s(I)
Detector resolution: 10.4167 pixels mm^{-1}	R_{int} = 0.030
rotation in f and w, 1°, shutterless scans	q_{max} = 72.1°, q_{min} = 4.0°
Absorption correction: multi-scan *SADABS* (Sheldrick, 2014)	h = -16®16
T_{min} = 0.820, T_{max} = 0.902	k = -22®21
50176 measured reflections	l = -17®17

Refinement

Refinement on F^2	Secondary atom site location: difference Fourier map
Least-squares matrix: full	Hydrogen site location: difference Fourier map
$R[F^2 > 2s(F^2)]$ = 0.038	H atoms treated by a mixture of independent and constrained refinement
$wR(F^2)$ = 0.101	$w = 1/[s^2(F_o^2) + (0.0485P)^2 + 1.1456P]$ where $P = (F_o^2 + 2F_c^2)/3$
S = 1.03	$(D/s)_{max}$ = 0.001
6387 reflections	$Dñ_{max}$ = 0.26 e $Å^{-3}$
424 parameters	$Dñ_{min}$ = -0.16 e $Å^{-3}$
2 restraints	Extinction correction: *SHELXL2014/7* (Sheldrick 2014, $Fc^* = kFc[1+0.001xFc^2l^3/sin(2q)]^{-1/4}$
Primary atom site location: dual	Extinction coefficient: 0.00070 (8)

7-(1-Butylcyclopentyl)-5-hydroxy-3-(o-tolyl)-2*H*-chromen-2-one – (40d-H)

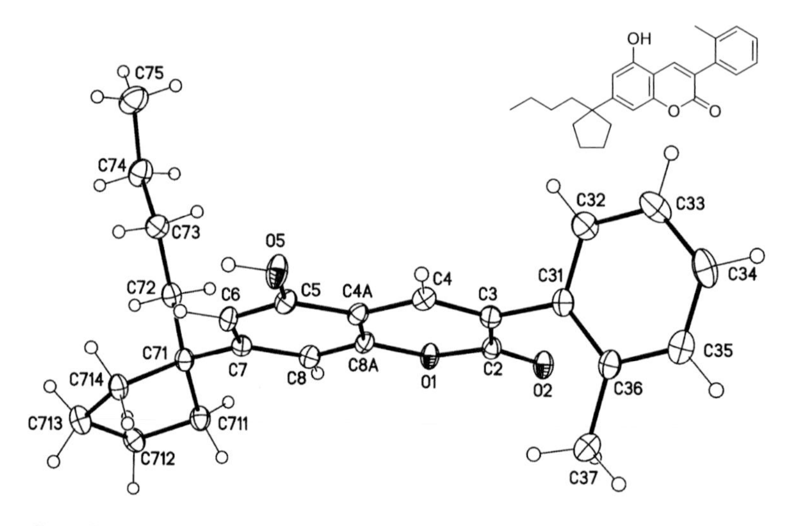

Crystal data

$C_{25}H_{28}O_3$	D_x = 1.242 Mg m^{-3}
M_r = 376.47	Cu $K\alpha$ radiation, λ = 1.54178 Å
Orthorhombic, $P2_12_12_1$ (no.19)	Cell parameters from 9743 reflections
a = 7.5849 (2) Å	θ = 3.8–72.2°
b = 15.4996 (4) Å	μ = 0.63 mm^{-1}
c = 17.1249 (5) Å	T = 123 K
V = 2013.25 (9) Å^3	Blocks, colourless
Z = 4	0.24 × 0.20 × 0.18 mm
F(000) = 808	

Data collection

Bruker D8 VENTURE diffractometer with Photon100 detector	3951 independent reflections
Radiation source: INCOATEC microfocus sealed tube	3857 reflections with $I > 2\sigma(I)$
Detector resolution: 10.4167 pixels mm^{-1}	R_{int} = 0.031
rotation in ϕ and ω, 1°, shutterless scans	θ_{max} = 72.3°, θ_{min} = 3.9°
Absorption correction: multi-scan *SADABS* (Sheldrick, 2014)	h = -9→9
T_{min} = 0.836, T_{max} = 0.915	k = -19→18
21481 measured reflections	l = -21→18

Refinement

Refinement on F^2	Hydrogen site location: difference Fourier map
Least-squares matrix: full	H atoms treated by a mixture of independent and constrained refinement
$R[F^2 > 2\sigma(F^2)] = 0.028$	$w = 1/[\sigma^2(F_o^2) + (0.0347P)^2 + 0.4679P]$ where $P = (F_o^2 + 2F_c^2)/3$
$wR(F^2) = 0.071$	$(\Delta/\sigma)_{max} < 0.001$
$S = 1.04$	$\Delta\rangle_{max} = 0.20$ e Å^{-3}
3951 reflections	$\Delta\rangle_{min} = -0.15$ e Å^{-3}
258 parameters	Extinction correction: *SHELXL2014/7* (Sheldrick 2014), $Fc^*=kFc[1+0.001xFc^2\lambda^3/\sin(2\theta)]^{-1/4}$
1 restraint	Extinction coefficient: 0.0026 (3)
Primary atom site location: structure-invariant direct methods	Absolute structure: Flack x determined using 1620 quotients [(I+)-(I-)]/[(I+)+(I-)] (Parsons, Flack and Wagner, Acta Cryst. B69 (2013) 249-259).
Secondary atom site location: difference Fourier map	Absolute structure parameter: -0.03 (6)

7-(1-butylcyclopentyl)-5-hydroxy-3-(m-tolyl)-2H-chromen-2-one – (40e-H)

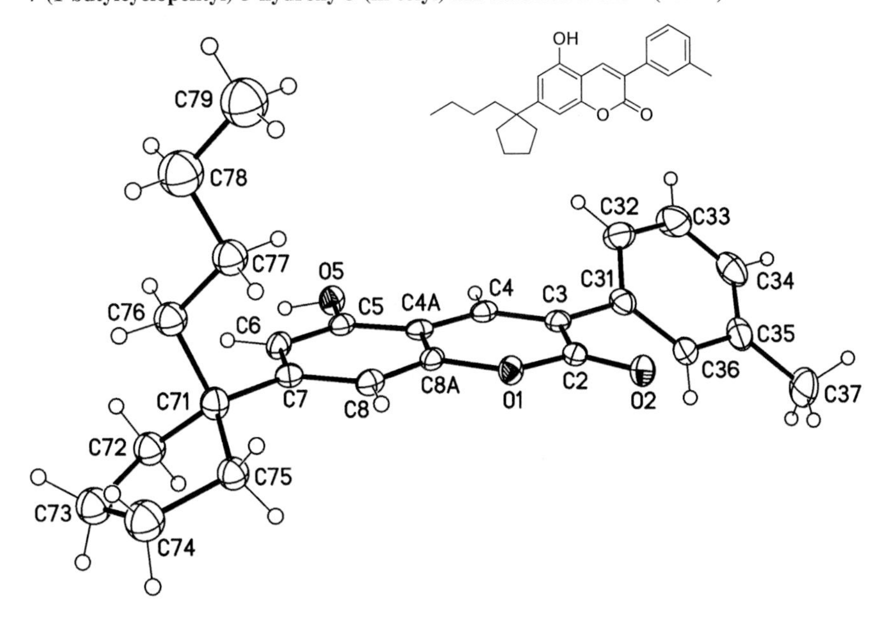

Crystal data

$C_{25}H_{28}O_3$	D_x = 1.232 Mg m^{-3}
M_r = 376.47	Cu *Ka* radiation, l = 1.54178 Å
Orthorhombic, *Pccn (no.56)*	Cell parameters from 9785 reflections
a = 21.6569 (9) Å	q = 2.8–72.1°
b = 22.1818 (8) Å	m = 0.63 mm^{-1}
c = 16.9012 (6) Å	*T* = 123 K
V = 8119.2 (5) $Å^3$	Blocks, colourless
Z = 16	0.20 × 0.10 × 0.06 mm
F(000) = 3232	

Data collection

Bruker D8 VENTURE diffractometer with Photon100 detector	8016 independent reflections
Radiation source: INCOATEC microfocus sealed tube	6575 reflections with *I* > 2s(*I*)
Detector resolution: 10.4167 pixels mm^{-1}	R_{int} = 0.050
rotation in f and w, 1°, shutterless scans	q_{max} = 72.2°, q_{min} = 2.9°
Absorption correction: multi-scan *SADABS* (SHeldrick, 2014)	*h* = -26®23
T_{min} = 0.887, T_{max} = 0.971	*k* = -25®27
58720 measured reflections	*l* = -19®20

Refinement

Refinement on F^2	Secondary atom site location: difference Fourier map
Least-squares matrix: full	Hydrogen site location: mixed
$R[F^2 > 2s(F^2)]$ = 0.047	H atoms treated by a mixture of independent and constrained refinement
$wR(F^2)$ = 0.121	$w = 1/[s^2(F_o^2) + (0.0511P)^2 + 6.3124P]$ where $P = (F_o^2 + 2F_c^2)/3$
S = 1.02	$(D/s)_{max}$ = 0.001
8016 reflections	$Dñ_{max}$ = 0.35 e $Å^{-3}$
503 parameters	$Dñ_{min}$ = -0.35 e $Å^{-3}$
369 restraints	Extinction correction: *SHELXL2014/7* (Sheldrick 2014), $Fc^*=kFc[1+0.001xFc^2l^3/sin(2q)]^{-1/4}$
Primary atom site location: dual	Extinction coefficient: 0.00016 (2)

7-(1-butylcyclopentyl)-5-hydroxy-3-(p-tolyl)-2H-chromen-2-one – (40f-H)

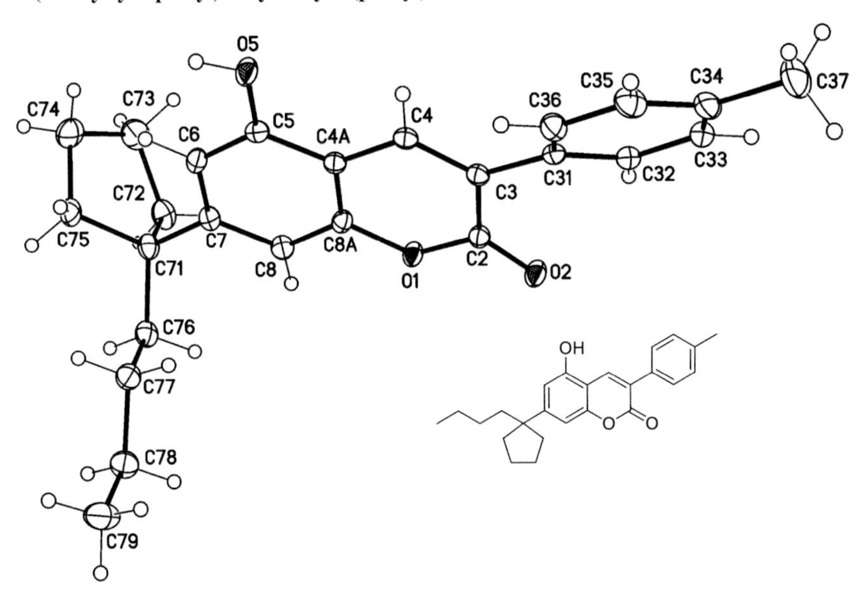

Crystal data

$C_{25}H_{28}O_3$	$F(000) = 1616$
$M_r = 376.47$	$D_x = 1.232$ Mg m^{-3}
Monoclinic, $P2_1/c$ *(no.14)*	Cu *K*a radiation, l = 1.54178 Å
$a = 16.4989$ (4) Å	Cell parameters from 9923 reflections
$b = 17.2141$ (4) Å	q = 2.8–72.0°
$c = 15.3705$ (4) Å	m = 0.63 mm^{-1}
b = 111.580 (1)°	$T = 123$ K
$V = 4059.44$ (17) Å^3	Blocks, colourless
$Z = 8$	0.20 × 0.12 × 0.08 mm

Data collection

Bruker D8 VENTURE diffractometer with Photon100 detector	7985 independent reflections
Radiation source: INCOATEC microfocus sealed tube	6868 reflections with $I > 2s(I)$
Detector resolution: 10.4167 pixels mm^{-1}	$R_{int} = 0.034$
rotation in f and w, 1°, shutterless scans	$q_{max} = 72.1°$, $q_{min} = 2.9°$
Absorption correction: multi-scan *SADABS* (Sheldrick, 2014)	h = -19®20
$T_{min} = 0.900$, $T_{max} = 0.958$	k = -20®21

44363 measured reflections	l = -17®18

Refinement

Refinement on F^2	Secondary atom site location: difference Fourier map
Least-squares matrix: full	Hydrogen site location: difference Fourier map
$R[F^2 > 2s(F^2)] = 0.037$	H atoms treated by a mixture of independent and constrained refinement
$wR(F^2) = 0.093$	$w = 1/[s^2(F_o^2) + (0.0404P)^2 + 1.815P]$ where $P = (F_o^2 + 2F_c^2)/3$
$S = 1.02$	$(D/s)_{max} = 0.001$
7985 reflections	$Dñ_{max} = 0.28$ e Å^{-3}
514 parameters	$Dñ_{min} = -0.18$ e Å^{-3}
2 restraints	Extinction correction: *SHELXL2014/7* (Sheldrick 2014, $Fc^* = kFc[1+0.001xFc^2l^3/sin(2q)]^{-1/4}$
Primary atom site location: dual	Extinction coefficient: 0.00071 (7)

6. List of Abbreviations

$	US-Dollar
(v/v)	volume/volume ratio
(w/w)	weight/weight ratio
°C	degree celsius
µg	microgram
µL	microliter
µmol	micromole
2-AG	2-arachidonoylglycerol
Ac	acetyl
ACN	acetonitrile
AD	anno domini
AIDS	acquired immune deficiency syndrome through human immunodeficiency virus infection
AR	adrenergic receptor
Ar	aromat
ATR	attenuated total reflection
b	broad
BC	before christ
Bn	benzyl
bs	broad singlet,
C. sativa	*Cannabis sativa*, species of the hemp plant
calc.	calculated
CB	cannabinoid

CB_1	cannabinoid receptor 1
CB_2	cannabinoid receptor 2
CBC	cannabichromene
CBD	cannabidiol
CBG	cannabigerol
CH	cyclohexane
CNS	central nervous system
d	day
d	doublet
DBU	1,8-diazabicyclo[5.4.0]undec-7-ene
DCM	dichloro methane
DEPT	distortionless enhancement by polarization transfer
DLS	dynamic light scattering
DMAP	4-dimethylaminopyridine
DMDO	dimethyldioxiran
DMF	dimethylformamide
DMSO	dimethylsulfoxid
DSC	differential scanning calorimetry
e.g.	exempli gratia (for example)
EE	encapsulation efficiency
equiv.	equivalents
Et	ethyl
EtOAc	ethyl acetate

FG	functional group
g	gram
GCMS	gas chromatography–mass spectrometry
GDP	guanosine diphosphate
gem.	geminal
GP	general procedure
GPCR	G-protein coupled receptor
GPP	geranyldiphosphat
GPR18	G-protein coupled receptor 18
GPR55	G-protein coupled receptor 55
G-Protein	guanine nucleotide-binding protein
GTP	guanosine triphosphate
h	hour
hept	heptet,
HPLC	high performance liquid chromatography
HRMS	high resolution mass spectrometry
Hz	hertz
i.e.	id est (that is)
IL	ionic liquid
in situ	latin for “on site”, without isolation
*i*Pr	isopropyl, prop-2-yl
IR	infrared
J	coupling constant

K	kelvin
kV	kilo volt
L	liter
Log *P*	octanol-water partition coefficient
LotNr.	charge number
m	meta
m	middle
M	molar
m	multiplet
m-cPBA	meta-chloroperoxybenzoic acid
Me	methyl
mg	milligram
MHz	mega hertz
min	minute
mL	milliliter
mM	milli molar
mmol	millimole
MS	molecular sieves
M_w	molecular weight
MWI	micro wave irradiation
N	normality/ equivalent concentration
n. i.	not isolated
n.d.	not determined

NAH	northern aliphatic hydroxy
*n*BuL	butyllithium
NHC	*N*-heterocyclic carbene
NMR	nuclear magnetic resonance
NOESY	nuclear overhauser enhancement and exchange spectroscopy
o	ortho
o/w	oil in water
p	para
PBL	peripheral blood leukocytes
PBS	phosphate-buffered saline
PBST	PBS + Tween® 80 surfactant
PCL	polycaprolactone
PH	phenolic hydroxy
Ph	phenyl
pH	potential of hydrogen, logarithm of the activity of hydrogen ions
PLA	poly lactic acid
PLGA	poly[(D,L-lactide-co-glycolide)]
PLGA-COOH	poly[(D,L-lactide-co-glycolide)] with free acid endgroups
PLGA-Et	poly[(D,L-lactide-co-glycolide)] with ethyl ester endgroups
ppm	parts per million
p-TsOH	*para*-toluonsulfonsäure
PVA	polyvinylalcohol
q	quartet

RO5	rule of five
rpm	rounds per minute
s	seconds
s	singlet
s	strong
s/o/w	solid in oil in water
SAH	southern aliphatic hydroxy
SC	side chain
SEM	Scanning Electron Microscopy
SLS	Static Light Scattering
t	triplet
T	transmission
*t*Bu	*tert*-butyl
TFA	trifluoroacetic acid
TFAA	trifluoracetic anhydride
Tg	glass transition temperature
THC	Δ^9-tetrahydrocannabinol
THCA	tetrahydrocannabinolic acid
THF	tetrahydrofuran
TLC	thin layer chromatogrphy
TMEDA	tetramethylethylenediamine
UNODC	United Nations Office on Drugs and Crime
US, USA	United States of America

UV	ultraviolet
V	volt
Vis	visible light
vs	very strong
vw	very weak
w	weak
W	watt
wt%	weight percent
δ	chemical shift

7. Literature

[1] T. Hurrle, Master Thesis, *Arbeiten zur formalen Totalsynthese von Δ9-Tetrahydrocannabinol und Synthese von Cannabinoid-Analoga*, Karlsruhe Institute of Technology (Kalrsruhe), **2014**.

[2] H. Peters, G. G. Nahas, in *Marihuana and medicine*, Springer, New York, **1999**, pp. 3-7, *A Brief History of Four Millennia (BC 2000—AD 1974).*

[3] M. Touw, *J. Psychoactive Drugs* **1981**, *13*, 23-34, *The Religious and Medicinal Uses of Cannabis in China, India and Tibet.*

[4] E. B. Russo, *Chem. Biodivers.* **2007**, *4*, 1614-1648, *History of cannabis and its preparations in saga, science, and sobriquet.*

[5] P. A. Matthioli, *Kreutterbuch des hochgelehrten und weitberühmten Herrn D. Petri Andrea Matthioli. Gemehrte und verfertigt durch Ioachimum Camerarium*, Faksimile, Frankfurt, **1626**.

[6] R. Mechoulam, in *Cannabinoids*, John Wiley & Sons, Ltd, Chichester, UK, **2014**, pp. 1-15, *Looking ahead after 50 years of research on cannabinoids.*

[7] L. Grinspoon, *Sci. Am.* **1969**, *221*, 17-25, *Marihuana.*

[8] D. F. Musto, *Arch. Gen. Psychiat.* **1972**, *26*, 101-108, *The marihuana tax act of 1937.*

[9] J. C. Anthony, F. Echeagaray-Wagner, *Alcohol Res. Health* **2000**, *24*, 201-208, *Epidemiologic analysis of alcohol and tobacco use.*

[10] D. E. Falk, H. Yi, S. Hiller-Sturmhofel, *Alcohol Res. Health* **2006**, *29*, 162-171, *An epidemiologic analysis of co-occurring alcohol and tobacco use and disorders.*

[11] United Nations Office on Drugs and Crime, *World Drug Report 2008*, **2008**.

[12] United Nations Office on Drugs and Crime, *World Drug Report 2017*, **2017**.

[13] A. C. Howlett, M. R. Johnson, L. S. Melvin, G. M. Milne, *Mol. Pharmacol.* **1988**, *33*, 297-302, *Nonclassical cannabinoid analgetics inhibit adenylate cyclase: development of a cannabinoid receptor model.*

[14] R. G. Pertwee, *Pharmacol. Therapeut.* **1997**, *74*, 129-180, *Pharmacology of cannabinoid CB1 and CB2 receptors.*

[15] R. G. Pertwee, *Life Sci.* **1999**, *65*, 597-605, *Evidence for the presence of CB1 cannabinoid receptors on peripheral neurones and for the existence of neuronal non-CB1 cannabinoid receptors.*

[16] United Nations Office on Drugs and Crime, *World Drug Report 2013*, **2013**.

[17] United Nations Office on Drugs and Crime, *World Drug Report 2014*, **2014**.

[18] S. Lake, T. Kerr, *Int. J. Health Policy Manag,* **2017**, *6*, 285, *The Challenges of Projecting the Public Health Impacts of Marijuana Legalization in Canada: Comment on" Legalizing and Regulating Marijuana in Canada: Review of Potential Economic, Social, and Health Impacts".*

[19] D. Piomelli, *Cannabis and Cannabinoid Research* **2016**, *Introduction to Cannabis and Cannabinoid Research.*

[20] Y. Gaoni, R. Mechoulam, *J. Am. Chem. Soc.* **1964**, *86*, 1646-1647, *Isolation, Structure, and Partial Synthesis of an Active Constituent of Hashish.*

[21] R. Mechoulam, Y. Shvo, *Tetrahedron* **1963**, *19*, 2073-2078, *Hashish—I: The structure of Cannabidiol.*

[22] Y. Gaoni, R. Mechoulam, *P. Chem. Soc. London* **1964**, 82, *Structure+ synthesis of cannabigerol new hashish constituent.*

[23] Y. Gaoni, R. Mechoulam, *Chem. Commun.* **1966**, 20-21, *Cannabichromene, a new active principle in hashish.*

[24] V. Di Marzo, L. D. Petrocellis, in *Cannabinoids*, John Wiley & Sons, Ltd, Chichester, UK, **2014**, pp. 261-289, *Fifty years of 'cannabinoid research' and the need for a new nomenclature.*

[25] R. G. Pertwee, in *Endocannabinoids* (Ed.: R. G. Pertwee), Springer International Publishing, Heidelberg, **2015**, pp. 1-37, *Endocannabinoids and Their Pharmacological Actions.*

[26] J. Gertsch, R. G. Pertwee, V. Di Marzo, *Brit. J. Pharmacol.* **2010**, *160*, 523-529, *Phytocannabinoids beyond the Cannabis plant – do they exist?*

[27] R. Mechoulam, S. Ben-Shabat, L. Hanus, M. Ligumsky, N. E. Kaminski, A. R. Schatz, A. Gopher, S. Almog, B. R. Martin, D. R. Compton, *Biochem. Pharmacol.* **1995**, *50*, 83-90, *Identification of an endogenous 2-monoglyceride, present in canine gut, that binds to cannabinoid receptors.*

[28] T. Sugiura, S. Kondo, A. Sukagawa, S. Nakane, A. Shinoda, K. Itoh, A. Yamashita, K. Waku, *Biochem. Biophys. Res. Commun.* **1995**, *215*, 89-97, *2-Arachidonoylgylcerol: A Possible Endogenous Cannabinoid Receptor Ligand in Brain.*

[29] W. A. Devane, A. Breuer, T. Sheskin, T. U. Jaerbe, M. S. Eisen, R. Mechoulam, *J. Med. Chem.* **1992**, *35*, 2065-2069, *A novel probe for the cannabinoid receptor.*

[30] L. Lemberger, in *Marihuana and Medicine* (Eds.: G. G. Nahas, K. M. Sutin, D. Harvey, S. Agurell, N. Pace, R. Cancro), Humana Press, Totowa, NJ, **1999**, pp. 561-566, *Nabilone.*

[31] C. Manera, T. Tuccinardi, A. Martinelli, *Mini-Rev. Med. Chem.* **2008**, *8*, 370-387, *Indoles and related compounds as cannabinoid ligands.*

[32] Biornica, *From seed to crystalline Cannabidiol: We leave no detail to chance.*, **30.10.2017**, http://www.cannabidiol-solutions.com/production-process/.

[33] R. Mechoulam, L. r. Hanuš, *Chem. Phys. Lipids* **2002**, *121*, 35-43, *Cannabidiol: an overview of some chemical and pharmacological aspects. Part I: chemical aspects.*

[34] R. Mechoulam, L. A. Parker, R. Gallily, *J. Clin. Pharmacol.* **2002**, *42*, 11S-19S, *Cannabidiol: An Overview of Some Pharmacological Aspects.*

[35] R. Kupper, Google Patents, **2006**, *Cannabinoid active pharmaceutical ingredient for improved dosage forms.*

[36] W. Horper, F.-J. Marner, *Phytochemistry* **1996**, *41*, 451-456, *Biosynthesis of primin and miconidin and its derivatives.*

[37] M. Fellermeier, M. H. Zenk, *FEBS Lett.* **1998**, *427*, 283-285, *Prenylation of olivetolate by a hemp transferase yields cannabigerolic acid, the precursor of tetrahydrocannabinol.*

[38] R. Mechoulam, Y. Gaoni, *JACS* **1965**, *87*, 3273-3275, *A Total Synthesis of dl-Δ1-Tetrahydrocannabinol, the Active Constituent of Hashish1.*

[39] K. E. Fahrenholtz, M. Lurie, R. W. Kierstead, *J. Am. Chem. Soc.* **1966**, *88*, 2079-2080, *Total Synthesis of dl-Δ9-Tetrahydrocannabinol and of dl-Δ8-Tetrahydrocannabinol, Racemates of Active Constituents of Marihuana.*

[40] K. E. Fahrenholtz, M. Lurie, R. W. Kierstead, *J. Am. Chem. Soc.* **1967**, *89*, 5934-5941, *Total synthesis of (.+-.)-. DELTA. 9-tetrahydrocannabinol and four of its isomers.*

[41] R. Mechoulam, P. Braun, Y. Gaoni, *J. Am. Chem. Soc.* **1967**, *89*, 4552-4554, *Stereospecific synthesis of (-)-.DELTA.1- and (-)-.DELTA.1(6)-tetrahydrocannabinols.*

[42] R. Mechoulam, P. Braun, Y. Gaoni, *J. Am. Chem. Soc.* **1972**, *94*, 6159-6165, *Syntheses of .DELTA.1-tetrahydrocannabinol and related cannabinoids.*

[43] R. K. Razdan, H. C. Dalzell, G. R. Handrick, *J. Am. Chem. Soc.* **1974**, *96*, 5860-5865, *Hashish. X. Simple one-step synthesis of (-)-.DELTA.1-tetrahydrocannabinol (THC) from p-mentha-2,8-dien-1-ol and olivetol.*

[44] T. H. Chan, T. Chaly, *Tetrahedron Lett.* **1982**, *23*, 2935-2938, *A biomimetic synthesis of Δ1-tetrahydrocannabinol.*

[45] W. E. Childers, H. W. Pinnick, *J. Org. Chem.* **1984**, *49*, 5276-5277, *A novel approach to the synthesis of the cannabinoids.*

[46] M. Moore, R. Rickards, H. Rønneberg, *Aust. J. Chem.* **1984**, *37*, 2339-2348, *Cannabinoid studies. IV. Stereoselective and regiospecific syntheses of Delta-9trans- and Delta-9cis-6a, 10a-Tetrahydrocannabinol.*

[47] L. Crombie, W. M. L. Crombie, S. V. Jamieson, C. J. Palmer, *J. Chem. Soc., Perkin Trans. 1* **1988**, 1243-1250, *Acid-catalysed terpenylations of olivetol in the synthesis of cannabinoids.*

[48] D. A. Evans, E. A. Shaughnessy, D. M. Barnes, *Tetrahedron Lett.* **1997**, *38*, 3193-3194, *Cationic bis(oxazoline)Cu(II) lewis acid catalysts. Application to the asymmetric synthesis of ent-Δ1-tetrahydrocannabinol.*

[49] D. A. Evans, D. M. Barnes, J. S. Johnson, T. Lectka, P. von Matt, S. J. Miller, J. A. Murry, R. D. Norcross, E. A. Shaughnessy, K. R. Campos, *J. Am. Chem. Soc.* **1999**, *121*, 7582-7594, *Bis(oxazoline) and Bis(oxazolinyl)pyridine Copper Complexes as Enantioselective Diels−Alder Catalysts: Reaction Scope and Synthetic Applications.*

[50] A. V. Malkov, P. Kočovský, *Collect. Czech. Chem. Commun.* **2001**, *66*, 1257-1268, *Tetrahydrocannabinol revisited: Synthetic approaches utilizing molybdenum catalysts.*

[51] A. D. William, Y. Kobayashi, *Org. Lett.* **2001**, *3*, 2017-2020, *A Method To Accomplish a 1,4-Addition Reaction of Bulky Nucleophiles to Enones and Subsequent Formation of Reactive Enolates.*

[52] A. D. William, Y. Kobayashi, *J. Org. Chem.* **2002**, *67*, 8771-8782, *Synthesis of Tetrahydrocannabinols Based on an Indirect 1,4-Addition Strategy.*

[53] S. P. Nikas, G. A. Thakur, D. Parrish, S. O. Alapafuja, M. A. Huestis, A. Makriyannis, *Tetrahedron* **2007**, *63*, 8112-8123, *A concise methodology for the synthesis of (−)-Δ9-tetrahydrocannabinol and (−)-Δ9-tetrahydrocannabivarin metabolites and their regiospecifically deuterated analogs.*

[54] B. M. Trost, K. Dogra, *Org. Lett.* **2007**, *9*, 861-863, *Synthesis of (-)-Δ9-trans-Tetrahydrocannabinol: Stereocontrol via Mo-Catalyzed Asymmetric Allylic Alkylation Reaction.*

[55] E. L. Pearson, N. Kanizaj, A. C. Willis, M. N. Paddon-Row, M. S. Sherburn, *Chem. Eur. J.* **2010**, *16*, 8280-8284, *Experimental and Computational Studies into an ATPH-Promoted exo-Selective IMDA Reaction: A Short Total Synthesis of Δ9-THC.*

[56] L. Minuti, E. Ballerini, *J. Org. Chem.* **2011**, *76*, 5392-5403, *High-Pressure Access to the Δ9-cis- and Δ9-trans-Tetrahydrocannabinols Family.*

[57] L.-J. Cheng, J.-H. Xie, Y. Chen, L.-X. Wang, Q.-L. Zhou, *Org. Lett.* **2013**, *15*, 764-767, *Enantioselective Total Synthesis of (−)-Δ8-THC and (−)-Δ9-THC via Catalytic Asymmetric Hydrogenation and SNAr Cyclization.*

[58] R. Mechoulam, Y. Gaoni, *Tetrahedron Lett.* **1967**, *8*, 1109-1111, *The absolute configuration of δ1-tetrahydrocannabinol, the major active constituent of hashish.*

[59] H. v. Pechmann, *Ber. Dtsch. Chem. Ges.* **1884**, *17*, 929-936, *Neue Bildungsweise der Cumarine. Synthese des Daphnetins. I.*

[60] R. G. Pertwee, *Pharmacol. Therapeut.* **1988**, *36*, 189-261, *The central neuropharmcology of psychotropic cannabinoids.*

[61] A. G. Gilman, *JAMA* **1989**, *262*, 1819-1825, *G proteins and regulation of adenylyl cyclase.*

[62] M. Rodbell, *Nature* **1980**, *284*, 17, *proteins in membrane transduction.*

[63] *The Nobel Prize in Physiology or Medicine 1994,* **1994**, 24.09.2014, http://www.nobelprize.org/nobel_prizes/medicine/laureates/1994/.

[64] A. C. Howlett, F. Barth, T. I. Bonner, G. Cabral, P. Casellas, W. A. Devane, C. C. Felder, M. Herkenham, K. Mackie, B. R. Martin, R. Mechoulam, R. G. Pertwee, *Pharmacol. Rev.* **2002**, *54*, 161-202, *International Union of Pharmacology. XXVII. Classification of Cannabinoid Receptors.*

[65] *The Nobel Prize in Chemistry 2012,* **2012**, 24.09.2014, http://www.nobelprize.org/nobel_prizes/chemistry/laureates/2012/.

[66] R. J. Lefkowitz, J. Roth, W. Pricer, I. Pastan, *P. Natl. Acad. Sci. USA* **1970**, *65*, 745-752, *ACTH Receptors in the Adrenal: Specific Binding of ACTH-125I and Its Relation to Adenyl Cyclase.*

[67] R. J. Lefkowitz, J. Roth, I. Pastan, *Science* **1970**, *170*, 633-635, *Radioreceptor Assay of Adrenocorticotropic Hormone: New Approach to Assay of Polypeptide Hormones in Plasma.*

[68] T. Frielle, S. Collins, K. W. Daniel, M. G. Caron, R. J. Lefkowitz, B. K. Kobilka, *P. Natl. Acad. Sci. USA* **1987**, *84*, 7920-7924, *Cloning of the cDNA for the human beta 1-adrenergic receptor.*

[69] R. Iyengar, L. Birnbaumer, *G-proteins*, Academic Pr., San Diego, **1990**.

[70] P. Kolb, D. M. Rosenbaum, J. J. Irwin, J. J. Fung, B. K. Kobilka, B. K. Shoichet, *P. Natl. Acad. Sci. USA* **2009**, *106*, 6843-6848, *Structure-based discovery of β2-adrenergic receptor ligands.*

[71] S. G. F. Rasmussen, B. T. DeVree, Y. Zou, A. C. Kruse, K. Y. Chung, T. S. Kobilka, F. S. Thian, P. S. Chae, E. Pardon, D. Calinski, J. M. Mathiesen, S. T. A. Shah, J. A. Lyons, M. Caffrey, S. H. Gellman, J. Steyaert, G. Skiniotis, W. I. Weis, R. K. Sunahara, B. K. Kobilka, *Nature* **2011**, *477*, 549-555, *Crystal structure of the [bgr]2 adrenergic receptor-Gs protein complex.*

[72] T. Hua, K. Vemuri, M. Pu, L. Qu, G. W. Han, Y. Wu, S. Zhao, W. Shui, S. Li, A. Korde, *Cell* **2016**, *167*, 750-762., *Crystal structure of the human cannabinoid receptor CB 1.*

[73] Z. Shao, J. Yin, K. Chapman, M. Grzemska, L. Clark, J. Wang, D. M. Rosenbaum, *Nature* **2016**, *High-resolution crystal structure of the human CB1 cannabinoid receptor.*

[74] S. G. F. Rasmussen, H.-J. Choi, D. M. Rosenbaum, T. S. Kobilka, F. S. Thian, P. C. Edwards, M. Burghammer, V. R. P. Ratnala, R. Sanishvili, R. F. Fischetti, G. F. X. Schertler, W. I. Weis, B. K. Kobilka, *Nature* **2007**, *450*, 383-387, *Crystal structure of the human [bgr]2 adrenergic G-protein-coupled receptor.*

[75] *The Nobel Prize in Chemistry 2012 - Popular Information,* **2014**, http://www.nobelprize.org/nobel_prizes/chemistry/laureates/2012/popular.html.

[76] W. Müller-Esterl, *Biochemie - Eine Einführung für Mediziner und Naturwissenschaftler*, Spektrum Akademischer Verlag, Heidelberg, **2009**.

[77] T. Kenakin, *Mol. Pharmacol.* **2004**, *65*, 2-11, *Efficacy as a Vector: the Relative Prevalence and Paucity of Inverse Agonism.*

[78] B. Jacobi, S. Partovi, *Basics Molekulare Zellbiologie*, 1. Aufl. ed., Elsevier, Urban & Fischer, München, **2011**.

[79] B. Lutz, in *Cannabinoids*, John Wiley & Sons, Ltd, Chichester, UK, **2014**, pp. 95-137, *Genetic dissection of the endocannabinoid system and how it changed our knowledge of cannabinoid pharmacology and mammalian physiology.*

[80] O. Aizpurua-Olaizola, I. Elezgarai, I. Rico-Barrio, I. Zarandona, N. Etxebarria, A. Usobiaga, *Drug Discovery Today* **2017**, *22*, 105-110, *Targeting the endocannabinoid system: future therapeutic strategies.*

[81] R. G. Pertwee, *Prog. Neurobiol.* **2001**, *63*, 569-611, *Cannabinoid receptors and pain.*

[82] K. Starowicz, N. Malek, B. Przewlocka, *Wiley Interdisciplinary Reviews: Membrane Transport and Signaling* **2013**, *2*, 121-132, *Cannabinoid receptors and pain.*

[83] J. M. Walker, A. G. Hohmann, W. J. Martin, N. M. Strangman, S. M. Huang, K. Tsou, *Life Sci.* **1999**, *65*, 665-673, *The neurobiology of cannabinoid analgesia.*

[84] K. Tsou, S. Brown, M. C. Sañudo-Peña, K. Mackie, J. M. Walker, *Neuroscience* **1998**, *83*, 393-411, *Immunohistochemical distribution of cannabinoid CB1 receptors in the rat central nervous system.*

[85] R. G. Pertwee, *Brit. J. Pharmacol.* **2007**, *152*, 984-986, *GPR55: a new member of the cannabinoid receptor clan?*

[86] S. P. H. Alexander, *Brit. J. Pharmacol.* **2012**, *165*, 2411-2413, *So what do we call GPR18 now?*

[87] D. McHugh, J. Page, E. Dunn, H. B. Bradshaw, *Brit. J. Pharmacol.* **2012**, *165*, 2414-2424, *Δ9-Tetrahydrocannabinol and N-arachidonyl glycine are full agonists at GPR18 receptors and induce migration in human endometrial HEC-1B cells.*

[88] J. Yu, E. Deliu, X.-Q. Zhang, N. E. Hoffman, R. L. Carter, L. A. Grisanti, G. C. Brailoiu, M. Madesh, J. Y. Cheung, T. Force, M. E. Abood, W. J. Koch, D. G. Tilley, E. Brailoiu, *J. Biol. Chem.* **2013**, *288*, 22481-22492, *Differential Activation of Cultured Neonatal Cardiomyocytes by Plasmalemmal Versus Intracellular G Protein-coupled Receptor 55.*

[89] C. M. Henstridge, N. A. B. Balenga, J. Kargl, C. Andradas, A. J. Brown, A. Irving, C. Sanchez, M. Waldhoer, *Molecular Endocrinology* **2011**, *25*, 1835-1848, *Minireview:*

Recent Developments in the Physiology and Pathology of the Lysophosphatidylinositol-Sensitive Receptor GPR55.

[90] C. M. Henstridge, *Pharmacology* **2012**, *89*, 179-187, *Off-target cannabinoid effects mediated by GPR55.*

[91] P. Zhao, M. E. Abood, *Life Sci.* **2013**, *92*, 453-457, *GPR55 and GPR35 and their relationship to cannabinoid and lysophospholipid receptors.*

[92] E. J. Rahn, A. G. Hohmann, *Neurotherapeutics* **2009**, *6*, 713-737, *Cannabinoids as Pharmacotherapies for Neuropathic Pain: From the Bench to the Bedside.*

[93] P. G. Baraldi, G. Saponaro, A. R. Moorman, R. Romagnoli, D. Preti, S. Baraldi, E. Ruggiero, K. Varani, M. Targa, F. Vincenzi, P. A. Borea, M. Aghazadeh Tabrizi, *J. Med. Chem.* **2012**, *55*, 6608-6623, *7-Oxo-[1,4]oxazino[2,3,4-ij]quinoline-6-carboxamides as Selective CB2 Cannabinoid Receptor Ligands: Structural Investigations around a Novel Class of Full Agonists.*

[94] R. A. Ross, *Trends Pharmacol. Sci.* **2011**, *32*, 265-269, *L-α-Lysophosphatidylinositol meets GPR55: a deadly relationship.*

[95] H. Sharir, M. E. Abood, *Pharmacol. Therapeut.* **2010**, *126*, 301-313, *Pharmacological characterization of GPR55, a putative cannabinoid receptor.*

[96] A. D. Khanolkar, S. L. Palmer, A. Makriyannis, *Chem. Phys. Lipids* **2000**, *108*, 37-52, *Molecular probes for the cannabinoid receptors.*

[97] J. E. Beal, R. Olson, L. Lefkowitz, L. Laubenstein, P. Bellman, B. Yangco, J. O. Morales, R. Murphy, W. Powderly, T. F. Plasse, K. W. Mosdell, K. V. Shepard, *J. Pain. Symptom Manag.* **1997**, *14*, 7-14, *Long-term efficacy and safety of dronabinol for acquired immunodeficiency syndrome-associated anorexia.*

[98] M. Lane, C. L. Vogel, J. Ferguson, S. Krasnow, J. L. Saiers, J. Hamm, K. Salva, P. H. Wiernik, C. P. Holroyde, S. Hammill, K. Shepard, T. Plasse, *J. Pain. Symptom Manag.* **1991**, *6*, 352-359, *Dronabinol and prochlorperazine in combination for treatment of cancer chemotherapy-induced nausea and vomiting.*

[99] F. Grotenhermen, *Cannabinoids* **2006**, *1*, 10-14, *Cannabinoids and the endocannabinoid system.*

[100] R. Q. Skrabek, L. Galimova, K. Ethans, D. Perry, *J. Pain* **2008**, *9*, 164-173, *Nabilone for the Treatment of Pain in Fibromyalgia.*

[101] M. P. Barnes, *Expert Opin. Pharmaco.* **2006**, *7*, 607-615, *Sativex®: clinical efficacy and tolerability in the treatment of symptoms of multiple sclerosis and neuropathic pain.*

[102] R. S. Padwal, S. R. Majumdar, *The Lancet* **1974**, *369*, 71-77, *Drug treatments for obesity: orlistat, sibutramine, and rimonabant.*

[103] C. B. Lee, *J. Korean Diabetes* **2013**, *14*, 58-62, *Weight Loss Drugs Recently Approved by the FDA.*

[104] P. Erkekoğlu, B. Giray, G. Şahin, *Fabad J. Pharm. Sci.* **2008**, *33*, 95-108, *The toxicological evaluation of rimonabant, taranabant, surinabant and otenabant in the treatment of obesity: Why the trials on endocannabinoid receptor antagonists and inverse agonists are suspended.*

[105] J. Toräng, S. Vanderheiden, M. Nieger, S. Bräse, *Eur. J. Org. Chem.* **2007**, *2007*, 943-952, *Synthesis of 3-Alkylcoumarins from Salicylaldehydes and α,β-Unsaturated Aldehydes Utilizing Nucleophilic Carbenes: A New Umpoled Domino Reaction.*

[106] V. Rempel, N. Volz, F. Glaser, M. Nieger, S. Brase, C. E. Muller, *J. Med. Chem.* **2013**, *56*, 4798-4810, *Antagonists for the orphan G-protein-coupled receptor GPR55 based on a coumarin scaffold.*

[107] D. Compton, K. C. Rice, B. R. De Costa, R. Razdan, L. S. Melvin, M. R. Johnson, B. R. Martin, *J. Pharmacol. Exp. Ther.* **1993**, *265*, 218-226, *Cannabinoid structure-activity relationships: correlation of receptor binding and in vivo activities.*

[108] A. Mahadevan, C. Siegel, B. R. Martin, M. E. Abood, I. Beletskaya, R. K. Razdan, *J. Med. Chem.* **2000**, *43*, 3778-3785, *Novel Cannabinol Probes for CB1 and CB2 Cannabinoid Receptors.*

[109] Y. Gareau, C. Dufresne, M. Gallant, C. Rochette, N. Sawyer, D. M. Slipetz, N. Tremblay, P. K. Weech, K. M. Metters, M. Labelle, *Bioorg. Med. Chem. Lett.* **1996**, *6*, 189-194, *Structure activity relationships of tetrahydrocannabinol analogues on human cannabinoid receptors.*

[110] L. E. Hollister, *Pharmacology* **1974**, *11*, 3-11, *Structure-Activity Relationships in Man of Cannabis Constituents, and Homologs and Metabolites of Δ9-Tetrahydrocannabinol.*

[111] N. Volz, Dissertation, *Sauerstoff-Heterocyclen als neue, selektive Liganden für die Cannabinoid-Rezeptoren*, Karlsruhe Institute of Technology (Karlsruhe), **2010**.

[112] D. P. Papahatjis, V. R. Nahmias, S. P. Nikas, T. Andreou, S. O. Alapafuja, A. Tsotinis, J. Guo, P. Fan, A. Makriyannis, *J. Med. Chem.* **2007**, *50*, 4048-4060, *C1'-Cycloalkyl Side Chain Pharmacophore in Tetrahydrocannabinols.*

[113] D. P. Papahatjis, S. P. Nikas, T. Kourouli, R. Chari, W. Xu, R. G. Pertwee, A. Makriyannis, *J. Med. Chem.* **2003**, *46*, 3221-3229, *Pharmacophoric Requirements for the Cannabinoid Side Chain. Probing the Cannabinoid Receptor Subsite at C1'.*

[114] F. Gläser, Dissertation, *Neuartige Cannabinoide - Synthese und biologische Evaluierung*, Karlsruhe Institute of Technology (Karlsruhe), **2014**.

[115] R. Sharma, S. P. Nikas, C. A. Paronis, J. T. Wood, A. Halikhedkar, J. J. Guo, G. A. Thakur, S. Kulkarni, O. Benchama, J. G. Raghav, *J. Med. Chem.* **2013**, *56*, 10142-10157, *Controlled-deactivation cannabinergic ligands.*

[116] Y. Jiang, W. Chen, W. Lu, *RSC Advances* **2012**, *2*, 1540-1546, *N-Heterocyclic carbene catalyzed conjugate umpolung reactions leading to coumarin derivatives.*

[117] J. G. Lombardino, J. A. Lowe, *Nat. Rev. Drug Discov.* **2004**, *3*, 853-862, *The role of the medicinal chemist in drug discovery [mdash] then and now.*

[118] S. Morgan, P. Grootendorst, J. Lexchin, C. Cunningham, D. Greyson, *Health Policy* **2011**, *100*, 4-17, *The cost of drug development: A systematic review.*

[119] K. H. Bleicher, H.-J. Bohm, K. Muller, A. I. Alanine, *Nat. Rev. Drug Discov.* **2003**, *2*, 369-378, *Hit and lead generation: beyond high-throughput screening.*

[120] J. Drews, *Science* **2000**, *287*, 1960-1964, *Drug Discovery: A Historical Perspective.*

[121] R. Santos, O. Ursu, A. Gaulton, A. P. Bento, R. S. Donadi, C. G. Bologa, A. Karlsson, B. Al-Lazikani, A. Hersey, T. I. Oprea, *Nat. Rev. Drug Discov.* **2017**, *16*, 19-34, *A comprehensive map of molecular drug targets.*

[122] C. A. Lipinski, F. Lombardo, B. W. Dominy, P. J. Feeney, *Adv. Drug. Deliver. Rev.* **1997**, *23*, 3-25, *Experimental and computational approaches to estimate solubility and permeability in drug discovery and development settings.*

[123] C. A. Lipinski, *Drug Discov. Today: Technologies* **2004**, *1*, 337-341, *Lead- and drug-like compounds: the rule-of-five revolution.*

[124] A. K. Ghose, V. N. Viswanadhan, J. J. Wendoloski, *J. Comb. Chem.* **1999**, *1*, 55-68, *A Knowledge-Based Approach in Designing Combinatorial or Medicinal Chemistry Libraries for Drug Discovery. 1. A Qualitative and Quantitative Characterization of Known Drug Databases.*

[125] C. Robert, C. S. Wilson, A. Venuta, M. Ferrari, C. D. Arreto, *J. Controlled Release* **2017**, *260*, 226-233, *Evolution of the scientific literature on drug delivery: A 1974–2015 bibliometric study.*

[126] L. Allen, H. C. Ansel, *Ansel's pharmaceutical dosage forms and drug delivery systems*, 10th Edition ed., Lippincott Williams & Wilkins, Baltimore, USA, **2013**.

[127] J. Folkman, D. M. Long, *J. Surg. Res.* **1964**, *4*, 139-142, *The use of silicone rubber as a carrier for prolonged drug therapy.*

[128] J. Folkman, D. M. Long, R. Rosenbaum, *Science* **1966**, *154*, 148-149, *Silicone rubber: a new diffusion property useful for general anesthesia.*

[129] A. S. Hoffman, *J. Controlled Release* **2008**, *132*, 153-163, *The origins and evolution of "controlled" drug delivery systems.*

[130] K. E. Uhrich, S. M. Cannizzaro, R. S. Langer, K. M. Shakesheff, *Chem. Rev.* **1999**, *99*, 3181-3198, *Polymeric Systems for Controlled Drug Release.*

[131] *Vitamin C + Zinc Capsules Depot,* **2017**, 10.10.2017, http://www.zeinpharma.com/vitamin-c-zinc-depot-capsules.

[132] P. M. Finch, L. J. Roberts, L. Price, N. C. Hadlow, P. T. Pullan, *Clin. J. Pain* **2000**, *16*, 251-254, *Hypogonadism in patients treated with intrathecal morphine.*

[133] P. P. Palmer, R. D. Miller, *Anesthesiology Clinics* **2010**, *28*, 587-599, *Current and developing methods of patient-controlled analgesia.*

[134] J. Weissberg-Benchell, J. Antisdel-Lomaglio, R. Seshadri, *Diabetes care* **2003**, *26*, 1079-1087, *Insulin pump therapy.*

[135] C. Wischke, S. P. Schwendeman, *Int. J. Pharm.* **2008**, *364*, 298-327, *Principles of encapsulating hydrophobic drugs in PLA/PLGA microparticles.*

[136] S. Mura, J. Nicolas, P. Couvreur, *Nat. Mater.* **2013**, *12*, 991, *Stimuli-responsive nanocarriers for drug delivery.*

[137] I. Bala, S. Hariharan, M. R. Kumar, *Crit. Rev. Ther. Drug.* **2004**, *21*, *PLGA nanoparticles in drug delivery: the state of the art.*

[138] V. V. Mody, A. Cox, S. Shah, A. Singh, W. Bevins, H. Parihar, *Applied Nanoscience* **2014**, *4*, 385-392, *Magnetic nanoparticle drug delivery systems for targeting tumor.*

[139] U. Prabhakar, H. Maeda, R. K. Jain, E. M. Sevick-Muraca, W. Zamboni, O. C. Farokhzad, S. T. Barry, A. Gabizon, P. Grodzinski, D. C. Blakey, in *Cancer Res.*, AACR, **2013**, *Challenges and key considerations of the enhanced permeability and retention effect for nanomedicine drug delivery in oncology.*

[140] E. Blanco, H. Shen, M. Ferrari, *Nat. Biotechnol.* **2015**, *33*, 941-951, *Principles of nanoparticle design for overcoming biological barriers to drug delivery.*

[141] V. Biju, *Chem. Soc. Rev.* **2014**, *43*, 744-764, *Chemical modifications and bioconjugate reactions of nanomaterials for sensing, imaging, drug delivery and therapy.*

[142] T. Garg, G. Rath, A. K. Goyal, *Drug deliv.* **2015**, *22*, 969-987, *Comprehensive review on additives of topical dosage forms for drug delivery.*

[143] V. P. Torchilin, *Nat. Rev. Drug Discov.* **2014**, *13*, 813, *Multifunctional, stimuli-sensitive nanoparticulate systems for drug delivery.*

[144] M. J. Alonso, R. K. Gupta, C. Min, G. R. Siber, R. Langer, *Vaccine* **1994**, *12*, 299-306, *Biodegradable microspheres as controlled-release tetanus toxoid delivery systems.*

[145] F. Esmaeili, M. H. Ghahremani, B. Esmaeili, M. R. Khoshayand, F. Atyabi, R. Dinarvand, *Int. J. Pharm.* **2008**, *349*, 249-255, *PLGA nanoparticles of different surface properties: Preparation and evaluation of their body distribution.*

[146] S. Freiberg, X. X. Zhu, *Int. J. Pharm.* **2004**, *282*, 1-18, *Polymer microspheres for controlled drug release.*

[147] C. Wischke, Y. Zhang, S. Mittal, S. P. Schwendeman, *Pharm. Res.* **2010**, *27*, 2063-2074, *Development of PLGA-Based Injectable Delivery Systems For Hydrophobic Fenretinide.*

[148] W. Jiang, R. K. Gupta, M. C. Deshpande, S. P. Schwendeman, *Adv. Drug. Deliver. Rev.* **2005**, *57*, 391-410, *Biodegradable poly(lactic-co-glycolic acid) microparticles for injectable delivery of vaccine antigens.*

[149] J. M. Anderson, M. S. Shive, *Adv. Drug. Deliver. Rev.* **2012**, *64*, 72-82, *Biodegradation and biocompatibility of PLA and PLGA microspheres.*

[150] R. K. Kulkarni, E. G. Moore, A. F. Hegyeli, F. Leonard, *J. of Biomed. Mater. Res.* **1971**, *5*, 169-181, *Biodegradable poly(lactic acid) polymers.*

[151] J. M. Brady, D. E. Cutright, R. A. Miller, G. C. Battistone, E. E. Hunsuck, *J. of Biomed. Mater. Res.* **1973**, *7*, 155-166, *Resorption rate, route of elimination, and ultrastructure of the implant site of polylactic acid in the abdominal wall of the rat.*

[152] D. E. Cutright, J. D. Beasley, B. Perez, *Oral Surg. Oral Med. O.* **1971**, *32*, 165-173, *Histologic comparison of polylactic and polyglycolic acid sutures.*

[153] A. J. Domb, W. Khan, *Polymeric Biomaterials: Structure and Function* **2013**, *1*, 135, *5 Biodegradable Polymers.*

[154] C. Wischke, J. Zimmermann, B. Wessinger, A. Schendler, H.-H. Borchert, J. H. Peters, T. Nesselhut, D. R. Lorenzen, *Int. J. Pharm.* **2009**, *365*, 61-68, *Poly (I: C) coated PLGA microparticles induce dendritic cell maturation.*

[155] S. Jhunjhunwala, G. Raimondi, A. W. Thomson, S. R. Little, *J. Controlled Release* **2009**, *133*, 191-197, *Delivery of rapamycin to dendritic cells using degradable microparticles.*

[156] C. Wischke, S. Mathew, T. Roch, M. Frentsch, A. Lendlein, *J. Controlled Release* **2012**, *164*, 299-306, *Potential of NOD receptor ligands as immunomodulators in particulate vaccine carriers.*

[157] J. T. Castaneda, A. Harui, S. M. Kiertscher, J. D. Roth, M. D. Roth, *J. Neuroimmune Pharm.* **2013**, *8*, 323-332, *Differential Expression of Intracellular and Extracellular CB2 Cannabinoid Receptor Protein by Human Peripheral Blood Leukocytes.*

[158] R. Mechoulam, *Curr. Pharm. Design* **2000**, *6*, 1313-1322, *Looking back at Cannabis research.*

[159] B. Lesch, J. Toräng, M. Nieger, S. Bräse, *Synthesis* **2005**, *2005*, 1888-1900, *The Diels-Alder approach towards cannabinoids.*

[160] B. Lesch, Dissertation, *Synthese von Benzo[b]pyranen und Diels-Alder-Strategien zur Synhtese von Cannabinoiden*, Friedrich-Wilhelms-Universität Bonn (Bonn), **2005**.

[161] M. C. Bröhmer, Dissertation, *Die Domino-oxa-Michael–Aldol-Reaktion in der Naturstoffsynthese: Asymmetrische Totalsynthesen von (–)-Diversonol, (+)-Lachnon C und Tetrahydrocannabinol-Analoga*, Karlsruhe Institute of Technology (Karlsruhe), **2011**.

[162] F. Gläser, M. C. Bröhmer, T. Hurrle, M. Nieger, S. Bräese, *Eur. J. Org. Chem.* **2015**, *2015*, 1516-1524, *The Diels-Alder Approach to Δ9-Tetrahydrocannabinol Derivatives.*

[163] T. Minami, Y. Matsumoto, S. Nakamura, S. Koyanagi, M. Yamaguchi, *J. Org. Chem.* **1992**, *57*, 167-173, *3-Vinylcoumarins and 3-vinylchromenes as dienes. Application to the synthesis of 3,4-fused coumarins and chromenes.*

[164] J. W. Huffman, X. Zhang, M. J. Wu, H. H. Joyner, *J. Org. Chem.* **1989**, *54*, 4741-4743, *Regioselective synthesis of (+-)-11-Nor-9-carboxy-.DELTA.9-THC.*

[165] J. W. Huffman, X. Zhang, M. J. Wu, H. H. Joyner, W. T. Pennington, *J. Org. Chem.* **1991**, *56*, 1481-1489, *Synthesis of (+-)-11-nor-9-carboxy-.DELTA.9-tetrahydrocannabinol. New synthetic approaches to cannabinoids.*

[166] J. B. Press, G. H. Birnberg, *J. Heterocyclic Chem.* **1984**, *22*, 561, *Heterocyclic-fused Benzopyrans as Cannabinoid Analogues.*

[167] V. Aggarwal, R. Grainger, P. Spargo, *J. Chem. Soc., Perkin Trans. 1* **1998**, 2771-2782, *(1 R, 3 R)-2-Methylene-1, 3-dithiolane 1, 3-dioxide: a highly reactive and highly selective chiral ketene equivalent in cycloaddition reactions with a broad range of dienes.*

[168] D. F. Taber, P. W. DeMatteo, R. A. Hassan, *Org. Synth.* **2013**, *90*, 350-357, *Simplified Preparation of Dimethyldioxirane (DMDO).*

[169] A. Behrenswerth, N. Volz, J. Toräng, S. Hinz, S. Bräse, C. E. Müller, *Biorg. Med. Chem.* **2009**, *17*, 2842-2851, *Synthesis and pharmacological evaluation of coumarin derivatives as cannabinoid receptor antagonists and inverse agonists.*

[170] V. Rempel, N. Volz, F. Gläser, M. Nieger, S. Bräse, C. E. Müller, *J. Med. Chem.* **2013**, *56*, 4798-4810, *Antagonists for the Orphan G-Protein-Coupled Receptor GPR55 Based on a Coumarin Scaffold.*

[171] R. Mechoulam, M. Peters, E. Murillo-Rodriguez, L. O. Hanuš, *Chem. Biodivers.* **2007**, *4*, 1678-1692, *Cannabidiol–recent advances.*

[172] S. A. Muthafer, Bachelor Thesis, *Synthese von Coumarinderivaten*, Karlsruhe Institute of Technology (Karlsruhe), **2016**.

[173] L. Kurti, B. Czakó, *Strategic applications of named reactions in organic synthesis*, Elsevier Academic Press, Amsterdam, **2005**.

[174] J. Clayden, *Organolithiums: selectivity for synthesis, Vol. 23*, Elsevier Science Ltd., Oxford, UK, **2002**.

[175] H. Gilman, A. Haubein, H. Hartzfeld, *J. Org. Chem.* **1954**, *19*, 1034-1040, *The cleavage of some ethers by organolithium compounds.*

[176] P. T. Lansbury, V. Pattison, J. Sidler, J. Bieber, *J. Am. Chem. Soc.* **1966**, *88*, 78-84, *Mechanistic aspects of the rearrangement and elimination reactions of α-metalated benzyl alkyl ethers.*

[177] J.-S. Friedrichs, Bachelor Thesis, *Synthese von Coumarin-Derivaten*, Karlsruhe Institute of Technology (Karlsruhe), **2017**.

[178] N. U. Hofslokken, L. Skattebol, *Acta Chem. Scand* **1999**, *53*, 258-262, *Convenient Method for the ortho-Formylation*.

[179] G. Sedlmeyer, **2016**, *unpublished results*.

[180] W. Perkin, *J. Chem. Soc.* **1868**, *21*, 53-63, *VI.—On the artificial production of coumarin and formation of its homologues*.

[181] M. Crawford, J. Shaw, *J. Chem. Soc. Pak.* **1953**, 3435-3439, *688. The course of the Perkin coumarin synthesis. Part I*.

[182] S. Rahmani-Nezhad, L. Khosravani, M. Saeedi, K. Divsalar, L. Firoozpour, Y. Pourshojaei, Y. Sarrafi, H. Nadri, A. Moradi, M. Mahdavi, A. Shafiee, A. Foroumadi, *Synth. Commun.* **2015**, *45*, 741-749, *Synthesis and Evaluation of Coumarin–Resveratrol Hybrids as 15-Lipoxygenaze Inhibitors*.

[183] W. Pu, Y. Lin, J. Zhang, F. Wang, C. Wang, G. Zhang, *Bioorg. Med. Chem. Lett.* **2014**, *24*, 5432-5434, *3-Arylcoumarins: Synthesis and potent anti-inflammatory activity*.

[184] M.-S. Schiedel, C. A. Briehn, P. Bäuerle, *J. Organomet. Chem.* **2002**, *653*, 200-208, *C-C Cross-coupling reactions for the combinatorial synthesis of novel organic materials*.

[185] K.-M. Kim, I.-H. Park, *Synthesis* **2004**, *2004*, 2641-2644, *A convenient halogenation of α, β-unsaturated carbonyl compounds with OXONE® and hydrohalic acid (HBr, HCl)*.

[186] D. Janssen-Müller, M. Schedler, M. Fleige, C. G. Daniliuc, F. Glorius, *Angew. Chem. Int. Ed.* **2015**, *54*, 12492-12496, *Enantioselective Intramolecular Hydroacylation of Unactivated Alkenes: An NHC-Catalyzed Robust and Versatile Formation of Cyclic Chiral Ketones*.

[187] J. Gordo, J. Avó, A. J. Parola, J. C. Lima, A. Pereira, P. S. Branco, *Org. Lett.* **2011**, *13*, 5112-5115, *Convenient Synthesis of 3-Vinyl and 3-Styryl Coumarins*.

[188] K. Waibel, Bachelor Thesis, *Eisenkatalysierte Kreuzkupplungen an Cumarinderivaten*, Karlsruhe Institute of Technology (Karlsruhe), **2015**.

[189] J. W. de Boer, W. R. Browne, S. R. Harutyunyan, L. Bini, T. D. Tiemersma-Wegman, P. L. Alsters, R. Hage, B. L. Feringa, *Chem. Commun.* **2008**, 3747-3749, *Manganese catalysed asymmetric cis-dihydroxylation with H2O2*.

[190] R. Gericke, J. Harting, I. Lues, C. Schittenhelm, *J. Med. Chem.* **1991**, *34*, 3074-3085, *3-Methyl-2H-1-benzopyran potassium channel activators*.

[191] C. N. Chiang, R. S. Rapaka, *NIDA Res. Monogr* **1987**, *79*, 173-188, *Pharmacokinetics and disposition of cannabinoids*.

[192] M. A. Huestis, E. J. Cone, *J. Anal. Toxicol.* **2004**, *28*, 394-399, *Relationship of Δ9-Tetrahydrocannabinol Concentrations in Oral Fluid and Plasma after Controlled Administration of Smoked Cannabis*.

[193] V. Rempel, N. Volz, S. Hinz, T. Karcz, I. Meliciani, M. Nieger, W. Wenzel, S. Brase, C. E. Muller, *J. Med. Chem.* **2012**, *55*, 7967-7977, *7-Alkyl-3-benzylcoumarins: a versatile scaffold for the development of potent and selective cannabinoid receptor agonists and antagonists*.

[194] A. K. Ghose, G. M. Crippen, *J. Chem. Inf. Comp. Sci.* **1987**, *27*, 21-35, *Atomic physicochemical parameters for three-dimensional-structure-directed quantitative structure-activity relationships. 2. Modeling dispersive and hydrophobic interactions*.

[195] D. Klose, F. Siepmann, K. Elkharraz, S. Krenzlin, J. Siepmann, *Int. J. Pharm.* **2006**, *314*, 198-206, *How porosity and size affect the drug release mechanisms from PLGA-based microparticles.*

[196] 20.09.2017, http://www.sigmaaldrich.com/catalog/product/aldrich/739952?lang=de®ion=DE&cm_sp=Insite-_-recent_fixed-_-recent5-1.

[197] 20.09.2017, http://www.sigmaaldrich.com/catalog/product/aldrich/719870?lang=de®ion=DE.

[198] C. Yan, J. H. Resau, J. Hewetson, M. West, W. L. Rill, M. Kende, *J. Controlled Release* **1994**, *32*, 231-241, *Characterization and morphological analysis of protein-loaded poly(lactide-co-glycolide) microparticles prepared by water-in-oil-in-water emulsion technique.*

[199] H. P. Schuchmann, T. Danner, *Chem. Ing. Tech.* **2004**, *76*, 364-375, *Emulsification: More than just comminution.*

[200] R. Jalil, J. Nixon, *J. Microencapsulation* **1990**, *7*, 25-39, *Microencapsulation using poly (L-lactic acid) II: Preparative variables affecting microcapsule properties.*

[201] R. Jalil, J. R. Nixon, *J. Microencapsulation* **1989**, *6*, 473-484, *Microencapsulation using Poly(L-Lactic Acid) I: Microcapsule Properties Affected by the Preparative Technique.*

[202] H. Zhao, J. Gagnon, U. O. Häfeli, *BioMagnetic Research and Technology* **2007**, *5*, 2, *Process and formulation variables in the preparation of injectable and biodegradable magnetic microspheres.*

[203] J. Dubochet, N. Sartori Blanc, *Micron* **2001**, *32*, 91-99, *The cell in absence of aggregation artifacts.*

[204] N. Brunacci, C. Wischke, T. Naolou, A. T. Neffe, A. Lendlein, *Eur. J. Pharm. Biopharm.* **2017**, *116*, 61-65, *Influence of surfactants on depsipeptide submicron particle formation.*

[205] R. T. Liggins, H. M. Burt, *Int. J. Pharm.* **2001**, *222*, 19-33, *Paclitaxel loaded poly(L-lactic acid) microspheres: properties of microspheres made with low molecular weight polymers.*

[206] F. Boury, H. Marchais, J. E. Proust, J. P. Benoit, *J. Controlled Release* **1997**, *45*, 75-86, *Bovine serum albumin release from poly(α-hydroxy acid) microspheres: effects of polymer molecular weight and surface properties.*

[207] F. Boury, T. Ivanova, I. Panaïotov, J. E. Proust, A. Bois, J. Richou, *J. Colloid Interface Sci.* **1995**, *169*, 380-392, *Dynamic Properties of Poly(DL-lactide) and Polyvinyl Alcohol Monolayers at the Air/Water and Dichloromethane/Water Interfaces.*

[208] G. Reich, *Eur. J. Pharm. Biopharm.* **1998**, *45*, 165-171, *Ultrasound-induced degradation of PLA and PLGA during microsphere processing: influence of formulation variables.*

[209] P. B. O'Donnell, J. W. McGinity, *Adv. Drug. Deliver. Rev.* **1997**, *28*, 25-42, *Preparation of microspheres by the solvent evaporation technique.*

[210] G. Odian, *Principles of polymerization*, 4th Edition ed., John Wiley & Sons, Hoboken, New Jersey, **2004**.

[211] P. Blasi, A. Schoubben, S. Giovagnoli, L. Perioli, M. Ricci, C. Rossi, *AAPS PharmSciTech* **2007**, *8*, E78-E85, *Ketoprofen poly(lactide-co-glycolide) physical interaction.*

[212] H. Okada, Y. Doken, Y. Ogawa, H. Toguchi, *Pharm. Res.* **1994**, *11*, 1143-1147, *Preparation of Three-Month Depot Injectable Microspheres of Leuprorelin Acetate Using Biodegradable Polymers.*

[213] M. M. Feldstein, G. A. Shandryuk, N. A. Platé, *Polymer* **2001**, *42*, 971-979, *Relation of glass transition temperature to the hydrogen-bonding degree and energy in poly(N-vinyl pyrrolidone) blends with hydroxyl-containing plasticizers. Part 1. Effects of hydroxyl group number in plasticizer molecule.*

[214] P. Di Martino, E. Joiris, R. Gobetto, A. Masic, G. F. Palmieri, S. Martelli, *J. Cryst. Growth* **2004**, *265*, 302-308, *Ketoprofen-poly (vinylpyrrolidone) physical interaction.*

[215] R. Nair, N. Nyamweya, S. Gönen, L. J. Martínez-Miranda, S. W. Hoag, *Int. J. Pharm.* **2001**, *225*, 83-96, *Influence of various drugs on the glass transition temperature of poly(vinylpyrrolidone): a thermodynamic and spectroscopic investigation.*

[216] A. Waugh, A. Grant, *Ross & Wilson Anatomy and Physiology in Health and Illness E-Book*, 11th ed., Elsevier Health Sciences, **2010**.

[217] A. K. Basak, A. S. Raw, L. X. Yu, *Adv. Drug. Deliver. Rev.* **2007**, *59*, 1-2, *Pharmaceutical impurities: Analytical, toxicological and regulatory perspectives.*

[218] R. Bodmeier, J. W. McGinity, *Pharm. Res.* **1987**, *4*, 465-471, *The Preparation and Evaluation of Drug-Containing Poly(dl-lactide) Microspheres Formed by the Solvent Evaporation Method.*

[219] E. Corey, *Angew. Chem. Int. Ed.* **2002**, *41*, 1650-1667, *Catalytic enantioselective Diels–Alder reactions: methods, mechanistic fundamentals, pathways, and applications.*

[220] C. Li, G. Xiao, Q. Zhao, H. Liu, T. Wang, W. Tang, *Org. Chem. Front.* **2014**, *1*, 225-229, *Sterically demanding aryl-alkyl Suzuki-Miyaura coupling.*

[221] H. E. Gottlieb, V. Kotlyar, A. Nudelman, *J. Org. Chem.* **1997**, 7512-7515, *NMR chemical shifts of common laboratory solvents as trace Impurities.*

[222] W. C. Still, M. Kahn, A. Mitra, *J. Org. Chem.* **1978**, *43*, 2923-2925, *Rapid chromatographic technique for preparative separations with moderate resolution.*

[223] M. Shiraishi, K. Kato, S. Terao, Y. Ashida, Z. Terashita, G. Kito, *J. Med. Chem.* **1989**, *32*, 2214-2221, *Quinones. 4. Novel eicosanoid antagonists: synthesis and pharmacological evaluation.*

[224] P. D. Knight, G. Clarkson, M. L. Hammond, B. S. Kimberley, P. Scott, *J. Organomet. Chem.* **2005**, *690*, 5125-5144, *Radical and migratory insertion reaction mechanisms in Schiff base zirconium alkyls.*

[225] T.-S. Jiang, J.-H. Li, *Chem. Commun.* **2009**, 7236-7238, *Palladium-catalyzed oxidative tandem reaction of allylamines with aryl halides leading to α, β-unsaturated aldehydes.*

[226] I. Šagud, F. Faraguna, Ž. Marinić, M. Šindler-Kulyk, *J. Org. Chem.* **2011**, *76*, 2904-2908, *Photochemical Approach to Naphthoxazoles and Fused Heterobenzoxazoles from 5-(Phenyl/heteroarylethenyl)oxazoles.*

[227] M. Zhu, M. H. Kim, S. Lee, S. J. Bae, S. H. Kim, S. B. Park, *J. Med. Chem.* **2010**, *53*, 8760-8764, *Discovery of Novel Benzopyranyl Tetracycles that Act as Inhibitors of Osteoclastogenesis Induced by Receptor Activator of NF-κB Ligand.*

8. Appendix

8.1 Curriculum Vitae

Thomas Hurrle

*24.04.1989 in Gernsbach, Deutschland

E-Mail: hurrle.thomas@gmail.com

Work Experience

01/2018 – 04/2018

PostDoc, Karlsruhe Institute of Technology (KIT), Karlsruhe, Institute of Organic Chemistry/Institute of Toxicology and Genetics in the work Group of Prof. Dr. Stefan Bräse

- Project development and organization

11/2014 – 12/2017

Ph.D. studies, KIT, Karlsruhe, Institute of Organic Chemistry/ Institute of Toxicology and Genetics in the research group of Prof. Dr. Stefan Bräse, graduation 1.0 "*very good* "(magna cum laude)

- Thesis: "*Synthesis of Cannabinoid Ligands – Novel Compound Classes, Routes and Perspectives*"
- Synthesis planning and implementation, analysis of synthesized compounds (NMR-spectroscopy, IR-spectroscopy, mass-spectrometry, HPLC, GC-MS)
- Supervision of students in practical courses and direct supervision of 6 theses and a DAAD trainee (USA)
- Responsibility for GC-MS and homepage administration

02/2016 – 03/ 2016

InterDoc, Helmholtz-Zentrum Geesthacht (HZG), Teltow, Department for Pharmaceutical Technology

- Focus on material science
- Development of micro- and nanoparticulate release systems for cannabinoids
- Method development and utilization of HPLC

Education

01/2015 – 12/2017

Graduate training in the BioInterfaces International Graduate School (BIF-IGS), KIT, Karlsruhe

- Transdisciplinary Graduate School

- Insight in the disciplines physics, chemistry, biology, engineering and information technology

10/2012 – 10/2014 **Study of chemistry (Master of Science), KIT,** Karlsruhe

Graduation mark 1,2

- Focus on organic chemistry with specialization in medicinal chemistry and biochemistry
- Master thesis in organic chemistry: Translation: "*Work on the formal synthesis of Δ9-Tetrahydrocannabinol and Synthesis of Cannabinoid-Analogs* "

10/2009 – 10/2012 **Study of chemistry (Bachelor of Science), KIT,** Karlsruhe

Graduation mark 1,5

- Bachelor thesis in organic chemistry Translation: "*Synthesis of Coumarin Derivatives*"

Internship abroad

10/2013 – 12/2013 **Scientific Internship, Queensland University of Technology (QUT),** Brisbane (Australia), Science and Engineering Faculty

- Supported by the PROMOS scholarship (DAAD)
- Focus on method development
- Title of the report: "*Investigations of 2,6-Disubstituted Substrates for the Acetal Method*"

Military Service

7/2008 – 03/2009 **Medical orderly, Lazarettregiment 41**, Horb am Neckar, Germany

- With additional training as CBRN-defense soldier

School Education

2005 – 2008 **Abitur (A-Level), Josef-Durler-Schule,** Rastatt, Germany

- Technical school with profile in information technology
- Award of the Gesellschaft Deutscher Chemiker (GDCh) for the best performance in chemistry

Professional Experience

10/2012 – 12/2013 **Student Employee, Fraunhofer Institute for Chemical Technology (ICT),** Berghausen, Germany, Department for Environmental Engineering

- Synthesis of flame retardants and their implementation as additives in polymers
- Extraction of renewable starting materials for polymerizations from biomass

05/2012 – 08/2012 **Research Assistant, KIT,** Karlsruhe, Institute for Organic Chemistry

- Synthetic, preparative and analytical operations

08/2010 – 09/2010 & 08/2011 – 09/2011 **Student Employee at Daimler,** Gaggenau, Germany, Department for Materials and Processes

- Quality control, data analysis and processing

Languages

German	native language
English	proficient
Spanish	basic
Italian	basic

8.2 Publications and Conference Contributions

Book chapter

S. Bräse, F. Gläser, **T. Hurrle** (**2015**), In *Privileged Scaffolds in Medicinal Chemistry* (pp. 287-311). Coumarins

Publications

P. Poonpatana, G. dos Passos Gomes, **T. Hurrle**, K. Chardon, S. Bräse, K.-S. Masters, I. Alabugin (**2017**). Formaldehyde-Extruding Homolytic Aromatic Substitution via C→ O Transposition: Selective 'Traceless-Linker'access to Congested Biaryl Bonds. *Chemistry-A European Journal*, *23*(38), 9091-9097.

F. Gläser, M. C. Broehmer, **T. Hurrle**, M. Nieger, S. Bräse, (**2015**). The Diels–Alder Approach to Δ9-Tetrahydrocannabinol Derivatives. *European Journal of Organic Chemistry*, 2015(7), 1516-1524.

Posters

T. Hurrle, S. Bräse, BioInterfaces International Graduate School Retreat, **2016**, Althütte, Germany. *Coumarin Derivatives as Potent Cannabinoid Analogues*

T. Hurrle, F. Mohr, D. Marcato, A.Hariharan, U. Strähle, R. Peravali, S. Bräse, *ICRS 2017 – 27th Annual ICRS Symposium on the Cannabinoids*, 22.-27. June **2017**, Montréal, Canada. *Novel cannabinoids based on the coumarin motif and fast evaluation through PMR-study on Embryonic Zebrafish*

F. Mohr, **T. Hurrle**, H. Jauch, A. Keil, S. Bräse, B. L. Fiebich, *ICRS 2017 – 27th Annual ICRS Symposium on the Cannabinoids*, 22.-27. June **2017**, Montréal, Canada. Modular Synthesis of Novel Cannabinoid Ligands Based on the Coumarin Motif as CB_1, CB_2, GPR55 Agonists and Antagonists

Oral Presentations

T. Hurrle, S. Bräse, BioInterfaces International Graduate School Retreat, **2017**, Frankfurt, Germany. *Novel Cannabinoids based on the Coumarin Motif,* price for one of the best 3 speakers

8.3 Acknowledgements

An dieser Stelle möchte ich mich bei allen bedanken, die zum Gelingen meiner Doktorarbeit beigetragen haben.

Zunächst möchte ich mich bei meinem Doktorvater Prof. Dr. Stefan Bräse bedanken. Vielen Dank für die interessante Aufgabenstellung und die Freiheit meine eigenen Ideen zu verwirklichen können, aber auch für die herrvoragende fachliche Betreuung und das große Vertrauen, dass du mir nicht nur während der Promotion sondern bereits während des Studiums engegengebracht hast.

Ebenso möchte ich mich bei Dr. Christian Wischke und seiner Gruppe am HZG Teltow für die freundiche aufnahme in die Gruppe bedanken. Die Exkursion hat meine Arbeit sehr bereichert und dazu beigetragen einen erweiterten Ausblick über das synthetische Thema hinaus zu erlangen.

Bei Prof. Dr. Michael Meier bedanke ich mich für die freundliche Übernahme des Korreferats.

Ein großes Dankeschön auch an Selin Samur, Christiane Lampert und Monique Antoniak für die Hilfe bei allen organisatorischen Angelegenheiten, die das IOC und den Campus Süd betrafen. Ebenfalls ein großes Dankeschön geht an die Administration am ITG, insbesondere Dr. Sandra Schneider und Irina Schierholz, für die Unterstützung bei den Campus Nord betreffenden Angelegenheiten.

Danke auch an die Arbeitskollegen am HZG, besonders Fabian Frieß, Nadia Brunacci und Andrea Pfeiffer für die freundliche Unterstützung während und auch nach meiner Zeit in Teltow. Danke auch an die analytische Abteilung für die DSC Messungen und besonders Frau Pieper und Frau Radzik für die schönen SEM Bilder.

Ein großer Dank geht an das Complat Team um Nicole Jung. Durch euch wird effektiv die Chance erhöht das unsere Verbindungen irgendwann mal Menschen helfen könnten.

Vielen Dank an Martin Nieger für die Kristallstrukturanalysen.

A big thanks goes to Alena Kalyakina, Florian Mohr for proofreading my work, I'll try to make it up to you. Also to Tim Wezemann and Isabelle Wessely for giving advice for the "finishing" touch.

Ein herzliches Dankeschön auch an die Analytikabteilung und die weiteren Personen, die das Rückgrad unseres Institut darstellen. Ohne diese Infrastruktur wäre es nicht möglich gewesen die Arbeit in diesem Umfang fertig zu stellen.

“Stand on the shoulders of giants.”

– *Google-scholar search screen.*

Ein großes Dankeschön an Manuel Bröhmer und Franzika Gläser für ihre Vorarbeit zur THC-Synthese sowie Nicole Volz, Jakob Toräng und Franziska Gläser für die Vorarbeiten zu den 3-Benzylcumarinen.

Dankeschön auch an den “AK Hurrle”: Eugen Dick, Kevin Waibel, Mareen Stahlberger, Leonora Nurcaj, Sarah Al Muthafer, Monica Theibault und Jan-Simon Friedrichs. Auch wenn nicht immer alles geklappt hat, es hat Spaß gemacht euch zu betreuen und war eine Bereicherung für meine Promotion.

Ganz besonderer Dank gilt natürlich dem Arbeitskreis Bräse, aktuelle und ehemalige Mitglieder eingeschlossen. Es war immer interessant und kurzweilig mit euch.

Vielen Dank auch an meine Familie und Freunde, ich weiß das ich immer auf euch zählen kann.

Zu guter Letzt möchte ich meinen Studienkollegen danken. Ihr habt großen Anteil daran, dass die acht Jahre mit Beginn der O-Phase 2009 etwas ganz besonderes waren.

“He was a dreamer, a thinker, a speculative philosopher…
or, as his wife would have it, an idiot.”

-Douglas Adams

Thank you for everything, Silvana.